Antibiotic Inhibition of Bacterial Cell Surface Assembly and Function

Antibiotic Inhibition of Bacterial Cell Surface Assembly and Function

Edited by:

Paul Actor
Smith Kline & French Laboratories, Philadelphia, Pennsylvania

Lolita Daneo-Moore
Temple University, Philadelphia, Pennsylvania

Michael L. Higgins
Temple University, Philadelphia, Pennsylvania

Milton R. J. Salton
New York University School of Medicine, New York, New York

Gerald D. Shockman
Temple University, Philadelphia, Pennsylvania

American Society for Microbiology
Washington, D.C. 20006

Library of Congress Cataloging-in-Publication Data

Antibiotic inhibition of bacterial cell surface
  assembly and function.

  Based on an ASM conference held at the College of
Physicians of Philadelphia on May 17–20, 1987.
Includes index.
    1. Bacterial cell walls—Congresses.   2. Bacterial
cell walls—Synthesis—Inhibitors—Congresses.   3. Beta
lactam antibiotics—Physiological effect—Congresses.
I. Actor, Paul.   II. American Society for Microbiology.
QR77.3.A56   1988   615'.329   87-30737

ISBN 0-914826-98-0

Contents

IV.   AUTOLYSIS AND PEPTIDOGLYCAN HYDROLASES

V.   PENICILLIN-BINDING PROTEINS

IX.   RESISTANCE

X.   CONCLUDING REMARKS

Contributors

Hiroyuki Adachi • Department of Agricultural Chemistry, The University of Tokyo, Bunkyo-ku, Tokyo 113, Japan

W. E. Alborn, Jr. • Lilly Research Laboratories, Indianapolis, IN 46285

N. E. Allen • Lilly Research Laboratories, Indianapolis, IN 46285

John S. Anderson • Department of Biochemistry, University of Minnesota, St. Paul, MN 55108

Anneleen Asmus • National Science Department, State University of New York College of Technology, Utica, NY 13504

M. E. Bayer • Fox Chase Cancer Center, Institute for Cancer Research, Philadelphia, PA 19111

M. H. Bayer • Fox Chase Cancer Center, Institute for Cancer Research, Philadelphia, PA 19111

Frank Beise • Robert Koch Institute of the Federal Health Office, D-1000 Berlin (West) 65, Federal Republic of Germany

Spencer A. Benson • Department of Molecular Biology, Princeton University, Princeton, NJ 08544

Terry J. Beveridge • Department of Microbiology, University of Guelph, Guelph, Ontario N1G 2W1, Canada

Lucas D. Bowler • Microbial Genetics Group, School of Biological Sciences, University of Sussex, Falmer, Brighton BN1 9QG, United Kingdom

Sandra Bräutigam • Institut für Mikrobiologie, Technische Hochschule, D-6100 Darmstadt, Federal Republic of Germany

Thomas Briese • Max-Planck-Institut für Molekulare Genetik, D-1000 Berlin 33, Federal Republic of Germany

Janice L. Brissette • Department of Biochemistry, Temple University School of Medicine, Philadelphia, PA 19140

Jenny K. Broome-Smith • Microbial Genetics Group, School of Biological Sciences, University of Sussex, Falmer, Brighton BN1 9QG, United Kingdom

Christine E. Buchanan • Department of Biology, Southern Methodist University, Dallas, TX 75275

Erlinda A. Cabacungan • Department of Biochemistry, Temple University School of Medicine, Philadelphia, PA 19140

Pietro Canepari • Istituto di Microbiologià dell'Università di Verona, 37134 Verona, Italy

Stephen Cooper • Department of Microbiology and Immunology,
University of Michigan Medical School, Ann Arbor, MI 48109-0620
Jacques Coyette • Service de Microbiologie, Université de Liège,
B-4000 Sart Tilman (Liège 1), Belgium
Lisa M. Cummings • Department of Chemotherapy, Roche Research
Center, Nutley, NJ 07110
L. Daneo-Moore • Department of Microbiology and Immunology,
Temple University School of Medicine, Philadelphia, PA 19140
E. Daub • Department of Biology, Massachusetts Institute of
Technology, Cambridge, MA 02139
Maria del Mar Lleò • Istituto di Microbiologià dell'Università di
Verona, 37134 Verona, Italy
Masaki Doi • Institute of Applied Microbiology, University of Tokyo,
Bunkyo-ku, Tokyo 113, Japan
D. Dolinger • Department of Microbiology and Immunology, Temple
University School of Medicine, Philadelphia, PA 19140
R. J. Doyle • Department of Microbiology and Immunology,
University of Louisville Health Sciences Center, Louisville, KY
40292
K. Duncan • Department of Chemistry, Massachusetts Institute of
Technology, Cambridge, MA 02139
Bernard Dussenne • Service de Microbiologie, Université de Liège,
B-4000 Sart Tilman (Liège 1), Belgium
Alex Edelman • Unité de Biochimie Microbienne, Institut Pasteur,
Paris 75724, Cedex 15, France
Aboubaker El Kharroubi • Service de Microbiologie, Université de
Liège, B-4000 Sart Tilman (Liège 1), Belgium
Heinz Ellerbrok • Max-Planck-Institut für Molekulare Genetik,
D-1000 Berlin 33, Federal Republic of Germany
H. Fletcher • Department of Microbiology and Immunology, Temple
University School of Medicine, Philadelphia, PA 19140
Roberta Fontana • Istituto di Microbiologià dell'Università di
Verona, 37134 Verona, Italy
Marita Franz • Robert Koch Institute of the Federal Health Office,
D-1000 Berlin (West) 65, Federal Republic of Germany
Jean-Marie Frère • Service de Microbiologie, Université de Liège,
B-4000 Sart Tilman (Liège 1), Belgium
Ernesto García • Centro de Investigaciones Biologicas, Consejo
Superior de Investigaciones Científicas, 28006 Madrid, Spain
José L. García • Centro de Investigaciones Biologicas, Consejo
Superior de Investigaciones Científicas, 28006 Madrid, Spain
Pedro García • Centro de Investigaciones Biologicas, Consejo
Superior de Investigaciones Científicas, 28006 Madrid, Spain

Kalle Gehring • Department of Microbiology and Immunology, University of California, Berkeley, CA 94720

Nafsika H. Georgopapadakou • Department of Chemotherapy, Roche Research Center, Nutley, NJ 07110

Jean-Marie Ghuysen • Service de Microbiologie, Université de Liège, B-4000 Sart Tilman (Liège 1), Belgium

Peter Giesbrecht • Robert Koch Institute of the Federal Health Office, D-1000 Berlin (West) 65, Federal Republic of Germany

B. Glauner • Max-Planck-Institut für Entwicklungsbiologie, D-7400 Tübingen, Federal Republic of Germany

E. W. Goodell • Natural Science Department, State University of New York College of Technology, Utica, NY 13504

Matthew J. Grossman • Waksman Institute of Microbiology, Rutgers University, The State University of New Jersey, Piscataway, NJ 08855-0759

Evan Gustow • Department of Biochemistry, Temple University School of Medicine, Philadelphia, PA 19140

S. Haberer • Fox Chase Cancer Center, Institute for Cancer Research, Philadelphia, PA 19111

Regine Hakenbeck • Max-Planck-Institut für Molekulare Genetik, D-1000 Berlin 33, Federal Republic of Germany

M. L. Higgins • Department of Microbiology and Immunology, Temple University School of Medicine, Philadelphia, PA 19140

Kim M. Hildebrandt • Department of Biochemistry, University of Minnesota, St. Paul, MN 55108

J. N. Hobbs, Jr. • Lilly Research Laboratories, Indianapolis, IN 46285

Joachim-Volker Höltje • Max-Planck-Institut für Entwicklungsbiologie, D-7400 Tübingen, Federal Republic of Germany

P. Huls • Department of Electron Microscopy and Molecular Cytology, University of Amsterdam, 1018 TV Amsterdam, The Netherlands

Akira Imada • Biology Laboratories, Takeda Chemical Industries, Ltd., Osaka, Japan

E. E. Ishiguro • Department of Biochemistry and Microbiology, University of Victoria, Victoria, British Columbia V8W 2Y2, Canada

Kohki Ishikawa • Faculty of Pharmaceutical Sciences, University of Tokyo, Bunkyo-ku, Tokyo 113, Japan

Fumitoshi Ishino • Institute of Applied Microbiology, University of Tokyo, Bunkyo-ku, Tokyo 113, Japan

Yoshiyasu Ito • Institute of Applied Microbiology, University of Tokyo, Bunkyo-ku, Tokyo 113, Japan

Philippe Jacques • Service de Microbiologie, Université de Liège, B-4000 Sart Tilman (Liège 1), Belgium

Peter W. Jeffs • Smith Kline & French Research Laboratories, Swedeland, PA 19479

Bernard Joris • Service de Microbiologie, Université de Liège, B-4000 Sart Tilman (Liège 1), Belgium

D. Karamata • Institut de Génétique et Biologie Microbiennes, 1005 Lausanne, Switzerland

Wolfgang Keck • Chemische Laboratoria der Rijksuniversiteit, 9747 AG-Groningen, The Netherlands

Judith A. Kelly • Department of Molecular and Cell Biology, University of Connecticut, Storrs, CT 06268

M. A. Kemper • Department of Microbiology and Immunology, University of Louisville Health Sciences Center, Louisville, KY 40292

James R. Knox • Department of Molecular and Cell Biology, University of Connecticut, Storrs, CT 06268

Tetsuo Kobayashi • Waksman Institute of Microbiology, Rutgers University, The State University of New Jersey, Piscataway, NJ 08855-0759

Arthur L. Koch • Department of Biology, Indiana University, Bloomington, IN 47405

Bogomira Korat • Max-Planck-Institut für Entwicklungsbiologie, D-7400 Tübingen, Federal Republic of Germany

Dominique Krüger • Robert Koch Institute of the Federal Health Office, D-1000 Berlin (West) 65, Federal Republic of Germany

W. Kusser • Department of Biochemistry and Microbiology, University of Victoria, Victoria, British Columbia V8W 2Y2, Canada

Harald Labischinski • Robert Koch Institute of the Federal Health Office, D-1000 Berlin (West) 65, Federal Republic of Germany

Götz Laible • Max-Planck-Institut für Molekulare Genetik, D-1000 Berlin 33, Federal Republic of Germany

J. Oliver Lampen • Waksman Institute of Microbiology, Rutgers University, The State University of New Jersey, Piscataway, NJ 08855-0759

Eugene R. LaSala • Department of Chemotherapy, Roche Research Center, Nutley, NJ 07110

Frederik Lindberg • Department of Microbiology, University of Umeå, S-901 87 Umeå, Sweden

Susanne Lindqvist • Department of Microbiology, University of Umeå, S-901 87 Umeå, Sweden

Rubens López • Centro de Investigaciones Biologicas, Consejo

Superior de Investigaciones Cientificas, 28006 Madrid, Spain

Yoshimi Maeda • Institute of Applied Microbiology, University of Tokyo, Bunkyo-ku, Tokyo 113, Japan

Shigefumi Maesaki • Institute of Applied Microbiology, University of Tokyo, Bunkyo-ku, Tokyo 113, Japan

Heinrich Maidhof • Robert Koch Institute of the Federal Health Office, D-1000 Berlin (West) 65, Federal Republic of Germany

Robert E. Marquis • Department of Microbiology and Immunology, The University of Rochester, Rochester, NY 14642

Christiane Martin • Max-Planck-Institut für Molekulare Genetik, D-1000 Berlin 33, Federal Republic of Germany

Hans H. Martin • Institut für Mikrobiologie, Technische Hochschule, D-6100 Darmstadt, Federal Republic of Germany

Stephen A. Martin • Department of Chemistry, Massachusetts Institute of Technology, Cambridge, MA 02139

Hans J. P. Marvin • Department of Biochemistry, University of Groningen, 9747 AG Groningen, The Netherlands

O. Massidda • Department of Microbiology and Immunology, Temple University School of Medicine, Philadelphia, PA 19140

Michio Matsuhashi • Institute of Applied Microbiology, University of Tokyo, Bunkyo-ku, Tokyo 113, Japan

Hiroshi Matsuzawa • Department of Agricultural Chemistry, The University of Tokyo, Bunkyo-ku, Tokyo 113, Japan

Thomas D. McDowell • Department of Microbiology, University of New Mexico School of Medicine, Albuquerque, NM 87131

Estelle J. McGroarty • Department of Biochemistry, Michigan State University, East Lansing, MI 48824

Neil H. Mendelson • Department of Molecular and Cellular Biology, University of Arizona, Tucson, AZ 85721

Enrique Méndez • Departamento de Endocrinología, Centro Ramón y Cajal, Carretera de Colmenar, 28034 Madrid, Spain

Claudia Metelmann • Institut für Medizinische Mikrobiologie der Freien Universität Berlin, D-1000 Berlin 45, Federal Republic of Germany

Helmut Mett • Research Laboratories, Pharmaceuticals Division, CIBA-GEIGY Ltd., CH-4002 Basel, Switzerland

Rajeev Misra • Department of Molecular Biology, Princeton University, Princeton, NJ 08544

Yukio Mitsui • Faculty of Pharmaceutical Sciences, University of Tokyo, Bunkyo-ku, Tokyo 113, Japan

H. L. T. Mobley • Department of Medicine, University of Maryland, Baltimore, MD 21201

R. C. Moellering, Jr. • Department of Medicine, New England

Deaconess Hospital, Harvard Medical School, Boston, MA 02215

Paul C. Moews • Department of Molecular and Cell Biology, University of Connecticut, Storrs, CT 06268

Jill Moring • Department of Molecular and Cell Biology, University of Connecticut, Storrs, CT 06268

N. Nanninga • Department of Electron Microscopy and Molecular Cytology, University of Amsterdam, 1018 TV Amsterdam, The Netherlands

Francis C. Neuhaus • Department of Biochemistry, Molecular Biology and Cell Biology, Northwestern University, Evanston, IL 60208

Robert A. Nicholas • Department of Biochemistry and Molecular Biology, Harvard University, Cambridge, MA 02138

Nancy J. Nicholls • Waksman Institute of Microbiology, Rutgers University, The State University of New Jersey, Piscataway, NJ 08855-0759

Hiroshi Nikaido • Department of Microbiology and Immunology, University of California, Berkeley, CA 94720

L. J. Nisbet • Xenova Ltd., Slough SL1 4EQ, England

Hiroshi Noguchi • Takarazuka Research Center, Biotechnology Laboratory, Sumitomo Chemical Co., Ltd., Takarazuka, Hyogo-ken, Japan

Staffan Normark • Department of Microbiology, University of Umeå, S-901 87 Umeå, Sweden

James L. Occi • Department of Molecular Biology, Princeton University, Princeton, NJ 08544

Takahisa Ohta • Department of Agricultural Chemistry, The University of Tokyo, Bunkyo-ku, Tokyo 113, Japan

Kenji Okonogi • Biology Laboratories, Takeda Chemical Industries, Ltd., Osaka, Japan

James T. Park • Department of Molecular Biology and Microbiology, Tufts University, Boston, MA 02111

E. Pas • Department of Electron Microscopy and Molecular Cytology, University of Amsterdam, 1018 TV Amsterdam, The Netherlands

Arthur A. Patchett • Merck Sharp & Dohme Research Laboratories, Rahway, NJ 07065

Ronald A. Pieringer • Department of Biochemistry, Temple University School of Medicine, Philadelphia, PA 19140

M. F. Suzanne Pinette • Department of Biology, Indiana University, Bloomington, IN 47405

Graziella Piras • Service de Microbiologie, Université de Liège, B-4000 Sart Tilman (Liège 1), Belgium

F. Pittaluga • Department of Microbiology and Immunology, Temple University School of Medicine, Philadelphia, PA 19140

H. M. Pooley • Institut de Génétique et Biologie Microbiennes, 1005 Lausanne, Switzerland

David L. Pruess • Department of Chemotherapy, Roche Research Center, Nutley, NJ 07110

Kelynne E. Reed • University of New Mexico School of Medicine, Albuquerque, NM 87131

Peter E. Reynolds • Department of Biochemistry, University of Cambridge, Cambridge CB2 1QW, England

Mildred Rivera • Department of Biochemistry, Michigan State University, East Lansing, MI 48824

H. J. Rogers • Biological Laboratory, University of Kent, Canterbury, Kent CT2 7NJ, United Kingdom

Concepción Ronda • Centro de Investigaciones Biologicas, Consejo Superior de Investigaciones Científicas, 28006 Madrid, Spain

Sabine Rosta • Research Laboratories, Pharmaceuticals Division, CIBA-GEIGY Ltd., CH-4002 Basel, Switzerland

I. Said • Department of Microbiology and Immunology, Temple University School of Medicine, Philadelphia, PA 19140

Anthony J. Salerno • Waksman Institute of Microbiology, Rutgers University, The State University of New Jersey, Piscataway, NJ 08855-0759

Barbara A. Sampson • Department of Molecular Biology, Princeton University, Princeton, NJ 08544

José M. Sanchez-Puelles • Centro de Investigaciones Biologicas, Consejo Superior de Investigaciones Científicas, 28006 Madrid, Spain

Giuseppe Satta • Istituto di Microbiologià dell'Università di Siena, Siena, Italy

Barbara Schacher • Research Laboratories, Pharmaceuticals Division, CIBA-GEIGY Ltd., CH-4002 Basel, Switzerland

Hans-Martin Schier • Max-Planck-Institut für Molekulare Genetik, D-1000 Berlin 33, Federal Republic of Germany

Bärbel Schmidt • Institut für Mikrobiologie, Technische Hochschule, D-6100 Darmstadt, Federal Republic of Germany

U. Schwarz • Max-Planck-Institut für Entwicklungsbiologie, D-7400 Tübingen, Federal Republic of Germany

Toshiya Senda • Faculty of Pharmaceutical Sciences, University of Tokyo, Bunkyo-ku, Tokyo 113, Japan

G. D. Shockman • Department of Microbiology and Immunology, Temple University School of Medicine, Philadelphia, PA 19140

Thomas Sidow • Robert Koch Institute of the Federal Health Office, D-1000 Berlin (West) 65, Federal Republic of Germany

Rabindra K. Sinha • Department of Biochemistry, Molecular Biology and Cell Biology, Northwestern University, Evanston, IL 60208

Min Dong Song • Institute of Applied Microbiology, University of Tokyo, Bunkyo-ku, Tokyo 113, Japan

Brian G. Spratt • Microbial Genetics Group, School of Biological Sciences, University of Sussex, Falmer, Brighton BN1 9QG, United Kingdom

Jack L. Strominger • Department of Biochemistry and Molecular Biology, Harvard University, Cambridge, MA 02138

R. E. Studer • Institut de Génétique et Biologie Microbiennes, 1005 Lausanne, Switzerland

Tomohiro Takahashi • Institute of Applied Microbiology, University of Tokyo, Bunkyo-ku, Tokyo 113, Japan

Akiko Takasuga • Department of Agricultural Chemistry, The University of Tokyo, Bunkyo-ku, Tokyo 113, Japan

P. E. M. Taschner • Department of Electron Microscopy and Molecular Cytology, University of Amsterdam, 1018 TV Amsterdam, The Netherlands

Fabienne Thonon • Service de Microbiologie, Université de Liège, B-4000 Sart Tilman (Liège 1), Belgium

J. J. Thwaites • Department of Engineering, Cambridge University, Cambridge CB2 1PZ, United Kingdom

Alexander Tomasz • The Rockefeller University, New York, NY 10021

Spassena Tornette • Max-Planck-Institut für Molekulare Genetik, D-1000 Berlin 33, Federal Republic of Germany

Kuni-hiro Ueda • Institute of Applied Microbiology, University of Tokyo, Bunkyo-ku, Tokyo 113, Japan

Joel Unowsky • Department of Chemotherapy, Roche Research Center, Nutley, NJ 07110

Jean van Heijenoort • Biochimie Moléculaire et Cellulaire, Université Paris-Sud, 91405 Orsay, France

Yveline van Heijenoort • Biochimie Moléculaire et Cellulaire, Université Paris-Sud, 91405 Orsay, France

H. S. Ved • Department of Biochemistry, Temple University School of Medicine, Philadelphia, PA 19140

Masaaki Wachi • Institute of Applied Microbiology, University of Tokyo, Bunkyo-ku, Tokyo 113, Japan

Stephen G. Waley • Sir William Dunn School of Pathology, University of Oxford, Oxford OX1 3RE, United Kingdom

Christopher T. Walsh • Department of Chemistry, Massachusetts Institute of Technology, Cambridge, MA 02139

Peter Welzel • Fakultät für Chemie der Ruhr-Universität, Bochum, Federal Republic of Germany

F. B. Wientjes • Department of Electron Microscopy and Molecular Cytology, University of Amsterdam, 1018 TV Amsterdam, The Netherlands

Bernard Witholt • Department of Biochemistry, University of Groningen, 9747 AG Groningen, The Netherlands

C. L. Woldringh • Department of Electron Microscopy and Molecular Cytology, University of Amsterdam, 1018 TV Amsterdam, The Netherlands

Paul W. Wrezel • Kraft, Inc., Glenview, IL 60025

E. Wu • Lilly Research Laboratories, Indianapolis, IN 46285

L. Zawadzke • Department of Chemistry, Massachusetts Institute of Technology, Cambridge, MA 02139

Allen R. Zeiger • Department of Biochemistry and Molecular Biology, Jefferson Medical College, Thomas Jefferson University, Philadelphia, PA 19107

Hai Ching Zhao • Department of Molecular and Cell Biology, University of Connecticut, Storrs, CT 06268

Ying Fang Zhu • Waksman Institute of Microbiology, Rutgers University, The State University of New Jersey, Piscataway, NJ 08855-0759

S. Zighelboim-Daum • Department of Medicine, New England Deaconess Hospital, Harvard Medical School, Boston, MA 02215

Preface

This volume is based on an ASM conference entitled "Antibiotic Inhibition of Bacterial Cell Surface Assembly and Function," held at the College of Physicians of Philadelphia on 17–20 May 1987. The aim of this conference was to communicate the latest advances, from basic studies of structure, function, assembly, and degradation of bacterial cell walls to the application of this and other information to learning more of how the group of "cell wall antibiotics," and especially β-lactams, work to inhibit or modify these processes. We are intrigued by the many applications of modern chemical, physical, molecular, and genetic technologies to these problems of both theoretical and practical interest. We believe that the contents of this volume will provide the reader with an up-to-date view of this active field of research and perhaps inspire the application of these new thoughts and methods to help solve many of the remaining and newly uncovered problems.

Paul Actor
Lolita Daneo-Moore
Michael L. Higgins
Milton R. J. Salton
Gerald D. Shockman

Acknowledgments

First, we would like to thank each of the participants and contributors to this volume for their excellent and prompt contributions; the ASM Meetings Department, and in particular Richard Bray and Caroline Polk, for their help in organizing the conference that led to this volume; and the ASM Publications Office, and especially Kirk Jensen, for their help in rapidly editing and publishing these proceedings. We also thank those companies and government agencies that provided financial support, including Hoffmann-LaRoche, Miles Pharmaceuticals, Lilly Research Laboratories, Upjohn Co., E. I. Du Pont de Nemours and Co., Glaxo USA, Merck Institute of Therapeutic Research, Pfizer Central Research, E. R. Squibb and Sons, American Cyanamid, Burroughs Wellcome & Co., CIBA-GEIGY, Schering Corporation, Smith Kline & French, Wyeth Laboratories, the National Science Foundation (grant no. DCB-8704227), and the National Institute of Allergy and Infectious Diseases (grant no. AI24920-01). Finally, we express special gratitude to Elizabeth Asklar and Gregory Harvey of the Department of Microbiology and Immunology of Temple University School of Medicine for their help in organizing and presenting the conference and this volume.

The Editors

I. STRUCTURE AND FUNCTION

Chapter 1

Wall Ultrastructure: How Little We Know

Terry J. Beveridge

One of the characteristic features of procaryotes is that they are the smallest known cellular life form. One of the prime reasons for this trait is that they depend on diffusion for their existence; whatever goes into or out of these cells must diffuse through their cell walls. Accordingly, a bacterium's ratio of surface area to volume is important, and I suspect that one reason for the diversity of bacterial shapes is to satisfy this requirement. There are no inner scaffoldings such as microtubules or microfilaments in these cells; their shape and form are determined by their encompassing walls. Of course, there are certain exceptions to this general rule since the mollicutes do not possess walls and *Spiroplasma*, somehow, manages to maintain a spiral shape (18, 19). But, with few other exceptions (10a), walls provide rigidity against cellular turgor pressure, and the protoplast conforms to the shape of the wall.

Basic Wall Formats

For almost a century, the Gram reaction has been used to differentiate bacteria into gram-negative and gram-positive varieties. Although many reasons for this fundamental staining difference have been entertained, Salton's work in 1963 pointed clearly to a wall-mediated mechanism (40), and our own work unequivocally verified his results (9, 22). Although Christian Gram did not realize it, his reaction was actually telling us of a fundamental difference in the chemical-structural organization of two types of eubacterial walls. Gram-

Terry J. Beveridge • Department of Microbiology, College of Biological Science, University of Guelph, Guelph, Ontario N1G 2W1, Canada.

3

negative walls (as represented by *Escherichia coli*) possess an outer membrane (OM) consisting of lipopolysaccharide (LPS), protein, and phospholipid which overlies a thin layer of peptidoglycan (PG; typically of A1γ chemotype), whereas gram-positive walls (as represented by *Bacillus subtilis*) are relatively thick (ca. 25 nm), amorphous matrices of intertwined PG (a number of chemotypes exist; see reference 41) and secondary polymers (often teichoic or teichuronic acid) (7). Obviously, this is a simplistic interpretation of the biochemical-structural relationships within these wall types since subtle differences occur between genera, species, and strains and more chemical constituents are involved.

In ultrastructural terms, the advent of cryoelectron microscopy as a method for physically preserving biological structure (24, 51) and the use of low-temperature plastic resins for maintenance of native conformation (17) offer new hope for obtaining the "correct" view of walls. In fact, freeze substitution, which offers the advantages of both techniques, has profoundly altered our perception of gram-negative and gram-positive cells (2, 3, 31). *E. coli* walls show a definite heavy metal staining asymmetry of the OM which conforms to the concept that LPS resides predominantly on the outer leaflet (Fig. 1). Between the OM and the plasma membrane (PM) is the so-called periplasmic gel (31) which we believe is primarily a mixture of PG, periplasmic proteins, and wall precursors. *B. subtilis* walls are no longer a featureless, amorphous matrix (3; T. J. Beveridge et al., *Proc. Microsc. Soc. Can.* 1985, p. 22–23), but show a surface fringe which may reflect wall turnover or exposed secondary polymer (Fig. 2).

As far as the actual Gram reaction goes for gram-positive and gram-negative eubacteria, it is the amount of PG in the wall which appears to determine the bacterium's response to the staining regimen. The crystal violet penetrates both cell types, and a simple metathetical anion exchange (chloride is replaced by iodide) induced by "Gram's iodine" produces large, purple stain aggregates within the cytoplasm (22). Although the ethanol of the decolorizing step solubilizes the PM of gram-positive organisms, the stain aggregate cannot escape through the wall and be washed out (9). The OM and PM of gram-negative organisms are both solubilized, and the stain aggregate punctures the thin PG layer to be washed away (9); the cell remnants are stained red by the counterstain. Gram-positive and gram-negative bacteria are designated as such by the chemical-structural format of their respective walls.

A class of bacteria whose wall structure has not been studied in detail with respect to the Gram reaction is the gram-variable variety. Most frequently, this variability occurs with certain gram-positive organisms which have a tendency to become gram negative during the later stages of growth or under growth stress (e.g., nutrient limitation or suboptimal growth temperature). Several *Bacillus, Clostridium,* and *Butyrivibrio* spp. are notorious for this staining phenomenon and can be represented by *C. thermosaccharolyticum*. Cultures of this bacterium stain gram positive during their lag and early exponential phases

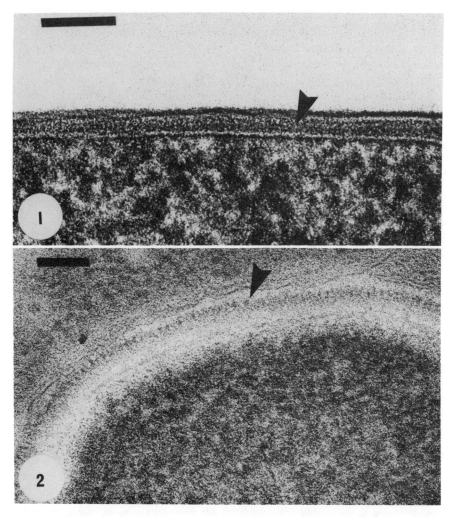

Figure 1. Cell envelope profile of *E. coli* K-12 by the freeze-substitution method. The arrowhead points to the "periplasmic gel." The outer face of the OM is intensely stained. Bar, 100 nm.
Figure 2. Cell wall of *B. subtilis* 168. Note the "picket fence" arrangement of the wall's surface (arrowhead). Bar, 100 nm. See T. J. Beveridge, B. J. Harris and Humphrey (*Proc. Microsc. Soc. Can.* 1985, p. 22–23) for more details of technique.

of growth, but become more gram negative as they age. Electron microscopy, using our modified Gram technique (22), revealed that the PG layer becomes increasingly thin with culture age (Fig. 3) until it can no longer withstand the stress of the decolorization step; these cells are now gram negative (Fig. 4). Cells of this type frequently possess a variable thickness of PG with a pro-

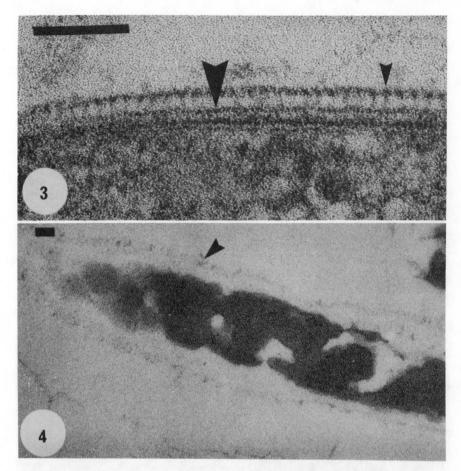

Figure 3. Cell envelope of *C. thermosaccharolyticum* by conventional fixation. The large arrow-head points to the thin PG layer, and the small one indicates the surface array. Bar, 100 nm.
Figure 4. A late-exponential-growth-phase cell of *C. thermosaccharolyticum* after ethanol de-colorization during our modified Gram stain (9, 22). Voids appear in the cytoplasm, and no platinum (from the stain) is detected. The arrowhead points to the remnants of the wall. Bar, 100 nm.

teinaceous surface array on top; obviously, this wall format cannot be considered a "typical" gram-positive variety.

Another group of gram-positive organisms, the *Mycobacterium-Coryne-bacterium-Proprionibacterium-Arthrobacter* group, have a different response to the Gram reaction. In cultures of these bacteria, a select percentage (often ca. 10%) of the culture will be gram negative. Electron microscopy has determined that it is those cells which have entered the septal division stage which become gram negative during staining. The growing septum seems to be a delicate region in the wall sacculus and is prone to "blow-out" during decolorization (Fig. 5).

So far, I have not mentioned archaebacteria and their reponse to the Gram stain. This is because their wall chemistry and structure are so profoundly different from those of eubacteria that no simple encompassing statement can be made. For example, murein is not a constituent of their walls, but a select few genera (i.e., *Methanobrevibacter, Methanobacterium,* and *Methanothermus*) possess pseudomurein (33). Walls of some genera (e.g., *Sulfolobus, Thermoproteus,* and *Methanococcus*) consist of only a single surface array, whereas other genera (e.g., *Methanospirillum* and *Methanothrix*) possess a wall and a complex sheath/spacer plug system (33). Cell envelope chemistry can be much more complicated; e.g., *Halococcus* and *Methanosarcina* spp. contain complex heteropolysaccharides (33), and phytanyl diethers are a major constituent of

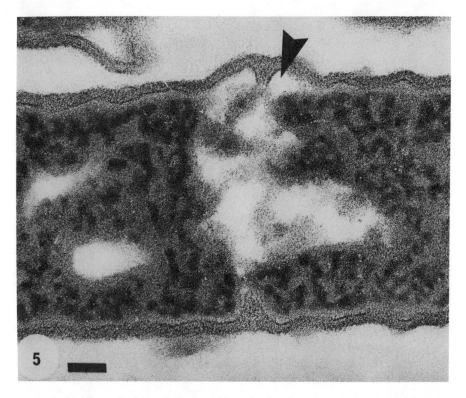

Figure 5. Dividing cell of *Mycobacterium phlei* which has been stained by our modified Gram stain (9, 22). The septum region has "blown-out," leaving a large void at the center of the cell. Only a weak platinum signal was obtained from this cell. Bar, 100 nm.

the membrane lipids (34). In fact, tetraether-based lipids have been found in several methanogens.

Obviously, gram-positive and gram-negative archaebacteria exist, but the wall chemical-structural basis for their staining response has not been widely studied. Intuitively, we suspect that those archaebacteria which possess only a surface array as their wall will be gram negative because the array cannot withstand the stress of decolorization. More difficult are those multienveloped archaebacteria such as *Methanospirillum hungatei*. This bacterium's envelope consists of a PM, a wall, and a sheath (Fig. 6); it is the sheath which is the shape-maintaining structure (10a). It possesses an extremely resilient proteinaceous array (15) which has a minute 2.8-nm spacing (42, 49) (Fig. 7). Clearly, this archaebacterial envelope should be at least as robust as a eubacterial gram-

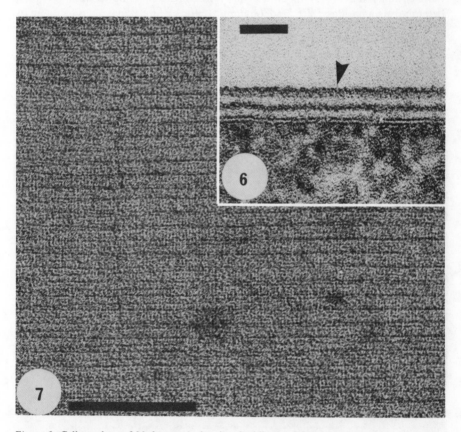

Figure 6. Cell envelope of *M. hungatei,* showing the bilayer of the plasma membrane, the overlying cell wall, and the sheath (arrowhead). Bar, 100 nm.
Figure 7. Negative stain of the sheath surface, which possesses a minute p2 spacing group of 5.6 by 2.8 nm (49). Bar, 100 nm.

positive wall and the bacterium should stain positive. In fact, *M. hungatei* stains gram negative because the sheath array has such a fine spacing that the crystal violet cannot get into the cells; it sticks to the outside of the sheath and is washed away.

The point, of course, is that although walls (and sheaths) in archaebacteria should control the Gram reaction, as they do in eubacteria, we cannot easily extrapolate from one "urkingdom" to the other. Staining differences reflect chemical-structural styles in walls but, ultimately, rely on penetration differences after ethanol decolorization. These, in fact, point out fundamental disparities in wall makeup which manifest themselves in a variety of ways, including sensitivity to antibiotics, to enzymes, and to mechanical disruption.

Extra Layers and Wall Extensions

Since I cannot spend much time on this section, it cannot be all inclusive, and I refer the reader to references 7, 43, and 44 for more information. It is important to recognize the existence of extra layers because we are detecting them with increasing frequency, especially on those bacteria which have been recently isolated from a natural horizon or which have been observed in situ. These layers are frequently lost after serial laboratory subculture.

Surface Arrays

Surface arrays (or S- or RS-layers) consist of protein or glycoprotein particles self-assembled into a planar crystal which completely encompasses the outside surface of the native wall and, accordingly, completely surrounds the bacterial cell (7, 43, 44). They are common to both archaebacteria and eubacteria (7, 43, 44) and can exist as single (e.g., *Thermoplasma* spp. [37]) or multiple layers (e.g., *Aquaspirillum metamorphum* [11] or *Aquaspirillum* "Ordal" [12]) as the outermost wall component. In addition, they may be closely associated with capsular material which, presumably, extends from the native wall through them (e.g., *Bacillus anthracis;* R. J. Doyle, T. Beveridge, M. Stewart, and J. Ezzell, *Abstr. Annu. Meet. Am. Soc. Microbiol.* 1986, J18, p. 191).

A number of lattice formats are available for these arrays. Most commonly, p6 (hexagonal) and p4 (tetragonal) symmetries are seen, but a few examples of p2 and at least one example of p1 have been described (43). The p1 symmetry of *B. anthracis* (Doyle et al., *Abstr. Annu. Meet. Am. Soc. Microbiol.* 1986) is especially exciting since the subunits presumably exist as single polypeptides (Fig. 8).

Most of the arrays on eubacterial walls are arranged so as to form aqueous channels ca. 2 to 3 nm in diameter, just the size which could exclude enzymes

Figure 8. Computer reconstruction of the surface array of *B. anthracis,* which has p1 symmetry. The light-gray regions represent the protein, whereas the dark areas are depressions filled with negative stain. Bar, 100 nm. This work was done in collaboration with M. Stewart and R. Doyle.

such as exomuramidases from entry to the underlying wall fabric. Accordingly, we speculated that one of the functions of these arrays could be to act as a sieving layer against harmful macromolecules (48). In addition, at least one of these layers seems somehow to be able to regulate its permeability; the A-layer of *Aeromonas salmonicida* is a virulence factor and exists in two separate patterns of the same protein, each having different channel diameters (50). Remarkably, Sleytr's group has been able to mount some of these arrays on synthetic support films for use in industry as ultrafiltration membranes (45).

The spacing of archaebacterial arrays is frequently larger (pore diameter, ca. 5 to 6 nm), as seen in *Thermoproteus* spp. (37) and *Sulfolobus acidocaldarius* (52). At the same time, the smallest array spacing is found on *Methanospirillum hungatei* and *Methanothrix concilii* (38, 42, 49). So once again, archaebacteria are full of surprises.

Capsules, Sheaths, and Wall Extensions

Capsules have also been called glycocalyces (20), but it is important to recognize that not all capsules possess a "glyco" chemical component; there is a wide range of chemotypes (e.g., *Bacillus licheniformis* ATCC 9945A has a polyglutamic acid capsule [7, 53]). Of all the constituents of bacterial en-velopes, capsules are the most difficult to preserve for ultrastructural deter-mination. We believe them to be thixotropic and to alternate between a gel and a liquid state. They are highly hydrated structures, and unless special precau-tions are taken, they collapse when conventional preparatory techniques for electron microscopy are used. Ruthenium red and alcian blue are electroposi-tive electron microscopy stains which have been used to stabilize anionic cap-sules, and specific capsular antibody has also proved successful as a stabilizing agent (6, 7, 20, 32). More recently (1985), Bayer has relied on dimethyl form-amide dehydration and Lowicryl K4M embedding of gelatin-enrobed *E. coli* to preserve the K-29 capsule (5). My own laboratory has had good success using the freeze-substitution method. For example, in our experience, the *E. coli* K-30 capsule is especially difficult to stabilize since only one-fourth of its mass contains chemically reactive groups [it is a branched polymer of mannose-glucuronic acid-(galactose)$_2$], yet freeze-substitution provides exquisite detail of its fibrous extensions from the cell surface (Fig. 9).

Several groups of bacteria form chains of cells and are encased in sheaths which resemble hollow cylinders when devoid of cells. For these cells, the

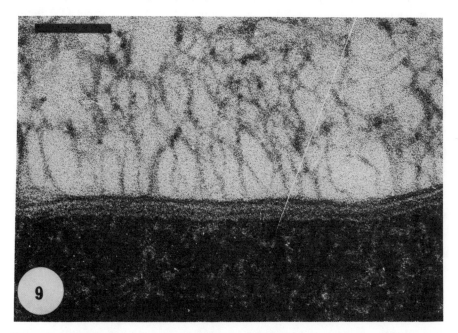

Figure 9. E. coli K-30 envelope processed by the freeze-substitution method, which has pre-served the capsule as thin fibrils. Bar, 100 nm. This work was done in collaboration with R. Harris and C. Whitfield.

sheath is the outermost surface layer and is in close contact with the underlying wall. They come in two structural varieties: those which are highly ordered (*Methanothrix* and *Methanospirillum* spp. are representative [Fig. 7]) and those which are fibrillar (e.g., in *Leptothrix* or *Sphaerotilus* spp. [Fig. 10]). The latter type are notorious for their ability to oxidize and precipitate iron and manganese from solution (30). Although chemical descriptions of *Methanothrix* and *Methanospirillum* sheaths have been published (they are mostly protein with small proportions of lipid and carbohydrate [15, 38]), very little is known about *Leptothrix* and *Sphaerotilus* sheaths except that they are composed mainly of acidic polysaccharides (30). The Mn^{2+}-oxidizing activity of *Leptothrix* spp. is due to an extracellular protein (1, 16).

Cell wall extensions are seen in some gram-negative eubacteria. By extension, I mean that the wall is forced out of its usual surface location to form a structure which occupies a position more remote from the cell. For example, in vibrioid cells such as *Vibrio cholerae* or *Bdellovibrio bacteriovorus,* the outer membrane forms a sheath which encompasses all but the distal tip of each single polar flagellum (Fig. 11) (28). These flagellar sheaths are physically parts of the wall, but their function remains unknown.

OM blebs are another wall extension which is common with some gram-negative organisms, especially members of the family *Pseudomonadaceae.* For these bacteria there appears to be a continuous excretion of small OM blebs during all growth phases, and although it is difficult to unequivocally state that

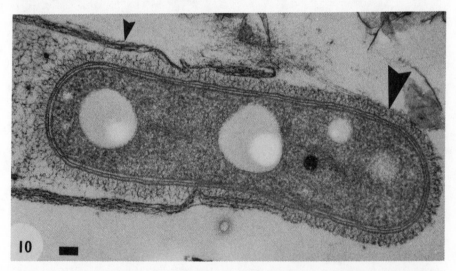

Figure 10. Freeze substitution of *L. discophora.* This is a swarmer cell which is escaping from the sheath (small arrowhead), and the cell possesses a capsule (large arrowhead) which is distinct from the sheath. The culture was supplied to us by W. Ghiorse.

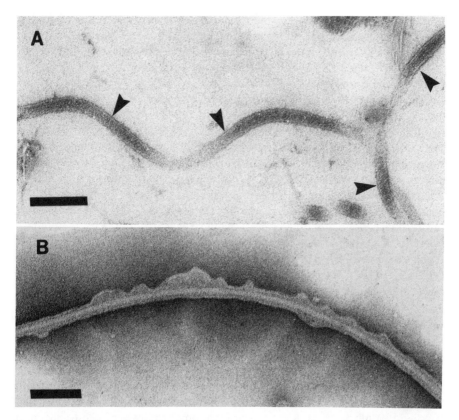

Figure 11. (A) Thin section of *V. cholerae* flagella. The arrowheads point to regions where the bilayer of the sheath can be seen. (B) Negative stain of a *V. cholerae* flagellum. The membranous sheath is disrupting from the flagellin shaft. Bars, 100 nm.

it is a natural phenomenon, intuitively the occurrence of blebbing under a variety of laboratory enrichment conditions suggests that it must be. Chemical agents such as EDTA, heavy metal cations, and aminoglycoside antibiotics tremendously increase the frequency of OM blebbing by interfering with constituent Mg^{2+} and Ca^{2+} which stabilize the membrane's lipid packing order (27, 35, 39; S. G. Walker and T. J. Beveridge, *Can. J. Microbiol.*, in press). When these metallic cations are removed, the packing order of select surface regions is so profoundly altered that the OM is forced into small, tensed vesicles of minimal free energy; in a thermodynamic sense, blebbing must occur (27; Walker and Beveridge, in press). The question, then, is whether or not natural OM blebbing is similarly a function of local, destabilized OM domains. If so, bacteria must have a regulatory mechanism which is able to modulate the lipid packing order to produce these local domains. Obviously, with bacteria, such as *Bacteroides succinogenes,* which grow on crystalline cellulose, OM bleb-bing is a necessity of life since the blebs contain all the exoenzymes required to digest the food source; the blebs adhere to and hydrolyze the cellulose (29).

Physicochemistry of Walls

Taken to their lowest common denominator, all bacterial walls, no matter their structural-chemical format, are an external interface with the local environment of the bacterium. For this reason, surface physicochemistry is important; it will be responsible for how the cells interact with their external milieu. Because interfacial chemistry is the controlling factor in this interaction, simple equilibrium solution chemistry is not the entire answer. The envelope can be regarded as a reaction surface where chemical reactions run more rapidly at a lower thermodynamic threshold than their soluble counterparts. At the same time, tremendous concentration gradients can be established. The physico-chemistry of the bacterial surface is built into the envelope fabric and is an essential design feature of procaryotic life.

Each wall type (whether it be gram positive or negative, or eubacterial or archaebacterial) has its own defined chemistry and structure which gives an individual "flavor" to its physicochemistry. For example, *B. subtilis* 168 walls, when grown in the presence of Mg^{2+} and PO_4^{3-}, are essentially a two-polymer system of approximately equal amounts of PG (A1γ) and teichoic acid. Both polymers are anionic and give the wall an overall electronegative charge density at neutral pH (7). When these walls are suspended in solutions of metallic salts, they bind so much metal that electron-scattering profiles are seen by electron microscopy (13). The COO^- groups of the peptidoglycan appear to dominate the reaction (14, 23, 36), and large metallic precipitates are grown (13, 14). *B. licheniformis* NCTC 6346 *his* walls are composed predominantly of teichoic and teichuronic acids (ca. 80% on a micromolar basis) when compared with constituent PG, and this biochemical difference ensures that they bind less metal than their *B. subtilis* counterparts (8, 10).

In actual fact, the physicochemistry of these two bacillus wall types differs much more than their chemistries would indicate. Electron microscopy of iso-lated walls treated with cationized ferritin to localize electronegative surface sites showed profound differences. Both the inner and outer surfaces of *B. licheniformis* walls were labeled by the probe, whereas only the outer surface of *B. subtilis* walls bound the cationized ferritin (Fig. 12) (47). The new inner surfaces of wall septa were also not labeled. The asymmetry of surface charge in *B. subtilis* is determined by the orientation of and combined charges of the PG and teichoic acid (47). Here is an example where the structural organization of the constituent wall polymers profoundly influences the surface effect. To complicate the issue further, very dilute concentrations of cationized ferritin

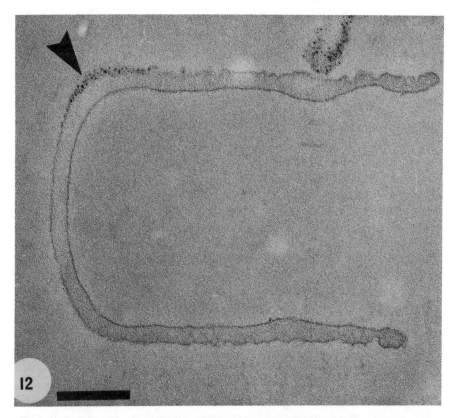

Figure 12. Thin section of a *B. subtilis* wall which has been treated with a very dilute concentration of cationized ferritin. For these walls only the outer surface is labeled, and because dilute cationized ferritin has been used, only the most electronegative areas of the wall have been labeled (arrowhead). See references 46 and 47 for further details. Bar, 100 nm. (Reproduced with permission of the authors and *Can. J. Microbiol.*)

bind to discrete surface patches on the wall (Fig. 12) (46); some surface regions are more highly charged than others.

Gram-negative walls, as represented by *E. coli* K-12, are obviously different from bacillus walls. Their PG is masked from the external milieu by the OM, and nonspecific solute access is controlled by means of a system of small aqueous channels composed of porin (7). As an external surface, it is the LPS face of the OM which is in immediate contact with the environment. The OM does not appear to have the same high charge density as bacillus walls (8), but metallic cations (primarily Mg^{2+} and Ca^{2+}) are essential for the existence of the membrane. They are especially important for LPS since this lipid contains numerous electronegative phosphoryl groups (21, 25, 27). Although 2-keto-3-deoxyoctonate accounts for three COO^- groups in this LPS, only one is available to react with soluble metals since the other two are neutralized by constituent amine functions within the inner core region (26). As mentioned previously, these LPS-associated metals are essential for bilayer stability in gram-negative OMs. For those bacteria which possess surface arrays above the wall, metallic ions are often required for both subunit-to-subunit and subunit-to-wall interaction (7).

How Little We Know

So far, I have been attempting to relate the diversity of wall types, their chemistry, and their structure with bacterial surface physicochemistry. Although a limitation on manuscript length has not allowed an in-depth treatment of all topics, it is apparent that most of our information is on only a few select wall systems. Currently, *E. coli* and *P. aeruginosa* systems are the most popular gram-negative models, whereas *Bacillus* spp., *Streptococcus* spp., and *Staphylococcus* spp. account for the majority of the work on gram-positive walls. Yet, we know there is a tremendous variety of wall types in the procaryotic kingdom, and one of the real questions is how representative our model systems are. Bacteria have inhabited the earth for at least 3.5 billion years and during this period have separated into two fundamentally different lines, the archaebacteria and the eubacteria. Given this tremendous timeframe, mutation, and natural selection, it is obvious that our model systems for walls cannot be all-encompassing.

Then, of course, there is the problem of laboratory versus natural conditions. In the laboratory, we pride ourselves on growing bacteria with rapid doubling rates (T_Ds of 20 to 30 min are possible) using monocultures; in nature, rapid growth is infrequent and monocultures are rarely, if ever, encountered. In the laboratory we are lacking the important natural relationships between organisms and their environment.

At the same time, we are completely neglecting the "individuality" of bacterial cells. Clearly, bacteria grow and divide, and if individually sampled from a culture, each cell would be in a specific growth state; some may be newly formed daughter cells, and others may be initiating septation. Obviously, each growth state requires a cell of a definite physical size and chemical composition. Since most of our physical and chemical analyses of walls have been derived from machinery requiring microgram, milligram, or even gram quantities of material (an *E. coli* cell weighs ca. 10^{-12} g), these analyses are enormously averaging. Subtle differences between individual cells are missed. For example, it is difficult to imagine that, as a bacillus grows and divides, there are not local differences within the wall matrix. Obviously, a growing septum must differ from cylindrical wall, and a completed septum must mature into a pole. There must also be differences between the pole matrix and that of the cylindrical wall which allows one, but not the other, to take on a hemispherical curvature. The point of junction between the two matrix types is of critical importance.

Certainly, the existence of unique shapes in bacteria requires local alterations to the wall fabric to manifest subtle folds and curves. We do not know exactly how cells such as spirilla and spirochetes determine their spiral shape, nor how *Microcyclus* becomes a ring, but in both cases it is a wall phenomenon. Likewise, stalk formation in *Caulobacter* and *Hyphomicrobium* spp. is due to wall extension.

Certainly, too, organelles such as flagella and pili must extend through and disturb the local wall fabric, yet (somehow) these regions resist the stress exerted on them by the turgor pressure in the protoplast; they do not "blowout." A similar point could be made for the "Bayer's patches" in gram-negative cells (4), which appear to be local sites of wall precursor transport. X-ray diffraction data suggest that the holes in the PG layer are so small that most macromolecules would have difficulty passing through (7). Is it possible that these patches do not contain PG, or is the PG in these patches of greater porosity?

For our perception of procaryotic cellular life, these are all fundamental questions which do not have a ready solution. "Averaging" techniques will not work since, unless we devise a means of isolation and purity, the subtle wall alterations we wish to examine will be a minor component of the entire wall matrix and will be overwhelmed by the major signal. To observe individual bacteria, we are forced to use some sort of microscopy. Bragg's law and the wavelength of visible light (0.7 to 0.4 μm) limit the resolving power of light microscopy and restrict it to detecting shape and the placement of associated large particles (e.g., developing endospores, flagella bundles, poly-β-hydroxybutyrate granules). To resolve, in detail, the structural components of individual bacteria, electron microscopy must be used. There are many problems associated with the technique, but a thorough understanding of the principles involved with imaging biological material by means of accelerated beams of electrons can provide a range of structural and compositional data (32). More than any other biological science, microbiology requires the use of electron microscopy for relating internal cell structure with cellular shape and form. Recent advances in cryoprocessing offer the potential to allow the examination of bacteria under conditions very similar to their native state.

ACKNOWLEDGMENTS. The work from my laboratory presented in this paper has been funded by grants from the Natural Sciences and Engineering Research Council of Canada and the Medical Research Council of Canada. Research on the gram-variable reaction proceeded in my laboratory with the help of M. Luckevich, who was partially funded by a Natural Sciences and Engineering Research Council of Canada Undergraduate Summer Research Award.

LITERATURE CITED

1. **Adams, L. F., and W. C. Ghiorse.** 1987. Characterization of extracellular Mn^{2+}-oxidizing activity and isolation of an Mn^{2+}-oxidizing protein from *Leptothrix discophora* SS-1. *J. Bacteriol.* **169:**1279–1285.

2. **Amako, K., K. Murata, and A. Umeda.** 1983. Structure of the envelope of *Escherichia coli* observed by the rapid-freezing and substitution-fixation method. *Microbiol. Immunol.* **27:**95–99.

3. **Amako, K., and A. Takade.** 1985. The fine structure of *Bacillus subtilis* revealed by rapid-freezing and substitution-fixation method. *J. Electron Microsc.* **34:**13–17.

4. **Bayer, M. E.** 1968. Areas of adhesion between wall and membrane of *Escherichia coli*. *J. Gen. Microbiol.* **53:**395–404.

5. **Bayer, M. E., E. Carlemalm, and E. Kellenberger.** 1985. Capsule of *Escherichia coli* K29: ultrastructure preservation and immunoelectron microscopy. *J. Bacteriol.* **162:**985–991.

6. **Bayer, M. E., and H. Thurow.** 1977. Polysaccharide capsule of *Escherichia coli:* microscope study of its size, structure, and sites of synthesis. *J. Bacteriol.* **130:**911–936.

7. **Beveridge, T. J.** 1981. Ultrastructure, chemistry and function of the bacterial wall. *Int. Rev. Cytol.* **72:**229–317.

8. **Beveridge, T. J.** 1984. Mechanisms of the binding of metallic ions to bacterial walls and the possible impact on microbial ecology, p. 601–607. *In* C. A. Reddy and M. J. Klug (ed.), *Current Perspectives in Microbial Ecology.* American Society for Microbiology, Washington, D.C.

9. **Beveridge, T. J., and J. A. Davies.** 1983. Cellular responses of *Bacillus subtilis* and *Escherichia coli* to the Gram stain. *J. Bacteriol.* **156:**846–858.

10. **Beveridge, T. J., C. W. Forsberg, and R. J. Doyle.** 1982. Major sites of metal binding in *Bacillus licheniformis* walls. *J. Bacteriol.* **150:**1438–1448.

10a. **Beveridge, T. J., B. J. Harris, and G. D. Sprott.** 1987. Septation and filament splitting in *Methanospirillum hungatei. Can. J. Microbiol.* **33:**725–732.

11. **Beveridge, T. J., and R. G. E. Murray.** 1975. Surface arrays on the cell wall of *Spirillum metamorphum. J. Bacteriol.* **124:**1529–1544.

12. **Beveridge, T. J., and R. G. E. Murray.** 1976. Superficial cell wall layers on *Spirillum* "Ordal" and their *in vitro* reassembly. *Can. J. Microbiol.* **22:**567–582.

13. **Beveridge, T. J., and R. G. E. Murray.** 1976. Uptake and retention of metals by cell walls of *Bacillus subtilis. J. Bacteriol.* **127:**1502–1518.

14. **Beveridge, T. J., and R. G. E. Murray.** 1980. Sites of metal deposition in the cell wall of *Bacillus subtilis. J. Bacteriol.* **141:**876–887.

15. **Beveridge, T. J., M. Stewart, R. J. Doyle, and G. D. Sprott.** 1985. Unusual stability of the *Methanospirillum hungatei* sheath. *J. Bacteriol.* **162:**728–737.

16. **Boogerd, F. C., and J. P. M. deVrind.** 1986. Manganese oxidation by *Leptothrix discophora. J. Bacteriol.* **169:**489–494.

17. **Carlemalm, E., R. M. Garavito, and W. Villiger.** 1982. Resin development for electron microscopy and an analysis of embedding at low temperature. *J. Microsc.* **126:**123–143.

18. **Charbonneau, D. L., and W. C. Ghiorse.** 1984. Ultrastructure and location of cytoplasmic fibrils in *Spiroplasma floricola* OBMG. *Curr. Microbiol.* **10:**65–72.

19. **Cole, R. M., J. G. Tully, T. J. Popkin, and J. M. Bové.** 1973. Morphology, ultrastructure, and bacteriophage infection of a helical mycoplasma-like organism (*Spiroplasma citri* gen. nov., sp. nov.) cultured from "stubborn" disease of citrus. *J. Bacteriol.* **115:**367–386.

20. **Costerton, J. W., R. T. Irvin, and K.-J. Cheng.** 1981. The bacterial glycocalyx in nature and disease. *Annu. Rev. Microbiol.* **35:**229–324.

21. **Couglin, R. T., S. Tonsager, and E. J. McGroarty.** 1983. Quantitation of metal cations bound to membranes and extracted lipopolysaccharide from *Escherichia coli. Biochemistry* **22:**2002–2007.

22. **Davies, J. A., G. K. Anderson, T. J. Beveridge, and H. C. Clarke.** 1983. Chemical mechanism of the Gram stain and synthesis of a new electron-opaque marker for electron microscopy which replaces the iodine mordant of the stain. *J. Bacteriol.* **156:**837–845.

23. **Doyle, R. J., T. H. Matthews, and U. N. Streips.** 1980. Chemical basis for the selectivity of metal ions by the *Bacillus subtilis* wall. *J. Bacteriol.* **143:**471–480.

24. **Dubochet, J., A. W. McDowall, B. Menge, E. N. Schmid, and K. G. Lickfeld.** 1983. Electron microscopy of frozen-hydrated bacteria. *J. Bacteriol.* **155:**381–390.

25. **Ferris, F. G., and T. J. Beveridge.** 1984. Binding of a paramagnetic metal cation to *Escherichia coli* K-12 outer membrane vesicles. *FEMS Microbiol. Lett.* **24:**43–46.

26. **Ferris, F. G., and T. J. Beveridge.** 1986. Site specificity of metallic ion binding in *Escherichia coli* K-12 lipopolysaccharide. *Can. J. Microbiol.* **32:**52–55.

27. **Ferris, F. G., and T. J. Beveridge.** 1986. Physicochemical roles of soluble metal cations in the outer membrane of *Escherichia coli* K-12. *Can. J. Microbiol.* **32:**594–601.

28. **Ferris, F. G., T. J. Beveridge, M. L. Marceau-Day, and A. D. Larson.** 1984. Structure and cell envelope associations of flagellar basal complexes of *Vibrio cholerae* and *Campylobacter fetus. Can. J. Microbiol.* **30:**322–333.

29. **Forsberg, C. W., T. J. Beveridge, and A. Hellstrum.** 1981. Cellulase and xylanase release from *Bacteroides succinogenes* and its importance in the rumen environment. *Appl. Environ. Microbiol.* **42:**886–896.

30. **Ghiorse, W. C.** 1984. Biology of iron- and manganese-depositing bacteria. *Annu. Rev. Microbiol.* **38:**515–550.

31. **Hobot, J. A., E. Carlemalm, W. Villiger, and E. Kellenberger.** 1984. The periplasmic gel: a new concept resulting from the re-investigation of bacterial cell envelope ultrastructure by new methods. *J. Bacteriol.* **160:**143–152.

32. **Holt, S. C., and T. J. Beveridge.** 1982. Electron microscopy: its development and application to microbiology. *Can. J. Microbiol.* **28:**1–53.

33. **Kandler, O., and H. König.** 1985. Cell envelopes of archaebacteria, p. 413–457. *In* C. R. Woese and R. S. Wolfe (ed.), *The Bacteria: a Treatise on Structure and Function,* vol. 8. *Archaebacteria.* Academic Press, Inc., New York.

34. **Langworthy, T. A.** 1985. Lipids of archaebacteria, p. 459–497. *In* C. R. Woese and R. W. Wolfe (ed.), *The Bacteria: a Treatise on Structure and Function,* vol. 8. *Archaebacteria.* Academic Press, Inc., New York.

35. **Martin, N. L., and T. J. Beveridge.** 1986. Gentamicin interaction with *Pseudomonas aeruginosa* cell envelope. *Antimicrob. Agents Chemother.* **29:**1079–1087.

36. **Matthews, T. H., R. J. Doyle, and U. N. Streips.** 1979. Contribution of peptidoglycan to the binding of metal ions by the cell wall of *Bacillus subtilis. Curr. Microbiol.* **3:**51–53.

37. **Messner, P., D. Pum, M. Sára, K. O. Stetter, and U. B. Sleytr.** 1986. Ultrastructure of cell envelope of the archaebacteria *Thermoproteus tenax* and *Thermoproteus neutrophilus. J. Bacteriol.* **166:**1046–1054.

38. **Patel, G. B., G. D. Sprott, R. W. Humphrey, and T. J. Beveridge.** 1986. Comparative analyses of the sheath structures of *Methanothrix concilii* GP6 and *Methanospirillum hungatei* strains GP1 and JF1. *Can. J. Microbiol.* **32:**623–631.

39. **Peterson, A. A., R. E. W. Hancock, and E. J. McGroarty.** 1985. Binding of polycationic antibiotics and polyamines to lipopolysaccharide of *Pseudomonas aeruginosa. J. Bacteriol.* **164:**1256–1261.

40. **Salton, M. R. J.** 1963. The relationship between the nature of the cell wall and the Gram stain. *J. Gen. Microbiol.* **30:**223–235.

41. **Schleifer, K. H., and O. Kandler.** 1972. Peptidoglycan types of bacterial cell walls and their taxonomic implications. *Bacteriol. Rev.* **36:**407–477.
42. **Shaw, P. J., G. J. Hills, J. A. Henwood, J. E. Harris, and D. B. Archer.** 1985. Three-dimensional architecture of the cell sheath and septa of *Methanospirillum hungatei. J. Bacteriol.* **161:**750–757.
43. **Sleytr, U. B., and P. Messner.** 1983. Crystalline surface layers on bacteria. *Annu. Rev. Microbiol.* **37:**311–339.
44. **Sleytr, U. B., P. Messner, M. Sára, and D. Pum.** 1986. Crystalline surface layers in archaebacteria. *Syst. Appl. Microbiol.* **7:**310–313.
45. **Sleytr, U. B., and M. Sára.** 1986. Ultrafiltration membranes with uniform pores from crystalline bacterial cell envelope layers. *Appl. Microbiol. Biotechnol.* **25:**83–90.
46. **Sonnenfeld, E. M., T. J. Beveridge, and R. J. Doyle.** 1985. Discontinuity of charge on cell wall poles of *Bacillus subtilis. Can. J. Microbiol.* **31:**875–877.
47. **Sonnenfeld, E. M., T. J. Beveridge, A. L. Koch, and R. J. Doyle.** 1985. Asymmetric distribution of charge on the cell wall of *Bacillus subtilis. J. Bacteriol.* **163:**1167–1171.
48. **Stewart, M., and T. J. Beveridge.** 1980. Structure of the regular surface layer of *Sporosarcina ureae. J. Bacteriol.* **142:**302–309.
49. **Stewart, M., T. J. Beveridge, and G. D. Sprott.** 1985. Crystalline order to high resolution in the sheath of *Methanospirillum hungatei:* a cross-beta structure. *J. Mol. Biol.* **183:**509–515.
50. **Stewart, M., T. J. Beveridge, and T. J. Trust.** 1986. Two patterns in the *Aeromonas salmonicida* A-layer may reflect a structural transformation that alters permeability. *J. Bacteriol.* **166:**120–127.
51. **Stewart, M., and G. Vigers.** 1986. Electron microscopy of frozen-hydrated biological material. *Nature* (London) **319:**631–636.
52. **Taylor, K. A., J. F. Deatherage, and L. A. Amos.** 1982. Structure of the S-layer of *Sulfolbus acidocaldarius. Nature* (London) **299:**840–842.
53. **Troy, F. A.** 1973. Chemistry and biosynthesis of the poly (γ-D-glutamyl) capsule in *Bacillus licheniformi*s: characterization and structural properties of the enzymatically synthesized polymer. *J. Biol. Chem.* **248:**316–324.

Chapter 2

Turgor Pressure, Sporulation, and the Physical Properties of Cell Walls

Robert E. Marquis

Turgor and Walls

At this stage of development of our knowledge of bacterial cell physiology, it is a truism to say that the physical properties of the peptidoglycan networks of cell walls are of importance for wall functions. The defined functions of the wall are primarily mechanical: to protect the cell against osmotic damage and to exclude potentially damaging macromolecules. The pertinent physical properties then include elasticity, tensile strength, and porosity. Other functions have been postulated for the wall, for example, the sequestering of minerals through ion exchange or buffering involving titration of carboxyl or amino groups in the wall (1, 8). However, these functions are less clearly defined than the mechanical functions, although certainly the wall or the peptidoglycan network does behave as an amphoteric ion-exchange resin with a capacity as high as 3.54 meq/g (dry weight) in organisms such as *Micrococcus luteus* (24). Bacterial walls are designed not only to protect the cell against catastrophic changes in environmental osmolality but also to maintain turgor pressure required for metabolism, growth, and cell division. The roles of turgor in cell morphology, surface growth, and cell division have recently been explored in novel ways by Koch and colleagues (9, 10) in the surface-stress theory. To maintain turgor at a nearly constant level, bacterial cells have adaptive systems for pooling

Robert E. Marquis • Department of Microbiology and Immunology, The University of Rochester, Rochester, New York 14642.

21

osmolites such as K^+, betaines, and other amino acids or polyols (12). The pooled osmolites are referred to as compatible solutes because, even at high concentrations, they are compatible with maintenance of the functional conformations of enzymes and other biopolymers. Clearly, cells have means to sense turgor and to respond by synthesizing osmolites or taking them up from the environment. A recent suggestion for a sensing apparatus is a pressure-sensitive ion channel in *Escherichia coli* described by Martinac et al. (20).

There has been some debate over the past few years about turgor, and there are still unresolved issues. That turgor exists seems no longer to be in question. For example, the studies by Reed and Walsby (26) with bacteria containing very sensitive, collapsible pressure indicators in the form of gas vacuoles have confirmed earlier estimations of turgor based on determinations of plasmolysis thresholds (18, 21). Gram-negative bacteria appear to have turgor pressures of some 0.5 MPa (5 atm). Their walls are complex, with an outer membrane and a thin layer of peptidoglycan which may not even be sufficient to envelope the cell completely. The elastic elements resisting the expansive force of turgor for cells of gram-negative bacteria include both the outer membrane and the murein, so that stretching and tension development would occur in all the layers of the wall. Even the inner or protoplast membranes of bacterial cells can withstand moderate tension. Generally, when protoplasts or spheroplasts burst osmotically, the damage is due to brittle fracture of the membrane following rapid osmotic shock with apparent crystallization within the membrane structure; slow stretching of the membrane is much less likely to lead to fracture (4).

Generally, the walls of gram-positive bacteria are cytologically simpler than those of gram-negative bacteria and are composed of thick networks of peptidoglycan, apparently in the form of a single huge molecule per cell, to which accessory polymers may be attached covalently. Enzymes or other polymers may also be enmeshed within this network. The large elastic envelope of a typical cell of a gram-positive bacterium allows for development of turgor pressures as great as 3 MPa (18, 21). It serves as a heteroporous, molecular sieve, as shown by the work of Scherrer and Gerhardt (29). The ultimate cell in regard to peptidoglycan elasticity is the bacterial endospore. During development, forespores are engulfed by the sporangial cell, in which they then undergo the process of differentiation leading to the fully dormant, resistant, mature spore. The forespore then has two membranes: an inner, right-side-out membrane and an outer, inside-out membrane. Between the two membranes, a thick envelope of peptidoglycan is laid down to form the cortex of the mature spore. There are claims that the turgor pressure within spores may be some tens of megapascals (34). However, these claims have not been substantiated by direct measurements, although spores can be made to explode by exposing them to certain acids (33). At present, the exploding of spores is only an amusement for laboratory sadists, and its physicochemical basis is not known.

How does turgor develop in a cell? If we were to take a bacterial cell, say, a cell of *Bacillus megaterium*, and throw it into distilled water, the cell would be in an environment of very low osmolality with water activity close to 1.0. Water would start to flow across the water-permeable membrane into the cytoplasm, where the water activity would be less than 1.0 because of cytoplasmic solutes. The cell would start to swell as water moved in. However, the elastic cell wall would resist this swelling, tension would build up in the wall, and the hydrostatic pressure within the cell would be increased.

To consider the final state of the cell, we need to consider the electrochemical potential of water within the cell (μ_w^{in}) and that outside the cell (μ_w^{out}).

$$\mu_w^{in} = \mu_w^0 + RT \ln a_w^{in} + \bar{V}P^{in}$$
$$\mu_w^{out} = \mu_w^0 + RT \ln a_w^{out} + \bar{V}P^{out}$$

Here μ_w^0 is the standard electrochemical potential of water, which would be the same for both the interior of the cell and the exterior; R is the gas constant; T is the Kelvin temperature; a_w is water activity; $\bar{V}$ is the molar volume of water; and P is the pressure. At equilibrium, $\mu_w^{in} = \mu_w^{out}$, and the turgor pressure across the cell wall is $P^{in} - P^{out} = (2.30 \, RT/\bar{V})(\log a_w^{out}/a_w^{in})$ or at 25°C approximately 311 MPa ($\log a_w^{out}/a_w^{in}$).

If one assumes on the basis of studies of stabilization of enzymes against heat damage through dehydration that the a_w within a bacterial spore is 0.7, then the turgor pressure for a spore in deionized water should be some 48 MPa (34). This remarkably high value provides ample need for the spore to construct a very thick, elastic peptidoglycan cortex. For a turgor pressure of 3 MPa for a gram-positive, vegetative cell, a_w^{out}/a_w^{in} would need to be about 1.023. If the cell were in deionized water, with a_w^{out} of 1.000, then a_w^{in} would be about 0.977. The difference in water activity across the membrane would be 0.023, and for an ideal solution, this difference would indicate an internal osmolality of about 1.25 osmol/kg, which is a very reasonable value in terms of what is known of the concentration of small solutes within bacterial cells.

Physical Structure and Hydration

Even though the view of cell turgor is becoming a little clearer, there is still much confusion about the physical structure and properties of the peptidoglycan networks that maintain turgor. Much of this confusion comes from incomplete consideration of hydration effects and of electrostatic interactions within the structures. In my own view, there is no compelling basis for not thinking of the fully hydrated, peptidoglycan network of the wall of a living, gram-positive bacterium as an elastic entanglement network with sufficient intramolecular flexibility to allow for nearly the maximal extent of electrostatic interactions among the charged groups of the network. Evidence for the highly ordered, and therefore dense, structures comes largely from studies of dehydrated walls or peptidoglycans. Even dried walls have relatively low densities. For example, Scherrer et al. (28) found that freeze-dried walls of *Bacillus megaterium* had relatively low density of 1.302 g/cm^3 determined by the helium-displacement method. Accompanying determinations of porosity indicated that the walls had collapsed during freeze-drying. Earlier determinations of buoyant densities of walls gave higher values for density of some 1.46 g/cm^3 (6) for mureins of *Lactobacillus plantarum* and *Spirillum serpens*. However, there are difficulties in interpreting apparent buoyant densities for structures such as walls which are insoluble polyelectrolytes. Density determinations for walls of *M. luteus* and *Staphylococcus aureus* with use of the CCl$_4$/ethanol-flotation technique (11) give intermediate values of 1.38 to 1.39 g/cm^3. The low values for density indicate that peptidoglycans in walls are not in the form of crystallites, of the sort formed by, say, cellulose, which has dry density of about 1.5 g/cm^3.

Of course, when we consider the functioning wall of a turgid cell, we are not considering a dried structure but rather a fully hydrated structure. Isolated cell walls have relatively low wet densities, less than 1.1 g/ml even for compact walls such as those of *S. aureus* (25, 31). Perhaps a more functional way to look at density is as the reverse in terms of milliliters of wall volume per gram (dry weight) determined by dextran exclusion or other techniques. Table 1 presents data for dextran-impermeable volumes of a variety of isolated peptidoglycan networks. Maximum and minimum measured volumes are presented to give an indication of the capacities of the structures to expand and contract electrostatically. When peptidoglycan networks are titrated with acid or base to pH values close to those for minimal net charge, the networks contract because of electrostatic attractions among negatively charged groups, mainly carboxyl groups, and positively charged amino groups. Acid titration to low pH values converts the structures to polycationic networks, and expansion results

Table 1. Dextran-impermeable volumes of isolated peptidoglycan networks

Organism	Minimum vol (ml/g [dry wt])	Maximum vol (ml/g [dry wt])	Reference
S. aureus	2.8	5.1	24
B. megaterium	5.2	10.0	
M. luteus	4.0	10.9[a]	25
B. megaterium (spore cortex)	3.7	13.3	16
B. cereus (spore cortex)	4.3	12.9	

[a]When 8 M urea was added to the networks, the maximum volume increased to about 16 ml/g.

from reduction in electrostatic attraction and increase in electrostatic repulsion among positively charged groups. Similarly, titration with base to high pH values converts the structures to polyanionic networks, and expansion results from reduction in electrostatic attraction and increase in electrostatic repulsion, now among negatively charged groups, mainly carboxyl but also phosphate groups in accessory polymers if they are present. Addition of salts such as KCl or NaCl to highly contracted or highly expanded walls results in expansion or contraction, respectively, because of the diminution of electrostatic forces associated with increased ionic strength of the suspending medium.

The extent of the volume changes indicates clearly that peptidoglycan networks are highly flexible, and the results of studies with *B. megaterium* (14) have shown that the walls of living cells also can undergo expansion and contraction due to changes in electrostatic interactions. The most compact network of Table 1 is that of the walls of *S. aureus* with minimum and maximum volumes of 2.8 and 5.1 ml/g (dry weight), respectively, and these compact structures can be related to the high degree of cross-linking in the peptidoglycan of the bacterium. At the other extreme, the cortical peptidoglycans of *B. megaterium* and *B. cereus* are much less compact with a maximum volume of some 13.3 ml/g (dry weight) for the cortex of *B. megaterium*. This looser structure can again be related to the structure of the peptidoglycan (32), which has a large net negative charge and a low degree of cross-linking.

The volume data lead to the conclusion that peptidoglycan networks are filled with water, much like microsponges. At the one extreme, the contracted peptidoglycan network of the wall of *S. aureus* had a volume in water suspension of 2.8 ml/g. The volume occupied by the polymer itself would be simply the reciprocal of the density or 0.77 ml/g. Therefore, the wall volume would be occupied by 2.0 ml of water and 0.8 ml of polymer. At the other extreme, the cortical peptidoglycan of *B. megaterium* would have in its expanded state about 12.5 ml of water and 0.8 ml of polymer. The water associated with cell walls does not appear to be abnormal, although at low levels of hydration, the water does have altered properties reflected by a shift in nuclear magnetic resonance relaxation (27). In essence, the water associates with the wall as one would expect from a knowledge of the chemical composition of peptidoglycans and accessory polymers, which have ionizable and other hydrophilic groups.

The interactions of water with peptidoglycan networks appear to have major effects on their physical properties, as shown, for example, by the very exacting work of Thwaites and Mendelson (30; Mendelson and Thwaites, this volume) on the mechanical properties of threads or macrofibers composed of cells of *B. subtilis* suppressed in cell division. The cells form long, helical filaments, which can associate to form larger macrofibers or still larger threads. The large structures can be used for macro-scale, at least from the perspective of a microbiologist, assessments of mechanical properties, primarily stress-strain relationships. Specimens were extended at a constant rate, and the resultant tension was recorded. At low relative humidities (RH), tension built up quickly in the threads, and they broke with little extension, only some 20% at RH of 65% but less than 1% at lower RH values. At higher RH values, tension buildup was less, and much greater extension was possible before breakage, up to 50% at RH of 88%. Behavior at 100% RH could be predicted only by extrapolation because the threads pulled out from the mounting cards of the stretch device at this high RH. There did appear to be sufficient tensile strength in the wall at high RH to allow for full development of turgor in vegetative cells (J. J. Thwaites, personal communication). Whether or not spore cortical structures are sufficiently strong to allow for the very high turgor suggested for the dormant spore remains an open question. However, there is no question that the state of hydration of peptidoglycan networks greatly affects their physical properties. Again, the state of the wall pertinent to bacterial cell physiology is the fully hydrated state. Water sorption data presented by Watt (35) for cortical peptidoglycans show that at RH of 65 and 90%, the networks were about 15 and 52% water, respectively, by weight. These values are well below the values for fully hydrated networks (Table 1), even in the most contracted state, i.e., 3.7 ml/g for the cortical peptidoglycan of *B. megaterium,* which would then be $(2.9/3.7)100 = 78\%$ water by volume or about 74% water by weight. At maximal expansion, the network would be $(12.5/13.3)100 = 94\%$ water by volume or about 93% water by weight. Overall, it is clear that mureins are porous networks of visco-elastic polymers impregnated with water and designed for maintenance of cell turgor.

Isolated Networks and Intact Cells

One concern in considering data such as those presented in Table 1 is that the peptidoglycan networks would of necessity be damaged during isolation, although the fragments used to obtain the data in Table 1 were all sufficiently large to be visible in a phase microscope. The phase images were faint because the refractive indices of the wall fragments were low, indicative of their low solids contents. In fact, the indices assessed by means of immersion refractometry were related to the solids contents of the various types of walls (15). Walls of intact cells also have faint images in positive phase contrast.

Comparisons of the electrical conductivities of wall fragments and of walls of intact cells of *M. luteus* (2) indicated little difference. However, the conductivity values did indicate high degrees of flexibility in the networks. Thus, for isolated fragments suspended in NaCl solutions, conductivities for low-frequency current predicted on the bases of the numbers of assessed anionic and cationic groups in the polymer networks were some 1.3 to 1.6 S/m. The actual, measured conductivities were between 0.8 and 1.0 S/m, and these lower measured values were more closely related to net charge densities rather than to total charge densities. In other words, the networks were sufficiently flexible to allow for internal neutralization of charge between network amino and carboxyl groups so that there was no need for mobile counterions, in this instance Na^+ and Cl^-, from the environment. It can be assumed that the charged groups of the networks did not conduct current because their mobilities would be extremely low. Current conduction would then involve only movements of small, mobile counterions associated with the fixed charges not neutralized by other network charges.

The conductivity values obtained with the isolated wall fragments were used to predict low-frequency conductivities for intact cells. The only conducting structure of the intact cell for low-frequency current would be the wall. The cell membrane would act as a complete insulator at frequencies below about 10 MHz to shield cytoplasmic ions from the electric field. The predicted value was some 0.15 S/m based on the assumption that the wall occupied 50% of the cell volume. This latter assumption was supported by determinations of the volume of the cell permeable to phosphate or to sucrose. The measured conductivity of the cells was 0.25 S/m, close to but somewhat higher than the predicted conductivity. Thus, it appeared that, if anything, walls of intact cells had somewhat higher conductivity than expected and so may have had a lower level of internal charge neutralization within the peptidoglycan network, possibly because of physical constraints due to turgor or to arrangement of the polymer strands. However, it does seem reasonable to take the view that the electrostatic interactions affecting the physical properties of isolated wall fragments also affect the physical properties of the walls of intact cells.

Cortical Peptidoglycans and Spores

The cortical peptidoglycans of spores appear to be designed for functions associated with the extreme dormancy and heat resistance of these cells. In fact, in the major theories for heat resistance of bacterial endospores the cortex plays a central role either as a contractile or an expansive element. Past work with forespores by us (17) and by Ellar and co-workers (5) has led to the conclusion that the cortical peptidoglycan serves to stabilize a state of osmotic dehydration which develops early in sporogenesis after engulfment by the large cell of the small cell formed during asymmetric cell division. This asymmetric division and the subsequent engulfment are integral processes in sporogenesis. The outer membrane of the forespore is inside out, and isolated forespores can hydrolyze exogenously added ATP, while vegetative protoplasts cannot. Because of this inversion, forespores lose potassium, and assessments of the levels of potassium and other minerals in isolated forespores (17) have indicated almost complete loss. For example, the potassium level in forespores was only about 0.01 μmol/mg (dry weight) compared with a value of 0.54 for the sporulating cells from which the forespores were isolated. In essence, the potassium in forespores is probably only sufficient to neutralize the net negative charges in cell polymers. When the cortex is laid down between the two spore membranes, it would act to stabilize this osmotic dehydration during subsequent mineralization of the spore and resist any associated movement of water into the cells. Of course, the mineralization of the spore is commonly associated with uptake of dipicolinic acid and possibly with precipitation of the accumulated mineral ions, mainly but not exclusively Ca, so that they would no longer be osmotically active. Certainly, the very low conductivities of spores even for high-frequency current for which the membranes would not act as insulators (19) indicate a high degree of electrostasis in mature spores. Isolated forespores also have low conductivities, but this is because they are nearly devoid of mineral ions of any type. Isolated forespores can be made to swell or shrink osmotically, but their responses are muted in comparison with those of vegetative protoplasts (36). Cortex formation further diminishes osmotic sensitivity. Even though forespores are dehydrated relative to vegetative cells, they are still as sensitive to heat. They retain full viability and can give rise to colonies, but only on media supplemented with an osmolite such as sucrose.

Cortex formation is required for development of heat resistance. A direct correlation has been established between cortex size and heat resistance (22), and cortex-deficient mutants do not become heat resistant even though they concentrate Ca and dipicolinic acid and become refractile. The cortical peptidoglycan of spores differs significantly from that of vegetative cells. The structural peculiarities of cortical peptidoglycan include low degrees of cross-linking (ca. 20%), presence of many (ca. 20%) muramyl residues without substituent peptides but with only a carboxy-terminal L-alanyl residue, δ-lactam formation in about half of the muramyl residues, and no substituents on the α-carboxyl of the D-glutamate residues. Only about 35% of the muramyl residues bear full peptides. This general structure was found to be common to all 14 organisms studied by Warth (32). The net result is a loose structure with a high, net negative charge.

If the model for spore cortical peptidoglycan structure is used to estimate charge density, the prediction is for a total negative charge of about 1.39 meq/g (dry weight), a positive charge of 0.30 meq/g, and a net negative charge of 1.09 meq/g. When isolated cortical fragments from spores of *B. megaterium* or *B. cereus* were soaked in concentrated KCl solution and then washed extensively with deionized water, the total K contents assayed with atomic absorption spectroscopy were 0.92 meq/g (dry weight) for *B. megaterium* and 1.00 meq/g for *B. cereus*. These values are sufficiently close to the predicted net negative charge to give confidence in the proposed structure.

The loose structure for cortical peptidoglycan is reflected also in shrinking-swelling behavior during acid/base titration (Table 1; 16). Cortical peptidoglycans are also very sensitive to salt-induced contraction, and exposure of isolated structures to dilute solutions of $CaCl_2$ results in contraction to the minimum volumes indicated in Table 1. This behavior provides ample basis for the contractile cortex hypothesis for heat resistance in which the cortex is viewed as squeezing water out of the core as a result of contraction (13). Of course, the same behavior could be applied to the hypothesis of the expanded, osmoregulatory cortex (7), or to variants of the hypothesis in which anisotropic swelling of the cortex in the radial direction is proposed (34).

Studies with isolated fragments of cortical peptidoglycans allow for an evaluation, albeit a preliminary one at this time, of the state of the cortex in living spores. Warth (34) suggested an expanded state: "The large volume of the cortex, compared with the peptidoglycan content, suggests an expanded rather than a contracted structure." Nakashio and Gerhardt (23) recently presented data for water contents of the cortices of spores lacking coats. The loss of coats is well known to have little or no effect on heat resistance. The water contents were estimated in terms of glucose-permeable volumes of the spores on the assumption that glucose can penetrate the water-filled pores of the cortex but cannot cross the inner membrane. If their data are converted to values for milliliters of total volume per gram (dry weight), with the assumption that the specific volumes of peptidoglycan and water are 0.8 and 1.0 ml/g, respectively, then the range of values is from 2.3 ml/g for *B. megaterium* ATTC 9885 to 7.1 ml/g for *B. cereus* 10^{LD}. The average value for eight spore types was 4.5 ml/g. Thus, it appears that the cortical peptidoglycan in the intact spore is more in the contracted than the expanded state. The values should perhaps be considered slightly low estimates because of possible problems with glucose penetration, but the uncertainty is not major.

A plausible sequence of events can be constructed for sporulation in terms of the acid/base physiology of the system and the properties of cortical peptidoglycans. After engulfment of the forespore compartment by the sporangial cell, the membrane ATPases of the two membranes of the forespore would act to acidify the cortical space. This acidification should be marked because of the small volume of the cortical space and would probably extend into the cytoplasm of the forespore. In fact, acidification could contribute to the low mineral content of the forespores because carboxyl groups of forespore biopolymers would become protonated and then not require counterions such as K^+. The capacity of the forespore to produce ATP would be diminished, and its capacity to maintain a Δ pH across the inner membrane would also diminish. The cortex would then be laid down in the acid environment of the cortical space. Because of the low pH value of the space, the cortex would contract. The osmotic dehydration of the forespore would then be stabilized, and the cortex would resist any subsequent influx of water. Initially, the turgor of the sporangial cell would act to compress the forespore through the cortical peptidoglycan. Based on assays of solute levels, the sporangial turgor pressure for

B. megaterium would be about 1 MPa, and that of the forespore would be less than one-tenth this value.

At stage IV, the sporangial cell develops new capacities to take up Ca and other minerals (5). Ordinarily, bacterial cells extrude Ca, possibly through a Ca/H antiport of the sort involved in mitochondrial Ca transport or by means of systems directly connected to ATP hydrolysis. The new uptake of Ca has been found to involve a system with a lower K_m than the extrusive system (5). The net result is that Ca flows into the vegetative cell faster than it can be extruded. Ca could flow into the acidified forespore by an antiport system, which is not electrogenic. Dipicolinic acid is concentrated by the forespores, somewhat later than Ca uptake. If the forespore became alkaline relative to the sporangial cell as a result of Ca uptake, dipicolinic acid would flow readily into the forespore because it is a weak acid with pK_a values of about 2.1 and 4.6 for the two carboxyl groups. Presumably, Ca and dipicolinic acid would precipitate in the forespore in association with proteins and nucleic acids. The cortex should remain in the contracted state because of salt effects. The addition of the coat layers would help to maintain minerals within the cortical space, and in fact, most spores have relatively high levels of mobile ions capable of moving in an electrical field within the cortical space (3).

There are still many aspects of this proposed scheme in need of experimental testing. However, it does give an orientation for future research based on knowledge of the acid/base physiology of bacterial cells involving proton-translocating ATPases and on the known physical and chemical properties of cortical peptidoglycans.

ACKNOWLEDGMENTS. The contributions of Gary R. Bender to the experimental work described and of Philipp Gerhardt and Ed Carstensen to the views presented are acknowledged with pleasure.

The work was supported by award DAAL 03-86-K-0075 from the U.S. Army Research Office.

LITERATURE CITED

1. **Beveridge, T. J.** 1981. Ultrastructure, chemistry and function of the bacterial wall. *Int. Rev. Cytol.* **72:**229–317.
2. **Carstensen, E. L., and R. E. Marquis.** 1968. Passive electrical properties of microorganisms. III. Conductivity of isolated bacterial cell walls. *Biophys. J.* **8:**536–548.
3. **Carstensen, E. L., R. E. Marquis, S. Z. Child, and G. R. Bender.** 1979. Dielectric properties of native and decoated spores of *Bacillus megaterium. J. Bacteriol.* **140:**917–928.
4. **Corner, T. R., and R. E. Marquis.** 1969. Why do bacterial protoplasts burst in hypotonic solutions? *Biochim. Biophys. Acta* **183:**544–558.
5. **Ellar, D. J.** 1978. Spore specific structures and their function, p. 295–325. *In* R. Y. Stanier,

H. J. Rogers, and J. B. Ward (ed.), *Relations between Structure and Function in the Prokaryotic Cell*. Cambridge University Press, Cambridge.

6. **Formanek, H., S. Formanek, and H. Wawra.** 1974. A three-dimensional atomic model of the murein layer of bacteria. *Eur. J. Biochem.* **46:**279–294.

7. **Gould, G. W., and G. J. Dring.** 1975. Heat resistance of bacterial endospores and concept of an expanded osmoregulatory cortex. *Nature* (London) **258:**402–405.

8. **Hughes, A. H., I. C. Hancock, and J. Baddiley.** 1973. The function of teichoic acids in cation control in bacterial membranes. *Biochem. J.* **132:**83–93.

9. **Koch, A. L.** 1985. Bacterial wall growth and division or life without actin. *Trends Biochem. Sci.* **10:**11–14.

10. **Koch, A. L., M. L. Higgins, and R. J. Doyle.** 1981. Surface tension-like forces determine bacterial shapes: *Streptococcus faecium. J. Gen. Microbiol.* **123:**151–161.

11. **Labischinski, H., G. Barnickel, H. Bradaczek, and P. Giesbrecht.** 1979. On the secondary and tertiary structure of murein. *Eur. J. Biochem.* **95:**147–155.

12. **Le Rudulier, D., A. R. Strøm, A. M. Dandekar, L. T. Smith, and R. C. Valentine.** 1984. Molecular biology of osmoregulation. *Science* **224:**1064–1068.

13. **Lewis, J. C., N. S. Snell, and H. K. Burr.** 1960. Water permeability of bacterial spores and the concept of a contractile cortex. *Science* **132:**544–545.

14. **Marquis, R. E.** 1968. Salt-induced contraction of bacterial cell walls. *J. Bacteriol.* **95:**775–781.

15. **Marquis, R. E.** 1973. Immersion refractometry of isolated bacterial cell walls. *J. Bacteriol.* **116:**1273–1279.

16. **Marquis, R. E., G. R. Bender, E. L. Carstensen, and S. Z. Child.** 1983. Electrochemical and mechanical interactions in *Bacillus* spore and vegetative mureins, p. 43–48. *In* R. Hakenbeck, J.-V. Holtje, and H. Labischinski (ed.), *The Target of Penicillin*. Walter de Gruyter and Co., Berlin.

17. **Marquis, R. E., G. R. Bender, E. L. Carstensen, and S. Z. Child.** 1983. Dielectric characterization of forespores isolated from *Bacillus megaterium* ATCC 19213. *J. Bacteriol.* **153:**436–442.

18. **Marquis, R. E., and E. L. Carstensen.** 1973. Electric conductivity and internal osmolality of intact bacterial cells. *J. Bacteriol.* **113:**1198–1206.

19. **Marquis, R. E., E. L. Carstensen, G. R. Bender, and S. Z. Child.** 1985. Physiological biophysics of spores, p. 227–240. *In* G. J. Dring, D. J. Ellar, and G. W. Gould (ed.), *Fundamental and Applied Aspects of Bacterial Spores*. Academic Press, Inc., London.

20. **Martinac, B., M. Buechner, A. H. Delcour, J. Adler, and C. Kung.** 1987. Pressure-sensitive ion channel in *Escherichia coli. Proc. Natl. Acad. Sci. USA* **84:**2297–2301.

21. **Mitchell, P., and J. Moyle.** 1956. Osmotic function and structure in bacteria, p. 150–180. *In* E. T. C. Spooner and B. A. D. Stocker (ed.), *Bacterial Anatomy*. Cambridge University Press, Cambridge.

22. **Murrell, W. G., and A. D. Warth.** 1965. Composition and heat resistance of bacterial spores, p. 1–24. *In* L. L. Campbell and H. O. Halvorson (ed.), *Spores III*. American Society for Microbiology, Ann Arbor, Mich.

23. **Nakashio, S., and P. Gerhardt.** 1985. Protoplast dehydration correlated with heat resistance of bacterial spores. *J. Bacteriol.* **162:**571–578.

24. **Ou, L.-T., and R. E. Marquis.** 1970. Electromechanical interactions in cell walls of gram-positive cocci. *J. Bacteriol.* **101:**92–101.

25. **Ou, L.-T., and R. E. Marquis.** 1972. Coccal cell-wall compactness and the swelling action of denaturants. *Can. J. Microbiol.* **18:**623–629.

26. **Reed, R. H., and A. E. Walsby.** 1985. Changes in turgor pressure in response to increases in external NaCl concentration in the gas-vacuolate cyanobacterium *Microcystis sp. Arch. Microbiol.* **143:**290–296.

27. **Resing, H. A., and R. A. Neihof.** 1970. Nuclear magnetic resonance relaxation of water adsorbed on bacterial cell walls. *J. Colloid Interface Sci.* **34:**480–487.
28. **Scherrer, R., E. Berlin, and P. Gerhardt.** 1977. Density, porosity, and structure of dried cell walls isolated from *Bacillus megaterium* and *Saccharomyces cerevisiae. J. Bacteriol.* **129:**1162–1164.
29. **Scherrer, R., and P. Gerhardt.** 1971. Molecular sieving by the *Bacillus megaterium* cell wall and protoplast. *J. Bacteriol.* **107:**718–735.
30. **Thwaites, J. J., and N. H. Mendelson.** 1985. Biomechanics of bacterial walls. Studies of bacterial thread made from *Bacillus subtilis. Proc. Natl. Acad. Sci. USA* **82:**2163–2167.
31. **Torbet, J., and M. Y. Norton.** 1982. The structure of the cell wall of *Staphylococcus aureus* studied with neutron scattering and magnetic birefringence. *FEBS Lett.* **147:**201–206.
32. **Warth, A. D.** 1978. Molecular structure of the bacterial spore. *Adv. Microb. Physiol.* **17:**1–45.
33. **Warth, A. D.** 1979. Exploding spores. *Spore Newslett.* **6:**4–6.
34. **Warth, A. D.** 1985. Mechanisms of heat resistance, p. 209–225. *In* G. J. Dring, D. J. Ellar, and G. W. Gould (ed.), *Fundamental and Applied Aspects of Bacterial Spores.* Academic Press, Inc., London.
35. **Watt, I. C.** 1981. Water vapor adsorption by *Bacillus stearothermophilus* endospores, p. 253–255. *In* H. S. Levinson, A. L. Sonenshein, and D. J. Tipper (ed.), *Sporulation and Germination.* American Society for Microbiology, Washington, D.C.
36. **Wilkinson, B. J., J. A. Deans, and D. J. Ellar.** 1975. Biochemical evidence for the reversed polarity of the outer membrane of the bacterial forespore. *Biochem. J.* **152:**561–569.

Chapter 3

Murein Structure Data and Their Relevance for the Understanding of Murein Metabolism in *Escherichia coli*

U. Schwarz
B. Glauner

For a long time investigators have used bacteria as an experimental system to study the generation of cellular shape and its maintenance, control, and modification on a molecular level, taking advantage of the assumed manageability of the system. In bacteria, one defined structural element in the envelope, the sacculus, maintains the integrity of the wall and the shape of the cell (for review, see reference 11). The chemistry of the polymer—peptidoglycan or murein—from which the sacculus is knitted, and the enzyme systems involved in its metabolism as well, appeared to be simple. With time, both of these notions had to be revised.

The long-held version of *Escherichia coli* murein is that of a single-layered net in which glycan strands with short peptide side chains are interlinked by peptide bridges making up a polymer of quite monotonous chemical composition (2). The only variation found was in the length of the peptide side chains, varying from two to five amino acids (3, 6, 20). The enlargement of such a simple structure and its modification during cell growth and division would essentially need only two types of synthetic activities: the elongation of glycan chains by the addition of new subunits would require a transglycosylase activ-

---

U. Schwarz and B. Glauner • Max-Planck-Institut für Entwicklungsbiologie, Abteilung Biochemie, D-7400 Tübingen, Federal Republic of Germany.

ity, and the cross-linkages between the peptide side chains of adjacent glycan strands would occur through transpeptidation. The chemical energy for the formation both of the glycosidic and peptide bonds would be carried in the low-molecular-weight precursors presynthesized in the cytoplasm (for review, see reference 21). This simple picture, however, changed with time. The enzyme systems involved turned out to be unexpectedly complex. There was not just one synthetic transglycosylase and one transpeptidase active in murein biosynthesis but a collection of proteins, the penicillin-binding proteins (22). Furthermore, a large set of murein hydrolases was revealed. For each type of chemical bond found in murein, at least one, and in some cases several, specific hydrolases have been identified. In the case of *Escherichia coli* the now probably complete list of murein hydrolases contains nine different proteins (for review, see reference 11).

For quite some time, there was a discomforting contrast between the complexity of the enzyme systems participating in the metabolism of murein and the simplicity of the known murein structure in *E. coli*. A careful reexamination of the chemical structure of murein with high-pressure liquid chromatography has finally solved this paradox. The separation of murein subunits obtained from isolated murein by complete enzymatic degradation with muramidase on a reversed-phase column yields a pattern of surprising complexity (4). Instead of only the few types of murein subunits known so far, around 80 different compounds were identified. They all were isolated and purified, and their chemical structure was determined. Their analysis showed very clearly that they all are indeed native components of murein (B. Glauner et al., manuscript in preparation).

Two types of novel subunits found are of special relevance for our conceptions about murein structure and murein metabolism. One of them is a group of oligomers, trimers (5), and tetramers consisting of three or four subunits in which all peptide side chains are cross-linked with one another. The other class of subunits encloses components with a new sort of peptide bond. In this case, unlike the normal type of cross-bridge in which the terminal alanine of one side chain is linked to the diaminopimelic acid (A_2pm) residue of the other side chain, the cross-linking bond is between the two A_2pm residues of the side chains. This type of peptide bridge has been identified by analysis of the products obtained by partial hydrolysis of the purified subunit. The link is not between the two D centers of the A_2pm residues, as we had assumed for some time, but between the L center of the donor A_2pm residue and the D center on the acceptor side. Conclusive evidence for this type of bond was obtained by Edman degradation (Fig. 1) and the characterization of the fragments obtained. This type of cross-bridge is sensitive to cleavage by LD-carboxypeptidase II from *E. coli*, but is stable when incubated with penicillin-sensitive DD-endopeptidase.

The existence of trimeric and tetrameric subunits forced us to reconsider

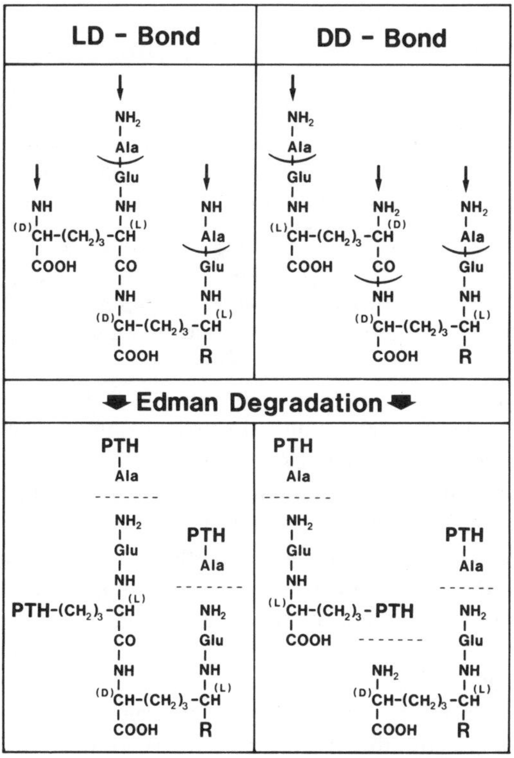

Figure 1. Determination of the configuration of the cross-linking bond in A_2pm-A_2pm dimers by Edman degradation. The peptide bridges were liberated from A_2pm-A_2pm dimers by amidase digestion (19) and were subjected to Edman degradation (13). In case of a DD bond a cleavage of the bridge between the two A_2pm residues should occur. The analysis of the Edman degradation products clearly excluded this possibility. This proves an LD configuration of the A_2pm-A_2pm bond. As an internal control the liberation of the alanine residues was determined.

the assumption that *E. coli* murein is a monolayered net. In a monolayered murein, trimeric subunits could exist only at chain ends if one accepts the data based on X-ray diffraction studies of murein, on infrared spectroscopy, and on ^{15}N-nuclear magnetic resonance relaxation analysis (1, 14–17). These data disagree with the assumption of a chitinlike structure of murein and suggest a helical arrangement of peptide side chains around the glycan strands. To say it another way, if murein were monolayered, trimers should occur exclusively at chain ends. This, however, is not what we found. On the contrary, about 80% of the trimers exist within glycan chains. For steric reasons this is possible only if one of the three glycan chains involved lies outside the plane of the two others (Fig. 2).

Also, pulse-chase experiments strongly support the assumption of a multilayered murein. They show that cross-bridges in the murein have different life times. The cross-links formed between newly incorporated glycan strands are relatively stable, whereas cross-links formed between new and preexisting glycan strands are cleaved and reformed during the aging of murein. The results are in good agreement with a multilayered murein growing according to an "inside-to-outside" mechanism as shown for gram-positive species (Fig. 3).

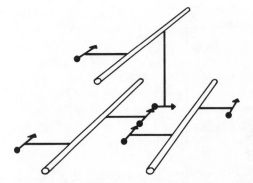

Figure 2. Assumed structure of a trimeric murein subunit cross-linking three different glycan chains. Stereochemical considerations make an arrangement of the three chains in one plane very unlikely (15, 16). The arrowheads mark the donor sites of the peptide side chain, and the spheres mark the respective receptor sites.

Other recent observations are in strong support of the concept that *E. coli* murein is a multilayered net. Electron microscopy of cell envelopes, using new techniques of fixation developed by Kellenberger's group, revealed murein as a rather thick gellike structure with a high water content rather than a very thin layer (10). Also, when envelopes are viewed after partial degradation of murein, as in studies done by van Heijenoort and colleagues, again a multilayered structure of murein is indicated (18). This concept, however, is valid only if there is enough murein available per cell to make a multilayered sacculus. This, in fact, is the case if one accepts stereochemical considerations and physicochemical data on murein structure from Labischinski's group. They indicate that the density of murein is lower than originally assumed, that murein is not a chitinlike polymer as mentioned already, that the peptide chains are helically arranged around the glycan chains, and finally that in stretched form the distance between adjacent sugar chains is considerably longer than assumed previously (1, 14–16).

In summary, all data available at present are very much in favor of a multilayered murein, also in the case of *E. coli*. According to our calculations,

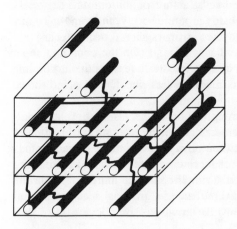

Figure 3. Inside (bottom)-to-outside growth model of murein (for review, see reference 12).

the murein found in a cell would suffice to make, on average, three layers. From an evolutionary point of view, the structure of the *E. coli* murein may be regarded as an improvement of the concept found in gram-positive species. The thickness of the murein seems to be optimized to the minimal number of layers necessary to guarantee the structural integrity of the wall.

Analysis of murein composition with high-pressure liquid chromatography not only gives us qualitative information but also allows a quantitation of the different components of murein with high accuracy. Thus, high-pressure liquid chromatography is a tool for a detailed study of the possible correlation between the composition of murein and its shape. For example, one can ask whether the hemispherical cell ends differ in chemical composition from the cylindrical part of the sacculus. The studies done so far are based on the availability of a series of mutants with altered cell shape, for example, cells growing and dividing more or less as spheres as a result of alterations in murein metabolism or, in contrast, as long filaments.

A collection of mureins from about 40 different mutants or strains growing under different conditions were analyzed and compared in both quantity and quality of their murein. The book-filling list of data is discussed here in very condensed form.

Mutants with modified murein metabolism show variations in the composition of their murein. These variations follow certain rules, the most important being a clear correlation between the length of the glycan chains and the extent of cross-linkage (Fig. 4). The shorter the glycan chains are, the more cross-bridges between them are found. The relative increase in cross-bridges is mainly due to an enrichment of trimeric and also tetrameric subunits cross-linking the glycan chains. This correlation can be interpreted in different ways; the most plausible for us is that there exist two systems which synthesize murein. One would be mainly busy with cell elongation, and the other would be mainly involved in cell septation. These two systems would operate in balance with each other. There is a host of experimental data from different laboratories which support this assumption (for review, see reference 11). According to our data, the quality of murein synthesized by the two systems is different. One produces murein with long, poorly cross-linked chains; the other would synthesize a polymer with highly cross-linked, but short chains and would be involved in cell division. Most spherical cell types which we have analyzed show the same composition of murein with short, highly cross-linked glycan chains. The results are in good agreement with data from Labischinski's group, postulating a different "topotype" for the murein of polar caps (15). This would mean that polar caps differ in chemical composition from the rest of murein.

If the addition of new murein in the growing cell is achieved through apposition to the preexisting sacculus, the participation of murein hydrolases in murein metabolism is strictly required. When new material is added at the inner side of the sacculus—inside-to-outside growth—the enlargement of the

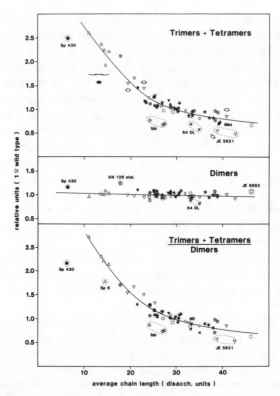

Figure 4. Correlation between the length of glycan strands and the extent of cross-linkage in murein from different *E. coli* strains. Strains growing more or less sphere shaped appear on the left side of the diagram.

surface area of murein requires the hydrolysis of bonds in the preexisting murein. In addition, part of it necessarily has to be removed at the outside to avoid the situation that after some generations the cell consists only of murein and nothing else.

A cleavage of cross-bridges and an excision of murein subunits during growth of *E. coli* indeed does occur. Goodell has shown that *E. coli* murein undergoes considerable turnover and recycling (7–9); half of the murein of the cell is cut out and reused during one generation. As postulated for its structure, the metabolism of murein also seems to be quite similar in gram-negative and gram-positive organisms.

The vital role of murein hydrolases in wall enlargement and cell division still awaits elucidation. There is good reason to hope, however, that this goal can be achieved in the near future. After many attempts, it was possible to clone the genes for the two lytic transglycosylases of *E. coli*, the soluble and the membrane-bound enzyme, and also for the penicillin-insensitive endopeptidase (W. Keck, A. Betzner, and E. W. Goodell, manuscript in preparation). This was achieved by using specific antibodies to screen clones overproducing the enzymes as a result of the presence of the gene on runaway plasmids. With these genes in hand, constructs can be envisaged in which the expression of the relevant genes can be manipulated by using the anti-sense RNA technique.

LITERATURE CITED

1. **Barnickel, G., D. Naumann, and H. Bradaczek.** 1983. Computer aided molecular modelling of the three-dimensional structure of bacterial peptidoglycan, p. 61–66. *In* R. Hakenbeck, J.-V. Höltje, and H. Labischinski (ed.), *The Target of Penicillin.* Walter de Gruyter, Berlin.
2. **Braun, V., H. Gnirke, U. Henning, and K. Rehn.** 1973. Model for the structure of the shape-maintaining layer of the *Escherichia coli* cell envelope. *J. Bacteriol.* **114:**1264–1270.
3. **De Pedro, M. A., and U. Schwarz.** 1981. Heterogeneity of newly inserted and preexisting murein in the sacculus of *Escherichia coli. Proc. Natl. Acad. Sci. USA* **78:**5856–5860.
4. **Glauner, B., and U. Schwarz.** 1983. The analysis of murein composition with high-pressure-liquid chromatography, p. 29–34. *In* R. Hakenbeck, J.-V. Höltje, and H. Labischinski (ed.), *The Target of Penicillin.* Walter de Gruyter, Berlin.
5. **Gmeiner, J.** 1980. Identification of peptide-cross-linked tridisaccharide peptide trimers in murein of *Escherichia coli. J. Bacteriol.* **143:**510–512.
6. **Gmeiner, J., and H.-P. Kroll.** 1981. N-acetylglucosaminyl-N-acetylmuramyl-dipeptide, a novel murein building block formed during the cell division of *Proteus mirabilis. FEBS Lett.* **129:**142–144.
7. **Goodell, E. W.** 1985. Recycling of murein by *Escherichia coli. J. Bacteriol.* **163:**305–310.
8. **Goodell, E. W., and U. Schwarz.** 1983. Cleavage and resynthesis of peptide cross bridges in *Escherichia coli* murein. *J. Bacteriol.* **156:**136–140.
9. **Goodell, E. W., and U. Schwarz.** 1985. Release of cell wall peptides into culture medium by exponentially growing *Escherichia coli. J. Bacteriol.* **162:**391–397.
10. **Hobot, J. A., E. Carlemalm, W. Villinger, and E. Kellenberger.** 1984. Periplasmic gel: new concept resulting from the reinvestigation of bacterial cell envelope ultrastructure by new methods. *J. Bacteriol.* **160:**143–152.
11. **Höltje, J.-V., and U. Schwarz.** 1985. Biosynthesis and growth of the murein sacculus, p. 77–119. *In* N. Nanninga (ed.), *Molecular Cytology of Escherichia coli.* Academic Press, Inc., London.
12. **Koch, A. L.** 1985. Inside-to-outside growth and turnover of the wall of gram-positive rods. *J. Bacteriol.* **117:**137–157.
13. **Konigsberg, W.** 1972. Subtractive Edman degradation. *Methods Enzymol.* **25:**326–332.
14. **Labischinski, M., G. Barnickel, and D. Naumann.** 1979. The state of order of bacterial peptidoglycan, p. 49–54. *In* R. Hakenbeck, J.-V. Höltje, and H. Labischinski (ed.), *The Target of Penicillin.* Walter de Gruyter, Berlin.
15. **Labischinski, M., G. Barnickel, D. Naumann, and P. Keller.** 1985. Conformational and topical aspects of the three-dimensional architecture of bacterial peptidoglycan. *Ann. Microbiol.* (Paris) **136A:**45–50.
16. **Labischinski, H., and L. Johannsen.** 1985. On the relationship between conformational and biological properties of murein, p. 37–42. *In* P. H. Seidl and K. H. Schleifer (ed.), *Biological Properties of Peptidoglycan.* Walter de Gruyter, Berlin.
17. **Lapidot, A., and L. Inbar.** 1983. Dynamic aspects of bacterial cell walls probed by *in vivo*

^{15}N NMR, p. 79–84. *In* R. Hakenbeck, J.-V. Höltje, and H. Labischinski (ed.), *The Target of Penicillin*. Walter de Gruyter, Berlin.

18. **Leduc, M., and J. Van Heijenoort.** 1985. Correlation between degradation and ultrastructure of peptidoglycan during autolysis of *Escherichia coli*. *J. Bacteriol*. **161:**627–635.

19. **Mollner, S., and V. Braun.** 1984. Murein hydrolase (N-acetyl-muramyl-L-alanine amidase) in human serum. *Arch. Microbiol*. **140:**171–177.

20. **Primosigh, J., H. Pelzer, D. Maass, and W. Weidel.** 1961. Chemical characterization of muropeptides released from the *E. coli* B cell wall by enzymic action. *Biochim. Biophys. Acta* **46:**68–80.

21. **Rogers, H. J., H. R. Perkins, and J. B. Ward.** 1980. *Microbial Cell Walls and Membranes*. Chapman & Hall, Ltd., London.

22. **Spratt, G. B.** 1977. Properties of the penicillin-binding proteins of *Escherichia coli* K12. *Eur. J. Biochem*. **72:**341–352.

II. MODELS OF CELL GROWTH

Chapter 4

The Sacculus, a Nonwoven, Carded, Stress-Bearing Fabric

Arthur L. Koch

Procaryotes generally are under turgor pressure (33). The turgor pressure is due to a higher osmotic pressure inside the cell than in the surrounding growth medium. This causes water to enter until counterbalanced by the elastic stretch of the murein sacculus. From estimates of the turgor pressure (34), it follows that in the typical gram-negative cell wall, the stress is 5.85×10^{-5} dynes/mol per average bond (14). This is large enough so that it would be impossible for the cell to patch a gap by pulling the edges together (A. L. Koch, *Microbiol. Rev.*, in press). Such a repair would require energy equivalent to dozens of ATP molecules. Consequently, a small defect that arose during growth would be expected to get larger and larger. It could be repaired by cross-linking a sufficiently long oligosaccharide crosswise to the gap. If caught in time, before the bonds in which the stress had become most concentrated had ripped, this could save the cell. Consequently, even if the peptidoglycan were originally laid down in parallel rows, imperfections and enzymological accidents would lead either to irregularities during repair or to further ripping and catastrophic death of the cell.

This need for a repair, or self-sealing, mechanism for the wall integrity argues against a regular structure for the wall fabric of murein. In addition, there are many other reasons for believing that the peptidoglycan is not arranged in a regular way. A review of the literature (Koch, in press) concerning X-ray diffraction, fine structure analysis by high-performance liquid chromatography, and end-group analysis of chain length has found that the evidence is solidly against a regular structure. In addition, it follows from thermodynamics that a regular structure requires more energy to produce than an irregular one for both entropic and free energy reasons. Finally, the manufacturing

Arthur L. Koch • Department of Biology, Indiana University, Bloomington, Indiana 47405.

of a regular structure requires more elaborate machinery than seemingly simple membrane-bound transpeptidases/transglycosylases may possess.

For these reasons, I discount the concept that the sacculus is like a fish net or that it is built of regular building blocks arranged in a regular way. But before rejecting a regular structure of repeating units, it is necessary to understand the stress relationships in differently shaped cells and reexamine four kinds of recent data seemingly in favor of a regular structure. I will also examine the consequences of the concept that the stress-bearing part of the wall is basically a "nonwoven" fabric. This does not mean that it has no order as might be expected for many man-made nonwoven fabrics; rather, it is expected to have some partial order because the stressed wall to which new peptidoglycan chains are added is subject to different stresses in different directions. Partial orientation of the fibers in the textile industry is called carding, hence the insertion of this term in the title of this chapter.

The Stresses in Pressure Vessels

From the engineering point of view, a bacterium is a pressure-containing vessel, subject to the same mechanical constraints as any other "boiler." The hoop stress, designated by N_θ, is in the tension in the surface, oriented in the circumferential direction. The meridional stress, designated by N_ϕ, is in the plane of the surface but oriented in the axial direction. For general shapes of a pressure-containing shell, engineers have developed methods to evaluate the stress in the surface of such boilers (8, 37).

The surface stresses in a hemispherical pole are the same in all directions and are mathematically expressed by $N_\theta = N_\phi = Pa/2$, where P is the turgor pressure and a is the radius of the cell. For the cylindrical part of the cell, the hoop stress, N_θ, is given by Pa while the meridional stress, N_ϕ, is equal to $Pa/2$ (see Fig. 1). The stress in the unsplit septum of a gram-positive organism should be zero or nearly zero because the surrounding external wall bears the stress generated by the hydrostatic pressure. However, there is compressive stress (i.e., N_θ becomes negative) on the portions of the developing constriction

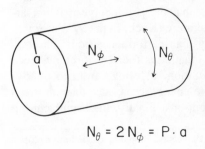

$$N_\theta = 2N_\phi = P \cdot a$$

Figure 1. Stress anisotropy in the cylindrical wall of a pressure-containing vessel. The axial or meridional stress is indicated by N_ϕ, and the circumferential or hoop stress is indicated by N_θ. Note that the circumferential stress is twice the axial stress. This means that sensitive bonds are more readily cleaved if oriented in the circumferential or hoop direction. See Fig. 2.

site in the gram-negative organism and in the partially split portion of the developing pole of the gram-positive organism. The hoop stress is tensile (i.e., N_θ is positive) in more pointed cells, like *Streptococcus faecium*, and of a smaller positive value in the more flattened *Bacillus subtilis* (3; Koch, in press).

Generation of Partial Order by Turgor in Asymmetric Cells

There are three ways that asymmetry of stress can influence the development of the wall structure. The first is differential autolysis, which is diagrammed in Fig. 2 and 3. Because of the anisotropism of stresses in the cylindrical portion of the gram-negative rod, bonds supporting stress in the hoop direction will be under greater tension and will be more liable to be hydrolyzed either chemically or by autolysins. It must be assumed that the autolysin responsible for wall enlargement is an endopeptidase attacking the same bridge bond as was formed by the transpeptidase. This follows because it is the only way known to convert a donor peptide into an acceptor peptide. Consequently, even though bonds are probably formed and linked in random orientations as the result of transpeptidase action, those in the hoop direction will have a lower energy of activation than those in the axial direction. Those at intermediate orientations will have an energy of activation decreased to an intermediary degree (Fig. 2). These effects theoretically can be very large. I calculated (14) that a lowering of the energy of activation due to 5 atm (506 kPa) of pressure would speed the hydrolysis 10^6-fold. Consequently, the twofold difference in stress should be effectively equivalent to an all-or-none response. The net effect of this is that endopeptidase-sensitive bonds should remain predominantly, but not exclusively, in the axial direction on the sidewalls of the cells (Fig. 3).

The second way giving rise to differential growth would be relevant only

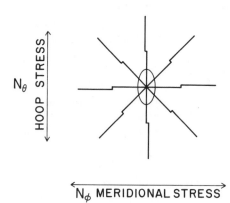

N_θ HOOP STRESS

N_ϕ MERIDIONAL STRESS

Figure 2. Stress in cylindrical wall as a function of orientation. The ellipse in the center of the figure symbolizes the fact that the stress on a bond in the functioning wall fabric is more sensitive if oriented in the hoop direction.

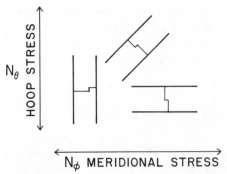

N_θ HOOP STRESS

N_ϕ MERIDIONAL STRESS

Figure 3. Cleavage as a function of orientation. Cross-linked pairs of disaccharides are depicted in various orientations. In the stressed wall fabric the bonds will be oriented in a more extended configuration than shown. For a gram-negative rod, the D-alanine-*meso*-muramic acid bond is probably the bond cleaved to permit enlargement, and thus the wall will be enriched for structures with the cross-linking peptides oriented in the axial direction. For the gram-positive organism in which the peptides are cleaved by amidases and the glycan chains are cleaved by glycoaminidases, the original wall, probably laid down at random, will be preferentially cleaved in the axial direction. Consequently, these cells need both types of enzymes. Note that the orientation of maximum stress is opposite to that of the gram-negative rod. This inversion arises for the gram-positive organism because the stress arises largely as a result of axial elongation. See Fig. 4.

to gram-positive walls growing by the inside-to-outside mechanism (23, 35, 36), where the nascent oligopeptidoglycan is laid down and is cross-linked in the unstressed state to other nascent oligopeptides and to the overlying extant wall at random. For this type of strategy the wall moves peripherally as a result of additions of younger layers underneath. In so doing it becomes slightly enlarged (strained) in the hoop direction, but because of elongation, it becomes strained in the axial direction to a greater degree, which tends to orient the polymer. This is a well-known phenomenon in polymer chemistry; e.g., when rubber is stretched near to the breaking point, it becomes paracrystalline (45). Sufficiently old lamina become cleaved by autolysin action. Here a combination of glycosaminidase and amidase action would cause the rupture because there would be no preferential orientation of the types of chemical bonds. The rupture would be akin to the fracture of a rock or other solid, the tearing of paper, etc. For the gram-positive case, the stress relieved on cleavage would be transferred to neighboring bonds that would then become much more susceptible to cleavage, and the fracture would continue. Because of the original stresses (Fig. 1), the fracture line will lie mainly in the hoop direction.

But there is an additional phenomenon that causes the fractures subsequently to become deeper. This is shown in Fig. 4, where the stresses through a cross section in the axial direction of a gram-positive rod are depicted. Two physical concepts are involved. One is that a chain breaks at its weakest link. Consequently, if the only stress arose from force applied at the ends of a chain, then the site of rupture is equi-probable along the length. However, if multiple additional forces are applied along the length of the chain, then the chain will break very nearly in the middle, as in a tug-of-war. For the gram-positive rod the situation is a combination of these two effects and is akin to that of the

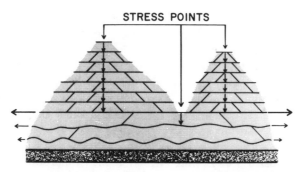

Figure 4. Formation of the external surface of the gram-positive rod. Depicted is a longitudinal section through the wall. The major axial stress is at the most peripheral intact lamella; inside this level the stress has not yet become maximal, and outside this level the stress has been partially dissipated by the fracturing process. Two features are indicated. First, the cracks tend to propagate inward in a circumferential pattern because a fraction of the stress is borne by the external fragmented layers, sparing the underlying wall. Second, the fragmented mass continues to become further fragmented as stresses build up underneath. These are transmitted because the underlying layer continues to expand. The cumulation of stresses acts to place the greatest stress in the longitudinal center of the mass. The relative effectiveness of these two processes depends on distribution of strand length, strand orientation, cross-linking pattern, autolytic enzymes, etc., and thus would give rise to varied patterns of surface topography.

case of the bark on a tree. It differs in that the expansion is mainly axial in the former and circumferential in the latter. As can be seen in Fig. 4, the partially fragmented portions of the external wall become partially restressed and come to share a portion of the load. There are two consequences: (i) the extant fissures tend to become deeper and (ii) new fissures are more likely to form in the middle of the masses of wall between the deeper fissures.

The third way that stresses affect the morphology has to do with stretch of the substratum during manufacture. As mentioned above, the septum of a gram-positive rod, such as *B. subtilis*, starts to be formed under stress-free conditions because all the stress is then supported by the outer wall. We have found (3, 20) that the septum is deformed to its definitive shape in becoming a pole as if the septum was entirely formed as a stress-free structure and then split, i.e., subject to the split-and-stretch mechanism (3, 20, 21). Now, this can be understood because the shape of the pole is that of a flattened hemisphere, which means that the hoop stresses are nearly zero or modestly compressive during formation. Consequently, the substratum of the annulus is under weak forces compared with, for example, *S. faecium*. In the latter case, the completed pole is elongated, the corresponding portion of the developing septum is thus subject to tensile stresses, and according to our measurements and calculations a lamina is laid down and becomes stretched slightly before further laminae are added (A. L. Koch and M. L. Higgins, unpublished data). This results in the more pointed pole.

There are yet additional forces to be considered in the case of aggregations of cells. Chains of cells form themselves into helices, cords, macrofibers, and threads (6, 28–32, 44). These clearly result from the attraction of cells for each other. The helix resulting from the hairpin foldback of a single, previously straight chain in some small part probably results from the torsional forces that develop when the foldback occurs. In the untwisted state one side of the folded back filament is under compressive stress and the other side is under tensile stress. These torsional forces resolve when the helix forms. A second factor concerns the biological effects of being surrounded by a large metabolizing biomass. This is discussed elsewhere (Koch, in press), but suffice it to say here that the force powering the helical twist of the macrofibers can be understood to result from a modest slowing of growth of those cells in the macrofiber that are remote from the nutrient surface. A third factor results from the increased amount of surface interaction possible with a helical arrangement or cables as in twisted ropes.

Partial Order Observed by Electron Microscopy

It has been shown with ultrasonic and endopeptidase treatment that the cylindrical portion of *Escherichia coli* is preferentially torn at right angles to the axis of the cell (46, 47). Even through the stresses due to turgor pressure were anisotropic during life (see above), no stress is acting during these treatments. Since the stress that develops in a cylinder under pressure is half as great in the meridional direction as in the hoop direction, preferential cleavage of bonds oriented in the circumferential direction during growth would occur, leaving the peptide bonds partially aligned in the axial direction. Because this pattern of partial orientation is imprinted on the covalent structure of the wall, it would remain even when the hydrostatic stress has been removed.

Similar electron micrographs have been observed with the thick-walled *B. subtilis*. The explanation proposed in the previous paragraph must be modified for this case and is given in more detail elsewhere (Koch, in press). It is quite different but basically goes back to the asymmetry of stresses in the wall, now, however, depending on the surface layers of partially autolyzed wall having more stress in the axial direction because of the elongation process.

Helices of Bacterial Filaments (Chains)

Mendelson's model for gram-positive bacteria (30) results from studies of mutants of *B. subtilis* which grow into filaments because of deficiencies in the autolysin system. These filaments, as well as some other naturally occurring organisms (6) and the wild-type strain under certain conditions (D. K. Mahnke and A. L. Koch, unpublished data), aggregate and fold back on themselves and grow in a complex system of helical cables. Such systems can be drawn into macroscopic threads (44). Mendelson presumes that the helical nature of the macroscopic threads results from some underlying regularity in the peptidoglycan structure within the individual cells; we disagree.

The helix clock model has been well reviewed (31). The observed helices require fiber bundles with at least two chains of cells. Single chains have little tendency to assume a helical shape until some time after they fold back on themselves. The helix-producing interaction depends, therefore, on the interaction between the external surfaces of the chains of cells, which we now know are quite rough (43). Mendelson presumes that these surfaces have a helical arrangement and that a second filament with the same arrangement will adhere and cause the formation of the twisted chain. In the first paper on the helical growth (28), he demonstrated that the outgrowth of spores of chain-forming organisms, with a second mutation that left the distal cell attached to the spore coat, developed into coils as if the cylindrical portion of the cell wall rotated as it was formed. In this case, the looped portion is under compressive strain on the inside arc of curvature and tensile strain on the outer arc. The strain resolves itself by the formation of a helix in such a way that the length of all original meridians is maintained. We believe that this is a significant factor in helix initiation.

Mendelson's later work showed that the outgrowth from spores was not needed (29). Rather, growth from the short filaments and single cells present in overnight cultures of the autolysin-deficient strains also gave rise to helical bundles. Note that the known mutant strains are not fully autolysin defective. This system requires alternative explanations. We have found that similar structures develop with autolysin-sufficient wild-type *B. subtilis* (Mahnke and Koch, unpublished data). Except for extreme dilutions of stationary cultures, the aggregates mainly arise by side-to-side interaction. One possibility for helix formation is that the macrofibers result from torsion between filaments of cells in the bundle adhering to each other and growing at slightly different rates. For example, the experimentally observed pitch would arise if the cells in the middle of the cable grew threefold slower than the ones on the surface of the macrofiber when the bundle contained 1,000 cell filaments (Koch, in press). Figure 5 shows the deficit in growth rate of the cells in the center of the bundle needed to account for the observed degree of helicity.

The Double-Stranded Helix of Burman and Park

The other model proposed recently positing a regular peptidoglycan structure is that of Burman and Park (4). Their point of departure was their measurements of the time course of the incorporation of diaminopimelic acid into

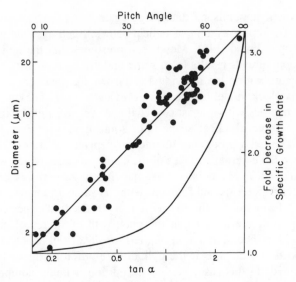

Figure 5. Relationship of helix pitch to the growth rate of cells internal or external to the cell mass. The data have been abstracted from Fig. 4 of reference 32. Superimposed on this is the prediction of a simple model for the presumed major force leading to helical formation. It was assumed that an aggregate of filaments (chains) of *B. subtilis* mutants continue to grow, but the cells at the center grow more slowly than do those on the periphery as a result of impoverished growth condition. An additional (heavy solid) line has been added to show the ratio of specific growth rates for the cells on the surface of the aggregate relative to those at the center needed to give the observed helix pitch.

cross bridges. The pentapeptide first serves as a donor, losing its terminal D-alanine in the transpeptidation process. Later the cross bridge is hydrolyzed (almost certainly requiring endopeptidase or a transpeptidase functioning in a different mode than usually envisioned) and then forms new cross bridges. In this latter event, the cross bridge must function as a recipient in the transpeptidation process. Their interpretation of the kinetics observed in pulse-chase experiments is that peptidoglycan is inserted in a double-strand helical way into the cylinder. They observed that in the initial phases of a pulse experiment, a significant fraction of the label was recovered in the form of the acceptor. To account for the data, they postulated that two new strands are inserted between an older pair of strands and both chains elongate processively. Assuming their model, it seems to me that helical elements could serve a structural role only if the helical strands ran the length of the cell and therefore had a contour length many times the length of a cell. Moreover, the Burman-Park model, as formulated, has the additional problem of maintaining the pitch angle, which should progressively decrease, as Mendelson has shown (30). Consequently, some additional provision to the published model is certainly needed.

We disagree with Burman and Park's interpretation because they have not taken into account the effects of incorporation into the walls of the cell poles. An alternative explanation of their data is that there is random insertion of new oligopeptidoglycan into the old wall all over the surface. However, at sites of constriction where the local activity is very high (38, 42), labeled donor peptides in the constricting regions quickly become cleaved and function as acceptor peptides. This conversion would take place in a much shorter time than the duration of the 10-min pulse. A simple model which includes the contributions of pole and side wall to predict the kinetics of the ratio of radioactivity in the acceptor molecules to that in the donor molecules is derived elsewhere (Koch, in press) and is a follows. Let F be the fraction of the wall that is pole, let T be the time, and let μ be the growth rate constant. Then the ADRR (acceptor donor radioactive ratio) is given by:

$$ADRR = [1 - \exp(-2\mu T) + F]/[\exp(-2\mu T) + F]$$

This function with $F = 0.25$ is shown in Fig. 6 superimposed on the experimental results of Burman and Park (4). The model is simplistic because it assumes that during the pulse part of the experiment the ADRR of the cylinder portion is zero. Actually, it should be a little higher because the pulse usually lasted 10 min. Also, the model assumes that the ADRR of the pole is 1, but actually it should be a little less because pole constriction could finish during the pulse period. More important, the chase is not perfect and, consequently, pool radioactivity continues to bleed into the murein, initially as donor peptides. Also, the presence of any cells that become quiescent for any reason during or shortly after the pulse will tend to keep the ADRR low.

Zonal Growth of Wall of Autolysin-Deficient Gram-Positive Rods

In several papers from Karamata's laboratory in Lausanne, Switzerland (40–42), a clear picture has emerged concerning the deposition of murein from an autolysin-deficient strain of *B. subtilis*. It was found that the label from uniformly radioactive wall ended up in relatively few zones after a five-generation chase. To put it differently, the wall that was synthesized during the chase separated the original label into isolated regions that were comparable to or a little longer than the normal poles of the parental cell. Even in these deficient cells there is some autolytic activity, and it is probable that this mutant, like its wild-type parent, grows by the inside-to-outside mechanism. In the deficient strain, the cleavages rarely liberate the original labeled side wall as small fragments into the medium; rather, the chased material ends up in the vicinity of the unsplit cross walls. This original side wall material and the

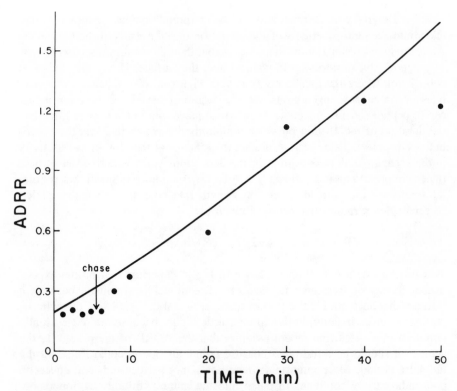

Figure 6. Fitting of the ADRR data of Burman and Park to a model that assumes random intercalation and corrects for the contribution of poles. See text.

conserved pole material then form a zone a little less than half the length of the cylindrical portion of a newborn parental cell.

This is exactly what is expected for the inside-to-outside mechanism and the considerations presented above concerning stress development in the growing gram-positive wall. If we designate the length of the cylindrical wall laid down in a newborn cell by L, then its length would be $2L$ when this lamina of wall reaches the outside of the cell if it was not ruptured but merely stretched (22). Most of the time, for the tug-of-war reasons stated above, when it is ruptured, it will be torn once near the middle. Depending on the relative degree of bonding between different laminae, the bonding within each lamina, and rate of hydrolytic activity, additional tears may develop. If the bonds holding it to other laminae of the wall are more rapidly breached, only one fracture will develop per original unit cell. The length of each portion will return to the original relaxed length of $L/2$. Our interpretation of the results reported by Karamata's group is that secondary ruptures occasionally take place, leading to more, but shorter, labeled zones. This interpretation is consistent with a diffuse (nonzonal) deposition of peptidoglycan over the entire area of the cylindrical portion. It is, of course, contrary to the replicon model (10), but consistent with our renovated replicon model (26).

How a Wall of Matlike Structure Could Be Made

On the basis of the above points and reconsideration of experimental data, it is concluded that wall synthesis is not due to a highly regular process. The biosynthesis of the growing sacculus is much easier to imagine if its structure is random because polymerization can then be assumed to be unregulated with respect to details. We assume that oligosaccharide chains of irregular, uncontrolled length would be secreted through the cell membrane. The composition and chain length distribution of the oligopeptides could depend on the species and on other circumstances. They would be polymerized largely to each other and to adjacent extant murein by membrane-bound transpeptidases and transglycosylases, again as the opportunity presents itself. This situation is consistent with the known fluidity of the membrane. The geometrical facts of pentapeptide oligosaccharide structure restrict, but do not eliminate, the cross-linking of a nascent chain to the same nascent chain or to chains not immediately adjacent in the existing wall. This follows since the peptides will vary in conformation as a result of thermal vibrations. The donor and acceptor peptides, even without the additional linking peptide portion present in some species, are certainly capable of ranging over a radius of 2 to 4 nm from their attachment to the muramic acid residue on the glycan chain (1, 2, 34). Consequently, a mat- or feltlike structure will be formed with little or no geometric order among the fibers. However, because the existent wall is under some stress, a limited order develops in the network, as discussed above. This would be different for the strategies of wall growth of different procaryotes. The cellular shape, as noted before, would depend on the turgor pressure, radius, free energy of synthesis, control of autolysis, and, for cylindrical growth, the existence of relatively rigid poles to support the side wall growth (13, 15). The linkage of wall material can only take place directly adjacent to the cytoplasmic membrane.

It might be expected that the wall would become, by chance, thicker in some regions of the side wall than in others. However, this would self-correct for the gram-negative organism because then the stress would be shared between layers and autolysis would decrease. For gram-positive organisms, an extra thickening would increase the thickness of the wall sufficiently so that additional peripheral walls would become autolyzed and sloughed.

In a series of papers (13–17, 19–24, 26; Koch, in press) we have developed the surface stress theory and have shown how such a random mechanism could function to give a stress-resistant layer capable of enlarging, dividing, and achieving nonspherical morphologies without invoking an energy-transducing mechanism, involving a protein like actin, to carry out mechanical work.

Nonwoven Fabrics, Rubbers, and Gels

There are two extreme types of nonwoven fabrics (7, 9, 27): (i) those in which all the bonds are substantially of the same strength and (ii) those which include a class of bonds that are much more fragile. In the case of the peptidoglycan in bacteria, the covalent bonds of the cross-linking peptides are made of comparable strength to the weakest bonds in the oligosaccharide chain.

The strength of covalent bonds is indeed large, but some of the bonds in the murein are not as resistant to hydrolysis in aqueous solution as are others. Even so, they remain intact unless the pH is very different from 7 and the temperature is much above the growth temperature of most bacteria.

Autolysin cleavage events split only particular bonds in the peptide chain, dependent on the specificity of the enzyme. Consequently, the sacculus of the bacteria in vivo should have the properties of the class of nonwoven fabrics which contain both less and more labile bonds. Thus, the wall fabric of living bacterial cells, in our view, is akin to a nonwoven fabric such as paper or thin slab gels of agarose or agar-agar. In all of these cases cross-linking bonds (usually hydrogen bonds or exchangeable salt linkages in these examples) are effectively weaker than the bonds holding the chain together.

It must be stressed again that the bacterial biopolymers belong to a special class of nonwoven fabrics because of differences in their mode of formation. The cell walls are generated by polymerizing the nascent material to a structure already under stress. Peptidoglycan is formed by transpeptidase-transglycosylase action to link the nascent murein to the extant structural framework of the cell. The cross-links are no doubt formed as a result of the simultaneous binding of donor and acceptor peptides to the membrane-bound enzyme. Thus, the salient difference is that the structure to which the newly secreted chains are bound has previously been deformed by the stress due to turgor pressure.

Mechanical Properties of the E. coli Sacculus In Vivo and In Vitro

Our recent studies (25) have shown that growing bacteria have a surface area 46% larger than the cell wall would have in the relaxed state. The amount of stretching of the wall fabric in vivo was measured by following the shrinkage of growing cell filaments when the turgor pressure was removed by disrupting the phospholipid membranes. Filaments of 7 to 27 μm in length were produced by temperature shifts for 1 to 2 h of *ftsA* and *ftsI* mutants. These filaments were long enough to be accurately measured by phase microscopy. They were mounted on the glass surfaces of polylysine-treated rectangular capillaries in an apparatus that allowed the bathing fluid to be replaced while the cells were viewed in a microscope. The degree of shrinkage was assessed by comparing the microphotographs obtained before disruption of the membrane with ones taken after the phase contrast was lost, signaling that the leakage had occurred and that the turgor pressure had been lost.

Although the physical dimensions of sacculi of gram-negative organisms can be assessed with an electron microscope when combined with low-angle shadowing of the preparations or other special techniques (48), the electron microscopy procedures have systematic errors because the cells are almost certainly shrunken by any known method of preparation and only reflect the properties in the dry state. For this reason, we developed a low-angle light-scattering apparatus and developed an appropriate theory to study the mean surface areas of sacculi under a variety of conditions (A. L. Koch and S. W. Woeste, in preparation). With this technique the sacculi were examined in aqueous suspension where they are hydrated and are not likely to be collapsed or wrinkled, as are samples prepared for electron microscopy. Since this is a technique unfamiliar to most microbiologists, it is necessary to state that the theory compares the linear dimension to the wavelength of light used for measurement (5, 11, 12, 18). The area and other physical dimensions of the sacculus can be found by comparing the intensity of light scattered over a range of very low angles (4 to 12 degrees) with the predictions of the appropriate theory for the relative intensity of light scattered by ellipsoidal shells of revolution. The theory requires only one parameter to be fitted, e.g., either the radius of the rod-shaped cells or the surface area.

It can be seen from molecular models (1, 2, 34) that regular, 100% cross-linked murein in a compact monolayer form could expand by 100 to 300% before covalent bonds must be broken. The expansion could be somewhat greater if the unstressed surface were dimpled or convoluted or if there were holes in the fabric. On the other hand, the possible expansion would be somewhat smaller because the configuration at which the energy is minimal is not quite the most compact one. The mean surface areas of pure sacculi preparations were estimated by our low-angle light-scattering method. For *E. coli* B/r H266 grown in rich medium, the area of sacculi near the isoionic point (pH 4.6) and in 1.0 M KCl was 6.1 μm^2. Lowering the ionic strength increased the area to 7.8 μm^2. This was further increased with alkali to 13.2 μm^2 at pH 12 and to 20.1 μm^2 with the further addition of saturated urea. Alternatively, it was raised to 24.5 μm^2 in low-ionic-strength medium by lowering the pH to 2 and saturating with urea. Thus, the maximum expansion, expressed as a percentage increase above the most compact structure, that we observed was 302%. This is consistent with a nearly complete fabric without major open gaps or spaces. None of the treatments mentioned should cleave the covalent bonds in the peptidoglycan. The changes in mean surface area due to pH shifts were shown to be reversible and no hysteresis effects were found (Woeste and Koch, unpublished data). The smaller expansion above the stress-free state of the in vivo sacculus found by the shrinkage of filaments of growing bacteria versus that found in studies of the in vitro sacculus suggests that the sacculus is not stretched to its elastic limit in vivo whereas the light-scattering data show that it could be stretched much more.

Because these methods are new, it is appropriate that they should be justified and related to other more familiar approaches. For this reason, the particular strain and culture conditions chosen were those extensively studied by Woldringh and collaborators in Amsterdam, The Netherlands. These workers found that the mean surface area of the organism is 11.9 μm^2 (48, 49). A similar estimate (10.9 μm^2) can be obtained from some of their other studies in which they used phase-contrast microscopy and examination of agar filtration samples by electron microscopy. These authors believe that the cells shrink during examination in the fixation procedure but are expanded in the grid plane as the cylinder becomes flattened during the filtration and drying, and they estimate that the two effects compensate for each other. Accepting this at face value, the area of the cell if it shrank when the phospholipid membrane was punctured would be 1.46-fold smaller, as with our filament studies reported above, or 7.8 μm^2. This value is in excellent agreement with the value of 7.8 μm^2 cited above.

The conclusion is that the growing sacculus is stretched somewhat above its relaxed state, but is capable of a good deal more expansion without recourse to autolytic enzyme activity.

Conclusions

The fact that fabric, factory, and manufacture all go back to a root word meaning "to accomplish something" is pertinent to the purpose of this paper. Whether an object is good and useful for a human or for a bacterium to make, the same engineering or husbandry concepts are needed and must be implemented. The fact that man's first cloth was woven implies that his mechanics were better than his chemical engineering. The reverse situation applies to the first inhabitants of this world, the procaryotes. Later, the eucaryotes developed a mechanics based on cytoskeletons and contractile proteins allowing mechanical movement. An extremely important movement is karyokinesis and cytokinesis. Here, we have scratched the surface of the equivalent techniques that the procaryotes have used to achieve the same goals to grow and divide.

ACKNOWLEDGMENTS. This new phase in the development of the surface stress theory was stimulated by discussion, action, and reaction by Ian Burdett,

Ron Doyle, Jean-Marie Ghuysen, Mike Higgins, Dimitri Karamata, Neil Mendelson, Nanne Nanninga, Ted Park, Gerry Shockman, Conrad Woldringh, and David White. It is dedicated to Lolito Daneo-Moore because, 10 years ago, she understood that maybe I could make a contribution if only I were taught a little more microbiology.

This work was supported by Public Health Service grant GM-34222 from the National Institutes of Health.

LITERATURE CITED

1. **Braun, V.** 1975. Covalent lipoprotein from the outer membrane of *Escherichia coli. Biochim. Biophys. Acta* **415:**335–377.
2. **Braun, V., H. Gnirke, U. Henning, and K. Rehn.** 1973. Model for the structure of the shape-maintaining layer of the *Escherichia coli* cell envelope. *J. Bacteriol.* **114:**1264–1270.
3. **Burdett, I. D. J., and A. L. Koch.** 1984. Shape of nascent and complete poles of *Bacillus subtilis. J. Gen. Microbiol.* **130:**1711–1722.
4. **Burman, L. G., and J. T. Park.** 1984. Molecular model for elongation of the murein sacculus of *Escherichia coli. Proc. Natl. Acad. Sci. USA* **81:**1844–1848.
5. **Cross, D. A., and P. Latimer.** 1972. Angular dependence of scattering from *Escherichia coli* cells. *Appl. Opt.* **11:**1225–1228.
6. **Davis, B. D., R. Dulbecco, H. N. Eisen, H. S. Ginsberg, W. B. Wood, Jr., and M. McCarty.** 1973. *Microbiology,* 2nd ed., p. 846–847. Harper & Row, Publishers, Inc., Hagerstown, Md.
7. **Emery, I.** 1980. *The Primary Structures of Fabrics.* The Textile Museum, Washington, D.C.
8. **Flugge, W.** 1973. *Stresses in Shells,* 2nd ed. Springer-Verlag, New York.
9. **Hearle, J. W. S., P. Grosberg, and S. Backer.** 1969. *Structure Mechanics of Fibers, Yarns, and Fabrics,* vol. 1. John Wiley & Sons, Inc., New York.
10. **Jacob, F., S. Brenner, and F. Cuzin.** 1963. On the regulation of DNA replication in bacteria. *Cold Spring Harbor Symp. Quant. Biol.* **28:**329–348.
11. **Koch, A. L.** 1961. Some calculations on the turbidity of mitochondria and bacteria. *Biochim. Biphys. Acta* **52:**429–441.
12. **Koch, A. L.** 1968. Theory of angular dependence of light scattered by bacteria and similar-sized biological objects. *J. Theor. Biol.* **18:**133–156.
13. **Koch, A. L.** 1982. On the growth and form of *Escherichia coli. J. Gen. Microbiol.* **128:**2527–2540.
14. **Koch, A. L.** 1983. The surface stress theory of microbial morphogenesis. *Adv. Microb. Physiol.* **24:**301–366.
15. **Koch, A. L.** 1984. How bacteria get their shapes: the surface stress theory. *Comment Mol. Cell. Biophys.* **2:**179–196.
16. **Koch, A. L.** 1984. Shrinkage of growing *Escherichia coli* cells through osmotic challenge. *J. Bacteriol.* **159:**914–924.
17. **Koch, A. L.** 1985. How bacteria grow and divide in spite of internal hydrostatic pressure. *Can. J. Microbiol.* **31:**1071–1083.
18. **Koch, A. L.** 1986. Estimation of size of bacteria by low-angle light-scattering measurements: theory. *J. Microbiol. Methods* **5:**221–235.
19. **Koch, A. L., and I. D. J. Burdett.** 1984. The variable T model for gram-negative morphology. *J. Gen. Microbiol.* **130:**2325–2338.

20. **Koch, A. L., and I. D. J. Burdett.** 1986. Normal pole formation during total inhibition of wall synthesis. *J. Gen. Microbiol.* **132:**3441–3449.
21. **Koch, A. L., and I. D. J. Burdett.** 1986. Biophysics of pole formation of Gram-positive rods. *J. Gen. Microbiol.* **132:**3451–3457.
22. **Koch, A. L., and R. J. Doyle.** 1985. Inside-to-outside growth and the turnover of the Gram-positive rod. *J. Theor. Biol.* **117:**137–157.
23. **Koch, A. L., M. L. Higgins, and R. J. Doyle.** 1981. Surface tension-like forces determine bacterial shapes: *Streptococcus faecium. J. Gen. Microbiol.* **123:**151–161.
24. **Koch, A. L., G. Kirchner, R. J. Doyle, and I. D. J. Burdett.** 1985. How does a *Bacillus* split its septum right down the middle? *Ann. Microbiol.* (Paris) **136A:**91–98.
25. **Koch, A. L., S. L. Lane, J. Miller, and D. Nickens.** 1987. Contraction of filaments of *Escherichia coli* after disruption of the cell membrane by detergent. *J. Bacteriol.* **166:**1979–1984.
26. **Koch, A. L., H. L. T. Mobley, R. J. Doyle, and U. N. Streips.** 1981. The coupling of wall growth and chromosome replication in Gram-positive rods. *FEMS Microbiol. Lett.* **12:**201–208.
27. **Lunenschloss, J., and W. Albrecht.** 1985. *Non-Woven Bonded Fabrics.* Halsted Press, Chichester, United Kingdom.
28. **Mendelson, N. H.** 1976. Helical growth of *Bacillus subtilis*: a new model for cell growth. *Proc. Natl. Acad. Sci. USA* **73:**1740–1744.
29. **Mendelson, N. H.** 1978. Helical *Bacillus subtilis* macrofibers: morphogenesis of a bacterial multicellular macroorganism. *Proc. Natl. Acad. Sci. USA* **75:**2478–2482.
30. **Mendelson, N. H.** 1982. The helix clock: a potential biomechanical cell cycle timer. *J. Theor. Biol.* **94:**209–222.
31. **Mendelson, N. H.** 1982. Bacterial growth and division: genes, structure, forces, and clocks. *Microbiol. Rev.* **46:**341–375.
32. **Mendelson, N. H., D. Favre, and J. J. Thwaites.** 1984. Twisted states of *Bacillus subtilis* macrofiber reflect structural states of the cell wall. *Proc. Natl. Acad. Sci. USA* **81:**3562–3566.
33. **Mitchell, P., and J. Moyle.** 1956. Osmotic function and structure of bacteria. *Symp. Soc. Gen. Microbiol.* **6:**150–180.
34. **Oldmixon, E. H., S. Glauser, and M. L. Higgins.** 1974. Two proposed general configurations for bacterial cell wall peptidoglycan shown by space filling models. *Biopolymers* **13:**2037–2060.
35. **Pauling, L.** 1944. *The Nature of the Chemical Bond.* Cornell University Press, London.
36. **Pooley, H. M.** 1976. Turnover and spreading of old wall during surface growth of *Bacillus subtilis. J. Bacteriol.* **125:**1127–1138.
37. **Pooley, H. M.** 1976. Layered distribution, according to age, within the cell wall of *Bacillus subtilis. J. Bacteriol.* **125:**1139–1147.
38. **Roark, R. J., and W. C. Young.** 1975. *Formulas for Stress and Strain,* 5th ed. McGraw-Hill Book Co., New York.
39. **Ryter, A., Y. Hirota, and U. Schwarz.** 1973. Process of cellular division in *Escherichia coli.* Growth pattern of *E. coli* murein. *J. Mol. Biol.* **78:**185–195.
40. **Schlaeppi, J.-M., and D. Karamata.** 1982. Cosegregation of cell wall and DNA in *Bacillus subtilis. J. Bacteriol.* **152:**1231–1240.
41. **Schlaeppi, J.-M., H. M. Pooley, and D. Karamata.** 1982. Identification of cell wall subunits in *Bacillus subtilis* and analysis of their segregation during growth. *J. Bacteriol.* **149:**329–337.
42. **Schlaeppi, J.-M., O. Schaefer, and D. Karamata.** 1985. Cell wall and DNA cosegregation in *Bacillus subtilis* studied by electron microscope autoradiography. *J. Bacteriol.* **164:**130–135.

43. **Schwarz, U., A. Ryter, A. Rambach, R. Hellio, and Y. Hirota.** 1975. Process of cellular division in *Escherichia coli*. Differentiation of growth zone in the sacculus. *J. Mol. Biol.* **98:**749–759.
44. **Sonnenfeld, E. M., T. J. Beveridge, A. L. Koch, and R. J. Doyle.** 1985. Asymmetric distribution of charges on the cell wall of *Bacillus subtilis*. *J. Bacteriol.* **163:**1167–1171.
45. **Thwaites, J. J., and N. H. Mendelson.** 1985. Biomechanics of bacterial walls: studies of bacterial thread made from *Bacillus subtilis*. *Proc. Natl. Acad. Sci. USA* **82:**2163–2167.
46. **Treloar, L. R. G.** 1958. *The Physics of Rubber Elasticity,* 2nd ed. Clarendon Press, Oxford.
47. **Verwer, R. W. H., E. H. Beachey, W. Keck, A. M. Stoub, and J. E. Poldermans.** 1980. Orientated fragmentation of *Escherichia coli* sacculi by sonication. *J. Bacteriol.* **141:**327–333.
48. **Verwer, R. W. H., N. Nanninga, W. Keck, and U. Schwarz.** 1978. Arrangements of glycan chains in the sacculus of *Escherichia coli*. *J. Bacteriol.* **136:**723–729.
49. **Woldringh, C. L., and N. Nanninga.** 1985. Structure of the nucleoid and cytoplasm in the intact cell, p. 161–197. *In* N. Nanninga (ed.), *Molecular Cytology of Escherichia coli.* Academic Press, Inc., London.

Chapter 5

Does *Escherichia coli* Use a Left-Handed Murein Endopeptidase and Transpeptidase to Assemble Its Sacculus?

James T. Park

Three years ago, Burman and I published a model for cell wall elongation in *Escherichia coli*. It was based on data obtained by following the fate of radioactive diaminopimelic acid as it became incorporated into the cross-linked muropeptide dimers of the growing sacculus (2, 9).The analysis was based on the fact that the donor and acceptor halves of cross-linked dimers isolated from the sacculus can be measured separately, and on the observation by dePedro and Schwarz (4) that newly incorporated disaccharide-pentapeptide units lose their terminal D-alanine very rapidly. Consequently, the average pentapeptide can serve as donor to drive the cross-linking transpeptidation reaction for a relatively short period of time (less than 30 s).

Without going into detail, I would like to summarize the main conclusions stemming from this work. There are four.

(i) About 90 to 100 membrane-bound enzyme complexes are involved which work independently of one another. Thus, insertion of murein into the sacculus is clearly diffuse while the number of sites is much larger than the minimum of three or four that could be estimated by autoradiographic studies. The number of complexes correlates with the measured quantity of several of the penicillin-binding proteins (PBPs) found in the cell.

(ii) The complexes move unidirectionally around the circumference of the cell about once every 8 min. This represents a unique example of directed

James T. Park • Department of Molecular Biology and Microbiology, Tufts University, Boston, Massachusetts 02111.

movement of membrane-bound enzymes. The effective rate of movement is about 5 nm/s.

(iii) During elongation the sacculus is in a very dynamic state (3, 5). One or two hundred thousand cross-links are broken per cell per generation to be replaced by new cross-links as new murein is incorporated.

(iv) Two strands of murein are synthesized and inserted by each complex as it progresses around the cell. This is perhaps the most unexpected result.

Why Are Two Strands Inserted at a Time?

Our initial rationalization of this observation (2) was that one enzyme (e.g., PBP 1a, a muropeptide polymerase and transpeptidase [8]) synthesized one glycan strand and cross-linked it to the existing sacculus to the right, and another enzyme (e.g., PBP 1b, a similar two-headed enzyme [8]) was needed to synthesize another strand and cross-link it to the existing strand on the left. This idea may still be correct. However, it is certainly not essential, as *E. coli* strains have been constructed in which the gene for either PBP 1a or PBP 1b has been deleted and each deletion mutant is viable (7, 10). This result would suggest that two PBP 1a or two PBP 1b enzymes can cooperate to insert two strands at a time. Why, then, are two strands inserted at a time? A much more attractive explanation arose from the report by Barnickel et al. (1), who, by determining the most favored conformations of the sugars and peptides, concluded that a strand of murein exists in the form of a helix with each repeating unit tilted roughly 90 degrees with respect to its neighbors. This idea is attractive because it explains why the extent of cross-linkage in gram-negative bacteria is usually 50% or less. Presumably, for cross-linking to occur, the peptides must be in the proper orientation and between neighboring helical strands; only every fourth set of peptides would be pointed toward each other. As shown in Fig. 1A, if we try to insert only one new strand between two existing helical strands, after cross-links are formed to the left, the donor peptides on the right will be facing a strand in which the acceptor peptide is pointing in the wrong direction. However, if two strands are inserted, all donor peptides are aligned with potential acceptors (Fig. 1B).

Does E. coli Use a Left-Handed Murein Endopeptidase and Transpeptidase to Assemble Its Sacculus?

We have repeatedly performed an experiment in which cells containing uniformly labeled sacculi are chased for several generations. The result is revealing but presents a dilemma, and this is the main point I wish to bring to your attention. As shown in Table 1, uniformly labeled sacculi would have an

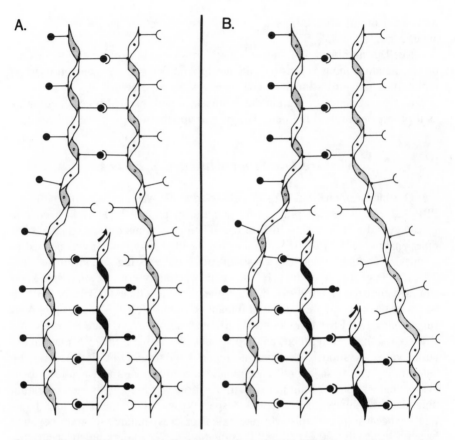

Figure 1. Cross-linking between helical strands of murein. (A) Misalignment of donor and acceptor peptides when only one donor strand is available. (B) Normal alignment of peptides when two strands are inserted. Closed circles, Donors; open arcs, acceptors. (Reprinted by permission from reference 8a).

equal amount of label in acceptors and donors, giving an initial acceptor/donor radioactivity ratio of 1.0, and for an average logarithmically growing cell of length 30, for every 22 counts present in the side walls there would be 8 counts in the poles of the cell. Now suppose that during a chase all cross-links are cleaved. This would result in loss of all counts from the donor half of dimers since once a cross-link is cleaved the former donor half now exists as a tetra-peptide and hence can serve only as an acceptor. The resulting acceptor/donor ratio would be infinity. If, as seems likely, poles once formed are relatively stable and only side wall cross-links are cleaved, complete cleavage of these cross-links would result in an acceptor/donor ratio of 6.5. If only half of the side wall cross-links are cleaved during growth, one would expect a final ratio of 2.15. These derivations are shown in Table 1. Experimentally, the ratio rises

Table 1. Theoretical values of the acceptor/donor (A/D) radioactivity ratio for various assumptions compared with the experimental value achieved during a chase of uniformly labeled cells

Assumption	Distribution of label		Final A/D ratio
	Poles	Side walls	
Uniformly labeled sacculi	8 A/8 D	22 A/22 D	1.0
All cross-links cleaved	16 A/0 D	44 A/0 D	∞
Side wall cross-links cleaved	8 A/8 D	44 A/0 D	6.5
Half of side wall cross-links cleaved	8 A/8 D	33 A/11 D	2.15
Experimental			1.55

to a figure of 1.55 within about one generation and remains essentially constant thereafter.

This result seems to clearly indicate that no more than half of the side wall cross-links are broken during growth. The experimental result may be too low because of one or more of the following factors. One is the remarkable observation of Goodell and Schwarz (6) that tripeptide from the murein is recycled by being incorporated preferentially to L-alanine at the nucleotide precursor level. This causes label to be reintroduced as donor, thereby maintaining the acceptor/donor ratio lower than it would otherwise be. The experiment has been repeated once with a strain that does not recycle, and the ratio was 1.81 or about 20% higher (unpublished data). Another effect that might cause the ratio to be deceptively low is gradual loss of counts from the sacculus. If 10% of the counts were lost per generation by the strain that does not recycle (a rough estimate) and all of these counts would otherwise have been counted in the acceptor fraction, the theoretical ratio for cleavage of half the sidewall cross-links (1.84) would about equal the preliminary experimental result (1.81). A third factor possibly contributing to an apparent low ratio is that the poles may be more highly cross-linked than the sidewalls.

Taking all this into consideration, it is most likely that approximately half of the side wall cross-links are cleaved and the remainder are conserved. What explanation can there be for only half of the side wall cross-links being cleaved and the other half remaining stable? The simplest explanation that suggests itself is that the relevant endopeptidase can only cleave cross-links of a certain orientation, e.g., left-handed cross-links. Figure 2 is an attempt to visualize the situation. Imagine that the endopeptidase, which is anchored in the cytoplasmic membrane, can cleave a left-handed cross-link via the shaded face of the donor muropeptide and the lighter face of the acceptor. If the endopeptidase now rotates 180 degrees and approaches a right-handed cross-link, the orientation of the sugars and cross-linked peptides is reversed; the enzyme will see the lighter face of the donor instead of the shaded face and will not fit. Presumably this argument, which depends on the helical conformation of the glycan strands, would also apply to the transpeptidases which form the cross-links.

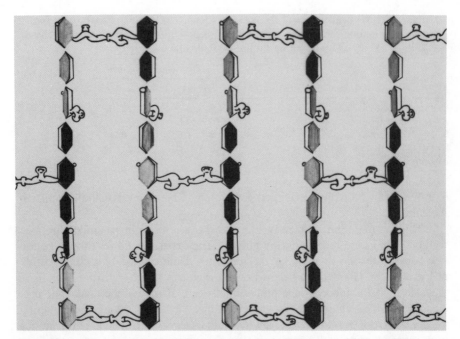

Figure 2. View of murein with cross-links between adjacent strands alternating between left handed and right handed. Note that the opposite surfaces of the cross-linked muropeptides would face the membrane depending on whether the linkage between helical strands was left handed or right handed.

It seems clear that *E. coli* uses the existing sacculus as a substrate, or template, and this implies that the membrane-bound enzymes involved must be able to distinguish left from right as these data suggest.

LITERATURE CITED

1. **Barnickel, G., D. Naumann, H. Bradaczek, H. Labischinski, and P. Giesbrecht.** 1983. Computer-aided molecular modeling of the three-dimensional structure of bacterial peptidoglycan, p. 61–66. *In* R. Hackenbeck, J. V. Holtje, and H. Labischinski (ed.), *The Target of Penicillin.* Walter de Gruyter, Berlin.
2. **Burman, L. G., and J. T. Park.** 1983. Molecular model for elongation of the murein sacculus of *Escherichia coli. Proc. Natl. Acad. Sci USA* **81:**1844–1848.
3. **Burman, L. G., J. Reichler, and J. T. Park.** 1983. Evidence for multisite growth of *Escherichia coli* murein involving concomitant endopeptidase and transpeptidase activities. *J. Bacteriol.* **156:**386–392.
4. **dePedro, M. A., and U. Schwarz.** 1981. Heterogeneity of newly inserted and preexisting murein in the sacculus of *Escherichia coli. Proc. Natl. Acad. Sci. USA* **78:**5856–5860.
5. **Goodell, E. W., and U. Schwarz.** 1983. Cleavage and resynthesis of peptide cross bridges in *Escherichia coli* murein. *J. Bacteriol.* **156:**136–140.

6. **Goodell, E. W., and U. Schwarz.** 1985. Release of cell wall peptides into culture medium by exponentially growing *Escherichia coli*. *J. Bacteriol.* **162:**391–397.

7. **Kato, J., H. Suzuki, and Y. Hirota.** 1985. Dispensability of either penicillin-binding protein -1a or -1b involved in the essential process for cell elongation in *Escherichia coli*. *Mol. Gen. Genet.* **200:**272–277.

8. **Matsuhashi, M., J. Nakagawa, S. Tomioka, F. Ishino, and S. Tamaki.** 1982. Mechanism of peptidoglycan synthesis by penicillin-binding proteins in bacteria and effects of antibiotics, p. 297–310. *In* S. Mitsuhashi (ed.), *Drug Resistance in Bacteria—Genetics, Biochemistry and Molecular Biology*. Japanese Scientific Societies Press, Tokyo.

8a. **Park, J. T.** 1987. The murein sacculus, p. 23–30. *In* F. C. Neidhardt, J. L. Ingraham, K. B. Low, B. Magasanik, M. Schaechter, and H. E. Umbarger (ed.), *Escherichia coli and Salmonella typhimurium: Cellular and Molecular Biology*, vol. 1. American Society for Microbiology, Washington, D.C.

9. **Park, J. T., and L. G. Burman.** 1985. Elongation of the murein sacculus of *Escherichia coli*. *Ann. Microbiol.* (Paris) **136A:**51–58.

10. **Yousif, S. Y., and J. K. Broome-Smith, and B. G. Spratt.** 1985. Lysis of *Escherichia coli* by β-lactam antibiotics: deletion analysis of the role of penicillin-binding proteins 1A and 1B. *J. Gen. Microbiol.* **131:**2839–2845.

Chapter 6

Autoradiographic Analysis of Peptidoglycan Synthesis in Shape and Cell Division Mutants of *Escherichia coli* MC4100

C. L. Woldringh

P. Huls

N. Nanninga

E. Pas

P. E. M. Taschner

F. B. Wientjes

The process of cell division in *Escherichia coli* and other cell systems is difficult to study because there is no direct assay for division other than an increase in the cell number of the population or a change in the individual cell shape. On the other hand, *E. coli* is a structurally simple organism, and it therefore seems possible that the mechanism underlying the universal relationship between DNA replication and cell division will eventually be resolved.

In *E. coli* this relationship is characterized by two seemingly opposed systems. The first enables the cell to divide under conditions of widely differing growth rates, cell sizes, and cell shapes (rods, filaments, and spheres). The second can inhibit, postpone, or dislocate the division process under unfavorable conditions (SOS or heat-shock response). This makes it difficult to characterize the role of division factors or proteins, which may be either a positive or a negative one.

C. L. Woldringh, P. Huls, N. Nanninga, E. Pas, P. E. M. Taschner, and F. B. Wientjes • Department of Electron Microscopy and Molecular Cytology, University of Amsterdam, 1018 TV Amsterdam, The Netherlands.

At present, the period between two major discontinuities occurring during cell growth, i.e., termination of DNA replication and initiation of the constriction process, has remained the least defined part of the cell cycle. For a better understanding of the two events and their relationship, we need more structural and biochemical information both on nucleoid positioning at termination and on rate of envelope synthesis in the cell's central compartment. To this end we studied nucleoid positions by fluorescence light microscopy and location of peptidoglycan synthesis by autoradiography under different conditions of division state and cell shape, i.e., in rods, filaments, and spheres. Especially in the last cells, there is, to our knowledge, theoretical (5) but no experimental information available on the direction of nucleoid segregation and on the mode of envelope synthesis during steady-state growth and division.

All strains used were derivatives of *E. coli* K-12 MC4100 *lysA,* which has been described previously (11). Strains carrying the *ftsZ*(Ts) and *pbpB2158*(Ts) genes have also been described (7, 12). *E. coli* MC4100 *lysA pbpA137*(Ts) *zbe*::Tn*10* was constructed from strain MC4100 *lysA* by transduction using P1 grown on strain SP177 [*pbpA137*(Ts) *zbe*::Tn*10*], obtained from B. G. Spratt (6). When cultured in glucose minimal medium, cells of this strain already grew as spheres at the permissive temperature (28°C). When cultured at a higher growth rate, for instance, in L-broth, the cells were rod shaped and became spherical only after a temperature shift to 42°C. The spherical growth at 28°C was not suppressible by a high salt concentration (300 mosM), nor were the cells resistant to mecillinam (2 μg/ml).

Labeling of the cells with radioactive diaminopimelic acid ([^{3}H]DAP, obtained from C.E.A., France, at a specific activity of 36.5 Ci/mmol) and autoradiography were performed as previously described (12). Pulse times extended for about 10% of the doubling time of the cells. For autoradiography whole cells were used rather than sacculi because the results were identical (P. Huls, unpublished data) and the procedure is less elaborate.

In experiments with the *pbpB* strain, pulse-labeling was carried out two doubling times after the shift to the nonpermissive temperature or after the addition of furazlocillin (2 μg/ml).

For fluorescence microscopy, cells were fixed with 0.1% OsO$_4$ and stained with the DNA-specific fluorochrome 33342 (Hoechst-Roussel Pharmaceuticals Inc.; 2 mM) as previously described (7).

Theoretical Predictions of Envelope Synthesis in Rods and Spheres

Geometrical considerations of envelope growth of rod-shaped cells, presented earlier (13), showed that the rate of envelope synthesis fluctuates during the cell cycle if we make the following assumptions: (i) cell shape is that of a cylinder with hemispherical polar caps, (ii) individual cell volume increases

exponentially during the cycle, and (iii) elongation proceeds at a constant diameter (see Fig. 1A). Depending on the duration of the constriction or T-period
relative to the doubling time, and with the above assumptions, the amount of
surface synthesized during this period (see hatched part of cell in Fig. 1) can
be calculated (12). Assuming that polar caps are entirely made out of newly
synthesized peptidoglycan, a rod-shaped cell like the one in Fig. 1A has (i) to
invest about 75% of its surface synthesized during the T-period into its polar
caps, the rest being available for lateral wall synthesis, and (ii) to increase its
rate of surface synthesis during constriction. Autoradiography of rod-shaped
E. coli MC4100 cells pulse-labeled with radioactive DAP confirmed the increased rate of envelope synthesis during polar cap formation (12). In addition,
the topography of DAP incorporation indicated that polar caps were indeed
made out of new material (12).

As a rough approximation for the situation in round cells, we assumed in
Fig. 1B that cells are born as a sphere and elongate at a constant diameter.
This implies that in a constricting cell the differential rate of surface synthesis
has to increase relative to surface synthesis rate in a nonconstricting cell and
that the constriction process extends over the entire cell cycle. If the T-period

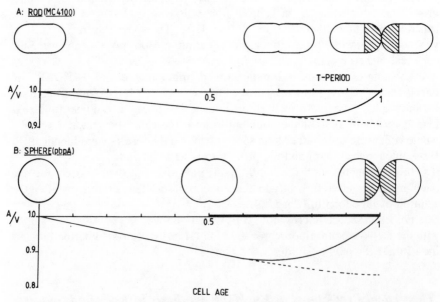

Figure 1. Comparison of growth and division of rod-shaped (A) and spherical (B) cells. Both
types of cells are assumed to have hemispherical polar caps and to elongate at constant diameter.
The surface-to-volume ratio (A/V), plotted as a function of cell age, was calculated with exponential volume increase during the cycle assumed. Hatched areas represent amount of cell
surface synthesized during the T-period, which may (in A) or may not (in B) be sufficient for
polar cap formation.

is shorter than the doubling time (see Fig. 1B), the differential rate of surface synthesis has to increase even more, while the situation may arise that surface-synthesis capacity during constriction is insufficient for polar cap formation. In other words, the cell also has to use old material for its polar caps. For the model cell in Fig. 1B, we can thus make the following two predictions: (i) all DAP incorporation during constriction occurs exclusively at the constriction site, and (ii) the rate of surface synthesis in constricting cells is much higher than in nonconstricting cells.

Topography and Rate of Surface Synthesis in Spherical Cells of *E. coli pbpA*

In glucose minimal medium at 28°C, we obtained a steady-state culture of *E. coli pbpA* with a mass doubling time of 77 min and a constant average cell mass of 4.3 (expressed as absorbance at 450 nm of a suspension of 10^9 cells per ml). About 44% of the cells were in the process of constriction, which matches a T-period of 40 min, i.e., only about half the cell cycle. Under the same conditions the parent strain grew with a doubling time of 72 min at an average cell mass of 2.15. The most striking difference, however, was that the *pbpA* cells were already spherical at the permissive temperature.

An autoradiogram of *pbpA* cells pulse-labeled with radioactive DAP at 28°C is shown in Fig. 2. Plots of grain distributions representing the amount of incorporated DAP along the normalized length axis of the cells (see reference 12) are given in Fig. 3. Already in cells which are slightly constricted (Fig. 3B) there is a peak in grain density in the middle of the cell. Although this central activity is maintained during further constriction (Fig. 3C), significant incorporation is still found all over the spherical cells, in contrast to our first prediction in the previous section.

Also, when plotting the grain number per cell averaged over length classes as a function of cell length (Fig. 4), we do not find the expected higher rate of surface synthesis in the constricting cells (hatched area in Fig. 4). We must therefore come to the tentative conclusion that, unlike the poles of rod-shaped cells (12), the polar caps of the round *pbpA* cells may not exclusively be made out of newly synthesized material and that the cells do not grow according to the assumptions made in Fig. 1B. The latter possibility is supported by the following observations obtained in two independent growth experiments. In contrast to the constant diameter of the parent strain, the width of the *pbpA* cells appeared to increase first, then to decrease at constriction, and to increase again during further growth (Fig. 5). This diameter decrease of about 14% may explain the bimodal length distribution in Fig. 4, which reflects a sudden increase in the length of constricting cells. Such a distribution has never been observed in the rod-shaped parent strain, which elongates at a virtually constant

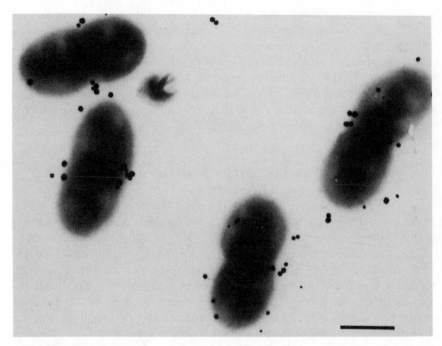

Figure 2. Autoradiogram of *E. coli pbpA* cells prepared by agar filtration. Bar, 1 μm.

diameter (Fig. 5). The diameter decrease, implying an increase in the surface-to-volume ratio, will require a drastic increase in the rate of surface synthesis. The results in Fig. 4 to not exclude such a rate increase in the middle of the cycle.

Nucleoid Segregation

We do not know how the growth mode of *pbpA* cells described above can serve the segregation of nucleoids in these cells. In Fig. 6 it can be seen that these slowly growing cells (doubling time, 77 min) contain on an average more than two nucleoids. Flow cytometric determination of the DNA content (J. A. C. Valkenburg, manuscript in preparation) showed that the *pbpA* cells contained 1.4 times the amount of DNA contained by the parent cells. From the position of the nucleoids in dividing and newborn cells it is not always clear whether the next constriction will be parallel or perpendicular to the previous plane of division. The latter mode of growth was reported for the gram-negative coccus *Neisseria gonorrhoeae* (10).

It seems as if the asymmetric position of nucleoids observed in many cells

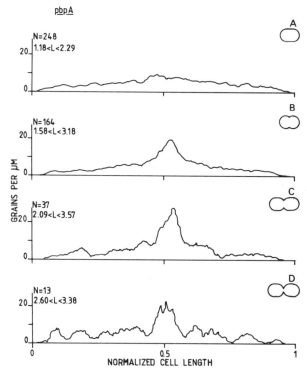

Figure 3. Silver grain distributions over a total of 462 *pbpA* cells measured from autoradiograms as shown in Fig. 2. Grain density is plotted as a function of normalized cell length (see reference 12). For nonconstricting (A), slightly constricting (B), halfway constricting, (C), and deeply constricting (D) cells, the total number of cells measured (N) and the range of cell length (L) considered are indicated.

is related to the asymmetric constrictions regularly observed in the *pbpA* cells (arrow, Fig. 6) and described previously for the round cells of *E. coli envB* (3). Experiments are in progress to determine whether the position of the nucleoid influences the nearby activity of peptidoglycan synthesis.

Topography and Rate of Surface Synthesis in Filamenting Cells

In a previous report (12) we presented observations on surface synthesis in *E. coli ftsZ* growing at the restrictive temperature as filaments which lack any persisting or newly initiated constrictions. These experiments indicated that lateral wall synthesis in growing filaments proceeds in a dispersed mode. We repeated these experiments with an isogenic *pbpB* mutant. Filaments were ob-

pbpA

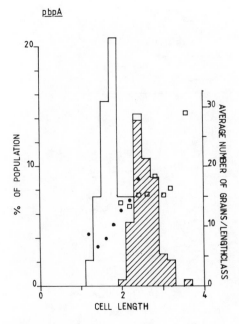

Figure 4. Length distribution of *pbpA* cells. Note the bimodal shape of the distribution. (• and □) Average number of grains per cell in each of the length classes of nonconstricting and constricting cells, respectively. Data are from the same population as in Fig. 3.

tained by inhibition of PBP 3 with furazlocillin, a β-lactam antibiotic known to inhibit PBP 3 specifically (1), or by a shift to the nonpermissive temperature (identical results not shown). Many of the constrictions which were initiated before the furazlocillin treatment were found to persist. After about two mass doublings these persisting constrictions have become extended and appear as blunt constrictions (4) or "necks" in the cell center (Fig. 7). In addition, new constrictions are initiated ("incomplete septa") as previously found at 42°C in *pbpB* filaments (9) and in *ftsA* filaments (2). We wondered whether these unfinished constrictions also resulted from a local increase in surface synthesis just as in normally dividing cells (12), in spite of the elimination of *pbpB* gene product (PBP 3) activity.

In nonconstricting filaments obtained with furazlocillin a diffuse pattern of DAP incorporation is evident (Fig. 8A), just as was found in *ftsZ* filaments obtained at 42°C (see Fig. 4 in reference 12). However, at all constrictions which must have been initiated during the inhibition, clear peaks in grain density are observed. The persisting and blunt constrictions in Fig. 8C and D show virtually no DAP incorporation. We interpret these observations as indicative for intact *ftsZ* gene product activity at new constriction sites and for complete inhibition of PBP 3 activity at the site of constriction.

Both kinds of filaments grew exponentially in mass (result not shown) for at least two mass doublings. This corresponds to the increasing rate of peptidoglycan synthesis found when plotting the average grain number over filaments

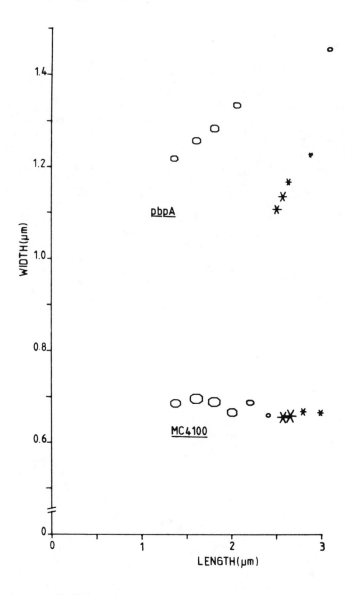

Figure 5. Average cell width per length class plotted versus cell length. Cells are the same as in Fig. 4. Symbols: ○, nonconstricting cells; *, constricting cells. The size of each symbol is proportional to the third root of the number of cells from which it is calculated, to show the different number of cells in each class (see reference 8). Average diameters of nonconstricting cells, of slightly constricting cells, and of the total population were 1.26, 1.12, and 1.22 μm, respectively.

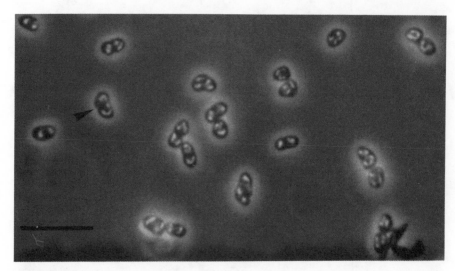

Figure 6. Fluorescence light micrograph of *E. coli pbpA* cells fixed with OsO₄ and stained with fluorochrome 33342. Arrow points to an asymmetrically constricting cell. Bar, 5 μm.

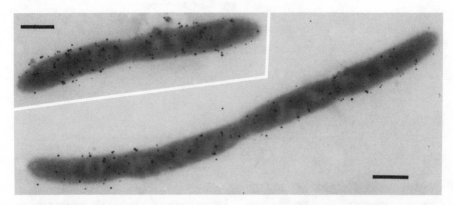

Figure 7. Autoradiogram of *E. coli pbpB* cells grown at 28°C and treated with furazlocillin. Note the central, blunt constrictions and the new constrictions on one and three-quarters of the cell length. Bar, 1 μm.

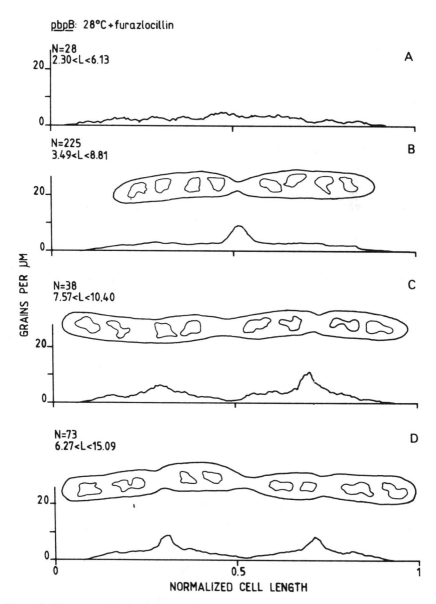

Figure 8. Silver grain distribution over *pbpB* cells as shown in Fig. 7. For nonconstricting filaments (A) and filaments showing a central (B), a central and a polar (C), and a central and two
polar (D) constrictions, the total number of cells (N) measured and the range of lengths (L)
considered are indicated. The central constrictions in C and D represent persisting, blunt constrictions.

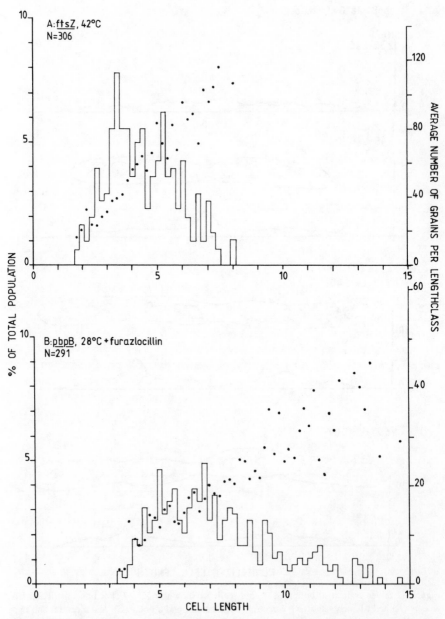

Figure 9. Length distributions of filaments of *ftsZ* grown at 42°C (A) and of *pbpB* treated with furazlocillin at 28°C (B). N, Number of measured filaments; ●, average number of grains per filament averaged over each length class.

per length class as a function of length (Fig. 9). Apparently filaments elongate exponentially, irrespective of inactivation of the gene products of either *ftsZ* or *pbpB*.

Conclusions

Our observations show that under conditions of inhibition of either elongation (as in *pbpA* cells) or cell division (as in *pbpB* filaments), the first stage in polar cap formation is always associated with a local increase in the rate of peptidoglycan synthesis. Although in rod-shaped cells the subsequent completion of the polar caps proceeded at an increasing rate of surface synthesis sufficient for the assembly of polar caps from new material (13), the situation appeared to be different in round cells.

A difference in the control of cell shape for spherical and rod-shaped cells has already been emphasized by Previc (5). In his geometrical model of spherical cells, the diameter first increases (22%) and gradually decreases again during a constriction period of 40% of the cycle. In this model, only half of the surface necessary for polar cap formation is synthesized during the T-period.

The relatively long constriction period (50% of the cycle) and the diameter increase in nonconstricting cells that we observed may help the cell in assembling its polar caps largely from new peptidoglycan. This growth mode implies (i) that little surface is made before constriction, reserving the surface synthesis capacity for polar cap formation, and (ii) that the rate of surface synthesis increases sharply halfway through the cycle. Although the results presented in Fig. 4 do not contradict such an interpretation, independent experiments with synchronized cells should show whether the necessary increase in surface synthesis rate at the onset of constriction does indeed take place.

The present observations in round cells are just the opposite of what we have observed thus far in certain *E. coli* B/r strains (8), which showed a diameter decrease during elongation and an increase during constriction. No DAP-incorporation studies have yet been performed on such cells.

A diameter increase, especially in nonconstricting *pbpA* cells, was also observed during rod-to-sphere conversion of *E. coli* SP45. In the first stage of these temperature-shift experiments (D. Buchnik, C. L. Woldringh, and A. Zaritsky, submitted for publication), the nonconstricting rods, which elongated at constant diameter under permissive conditions, increased much more in width than the constricting cells.

The observations on furazlocillin filaments obtained at 28°C from *E. coli pbpB* confirm the dispersed mode of envelope synthesis along the length of nonconstricting cells (Fig. 7A) reported for 42°C filaments of *ftsZ* (12). In the *pbpB* filaments newly initiated constrictions, not found in *ftsZ* filaments, coincided with a locally increased DAP incorporation, which may be ascribed to

FtsZ protein-aided activity. Apparently, the increased rate of peptidoglycan synthesis at the first stage of polar cap formation can take place when PBP 3 activity is inhibited.

ACKNOWLEDGMENTS. We thank Miriam Aarsman for measurements, Edwin van Spronsen and Fred Brakenhoff for computer programs, Jacques Valkenburg for flow cytometric measurements, and Christine van Wijngaarden for preparing the manuscript.

LITERATURE CITED

1. **Botta, G., and J. T. Park.** 1981. Evidence for involvement of penicillin-binding protein 3 in murein synthesis during septation but not during constriction. *J. Bacteriol.* **145**:333–340.
2. **Donachie, W. D., K. J. Begg, and N. F. Sullivan.** 1984. Morphogenes of *Escherichia coli*, p. 27–62. *In* R. Losick and L. Shapiro (ed.), *Microbial Development*. Cold Spring Harbor Laboratory, Cold Spring Harbor, N.Y.
3. **Iwaya, M., R. Goldman, D. J. Tipper, B. Feingold, and J. L. Strominger.** 1978. Morphology of an *Escherichia coli* mutant with a temperature-dependent round cell shape. *J. Bacteriol.* **136**:1143–1158.
4. **Olijhoek, A. J. M., S. Klencke, E. Pas, N. Nanninga, and U. Schwarz.** 1982. Volume growth, murein synthesis, and murein cross-linkage during the division cycle of *Escherichia coli* PA3092. *J. Bacteriol.* **152**:1248–1254.
5. **Previc, E. P.** 1970. Biochemical determination of bacterial morphology and the geometry of cell division. *J. Theor. Biol.* **27**:471–497.
6. **Spratt, B. G., A. Boyd, and N. Stoker.** 1980. Defective and plaque-forming transducing bacteriophage carrying penicillin-binding protein-cell shape genes: genetic and physical mapping and identification of gene products from the *lip-dacA-rodA-pbpA-leuS* region of the *Escherichia coli* chromosome. *J. Bacteriol.* **143**:569–581.
7. **Taschner, P. E. M., J. G. J. Verest, and C. L. Woldringh.** 1987. Genetic and morphological characterization of *ftsB* and *nrdB* mutants of *Escherichia coli*. *J. Bacteriol.* **169**:19–25.
8. **Trueba, F. J., and C. L. Woldringh.** 1980. Changes in cell diameter during the division cycle of *Escherichia coli*. *J. Bacteriol.* **142**:869–878.
9. **Walker, J. R., A. Kovarik, J. S. Allen, and R. A. Gustafson.** 1975. Regulation of bacterial cell division: temperature-sensitive mutants of *Escherichia coli* that are defective in septum formation. *J. Bacteriol.* **123**:693–703.
10. **Westling-Häggström, B., T. Elmros, S. Normark, and B. Winblad.** 1977. Growth pattern and cell division in *Neisseria gonorrhoeae*. *J. Bacteriol.* **129**:333–342.
11. **Wientjes, F. B., E. Pas, P. E. M. Taschner, and C. L. Woldringh.** 1985. Kinetics of uptake and incorporation of *meso*-diaminopimelic acid in different *Escherichia coli* strains. *J. Bacteriol.* **164**:331–337.
12. **Woldringh, C. L., P. Huls, E. Pas, G. J. Brakenhoff, and N. Nanninga.** 1987. Topography of peptidoglycan synthesis during elongation and polar cap formation in a cell division mutant of *Escherichia coli* MC4100. *J. Gen. Microbiol.* **133**:575–586.
13. **Woldringh, C. L., J. A. C. Valkenburg, E. Pas, P. E. M. Taschner, P. Huls, and F. B. Wientjes.** 1985. Physiological and geometrical conditions for cell division in *Escherichia coli*. *Ann. Microbiol.* (Paris) **136**:131–138.

Chapter 7

Where and When Is the Cell Wall Peptidoglycan of Gram-Negative Bacteria Synthesized?

Stephen Cooper

What is the rate of peptidoglycan synthesis during the division cycle? Most experiments on this question have observed an increase in peptidoglycan synthesis during the division cycle although none of the experiments was precise enough to distinguish among exponential, linear, bilinear, or other, more complex descriptions. Meanwhile, most of the theoretical discussions have considered only a small set of models that are exponential, linear, bilinear, or variations and combinations of these (reviewed by S. Cooper, submitted for publication).

The work presented here provides an unanticipated answer to this question. The precise answer is less important than the conclusion that the rate of peptidoglycan synthesis is neither exponential nor linear, but is a complex pattern that, paradoxically, is quite easy to describe (Cooper, submitted).

The experiments described here were stimulated by the observations of Woldringh et al. (13). Using refined autoradiographic techniques, they demonstrated a decrease in the rate of cell wall synthesis in the cylindrical portion of the cell after the start of cell constriction. Here I propose a model for the partition of wall synthesis between the pole and lateral wall during the constriction period that explains this unexpected finding. This model is consistent with the constant density of the cell and explains the decrease in the rate of cylinder extension in constricting cells. The predictions of the model for the rate of peptidoglycan synthesis during the division cycle are here confirmed by measurements, made by the membrane elution ("baby machine") technique, of the rate of synthesis of peptidoglycan during the division cycle.

Stephen Cooper • Department of Microbiology and Immunology, University of Michigan Medical School, Ann Arbor, Michigan 48109–0620.

The Model

Consider a rod-shaped gram-negative bacterial cell that is approximated as a right circular cylinder capped with two hemispheres (Fig. 1, age 0). Before constriction starts, growth of the cell surface occurs only by extension and elongation of the cylindrical lateral wall (Fig. 1, ages 0.0 to 0.5). If the cell diameter is constant, for the extension of the cylinder to produce a given increment of volume, there is a corresponding increment of surface area. The ratio of area increase to volume increase is constant during the period before constriction starts.

When invagination and constriction start, the rates of volume and area

CELL AGE

0

.1

.2

.3

.4

.5

.6

.7

.8

.9

1.0

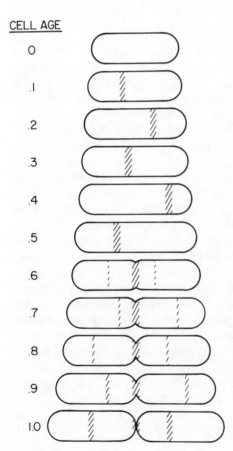

Figure 1. Illustration of the proposed model of cell growth. Newborn cells have a cylinder length (L) of 2 and a radius (r) of 0.5. Prior to invagination, growth of the cell proceeds only by cylinder growth. The shaded regions in the cell indicate the amount and location (whether in the pole or the cylinder) of extension since the previous cell. The width of the shaded area is drawn to scale. It should be considered that cell growth is occurring throughout the cylinder and not in a narrow contiguous band. Each shaded sector indicates the increase in volume (ΔV) and surface area (ΔA). During cylinder growth the ratio of these two factors is constant. When pole synthesis starts (at age α of 0.5 in the cells illustrated here), there is an increase in volume and area in the region of the new poles. Any volume increase, determined by the increase in mass, that is not accommodated by new poles is accommodated by additional cylinder growth. Cylinder growth after constriction starts is slower than before constriction. This is schematically illustrated by the thinner sector in the expanding side wall immediately after the start of constriction. As the new pole increases, in increments of equal area between the indicated ages, the volume accommodated by the new poles is continuously decreasing. Therefore, the growth rate in the cylindrical portion increases continuously during the constriction period. At the end of the division cycle the rate of synthesis in the cylinder is the same as the rate for a newborn cell. There is no sharp change in the rate of cylinder elongation at the instant of division.

increase are no longer proportional. This can be seen by considering that the surface area enclosing a unit of volume in a spherical pole is greater than the surface lateral wall area enclosing that unit of volume in a cylindrical portion of a cell.

The actual area-to-volume ratio during the division cycle depends on three factors: the age during the division cycle when constriction starts, the initial shape of the cells (length/diameter), and the presumed mode of increase of the polar hemispheres.

The precise predictions have been derived, and the main results are presented here. (The complete analysis is presented elsewhere [Cooper, submitted].) Assume that (i) the rate of mass and volume increase is exponential during the division cycle, (ii) the cell density is constant during the division cycle, (iii) constriction starts at a particular age during the division cycle, (iv) the cell can be approximated by a cylinder capped with two hemispheres, (v) the new pole grows at a constant rate of area increase after the start of constriction, and (vi) any volume increase in the cell that is not accommodated by the increase in new pole volume is accommodated by an increase in the cylindrical wall of the cell. The total cell area at any time during the division cycle is:

$$A_{\alpha_{tot}} = \frac{2}{r}\left\{ 2^\alpha\left(\frac{4}{3}\pi r^3 + \pi r^2 L\right) - \frac{4}{3}\pi r^3 \right.$$
$$\left. - \left(\frac{4}{3}\pi r^3 - \frac{2}{3}\pi[(r - h_\alpha)^2(2r + h_\alpha)]\right)\right\} \tag{1}$$
$$+ 4\pi r^2 + 4\pi r h_\alpha$$

which by rearrangement and reduction gives:

$$A_{\alpha_{tot}} = \frac{3}{3}(2^\alpha)\pi r^2 + 2(2^\alpha)\pi r L + \frac{4}{3}\pi r^2 + \frac{4}{3}\pi\frac{h_\alpha^3}{r} \tag{2}$$

with L the total length of the cell, α the age during the division cycle, and r the radius of the cell. For the particular model and assumptions presented here, one can differentiate and determine the rate of area increase during the division cycle:

$$\left(\frac{dA}{d\alpha}\right)_{0<\alpha<1} = \frac{8}{3}(2^\alpha)\ln 2\pi r^2 + 2(2^\alpha)\ln 2\pi L + \frac{4}{r}\pi\left[h_\alpha^2\left(\frac{r}{1 - T_c}\right)\right] \tag{3}$$

with T_c the cell age at the start of the invagination. As the differential rate of mass increase is given by:

$$\frac{dM}{d\alpha} = 2^{\alpha} \ln 2M_{0_{\text{tot}}} \tag{4}$$

one can divide equation 3 by equation 4 and get:

$$\left(\frac{dA}{dM}\right)_{0 \le \alpha \le 1} = 1 + \frac{6h_{\alpha}^{2}}{2^{\alpha} \ln 2(1 - T_{c})(4r^{2} + 3L)} \tag{5}$$

Consider the growth of a cell in steady-state conditions with a given initial and constant radius, an initial cylindrical wall length, and a time for the start of initiation during the cycle. By combining these in a new constant, κ, we arrive at:

$$\left(\frac{dA}{dM}\right)_{0 \le \alpha \le 1} = 1 + \kappa\left(\frac{h_{\alpha}^{2}}{2^{\alpha}}\right) \tag{6}$$

which indicates that the rate of increase of cell wall, compared with the rate of mass increase, is constant before constriction (as $h_{\alpha} = 0$) and increases after the start of constriction.

The central idea of this model is how cell wall synthesis is partitioned between the new pole and the lateral wall during the constriction period. The model proposed here assumes that after pole synthesis starts, whatever remains of the increased volume that is not accommodated by the volume increase in new pole is accommodated by an increase in the volume of the cylindrical portion of the cell. An analogy would be that new pole increase is a pressure relief system that relieves the stress on the side wall and so lowers the rate of cylindrical growth.

It is extremely difficult to distinguish the proposed rate of cell wall or cell area increase (equation 3) from a simple exponential (2), but if the rate of peptidoglycan synthesis is normalized to the rate of increase in cell mass (as in equation 6), then the predictions of the model can be tested experimentally. The model proposed here predicts a constant ratio of the rates of area increase to mass increase during the first part of the division cycle and an increase in the ratio during the latter part of the division cycle, after the start of invagination.

Experimental Details

Salmonella typhimurium was chosen for this study because it can be analyzed by the membrane elution technique (5), and, more important, the incorporation of diaminopimelic acid into *S. typhimurium* is much more efficient than incorporation into *Escherichia coli* (4), the organism used in most of the earlier studies. Compared with *S. typhimurium, E. coli* is relatively imperme-able to diaminopimelic acid, even in diaminopimelic acid-requiring strains. *N*-Acetylglucosamine is a specific cell wall label in both species and was used to analyze wall synthesis in *E. coli*. All of the results on the pattern of wall synthesis during the division cycle are consistently the same, independent of the peptidoglycan label or the particular gram-negative rod-shaped organism used. The experimental details are presented elsewhere (Cooper, submitted; S. Cooper and M. L. Hsieh, manuscript in preparation).

Membrane Elution Method

The membrane elution technique is a method for measuring the rate of synthesis of a particular macromolecule during the division cycle (3, 5, 9, 10; Cooper, submitted). Exponentially growing, unperturbed cells are labeled for a short time with a label specific for the molecule of interest. The cells are filtered onto a cellulose nitrate membrane, the membrane is inverted, and fresh medium is pumped through the membrane. A fraction of the labeled cells bind to the membrane and grow in the presence of the medium pumped over the cells. Only newborn cells arising by division are released from the membrane. The first cells released by division arise from the oldest cells in the labeled culture, i.e., those cells that were just about to divide at the time the cells were labeled. With further incubation the newborn cells are released from cells that were increasingly younger, in terms of the division cycle, at the time of la-beling. Cells are released from the membrane with label which reflects the labeling of the cells, during balanced and unperturbed growth, as a function of the division cycle. By considering cells in reverse order of elution, and over one generation of elution, the rate of incorporation of a radioactive label as a function of the division cycle can be determined.

It is important to note that the membrane elution technique, while pro-ducing synchronized populations of cells by elution, is not a synchrony tech-nique. The cells are labeled in unperturbed, exponential growth and analysis of the cells occurs after the label has been incorporated. Any perturbation of the cells by filtration or subsequent elution is irrelevant to the determination of the rate of incorporation of the compounds during the division cycle. All that is required is that the cells divide in order, with only newborn cells being eluted from the membrane.

A Comment on the Methodology

There have been many different approaches to the problem of bacterial cell growth—biochemical, biophysical, microscopical, and theoretical—and there have been many conflicting conclusions regarding the pattern of cell growth

(2). The results of the experiments described here are consistent with a model that is quite different from many previous models. Can one make a choice as to the proper methodology for analyzing biosynthetic patterns during the division cycle? The method used here has had a major success, and the model of DNA synthesis which was derived by using this methodology (3) has been confirmed by an entirely independent approach (12). I therefore suggest that the membrane elution method is a suitable candidate for revealing the rate of cell surface and mass synthesis during the division cycle.

Rate of Peptidoglycan and Protein Synthesis During the Division Cycle

S. typhimurium was labeled with radioactive leucine and diaminopimelic acid and was analyzed by the membrane elution technique. Typical results are presented in Fig. 2 and 3. The ratio of diaminopimelic acid to leucine starts relatively high and decreases to a constant level during the first generation of elution. The same data can be considered in reverse (i.e., reading backwards from approximately 70 min on the graph), to analyze the change in the ratio as the cells progress from young to old. The ratio is constant during the first part of the division cycle and increases toward the end of the division cycle (Fig. 2). The crests and troughs in the cell elution curve are used to identify the different generations of elution (9, 10). Similar results can be obtained with *N*-acetylglucosamine as a specific label for cell wall peptidoglycan (Fig. 4).

Examination of the curves of radioactivity per cell for the different labels reveals that the leucine-per-cell curves are straight decreasing exponential curves without observable breaks, while the diaminopimelic acid and *N*-acetylglucosamine curves have variations that lead to the observed ratio curves. The leucine curve indicates that the rate of incorporation of leucine into the cycle is exponential. Therefore, the rate of synthesis of protein during the division cycle is exponential.

Extended Elution Times: Analysis of the Stability and Segregation of Peptidoglycan

Elution experiments have been carried out for up to six generations, and the results consistently indicate that after the initial two generations the ratio is constant, indicating that there is no turnover of the peptidoglycan in *S. typhimurium* or *E. coli* and that the lateral wall material is partitioned equally for at least six generations with no preferential synthesis in any portion of the cylinder.

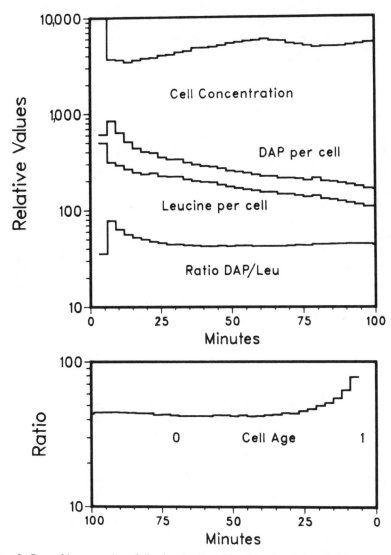

Figure 2. Rate of incorporation of diaminopimelic acid and leucine during division cycle determined by the membrane elution technique. *S. typhimurium* cells (50 ml at 2.5 × 10⁸/ml) were
labeled for 3 min with 2 μCi of [¹⁴C]leucine and 200 μCi of [³H]diaminopimelic acid. The cells
were then bound to a membrane and analyzed as described in the text. The lower panel plots the
ratio in reverse. Reading from left to right, from age 0 at the time of labeling to age 1 at the
time of labeling, there is a constant ratio that increases in the older cells.

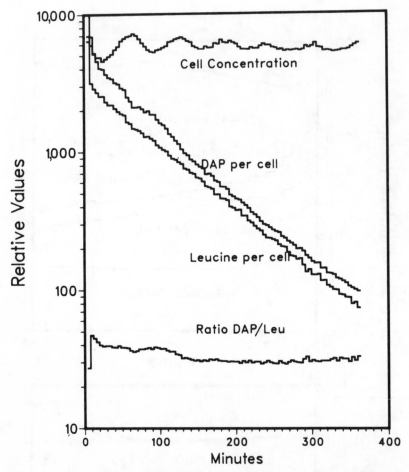

Figure 3. Rate of incorporation of diaminopimelic acid and leucine during the division cycle determined by the membrane elution technique. *S. typhimurium* cells (100 ml at 2.4 × 10⁸/ml) were labeled for 4 min with 2 μCi of [¹⁴C]leucine and 100 μCi of [³H]diaminopimelic acid and were analyzed as described in the text.

Discussion

Prior to the start of constriction, when cylindrical extension is the only means of cell growth, the rate of peptidoglycan synthesis appears exponential because the differential rate of wall increase is similar to the differential rate of mass and volume increase. But it is incorrect to say that synthesis is exponential. Exponential synthesis means not only that the rate of synthesis is exponential, but also that the total amount of material increases exponentially. This is not the case for cell wall material. After constriction starts, peptido-

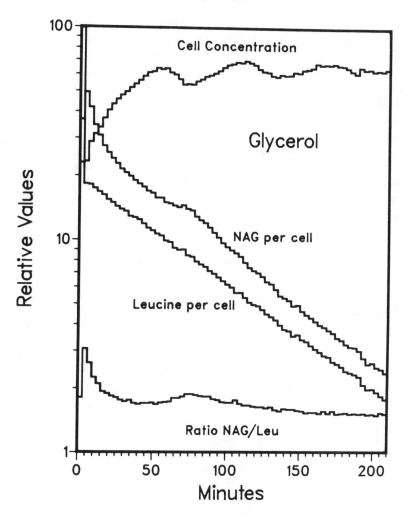

Figure 4. Rate of incorporation of *N*-acetylglucosamine and leucine during the division cycle of *S. typhimurium* growing in minimal medium supplemented with glycerol and lysine. The procedure was the same as described in the Fig. 3 legend except that 50 ml of cells at $1.2 \times 10^8/$ ml was labeled for 1 min with 2 μCi of [^{14}C]leucine and 20 μCi of *N*-[^{3}H]acetylglucosamine. Three-minute fractions were collected during the elution.

glycan synthesis is composed of both cylindrical extension and pole synthesis. In the model proposed here, the pole extends as mass increases but does not allow enough of a volume increase to accommodate the increased mass and volume. The extra volume required to enclose the new material is made up by an increase in the length of the cylindrical lateral wall (Fig. 1, ages 0.6 to 1.0).

Thus, the answer to the question, What is the rate of peptidoglycan synthesis during the division cycle? is that the rate is similar to the rate of mass and volume increase before constriction starts, and the rate increases compared with mass and volume synthesis after constriction starts.

Even if one had a perfect experimental result and exact measurements describing the rate of cell wall synthesis during the division cycle, the results on peptidoglycan synthesis would, by themselves, be unintelligible because the equation for the surface area is extremely complex (see equations 1, 2, and 3) but very similar to exponential. To understand the varying rate of cell wall synthesis, the rate of wall synthesis during the division cycle must be compared with the rate of mass synthesis or volume increase (equation 6). Control and reconstruction experiments reveal that the changes in the ratio are well within the ability of the membrane elution technique to determine such differential rates of synthesis.

Stability and Turnover of Peptidoglycan

Turnover of peptidoglycan during the growth of *E. coli* has been reported (1, 7, 8). The results presented here suggest that no measurable or significant turnover of peptidoglycan—i.e., processes that lead to the loss of material from cell—occurs in *S. typhimurium* or *E. coli*. If there was turnover, it would be expected to be more readily observed in this experimental situation, as the cells are continuously being washed with fresh medium. Turnover in which material is rapidly reincorporated, and not lost from the cell, is not eliminated by these experiments. If any turnover occurs, it must be minimal and not greater than the turnover of protein in the cell. As the bulk of the cell protein is stable in exponentially growing cells (11), this suggests that there is no measurable turnover of peptidoglycan in *S. typhimurium* or in *E. coli*.

It may be that the observed differences in turnover between *E. coli* and *S. typhimurium* are related to the fact that almost all the experiments on turnover in *E. coli* were carried out with a diaminopimelic acid auxotroph, strain W7. *E. coli* is rather impermeable to diaminopimelic acid (4), and large amounts of diaminopimelic acid must be present in the medium to ensure growth. It is possible that the psysiology of the organism is disturbed by diaminopimelic acid limitation. This is supported by Driehuis and Wouters (6), who noted that limiting diaminipimelic acid led to the production of abnormal peptidoglycan.

Rate of Cylinder Extension

Woldringh et al. (13) noted that the incorporation of diaminopimelic acid into the lateral walls of cells with constrictions was significantly less than into the lateral walls of cells without constrictions. What is the mechanism of this redistribution of synthesis between the pole and the side wall? When pole synthesis starts, the increase in cell volume by pole growth relieves the stress in the cylinder area. Because of this reduction in stress, the rate of insertion of peptidoglycan in the lateral wall is reduced.

Caveats and Generalizations

Although a precise model has been presented here based on a simple geometrical analog of the cell shape, it should be emphasized that the data do not yet allow differentiation between this model and many other equally likely models. For example, the cell may not be a cylinder with hemispherical ends; there may be bulges, and the shape of the poles may be more complex. Also, the new poles may not grow by adding equal areas in equal times but by other modes of increase. Nevertheless, the important point is that even if the actual model is different from the one presented here, it will still be true that the cell density does not have to change during the cycle and there will be an increase in the ratio of area increase to mass increase during invagination. Whether the model proposed here can ever be differentiated from other very similar models by any reasonable experimental procedure or measurement is a question which is difficult to answer at this time.

ACKNOWLEDGMENT. This work was supported by grant DMB 8417403 A01 from the National Science Foundation.

LITERATURE CITED

1. **Chaloupka, J., and M. Strnadovà.** 1972. Turnover of murein in a diaminopimelic acid dependent mutant of *Escherichia coli. Folia Microbiol.* (Prague) **17:**446–455.
2. **Cooper, S.** 1969. Cell division and DNA replication following a shift to a richer medium. *J. Mol. Biol.* **43:**1–11.
3. **Cooper, S, and C. E. Helmstetter.** 1968. Chromosome replication and the division cycle of *Escherichia coli. J. Mol. Biol.* **31:**519–540.
4. **Cooper, S., and N. Metzger.** 1986. Efficient and quantitative incorporation of diaminopimelic acid into the peptidoglycan of *Salmonella typhimurium. FEMS Microbiol. Lett.* **36:**191–94.
5. **Cooper, S., and T. Ruettinger.** 1973. Replication of deoxyribonucleic acid during the division cycle of *Salmonella typhimurium. J. Bacteriol.* **111:**966–973.
6. **Driehuis, F., and J. T. M. Wouters.** 1985. Effect of diaminopimelic acid-limited growth on the peptidoglycan composition of *Escherichia coli* W7. *Antonie van Leeuwenhoek J. Microbiol. Serol.* **51:**556.
7. **Goodell, E. W.** 1985. Recycling of murein by *Escherichia coli. J. Bacteriol.* **163:**305–310.
8. **Goodell, E. W., and U. Schwarz.** 1985. Release of cell wall peptides into culture medium by exponentially growing *Escherichia coli. J. Bacteriol.* **162:**391–397.

9. **Helmstetter, C. E.** 1967. Rate of DNA synthesis during the division cycle of *E. coli* B/r. *J. Mol. Biol.* **24:**417–427.
10. **Helmstetter, C. E., and S. Cooper.** 1968. DNA synthesis during the division cycle of rapidly growing *E. coli* B/r. *J. Mol. Biol.* **31:**507–518.
11. **Koch, A. L., and H. R. Levy.** 1955. Protein turnover in growing cultures of *Escherichia coli. J. Biol. Chem.* **217:**947–957.
12. **Skarstad, K., H. B. Steen, and E. Boye.** 1985. *Escherichia coli* DNA distribution measured by flow cytometry and compared with theoretical computer simulations. *J. Bacteriol.* **163:**661–668.
13. **Woldringh, C. L., P. Huls, E. Pas, G. J. Brakenhoff, and N. Nanninga.** 1987. Topography of peptidoglycan synthesis during elongation and polar cap formation in a cell division mutant of *Escherichia coli* MC4100. *J. Gen. Microbiol.* **133:**575–586.

Chapter 8

Surface Growth of Streptococci in the Presence and Absence of β-Lactam Antibiotic

M. L. Higgins

L. Daneo-Moore

The cell wall surface of *Streptococcus faecium* is assembled in discrete growth sites which can be seen in electron micrographs (Fig. 1 in reference 9). Each site is flanked by a pair of raised bands. A site is initiated by a ridge of wall being assembled under each equatorial band which forms a nascent cross wall. Subsequently, a division furrow bilaterally cleaves the band and a portion of the underlying cross wall. This results in the production of two new bands and allows the cleaved portion of the cross wall to separate and expand into two layers of peripheral wall (Fig. 1D). In the remainder of the process, the cross wall grows centripetally into the cytoplasm until it closes, and the division furrow continues to constrictively cleave the cross wall until the cross wall is completely severed and two new polar caps result. Thus, two poles are produced for every equatorial band that is split.

For the past few years, we have been studying this process by electron microscopy. The method is quite simple (3). Platinum-carbon replicas are made of populations of cells, electron micrographs are taken, a digitizer is employed to convert the perimeter of these replicas into a series of x,y coordinates, and then a computer is used to mathematically rotate these coordinates around the central axis of the cell to form a three-dimensional reconstruction of the cell surface. By reconstructing a large number of cells from a given population, one can predict how a site will change in shape as a cell increases in size during its cell cycle.

M. L. Higgins and L. Daneo-Moore • Temple University School of Medicine, Philadelphia, Pennsylvania 19140.

Using this technique, we found that the size of the two polar caps made by a site was invariant with growth rate (3). In this way, the process of cell wall assembly appeared to be similar to chromosome replication in that two poles of constant size were made during a round of synthesis. However, unlike chromosome replication, the time for a site to produce two polar caps appeared to increase with the mass doubling time of the culture. This conclusion was drawn from observations for which calculations showed that new sites were made, on the average, a few minutes before division in cultures with doubling times of less than about 80 min, while in cultures with doubling times greater than 80 minutes new sites were initiated at or just after division (4, 10). These calculations are based on the assumption that the volume of cells increases at a constant rate during the cell cycle. This suggested that while the time for a round of envelope synthesis might be a few minutes more or less than the mass doubling time of the culture, the overall rate of cell surface assembly in a site seemed to increase with growth rate.

With these observations in hand, we then asked whether chromosome replication might time the initiation of rounds of cell wall growth. In opposition to this idea, we could not show any correlation between the number of replication forks, or timing of the initiation and termination of rounds of chromosome synthesis, and the initiation of rounds of cell wall synthesis (6, 10). Furthermore, in cultures in which chromosome synthesis was virtually completely inhibited with mitomycin C, new sites of wall synthesis continued to be initiated at about the rate of the control cells over a 60-min period of observation (5).

While these data did not support the idea that chromosome replication timed the initiation of cell wall growth sites, we did observe that the inhibition of chromosome replication inhibited the terminal stages of site developement (i.e., cell division). Therefore, communication between chromosome synthesis and cell wall surface growth appeared to occur in the terminal stages of envelope growth, probably where some SOS-like process ensured that DNA-less cells would not result.

If chromosome synthesis did not regulate the formation of cell wall growth sites, what other mechanism might be responsible? One model supported by recent data is that sites are formed in response to increasing cytoplasmic density. When exponential-phase cells are separated on Percoll gradients (2), the densest cell-containing fractions are enriched with streptococci in the terminal stages of division and with cells which have small, newly made cell wall growth sites. Small sites are observed in large dividing cells at lateral positions (Fig. 1D) and in the center of small newborn cells (Fig. 1A). The interpretation of these results based on the above density model is that as the division furrow reduces the diameter cross wall (Fig. 1A–D), less and less new cell volume is produced as each portion of cross wall is cleaved and expands into a polar cap. If the growth of cytoplasmic constituents proceeds exponentially through the

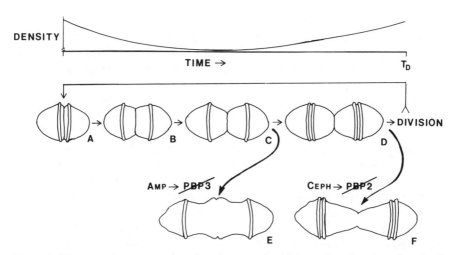

Figure 1. Diagrammatic representation of our interpretation of data gathered on the cell cycle of *Streptococcus faecium*. (A–D) Surface views of cells showing a round of envelope growth in a culture with a rapid doubling time. A new cell wall growth site is initiated independent of chromosome synthesis by the formation of a nascent cross wall and the subsequent splitting of the overlying band by a division furrow (D). The constrictive splitting of the cross wall allows the site to increase in volume (probably exponentially) until two poles are made. In cells with mass doubling times (T_Ds) of <80 min (as shown here), new sites are made just before division, whereas in cells with T_Ds of >80 min, sites are initiated at or a few minutes after division (i.e., the time to make two poles by a site increases with T_D). While the initiation of a new site appears to be independent of chromosome replication, division appears to require completion of a round of chromosome replication. As a site approaches its maximum size, an increase in buoyant density is observed which does not decrease until after new sites have begun to increase in size. If a drug is added having a high affinity for PBP 3 (e.g., ampicillin), the initiation of new sites is inhibited and old sites bulge, whereas if PBP 2 is preferentially saturated by a drug such as cephalothin, new sites are made, but old sites elongate, apparently as a result of an inhibition of some terminal stages in growth site development.

cycle, this would mean that in the latter phases of the cycle the density of cells would increase as a result of the cell surface growing slower than cytoplasmic mass (10). This increase in density would produce very large increases in the stresses placed on the portions of the wall jointed by the equatorial bands, because these bands join surfaces which have different curvatures (E. H. Old-mixon, M.S. thesis, Temple University, Philadelphia, Pa., 1973). In this view, increases in density which result in increased stress being placed on the equatorial bands could be the signal that activates the enzymes required to make new sites. After a new site is formed, some time would be necessary before such a site could grow enough to lower the internal density-pressure of the cell (Fig. 1B–D).

This scheme proposes an interplay between enzymes that carry out wall synthesis and physical density-pressure elements. What would happen to such a scheme if β-lactam antibiotics were added? We have begun to answer this question by asking how surface growth of these cells is disturbed by the addition of cephalothin and ampicillin to a final concentration of 80 and 2 μg/ml, respectively, to exponential-phase cultures (7). These concentrations were selected because they were the highest levels which had little effect on increases in culture turbidity. Cephalothin and ampicillin were chosen because they have selective high binding affinities for penicillin-binding proteins (PBPs) 2 and 3, respectively (1).

While the turbidity of the cultures treated separately with these concentrations of drug increased at essentially the rate of the controls, the rate of cell division was greatly inhibited (7). This resulted in a doubling of average mass of the treated cells over 60 min. When three-dimensional reconstructions of these cells were made from electron micrographs, measurements showed that the volume of these replicas also doubled over the same period. The measurements from the replicas were then used to determine what portions of the cell envelope (i.e., polar caps, growth site) underwent enlargement to allow the volume of the cells to double in 60 min.

This analysis showed that the increases in cell size came from increases in volume of the cell's growth sites rather than its polar caps. As described above, growth sites can appear at central locations, which we term primary sites (Fig. 1A), and at lateral locations, which we call secondary sites (Fig. 1D). Using these designations, we found that the treated primary sites showed greater increases in volume than did secondary sites, but with time, both became abnormally large. Some differences between treatments were seen. For example, the primary sites of the ampicillin-treated cells increased more, and their secondary sites less, than did the cephalothin-treated cells. More important, while new sites were made at close to the rate of the control in cephalothin-treated cells, the initiation of such new sites was greatly inhibited by ampicillin. This suggested that while these two types of treated cells were increasing in mass at about the same rate, this mass was being housed in the two types of treated cultures in slightly different patterns. Thus, in both cultures growth sites were growing to abnormally large sizes, but differences in the rates of growth of primary and secondary growth sites, and differences in the rate of initiation of sites, could be detected.

We then asked this question: What effect do these two drugs have on cell shape? This simple question proved to be very difficult to answer. When we examined the morphology of the replicas of the two types of treated cells, there appeared to be a great diversity of shapes which fit no predictable pattern. After several attempts to computer model these shape changes, we resorted to a "brute force" approach. Each set of electron micrographs representing a population of treated or untreated cells was placed in one of a series of increasing volume classes. From our measurements of cell mass and number we could predict the average increase in volume that cells in each volume class would undergo during a given interval of treatment. We then could tentatively identify the volume classes which treated cells would grow into with time and trace the changes in cell shape. By using this technique, we discovered a pattern of shape change which seemed logical and fit our contour measurements. The results of this study are summarized diagrammatically in Fig. 2. As stated above, the poles of cells present at the time of drug addition did not change in size, and Fig. 2 indicates that no change in shape was observed either. The central growth sites present at the time of treatment usually were able to divide but often produced misshapen poles. Even though these poles were frequently misshapen, their average volumes were indistinguishable from those of the untreated cells. The large abnormal increases in cell size observed in the presence of either drug were due to the enlargement of sites that were initiated just before or after drug addition. With time, these sites elongated or bulged, or both. Comparison of the two treatments showed that the sites of the ampicillin-treated cells bulged more than those of the cephalothin-treated cells; the ampicillin-treated cells also initiated fewer new sites.

Two models are proposed to explain these shape changes. The first model argues that both ampicillin and cephalothin result in the formation of wall of lowered tensile strength. The greater amount of bulging in the ampicillin-treated sites is due to a greater reduction in tensile strength of their walls. As a site bulges, a unit cell wall surface found in this site will contain more cytoplasmic volume than a comparable unit of wall surface in an untreated site. Hence, according to this scheme, the ampicillin-treated cells produce fewer new sites than cephlothin-treated cells because their bulged sites can hold more cytoplasmic mass before the formation of new sites is triggered by an increase in density.

The second model states that specific PBPs play critical roles at given times in the cell cycle in the formation of wall growth sites. Recently, it has been shown that β-lactam antibiotics such as ampicillin and cephalothin which have a selective binding affinity for PBP 3 and PBP 2 introduce blocks into the cell cycle about 40 and 10 min before division, respectively (11). These observations suggest that fewer new sites are formed in the ampicillin-treated cells because this drug inactivates PBP 3, which is required for new site production about 40 min before division (Fig. 1E). In cephalothin-treated cells new sites could be made at or near normal rates, because levels of PBP 3 are sufficient for near normal rates of new site production. However, the terminal stages of division which require PBP 2 are inhibited by the selective binding of cephalothin to this PBP.

At this time we cannot affirm or deny either of these models. This is the goal of work in progress. However, it seems that the central question is whether cell density can regulate the formation of new sites of surface growth or is simply an indirect manifestation of another mechanism(s) of surface growth regulation.

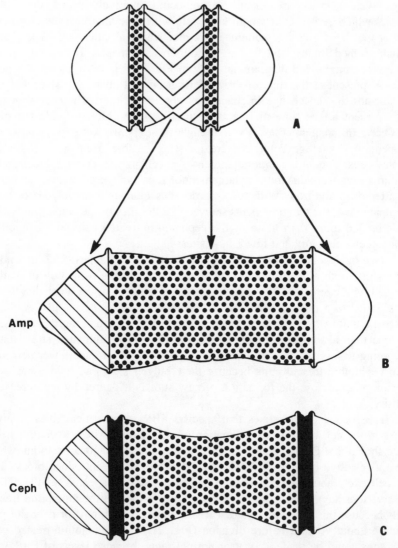

Figure 2. Diagrammatic sketch of the morphological effect of the addition of ampicillin (B) and cephalothin (C) to an untreated cell (A) for 60 min. Upon the addition of drugs, the primary growth site of the untreated cell (hatched area in figure), being relatively well developed, will go on to divide to give rise to two daughter cells, of which only one is shown in panels B and C. Therefore, on division, the primary site in panel A produces one of the poles seen in panel B or C. In contrast, the secondary growth sites of the untreated cell (stippled areas), being much less developed than the primary site, do not divide upon treatment but rather increase in length and girth (B and C). The ampicillin-treated cell differs from that treated with cephalothin in that its stippled zones are, on the average, greater in length and girth (B), whereas the cephalothin-treated cells have more new sites (C).

ACKNOWLEDGMENT. This work was supported in part by Public Health Service grant AI10971 from the National Institute of Allergy and Infectious Diseases.

LITERATURE CITED

1. **Coyette, J., J. M. Ghuysen, and R. Fontana.** 1980. The penicillin binding proteins in *Streptococcus faecalis* ATOC 9790. *Eur. J. Biochem.* **110:**445–456.
2. **Dicker, D. T., and M. L. Higgins.** 1987. Cell cycle changes in the buoyant density of exponential-phase cells of *Streptococcus faecium. J. Bacteriol.* **169:**1200–1204.
3. **Edelstein, E. M., M. S. Rosenzweig, L. Daneo-Moore, and M. L. Higgins.** 1980. Unit cell hypothesis for *Streptococcus faecalis. J. Bacteriol.* **143:**499–505.
4. **Gibson, C. W., L. Daneo-Moore, and M. L. Higgins.** 1983. Initiation of wall assembly sites in *Streptococcus faecium. J. Bacteriol.* **154:**573–579.
5. **Gibson, C. W., L. Daneo-Moore, and M. L. Higgins.** 1983. Cell wall assembly during inhibition of DNA synthesis in *Streptococcus faecium. J. Bacteriol.* **155:**351–356.
6. **Gibson, C. W., L. Daneo-Moore, and M. L. Higgins.** 1984. Analysis of initiation of sites of cell wall growth in *Streptococcus faecium* during a nutritional shift. *J. Bacteriol.* **160:**935–942.
7. **Higgins, M. L., M. Ferrero, and L. Daneo-Moore.** 1986. Relationship of shape to initiation of new sites of envelope growth in *Streptococcus faecium* cells treated with β-lactam antibiotics. *J. Bacteriol.* **167:**562–569.
8. **Higgins, M. L., A. L. Koch, D. T. Dicker, and L. Daneo-Moore.** 1986. Autoradiographic studies of chromosome replication during the cell cycle of *Streptococcus faecium. J. Bacteriol.* **108:**541–547.
9. **Higgins, M. L., and G. D. Shockman.** 1970. Model for cell wall growth of *Streptococcus faecalis. J. Bacteriol.* **101:**643–648.
10. **Koch, A. L., and M. L. Higgins.** 1984. Control of wall band splitting in *Streptococcus faecium* (ATCC9790). *J. Gen. Microbiol.* **130:**735–745.
11. **Pucci, M. J., E. T. Hinks, D. T. Dicker, M. L. Higgins, and L. Daneo-Moore.** 1986. Inhibition by β-lactam antibiotics at two different times in the cell cycle of *Streptococcus faecium* ATCC 9790. *J. Bacteriol.* **165:**682–688.

Chapter 9

How Do Bacilli Elongate?

M. A. Kemper
H. L. T. Mobley
R. J. Doyle

Some Difficulties in Assessing How Bacilli Elongate

In this chapter we review the literature concerned with the surface assembly of cell walls of *Bacillus subtilis* and conclude that surface extension in bacilli is by random intercalation of wall precursors. This conclusion draws heavily from the surface stress theory for microbial morphogenesis and from direct experimental details defining the sites of surface insertion and the subsequent fate of the sites during division.

It is conceptually appealing to consider that bacilli elongate by narrow zones of wall growth. A streptococcus enlarges its surface by wall extension at a defined and predictable zone. In a unit cell of *Streptococcus* spp., one-half of the cell wall is derived from the mother cell, whereas the other half is synthesized prior to division (19). Bacilli also possess two cell poles, one of which may be very old and the other of which may be newly assembled. Bacilli, however, possess a cylindrical side wall (4) interposed between the two cell poles. Length extension in gram-positive bacilli must therefore take into account growth of the cell cylinder (14, 31, 38).

There has been considerable controversy for a number of years regarding the mechanism of elongation of members of the genus *Bacillus*. One view holds that surface extension is by random addition of new wall at many sites on the cell surface (20, 29). In this view, old wall is diluted with new wall, resulting in an increase in the cell surface. Some authors subscribe to the notion that a

M. A. Kemper and R. J. Doyle • Department of Microbiology and Immunology, University of Louisville Health Sciences Center, Louisville, Kentucky 40292.   **H. L. T. Mobley** • Department of Medicine, University of Maryland, Baltimore, Maryland 21201.

bacillus possesses narrow growth zones in cell cylinders (7, 21). In recent years, Pooley and co-workers (36, 39) have suggested that wall growth occurs at several zones in the cell cylinder. Work in Archibald's laboratory (1–3, 8, 41) has shown that, under conditions favoring the synthesis of bacteriophage binding sites, adsorption of the phage particles is uniform over the cell surface. A switch to conditions favoring loss of phage binding sites resulted in a random loss of sites from the cell cylinder, but a retention of sites in cell poles. More recently, Mobley et al. (33) showed that when a temperature-sensitive *gtaC* mutant was shifted from a nonpermissive to a permissive temperature for α-D-glucosylation of cell wall teichoic acid, then concanavalin A (ConA) receptor sites appeared on the cell cylinder nearly one generation later. In addition, Mobley et al. (33) observed the loss of ConA receptors when cells were shifted from permissive to nonpermissive conditions. The loss of lectin-reactive sites, on both wild-type and autolysin-deficient cells, was random in the cell cylinder. Mobley et al. (33) suggested that bacilli elongate in the cylinders by random or diffuse intercalation of wall materials, but that cell poles are formed separately, probably by annular growth (see also reference 5).

Any attempt to establish a pattern for surface assembly must take into account cell wall turnover. Turnover is the loss of wall components into the growth medium (6, 16, 29) during cell division. The shed components are not reutilized for growth. Furthermore, only old wall (nearly one generation) is turned over (9, 30) and turnover is a natural consequence of growth (10, 17). The difficult problem is to distinguish between sites of new net wall synthesis and sites representing turned-over wall. Figure 1 describes the fate of a cell wall precursor, *N*-acetyl-D-glucosamine (GlcNAc), when added to a growing culture of *B. subtilis*. The GlcNAc is taken up by a high-affinity transport

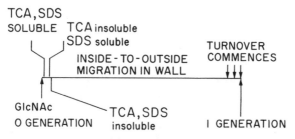

Figure 1. Metabolism of GlcNAc in *B. subtilis*. When GlcNAc is added to an exponential culture of *B. subtilis*, the sugar is quickly taken up, converted first into trichloroacetic acid (TCA)-, sodium dodecyl sulfate (SDS)-soluble precursors, and then into a TCA-insoluble, SDS-soluble form. The GlcNAc, now at approximately 0.1 generation, becomes TCA and SDS insoluble. As one generation of growth approaches, the GlcNAc in the cell wall becomes susceptible to turnover. Exponential turnover can then be measured for two to three generations. The GlcNAc remaining after three to four generations after its uptake is now found predominantly in cell poles, which turn over very slowly, or in a non-turning over compartment. See Koch and Doyle (25) for discussion of turning over and non-turning over compartments for GlcNAc.

system (32) and rapidly converted into wall intermediates. The newly added wall oligomer, once cross-linked, then is pushed to the cell periphery by even more recently added wall. As the wall material reaches the cell surface, it becomes susceptible to autolysins (18, 37), resulting in turnover.

Recent results from this laboratory seem to rule out zonal growth in the cell side walls. The results, coupled with the surface stress theory for microbial morphogenesis, strongly support a model which has surface extension occurring at many, randomly placed sites. The approach to the problem is based on the following considerations. (i) ConA interacts specifically and reversibly with α-D-glucose residues on cell wall teichoic acids (12). (ii) Teichoic acid is synthesized concomitantly with peptidoglycan (30). (iii) Glucosylation of the teichoic acid occurs at the same time as wall synthesis and is not delayed until after teichoic acid incorporation into peptidoglycan (33). (iv) Phosphoglucomutase-deficient mutants (*gtaC*) may glucosylate their teichoic acids at 35°C, but not at 46°C; ConA may therefore be used to study the fate of glucosylated teichoic acid (growth at 35°C, followed by 46°C) or the insertion of teichoic acid (growth at 46°C, followed by growth at 35°C). (v) Both autolysin-sufficient (LYT⁺) and autolysin-deficient (LYT⁻) (15) backgrounds may be used with the *gtaC* mutation.

Strategy for Determining the Mechanism of Surface Elongation in B. subtilis

The experimental methods employed in this paper have been described in detail by Mobley et al. (33), Koch and Doyle (25), Doyle and Birdsell (12), Doyle and Koch (*Crit. Rev. Microbiol.*, in press), and Doyle et al. (13). The present study extends earlier observations and at the same time attempts to define more accurately whether young wall or old wall is under more tension.

Figure 2 shows that when a culture of *B. subtilis* (*gtaC33*) is shifted from 35 to 46°C, there is a loss in the ability of the cells to bind fluorescein-labeled ConA. At the end of a single generation at the nonpermissive temperature, the cell cylinders seem to lose ConA-reactive sites rather uniformly. It is as if there is a dilution of old wall with new wall. Binding of ConA at old cell poles is retained at the nonpermissive temperature (Fig. 2B and C). This is consistent with previous reports (1–3, 11, 13, 41). When *B. subtilis* BUL114 (a *gtaC* and partially LYT⁻ transformant) was subjected to the same temperature change (from 35 to 46°C), there was again a "dilution" of ConA-reactive sites with nonreactive sites (Fig. 3). Cell poles in this LYT⁻ strain also retained their capacity to complex with the lectin after the temperature shift. In neither the LYT⁺ strain (Fig. 2) nor the LYT⁻ strain (Fig. 3) was there evidence for a growth zone (or a few zones) in the cell cylinder.

The shift of a *gtaC* culture from 46 to 35°C to permit the glucosylation

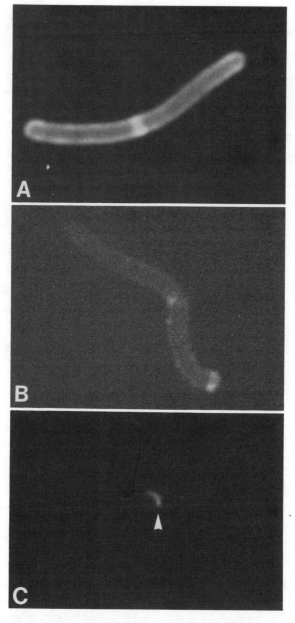

Figure 2. Binding of fluorescein-labeled ConA by *B. subtilis* (*gtaC33*). The *B. subtilis* strain was shifted from exponential growth at 35°C to 46°C. The nonpermissive temperature prevented glucosylation of newly synthesized teichoic acid. (A) Cells from 35°C culture; (B and C) cells from 46°C culture after 1.0 generation (B) and after 3 generations (C). Conditions for staining, washing, fixing, and photography were described by Mobley et al. (33).

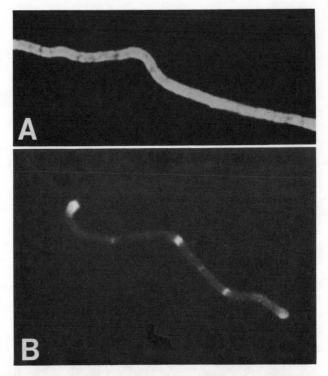

Figure 3. Distribution of ConA receptors on *B. subtilis* BUL114. Conditions were as described in the Fig. 2 legend. (A) Cells from 35°C culture; (B) cells after 1.0 generation at 46°C.

of teichoic acids (only newly synthesized teichoic acids are glucosylated, not older teichoic acids already bound to peptidoglycan) also gave rise to predictable ConA binding patterns. Cell cylinders began to bind the lectin after about one generation, whereas cell poles would not yet stain with fluorescein-labeled ConA (Fig. 4). There was no evidence for localized zones of insertion of the ConA receptors, other than on newly synthesized poles.

The surface stress theory (24, 26, 27) states that in gram-positive bacilli the very youngest side wall is much less stressed than the very oldest (see also Fig. 1). The theory would accept the view that there is a gradient of tension in side walls, from the inner wall face to the outer. A culture of *B. subtilis* 168 was pulsed for 0.1 generation with [^{3}H]GlcNAc and then subjected to cellular autolysis by addition of sodium azide (22). Another part of the pulsed culture was chased for 0.9 generation with nonradioactive GlcNAc. According to surface stress theory, the youngest wall should be the most resistant to solubilization by autolysins, whereas the oldest, more stressed wall should be the most susceptible. The results (Fig. 5) support this view. In several experiments

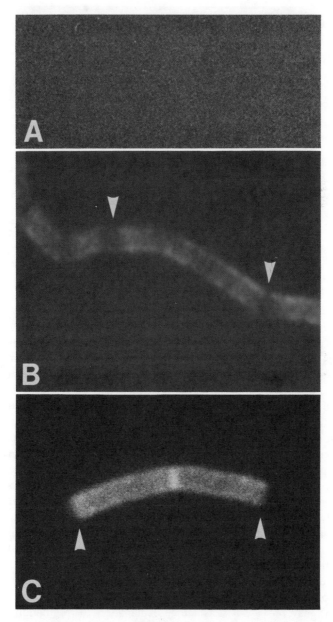

Figure 4. Pattern of binding of fluorescein-labeled ConA when *B. subtilis* (*gtaC33*) was shifted from a nonpermissive to a permissive temperature. Cells grown at 46°C for several generations did not bind ConA. Binding of the lectin was observed when the culture was shifted from 46 to 35°C. (A) Several generations at 46°C; (B) 1.0 generation at 35°C; (C) 2.0 generations at 35°C.

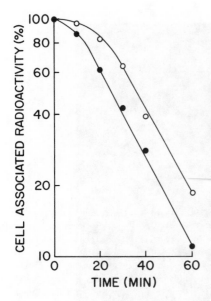

Figure 5. Loss of N-acetyl-D-[³H]glucosamine from *B. subtilis* 168. An exponential culture of *B. subtilis* in Penassay broth at 35°C was pulsed for 0.1 generation with N-acetyl-D-[1-³H]glucosamine (final activity, 2 μCi/ml), and then sodium azide (final concentration, 100 mM) was added. Samples of cell suspension were taken at intervals by use of 0.45-μm membrane filters (Millipore Corp.) and were washed with 5% TCA (●). Another part of the culture was pulsed for 0.1 generation with the labeled GlcNAc, but was chased for 0.9 generation with 500 K_ms of GlcNAc (32) before the addition of sodium azide (○).

in which the "age" of the wall was controlled by pulse-chase sequences, the older wall always was more susceptible to solubilization during autolysis. We believe this is strong support for the premise that highly stressed chemical bonds are more readily hydrolyzed than nonstressed bonds of a similar chemical nature (24).

Figure 6 contains a composite interpretation of the foregoing results. It is suggested that newly inserted wall materials create tension on preexisting wall. This tension causes a stretching of the wall matrix, resulting in elongation of the cell cylinder. The most stressed wall may be cleaved by autolysins and wall turnover may result. An axiom of the model is that only growing cells will be cleaved by the autolysins. This is consistent with literature which shows that division-arrested *B. subtilis* does not rapidly turn over its cell wall (17, 42). The model also would predict that cell wall would spread during growth. Pooley (34, 35) was the first to describe the "spreading" of cell wall during cell division. The model also embraces the view that autolysin is required for normal cell growth. It does not, however, state that turnover must accompany growth. Finally, there is an assumption that autolysins are regulated in such a manner that only old cell wall is cleaved. The influence of energized membrane on autolysin(s) may provide a satisfactory explanation for regulation of the enzyme(s) (22, 23, 28). Autolysin near the energized membrane or embedded within the wall matrix may be inhibited by low pH due to the extrusion of protons. Autolysin on the cell surface would not be influenced by the proton motive force and would have an active conformation.

INSERTION,DEVELOPMENT
OF TENSION
(cell elongates)

AUTOLYSIN RELIEVES
TENSION

TURNOVER

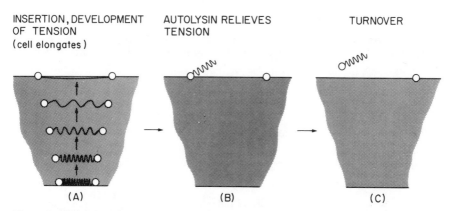

(A)   (B)   (C)

Figure 6. Model depicting the elongation of the side wall of *B. subtilis.* A cell wall precursor is secreted by the cytoplasmic membrane and then cross-linked into the wall matrix facing the membrane (A). The wall segment is then pushed out by more recently added wall. This causes the wall segment to stretch. By the time it reaches the outer wall face (about one generation) it has become highly susceptible to the actions of autolysins. An autolysin cleaves one site on the segment, thereby alleviating tension (B), and another autolysin finally cleaves the other bond holding the segment onto the wall matrix (turnover) (C). Addition of new wall with subsequent stretching will result in the elongation of the cell cylinder. Autolysin-deficient cells may have only enough enzyme to clip the most highly stressed bonds. Turnover in these cells will therefore be minimal and their surfaces may be "rough" (40).

Conclusions

It now seems reasonably clear that *B. subtilis* elongates by the inside-to-outside growth of its cell wall. Cell wall precursors are added to the face of the wall near the cytoplasmic membrane. Shortly thereafter, the newly inserted wall components are cross-linked. Upon cross-linking, the peptidoglycan becomes stretched due to cellular turgor pressure. The addition of more wall results in increased tension on the peptidoglycan as it is pushed away from the plasma membrane. The process of addition of wall precursors, their cross-linking, and their assuming tension, if repeated at very many sites in the wall, will cause elongation. The growth of bacilli is therefore from many sites in the side wall. This type of elongation of the cell during growth does not require a zone originating at the septum. The only zone of growth for cylinder extension is represented by the inside-to-outside movement of wall as the cell elongates. The foregoing conclusions were drawn from results with fluorescein-labeled ConA and two (one LYT[+]; the other partially LYT[-]) temperature-sensitive phosphoglucomutase mutants of *B. subtilis.* At the permissive temperature, labeled lectin bound evenly over cell surfaces. After a chase at the nonpermissive temperature at which cell wall glucosylation was inhibited, ConA receptor sites were lost randomly over the side walls. In contrast, when the temperature shift was from nonpermissive to permissive, then the binding of the lectin was evenly distributed in the cell cylinders after about one generation. Results from LYT⁻ or LYT⁺ cells were similar. In addition, younger, less stretched wall appears more resistant to solubilization than older, more stretched wall, during cellular autolysis.

ACKNOWLEDGMENTS. This work was supported in part by grant PCM 78-08903 from the National Science Foundation.
A. L. Koch conceived the idea that peptidoglycan components assume stress when cross-linked in growing bacteria.

LITERATURE CITED

1. **Anderson, A. J., R. S. Green, A. J. Sturman, and A. R. Archibald.** 1978. Cell wall assembly in *Bacillus subtilis:* location of wall material incorporated during pulsed release of phosphate limitation, its accessibility to bacteriophages and concanavalin A, and its susceptibility to turnover. *J. Bacteriol.* **136:**886–899.
2. **Archibald, A. R.** 1976. Cell wall assembly in *Bacillus subtilis:* development of bacteriophage-binding properties as a result of the pulsed incorporation of teichoic acid. *J. Bacteriol.* **127:**956–960.
3. **Archibald, A. R., and H. E. Coapes.** 1976. Bacteriophage SP50 as a marker for cell wall growth in *Bacillus subtilis. J. Bacteriol.* **125:**1195–1206.
4. **Beveridge, T. J., and R. G. E. Murray.** 1979. How thick is the *Bacillus subtilis* cell wall? *Curr. Microbiol.* **2:**1–4.
5. **Burdett, I. D. J., and M. L. Higgins.** 1978. Study of pole assembly in *Bacillus subtilis* by computer reconstruction of septal growth zones seen in central, longitudinal thin sections of cells. *J. Bacteriol.* **133:**959–971.
6. **Chaloupka, J., L. R. Kreckova, and P. Kreckova.** 1964. Degradation and turnover of bacterial cell wall mucopeptide in growing bacteria. *Folia Microbiol.* (Prague) **9:**9–15.
7. **Chung, K. L., R. Z. Hawirko, and P. K. Isaac.** 1964. Cell wall replication. I. Cell wall growth of *Bacillus cereus* and *Bacillus metaterium. Can. J. Microbiol.* **10:**43–48.
8. **Clark-Sturman, A. J., and A. R. Archibald.** 1982. Cell wall turnover in phosphate and potassium limited chemostat cultures of *Bacillus subtilis* W23. *Arch. Microbiol.* **131:**375–379.
9. **de Boer, W. R., F. J. Kruyssen, and J. T. M. Wouters.** 1981. Cell wall turnover in batch and chemostat cultures of *Bacillus subtilis. J. Bacteriol.* **145:**50–60.
10. **de Boer, W. R., P. D. Meyer, C. G. Jordens, F. J. Kruyssen, and J. T. M. Wouters.** 1982. Cell wall turnover in growing and nongrowing cultures of *Bacillus subtilis. J. Bacteriol.* **149:**977–984.
11. **DeChastellier, C., C. Frehel, and A. Ryter.** 1975. Cell wall growth of *Bacillus megaterium:* cytoplasmic radioactivity after pulse-labeling with tritiated diaminopimelic acid. *J. Bacteriol.* **123:**1197–1207.
12. **Doyle, R. J., and D. C. Birdsell.** 1972. Interaction of concanavalin A with the cell wall of *Bacillus subtilis. J. Bacteriol.* **109:**652–658.
13. **Doyle, R. J., H. L. T. Mobley, L. K. Jolliffe, and U. N. Streips.** 1980. Restricted turnover of the cell wall of *Bacillus subtilis. Curr. Microbiol.* **5:**19–22.
14. **Fan, D. P., B. E. Beckman, and H. L. Gardner-Eckstrom.** 1975. Mode of cell wall synthesis in gram-positive bacilli. *J. Bacteriol.* **123:**1157–1162.

15. **Fein, J. E., and H. J. Rogers.** 1976. Autolytic enzyme-deficient mutants of *Bacillus subtilis* 168. *J. Bacteriol.* **127**:1427–1442.
16. **Frehel, C., C. DeChastellier, and A. Ryter.** 1980. Peptidoglycan turnover during growth recovery after chloramphenicol treatment in a DAP-LYS-mutant of *Bacillus megaterium.* *Can. J. Microbiol.* **26**:308–317.
17. **Glaser, L., and B. Lindsay.** 1977. Relation between cell wall turnover and growth in *Bacillus subtilis. J. Bacteriol.* **130**:610–619.
18. **Herbold, J. R., and L. Glaser.** 1975. *Bacillus subtilis* N-acetylmuramic acid L-alanine amidase. *J. Biol. Chem.* **250**:1676–1682.
19. **Higgins, M. L., and G. D. Shockman.** 1970. Model for cell wall growth of *Streptococcus faecalis. J. Bacteriol.* **101**:643–648.
20. **Highton, P. J., and D. G. Hobbs.** 1972. Penicillin and cell wall synthesis: a study of *Bacillus cereus* by electron microscopy. *J. Bacteriol.* **109**:1181–1190.
21. **Hughes, R., and E. Stokes.** 1971. Cell wall growth in *Bacillus licheniformis* followed by immunofluorescence with mucopeptide-specific antiserum. *J. Bacteriol.* **106**:694–696.
22. **Jolliffe, L. K., R. J. Doyle, and U. N. Streips.** 1981. Energized membrane and cellular autolysis in *Bacillus subtilis. Cell* **25**:753–763.
23. **Kirchner, G., A. L. Koch, and R. J. Doyle.** 1984. Energized membrane regulates cell pole formation in *Bacillus subtilis. FEMS Microbiol. Lett.* **24**:143–147.
24. **Koch, A. L.** 1983. The surface stress theory of microbial morphogenesis. *Microb. Physiol.* **24**:301–366.
25. **Koch, A. L., and R. J. Doyle.** 1985. Inside-to-outside growth and turnover of the Gram-positive rod. *J. Theor. Biol.* **117**:137–157.
26. **Koch, A. L., M. L. Higgins, and R. J. Doyle.** 1981. Surface tension-like forces determine bacterial shapes: *Streptococcus faecium. J. Gen. Microbiol.* **123**:151–161.
27. **Koch, A. L., M. L. Higgins, and R. J. Doyle.** 1982. The role of surface stress in the morphology of microbes. *J. Gen. Microbiol.* **128**:925–945.
28. **Koch, A. L., G. Kirchner, R. J. Doyle, and I. D. J. Burdett.** 1985. How does a *Bacillus* split its septum right down the middle? *Ann. Microbiol.* (Paris) **136A**:91–98.
29. **Mauck, J., L. Chan, L. Glaser, and J. Williamson.** 1972. Mode of cell wall growth of *Bacillus megaterium. J. Bacteriol.* **109**:373–378.
30. **Mauck, J., and L. Glaser.** 1972. On the mode of *in vivo* assembly of cell wall of *Bacillus subtilis. J. Biol. Chem.* **247**:1180–1187.
31. **Mendelson, N., and J. Reeve.** 1973. Growth of the *Bacillus subtilis* cell surface. *Nature* (London) *New Biol.* **243**:62–64.
32. **Mobley, H. L. T., R. J. Doyle, U. N. Streips, and S. O. Langemeier.** 1982. Transport and incorporation of *N*-acetyl-D-glucosamine in *Bacillus subtilis. J. Bacteriol.* **150**:8–15.
33. **Mobley, H. L. T., A. L. Koch, R. J. Doyle, and U. N. Streips.** 1984. Insertion and fate of the cell wall in *Bacillus subtilis. J. Bacteriol.* **158**:169–179.
34. **Pooley, H. M.** 1976. Turnover and spreading of old wall during surface growth of *Bacillus subtilis. J. Bacteriol.* **125**:1127–1138.
35. **Pooley, H. M.** 1976. Layered distribution, according to age, within the cell wall of *Bacillus subtilis. J. Bacteriol.* **125**:1139–1147.
36. **Pooley, H. M., J.-M. Schlaeppi, and D. Karamata.** 1978. Localized insertion of new cell wall in *Bacillus subtilis. Nature* (London) **274**:264–266.
37. **Rogers, H. J., C. Taylor, S. Rayter, and J. B. Ward.** 1984. Purification and properties of autolytic endo-β-N-acetylglucosaminidase and the N-acetylmuramyl-L-alanine amidase from *Bacillus subtilis* strain 168. *J. Gen. Microbiol.* **130**:2395–2402.
38. **Sargent, M. G.** 1978. Surface extension and the cell cycle in prokaryotes. *Adv. Microb. Physiol.* **18**:106–176.
39. **Schlaeppi, J.-M., H. M. Pooley, and D. Karamata.** 1982. Identification of cell wall sub-

units in *Bacillus subtilis* and analysis of their segregation during growth. *J. Bacteriol.* **149:**329–337.

40. **Sonnenfeld, E. M., T. J. Beveridge, A. L. Koch, and R. J. Doyle.** 1985. Asymmetric distribution of charge on the cell wall of *Bacillus subtilis. J. Bacteriol.* **163:**1167–1171.

41. **Sturman, A. J., and A. R. Archibald.** 1978. Conservation of phage receptor material at polar caps of *Bacillus subtilis* W23. *FEMS Microbiol. Lett.* **4:**255–259.

42. **Vitkovic, L., H.-Y. Cheung, and E. Freese.** 1981. Absence of correlation between rates of cell wall turnover and autolysis shown by *Bacillus subtilis* mutants. *J. Bacteriol.* **157:**318–320.

Chapter 10

Studies of *Bacillus subtilis* Macrofiber Twist States and Bacterial Thread Biomechanics: Assembly and Material Properties of Cell Walls

Neil H. Mendelson

J. J. Thwaites

New information about the organization, assembly, and material properties of bacterial cell walls is being obtained by the study of two unique *Bacillus subtilis* systems, the macrofiber and thread systems, that provide highly amplified views of individual cells (9, 20). In the macrofiber system the establishment and maintenance of twist states are governed, respectively, by factors concerned with the assembly of cell wall and the properties of cell wall polymers within the fabric of the wall. The involvement of peptidoglycan was previously suggested by the facts that penicillin G blocks macrofiber formation and that lysozyme digestion of intact live macrofibers results in relaxation motions in which twist states change before the entire macrofiber is converted to spheroplasts (5, 23). Recently, it was found that establishment of twist states can be

Neil H. Mendelson • Department of Molecular and Cellular Biology, University of Arizona, Tucson, Arizona 85721.   **J. J. Thwaites** • Department of Engineering, Cambridge University, Cambridge CB2 1PZ, United Kingdom.

regulated by D-alanine and D-cycloserine (U. C. Surana, A. J. Wolfe, and N. H. Mendelson, personal communication). Changes in twist that were brought about by growth in D-alanine were always in a direction opposite to those caused by the presence of D-cycloserine. Thus, the two factors antagonized each other with respect to twist in the same manner that they do with respect to growth inhibition, suggesting that the same biochemical targets are involved. The hypothesis that twist may be regulated by the D-Ala-D-Ala dipeptide pool size operating by control of peptidoglycan cross-linking will be discussed.

Both covalent and electrostatic properties of the cell wall polymers are likely to contribute to the maintenance of twist states in macrofibers, as well as to the material properties of cell walls in all cells (14). Therefore, we have used the bacterial thread system to study the mechanical properties of cell walls. Although it has been known for some time that cell walls, in particular the load-bearing polymer, peptidoglycan, stretch appreciably and recover (6, 7, 16), this knowledge is difficult to obtain, even indirectly, in normal cultures. Bacterial thread can be measured by standard techniques used on textile threads, which allow direct measurement of mechanical properties. Such measurements showed that in general the walls behaved as typical viscoelastic polymers. The stress induced by stretching relaxed, though not thoroughly, if the structures were held at constant length. When stretched structures were made slack, the resulting bowed shape eventually straightened as the original length was regained, indicating a rearrangement of the wall polymers. The strength of cell wall material determined by straining to break was found to be highly dependent upon relative humidity. The results suggest that wall polymers must be able to change configurations, possibly unfolding in a manner that permits a return to something approximating, if not identical to, the initial starting structure when the stretching force is relieved. The mechanical properties observed place limitations on the kinds of cell wall polymer superstructure arrangements that are feasible.

Approaches Used in the Study of Macrofibers

Studies of macrofibers utilized standard protocols and growth conditions, as previously described (9, 10). Effects of twist-modulating compounds were assessed by comparison of structures produced in six-well tissue culture plates. Each well contained a different concentration of the compound being tested for twist-modulating activity. Twist states were determined by phase-contrast microscopy. Twist values representing the spectrum from maximum left handedness to maximum right handedness were assigned by using the convention −6 to 0 to +6 as described by U. C. Surana (Ph.D. dissertation, University of Arizona, Tucson, 1987).

Approaches Used in the Study of Bacterial Thread

Thread specimens were produced from cultures of the same separation-suppressed (*lyt*) mutants of *B. subtilis* 168 (strain FJ7) as were used for macrofiber experiments. Cultures grown in TB medium (20) produced many long filaments that resembled textile webs. Threads were produced by drawing out the web from the culture at a rate of 22 mm/min by use of a quartz clock movement. Individual filaments were drawn radially into the forming thread and, apparently, were compressed by surface tension into a compact bundle (20). Uniform threads of up to 1 m in length at 180-μm diameter were produced. Such threads have about 50,000 filaments in the cross section and contain in excess of 10^{10} cells.

Scanning electron micrographs show the cellular filaments aligned parallel to the thread axis and also show their close packing and deformation into hexagonal cross-sectional shape (20). The shrinkage associated with this deformation must be accompanied by substantial transverse compressive stress. This and possibly the gluing action of the residual growth medium maintain the integrity of the threads during tensile testing. No sign of slippage between filaments can be detected right up to break, even at high relative humidity. It is clear that the properties measured are those of cell wall material.

Some thread specimens were rehydrated in water to remove the residual growth medium and then, with difficulty, redrawn. Scanning electron micrographs of redrawn threads showed surface faults, i.e., gaps between the ends of filaments in the same line which were not apparent in standard threads.

The instrument used to measure bacterial thread was similar to standard machines used in textile testing (e.g., Instron). The specimens, of gauge length 30 or 100 mm, were mounted between jaws. The upper jaw was made to traverse at one of a set of fixed speeds, thus extending the thread. The lower jaw was fixed to a load cell. The variations in load with extension or simply as a function of time were recorded with a standard pen/chart recorder. The system was enclosed at constant relative humidity within an Atmosbag.

The Macrofiber System

Factors Affecting the Establishment of Twist

Previous work showed that many factors influence the development of twist in macrofibers (Table 1). The findings indicate that both genetic and physiological factors play a role in the development of twist states which must correspond to the establishment of cell surface conformational states at the time of cell wall assembly. Potential targets for twist-modulating factors are likely to be, therefore: (i) proteins concerned with the metabolism of peptidoglycan

Table 1. Factors that influence the establishment of twist in *B. subtilis* macrofibers

Factor	Twist bias with increasing value[a]	Conditions	Zero twist found at[b]:	Reference
Genes	LH	C6D, TB medium, 20 or 48°C	NA	9
	LH	PS6μB, TB medium, 20–48°C	NA	4
	LH	PS5, TB medium, 20–48°C	NA	4
	RH	84A, TB medium, 20–48°C	NA	Favre[c]
	RH	810A, TB medium, 20–48°C	NA	4
Temperature	LH	FJ7, TB medium	39°C	12, 13
	RH	FJ7, S1 medium + MgSO$_4$	ND	Mendelson[d]
Nutrition	RH	FJ7, TB medium, 20°C	TB + S1[e]	13
	LH	FJ7, S1 medium + MgSO$_4$, 18°C	TB + S1[e]	Wolfe[f]
	LH	FJ7, T medium, 20°C	NA	11
	RH	FJ7, Be medium, 20°C	NA	11
Ions				
MgSO$_4$	RH	FJ7, T medium, 20°C	10^{-3} M	11
	RH	734, T medium, 20°C	5×10^{-4} M	11
	LH	CGD, TB medium, 39°C	ND	11
	LH	FJ7, Be medium, 20°C	ND	11
(NH$_4$)$_2$SO$_4$	LH	FJ7, Be or TB, 20°C	10^{-2} M	11
	LH	734, Be or TB, 20°C	6×10^{-3} M	11
NaCl	RH	FJ7, T medium, 20°C	0.08 M	Mendelson[d]
Glycerol	LH	FJ7, T + MgSO$_4$, 20°C	0.3 mol/kg	Mendelson[d]
Protease	RH	FJ7, TB medium, 48°C	ND	4
	RH	PS5, TB medium, 20°C	ND	4
	RH	PS6μB, TB medium, 20°C	ND	4
Alkaline phosphate (intestinal)	RH	FJ7, TB medium, 20°C	NA	Surana[g]
pH	RH/LH	Variations with pH depend on medium or strains used	NA	Wolfe[f]
D-Alanine	RH	FJ7, T medium, 20°C	1.2×10^{-3} M	Mendelson[d]
	RH	C6D, TB medium, 20°C	2×10^{-3} M	Mendelson[d]
	LH	FJ7, TB + MgSO$_4$, 20°C	3×10^{-2} M	Mendelson[d]
	LH	C6D, TB medium, 20°C	2×10^{-2} M	Mendelson[d]
D-Cycloserine	LH	FJ7, TB medium, 20°C	$<2 \times 10^{-5}$ M	Mendelson[d]
	LH	C6D, TB + D-Ala, 20°C	3×10^{-4} M	Mendelson[d]
	RH	C6D, TB + D-Ala, 20°C	10^{-4}–10^{-3} M	Mendelson[d]
β-Chloro-D-alanine	LH	FJ7, TB + MgSO$_4$, 20°C	2.5×10^{-3} M	Mendelson[d]
	RH	C6D, TB + D-Ala, 20°C	1.2×10^{-4} M	Mendelson[d]

[a]LH, Left handed; RH, right handed.
[b]NA, Not applicable; ND, not determined.
[c]D. Favre, Ph.D. dissertation, University of Lausanne, Lausanne, Switzerland, 1984.
[d]N. H. Mendelson, unpublished data.
[e]In mixtures of TB + S1.
[f]A. J. Wolfe, Ph.D. dissertation, University of Arizona, Tucson, 1985.
[g]U. C. Surana, Ph.D. dissertation, University of Arizona, Tucson, 1987.

and teichoic acid, (ii) proteins concerned with the assembly of the cell wall polymers which may play a structural role during cell wall assembly, and (iii) possibly the cell wall polymers themselves which carry charged groups that may be influenced by counterions, pH, waters of hydration, and so forth. For most of the factors shown in Table 1 it is impossible, given present information, to identify the relevant target. D-Alanine and D-cycloserine are clearly exceptions. The focus of this section is, therefore, on new results obtained with these two twist modulators.

Twist Modulation by D-Alanine

The fact that D-alanine alone among components of cell wall polymers was able to influence the establishment of macrofiber twist when added to the culture medium was discovered by A. J. Wolfe (Ph.D. dissertation, University of Arizona, Tucson, 1985). Details to be reported elsewhere (U. C. Surana, A. J. Wolfe, and N. H. Mendelson, manuscript in preparation) establish that in most systems there is a concentration-dependent rightward shift in twist produced by D-alanine, whereas L-alanine in the same concentration range is without activity. In the case of strain C6D, one of the original macrofiber-producing strains previously described (9, 11) that normally produces left-handed structures in TB medium at low temperature, increasing concentrations of D-alanine were found to bring about two transitions: the first (at about 2×10^{-3} M) from left to right handedness and the second (at about 2×10^{-2} M) from right to left handedness (Fig. 1). All of the structures produced by this strain over the concentration range of D-alanine shown in Fig. 1 were excellent, highly organized macrofibers in contrast to the tightly coiled, rigid, single-filament helices (similar to those described by Tilby [21]) found in macrofibers produced by strain FJ7 and its derivatives at very high D-alanine concentrations. Strain C6D was therefore used initially to examine the interaction between D-alanine and D-cycloserine with respect to the establishment of twist states. In the macrofiber system we searched for antagonism of the D-alanine-induced twist modulation by D-cycloserine in a manner analogous to a previous approach in which D-alanine was used to antagonize the growth inhibition produced by D-cycloserine (1, 18).

Twist Modulation by D-Cycloserine

The experimental protocol generally followed involved the production of three sets of strain C6D cultures, each in six replicates resulting in 18 cultures used for comparison of twist states. One set of three was grown in TB medium alone, a second set contained sufficient D-alanine to result in right handedness,

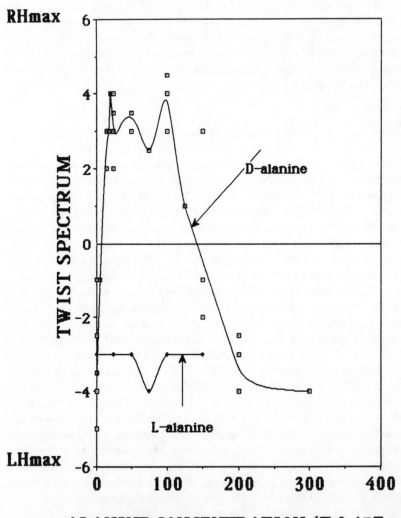

Figure 1. Effect of alanine on establishment of *B. subtilis* macrofiber twist. Strain C6D macrofibers were cultured in TB medium at 20°C in the standard manner. D-Alanine or L-alanine was added at the concentrations indicated. Twist states were determined by phase-contrast microscopy. Twist state values were assigned as described in the text. LH, Left handedness; RH, right handedness.

and a third set contained a high enough concentration of D-alanine to result in left handedness. For each of the three sets, one of the six subcultures served as a control, whereas the other five contained increasing amounts of D-cyclo-serine over a concentration range reaching that required for growth inhibition. All of the cultures were seeded from the same source and incubated at 20°C for 18 h until mature structures were present in the controls. The twist state of structures produced in each culture was then determined by microscopy. The findings are presented in diagrammatic form in Fig. 2a, which also shows the relationship of D-alanine and D-cycloserine twist modulation effects to that pro-duced by temperature (13).

Figure 2a indicates that all of the results can be interpreted in a consistent manner if the twist spectrum is drawn as a circle ranging from maximum left to maximum right handedness by two routes: going clockwise or counterclock-wise. In this context the influence of D-alanine is drawn as taking the clockwise route from left to right handedness and then to left handedness; that of D-cy-closerine is in the opposite direction, going counterclockwise. Interactions be-tween D-alanine and D-cycloserine were interpreted by using this convention concerning the direction in which each twist modulator causes change in twist. Thus, the left handedness produced in high concentrations of D-alanine was antagonized, resulting in right-handed structures when D-cycloserine was pres-ent in the growth medium in addition to the high concentration of D-alanine. Similarly, the right-handed twist state normally found when D-alanine was pres-ent in the concentrations shown in Fig. 1 did not arise when D-cycloserine was also present. Instead, left-handed structures were produced. Assuming the D-cycloserine influence again to be a counterclockwise shift in twist as shown in Fig. 2a, the left-handed structures produced under these conditions should be equivalent to those produced in TB medium in the absence of twist modulators. When D-cycloserine was added to TB cultures without D-alanine addition, no alteration of the left-handed twist state was noted, even in cultures containing concentrations of D-cycloserine sufficient to inhibit growth. The molar ratio of D-alanine to D-cycloserine required to convert structures to zero twist ranged from about 10 to 300, depending upon the D-alanine concentration. The ab-solute concentration of D-cycloserine required to convert twist to zero was in the range of 10^{-4} to 10^{-3} M.

Production of strain C6D macrofibers in TB medium at various temper-atures between 20 and 48°C results in two helix orientation inversions (from left handedness at 20°C, to right handedness at 37°C, to left handedness at 48°C), as previously described (11). The direction of the temperature route around the twist spectrum was chosen as clockwise, therefore, in view of the similarity of the temperature effect to the effect of D-alanine. Temperature in-teractions with D-alanine and D-cycloserine were examined at three points on the temperature spectrum corresponding to left handedness at low and high temperatures and to right handedness at intermediate temperatures. Neither of

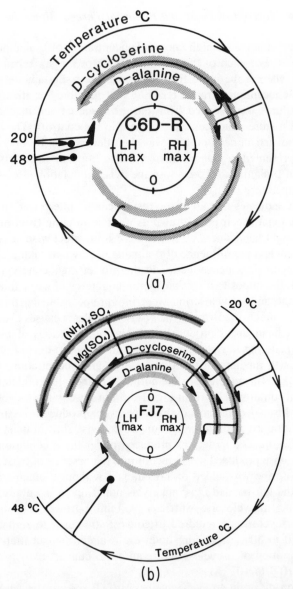

Figure 2. Influence of D-alanine and D-cycloserine on the establishment of macrofiber twist. (a) Results obtained with strain C6D. The twist spectrum is shown as a circle in the center ranging from maximum left handedness to maximum right handedness. Effects of D-alanine, D-cycloserine, and temperature are shown in concentric circles. Interactions between two systems are shown by connecting arrows. Arrows point in the direction of twist modification resulting from increasing values of the twist-modulating factor. Arrows pointing to dots represent conditions where no change of twist resulted from addition of the modulator. (b) Results obtained with strain FJ7. Details as for panel a. Twist modulation by magnesium and ammonium sulfate is also shown. Conventions concerning the direction of twist modulation obtained from studies of C6D macrofibers were followed in plotting of the FJ7 results as well.

116

the left-handed states was perceptibly changed by the presence of D-cycloser-ine, but both were influenced by D-alanine. The right-handed twist state was responsive to both D-cycloserine and D-alanine. As shown in Fig. 2a, all of the results are compatible with the routes assigned to twist changes produced by both modulators on the basis of initial results described above.

Interactions between D-Alanine/D-Cycloserine and Other Twist Modulators

Interactions between D-alanine and D-cycloserine and other systems of twist modulation are shown for strain FJ7 in Fig. 2b. The same conventions used in Fig. 2a were followed. As in strain C6D, D-alanine influenced twist of FJ7 toward the right. At high concentrations, however, the second transition to left handedness was not observed. Instead, highly coiled, single filaments similar to those described by Tilby (21) were produced. The temperature-controlled portion of the FJ7 twist spectrum, which spanned from right to left handedness with increasing temperature, appears to lie between 2 o'clock and 8 o'clock on the twist spectrum shown in Fig. 2b. D-Alanine and D-cycloserine inter-actions with temperature resulted in structures of appropriate twist, as indicated when the convention established with strain C6D was followed.

The magnesium-ammonium ion system previously studied (11) has also been used to establish initial twist states for challenge by D-alanine and D-cycloserine. When magnesium was used to produce right-handed structures, the presence of D-cycloserine antagonized the rightward influence of the ion, resulting in left-handed structures. D-Cycloserine also was able to change the twist of FJ7 structures produced in TB medium at 20°C in the absence of other known twist modulators from right to left handedness. In addition, the missing portion of the twist spectrum (left handedness produced at high D-alanine con-centrations) was found if D-alanine was added to structures made right handed by magnesium. Consistent results were also found when ammonium ions were used to produce left-handed structures that were challenged by D-alanine and D-cycloserine addition.

Similar experiments were carried out with the macrofiber-producing strain 2C8, which carries a defect in the alanine racemase and consequently requires addition of D-alanine for growth (data not shown). Rightward shift of twist was demonstrated by using T medium in which a low concentration of D-alanine sufficient to support growth resulted in left-handed structures. Leftward shift of twist was more difficult to document because of the high sensitivity of the strain to D-cycloserine but could definitely be observed at low concentrations of the inhibitor. The response of this alanine racemase-deficient strain sug-gested that the D-Ala-D-Ala dipeptide pool size may be the key to twist reg-ulation, possibly acting by control of peptidoglycan cross-linking.

β-Chloro-D-Alanine, Vancomycin, and Penicillin G Effects on Twist

Preliminary experiments using β-chloro-D-alanine, vancomycin, and penicillin G have also been conducted with strains C6D and FJ7 (details will be
presented elsewhere). The haloalanine behaved like D-cycloserine in the ion-
regulated twist system of FJ7 and at high concentrations of D-alanine in the
C6D system, but gave variable results with C6D structures made right handed
with intermediate amounts of D-alanine. Minimal effect on twist was noted with
vancomycin, but there appeared to be a differential sensitivity with respect to
growth inhibition of left-handed FJ7 structures produced in TB medium plus
ammonium ions in comparison with right-handed structures produced in TB
medium plus magnesium. A similar differential sensitivity was also noted in
this system when penicillin G, D-cycloserine, and β-chloro-D-alanine were used
as inhibitors. In all cases the cultures growing as left-handed structures were
more sensitive to growth inhibition than were those growing as right-handed
structures. Left- and right-handed twist states produced by effectors other than
ammonium and magnesium have not yet been examined, however, with respect
to their sensitivities to these growth inhibitors.

Macrofiber Twist as an Indication of Cell Wall Polymer Conformation

Macrofiber twist was previously shown to arise as a compensatory shape
deformation resulting from the impedance of rotation generated during the elongation of each cell within the structure (8, 9, 12). The forces responsible for
this deformation trace to the assembly of cell wall polymers during growth,
and it has been possible, therefore, to use macrofiber twist to reveal aspects
of cell growth hitherto undetected in individual cells. For example, the discovery that macrofibers can assume a spectrum of twist states from tight left
handedness to tight right handedness suggests that the cell wall polymers must
be able to assume on average a corresponding spectrum of conformational states.
For any given twist state it is possible to calculate a corresponding helical state
of the cell cylinder which defines the geometry that must be followed during
cell surface assembly and, consequently, the dynamics of surface growth from
which helical cell shape is derived when these normal motions become mechanically blocked. Central issues that remain to be explored in the macrofiber
system are (i) the cell wall molecular conformations that correspond to twist
states, (ii) the factors that regulate cell wall assembly leading to twist states,
and (iii) the biomechanics of force interactions in the cell wall responsible for
conversion of helical assembly of cell surfaces into helical cell shape of a particular twist in macrofibers. Here we have addressed points ii and iii, focusing
for the time being in the latter case on measuring the material properties of
cell walls inasmuch as such information is needed for all but the most super-

ficial analysis of the relevant mechanical principles.

The variety of factors found to regulate the establishment of macrofiber twist states suggests that gene products operating at the time of cell wall assembly influence the assembly process and that the assembly process is very sensitive to a number of environmental conditions. The findings show further that charge interactions are important in twist determination and that a cell surface protein(s) is needed to establish left-handed twist states. A direct link to peptidoglycan metabolism is now evident based upon the finding that D-alanine and D-cycloserine are effective twist modulators.

Peptidoglycan Metabolism and Twist Determination

The twist modulation effects of D-alanine, initially found to be a rightward shift proportional to concentration, are shown here to also entail a shift from right to left at high concentrations of D-alanine in strain C6D and in the presence of magnesium sulfate in strain FJ7. All of the observations can be interpreted in a consistent manner if the twist spectrum is drawn as a continuous circle, as shown in Fig. 2. When plotted in this fashion, there are two routes from left-handed maximum to right-handed maximum, going clockwise or counterclockwise around the circle. Twist modulation by D-alanine has arbitrarily been chosen as taking the clockwise route. The directions taken by other twist modulators have then been assigned by reference to the D-alanine route. The data indicate that whatever starting position on the twist spectrum one chooses, addition of D-alanine will drive the twist state from that position to one farther clockwise around the twist spectrum circle.

There are three known cell surface components that contain D-alanine and which may therefore be targets for the twist modulation effect of D-alanine: (i) the peptidoglycan, (ii) the teichoic acid, and (iii) two membrane proteins described by Clark and Young (3). The observation that D-cycloserine antagonizes the D-alanine effect on twist makes it likely that peptidoglycan rather than teichoic acid is the target. Nothing can be said about the role of the D-alanine-containing proteins because no further work has been reported on them. (We will report elsewhere [U. C. Surana and N. H. Mendelson, manuscript in preparation] a cell surface protein of very high molecular weight found in macrofiber strains that may be the left-twist protein which appears to have the same molecular weight as one of the two D-alanine-containing membrane proteins, but we do not yet know whether this protein contains D-alanine.) If the D-alanine metabolism with respect to peptidoglycan synthesis is indeed the target for twist modulation, then the relevant pathway must involve the well-known three-step sequence: (i) conversion of L-alanine to D-alanine catalyzed by a racemase that can operate in either direction, (ii) linkage of two D-alanine residues to form the dipeptide D-alanyl-D-alanine, catalyzed by a ligase, and

(iii) attachment of the dipeptide to the *meso*-diaminopimelic acid of *N*-acetyl muramic acid that carries the tripeptide, L-alanine-D-glutamic acid-*m*-diaminopimelic acid. The uptake of D-alanine could then enter this pathway and flow either to L-alanine or to the dipeptide. Twist modulation is not likely to be driven by increasing the L-alanine pool since addition of L-alanine to macrofibers did not affect twist and a racemase-deficient strain also responded to D-alanine by change of twist. It appears, therefore, that either D-alanine itself or the dipeptide pool size may be responsible for twist alteration.

Implications of the D-Alanyl-D-Alanine Pool Size in Twist Determination

Figure 2 shows that D-cycloserine can also influence twist in the absence of added D-alanine. In all cases a counterclockwise direction of twist modulation was observed. In peptidoglycan metabolism D-cycloserine has been shown to inhibit both the alanine racemase and the D-alanine ligase. Taken together with the above, it seems likely that in twist regulation the D-alanyl-D-alanine pool size may be the essential factor. If so, reactions that might be sensitive to variations in the dipeptide pool size could involve (i) the production of *N*-acetyl muramic acid pentapeptide, (ii) cross-linking reactions that involve the dipeptide when attached to the pentapeptide (i.e., enzymes such as the transpeptidase that have sites for a D-alanyl-D-alanine dipeptide), and possibly (iii) reactions that remove the terminal D-alanine from the peptapeptide catalyzed by the carboxypeptidase. Variations in the quantity or quality of peptidoglycan cross-linking should therefore be examined as a function of twist states to test these assumptions. Such variations would be expected to alter peptidoglycan structure, which in turn could affect the geometry of cell wall assembly.

The Bacterial Thread System

Test Protocol Used

The results of preliminary studies of the mechanics of bacterial thread specimens using a prototype system (20) provided sufficient information for the construction of a tensile testing instrument of appropriate sensitivity. The standard test protocol developed for use with this instrument is shown in Fig. 3. A thread was first extended at constant speed. The extension was then stopped, and the relaxation in stress in the extended state was observed over a period of time. The specimen was then rapidly returned to its original length and allowed to recover. Any residual extension was then noted in a final extension

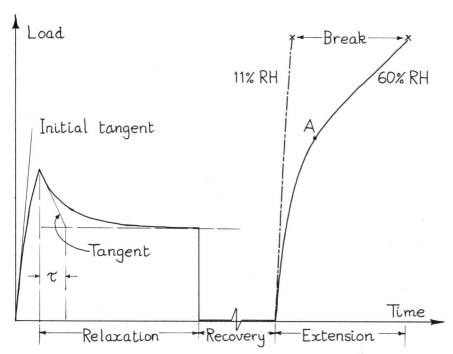

Figure 3. Record of a typical thread tensile test. The difference in behavior between high and low relative humidity (RH) (———-— with compressed "load scale") is shown only for final loading. Point A represents the breaking point for hydrated threads at the corresponding RH.

at fixed speed during which the stress was recorded up to break. A variation involved making repeated extension/retraction cycles to investigate recovery from repeated loading.

Cell wall properties were measured by observing the extensibility (geo-metrical strain at break), tensile strength (stress at break), initial (Young's) modulus (stress/strain ratio derived from initial tangent, Fig. 3), "final" mod-ulus (ratio of equilibrium stress after relaxation to strain imposed), residual strain after recovery upon removal of load, and the time scale of stress relax-ation. In calculating the wall stress it was assumed that the walls bear the whole load, and although it must vary with relative humidity, the ratio of wall to thread cross-section area was taken to be 0.2. The conclusions which follow are based upon the results of 120 tests in atmospheres ranging from 11 to 98% relative humidity, including 20 threads which were never allowed to dry before testing, 20 threads which were redrawn after hydration in either water or the original growth medium, and 12 which were treated in the culture with lyso-zyme (final concentration, 40 μg/ml). The test relative humidity range for hydrated threads was 50 ± 5%.

Cell Wall Properties Observed

For "standard" threads the results are summarized in Table 2. Extensibility, except at low humidity, was comparable with that of cellulose, but much less than that of elastin. Tensile strength, except for dry walls, was much less than that of cellulose (15 times weaker) and chitin (5 times weaker) (19). The variation in modulus, about 2,000 times over the humidity range, was much greater than that for other biological materials. When wet, cell wall modulus approached that of resilin and elastin (22). The figures for strength and, to a lesser extent, for modulus were higher than previously reported (20). This, no doubt, is due to the better experimental technique, which also resulted in less, but still substantial, spread of data points.

The properties of threads made, stored, and tested at the same relative humidity were no different from standard threads. More surprisingly perhaps, those of lysozyme-treated threads, over the whole relative humidity range, were no different. Nor, except for extensibility, were those of threads redrawn after hydration in TB medium. At essentially constant relative humidity there was apparently a large variation in extensibility from the standard value down to less than 1%. For water-hydrated threads it was always below 1% and the stress/strain graph appears to end at a point such as A in Fig. 3. We suspect that this is an artifact due to breakage at faults, for in almost all cases the thread diameter at break was significantly greater than the minimum diameter (which was never observed with "standard" threads). For this reason the strength (then calculated from minimum diameter) may be underestimated by, for water-hydrated threads at ~50% relative humidity, approximately four times. The initial modulus, which is not affected by this problem, was about five times standard.

The general shape of the load/extension graph for thread was very similar to those of other polymers, including man-made, the dry specimens behaving like glassy polymers. The load also depended on the rate of strain in the same way. Load relaxation in the extended state is a feature of viscoelasticity. The ratio of final to initial modulus for thread was quite typical (Table 2). The time

Table 2. Approximate mechanical property[a] values derived from standard threads at different relative humidities

Relative humidity (%)	Approx value of:				
	ϵ_b (%)	δ_b (N/mm^2)	E_i (kN/mm^2)	E_f/E_i	τ (min)
11	<1	300	20	0.8	1.4
65	20–60	20	0.2	0.25	0.7
98	50–70	>3	0.01	0.4	2.3

[a]Properties: ϵ_b, extensibility; δ_b, strength; E_i, initial modulus; E_f, final modulus; τ, time scale of load relaxation.

scale for relaxation cannot be represented by a single time constant (nonlinear viscoelasticity), but using the initial rate of decay, one can define a time τ as if there were a single constant as shown in Fig. 3. For thread, τ is of the order of 1 or 2 min and varies with relative humidity in approximately the same way as the ratio of final to initial modulus.

Recovery, even after several load cycles at a substantial fraction of the breaking load, can approach total at 65% relative humidity. Even at 90% relative humidity there was very little residual strain. It is not yet possible to make good quantitative assertions, but it is clear that cell wall material is better than many other polymers in this respect.

Cell Wall Mechanical Properties and Cell Wall Structures

Our studies of the mechanical properties of cell walls from the same strain as that used in studies of twist regulation confirm Marquis' supposition that peptidoglycan, because of its shorter-chain, noncrystalline structure, is much less stiff than cellulose or chitin (7, 16). However, the kinetic aspects of this behavior, i.e., stress relaxation and recovery from extension, together with the observed substantial extensibility, are little different from those of other polymers. The likely structure of cell wall is then, like theirs, one of molecular entanglement, with the (substantially reversible) extensions being associated with chain unfolding. It is unlikely that a cell wall model in which peptidoglycan is arranged in a regular pattern could account for such extension (2). This is not to say that there is not order in the structure. The observation of rotation in twisted macrofibers during growth implies some average orientation of the peptidoglycan backbone. Furthermore, because lysozyme cuts the backbone, the insensitivity of thread properties to its attack supports the view that this average orientation is circumferential or, for some states, as in macrofibers, at a small helix angle to this direction (12).

The strong dependence of wall properties on humidity and the changes due to water hydration could be a result of differences in ion binding by charged polymer end groups (7, 16). An alternative explanation, supported by the evidence about the effects of neutral salts and changes in pH on macrofibers, is that, like that of proteins, the structure of wall polymers is dependent on the ordering of water molecules within the cell wall (17). This also bears on the question of what the wall properties are in vivo.

The Cell Wall Is Stronger than Previously Reported

Standard thread properties at relative humidity approaching 100% are substantially greater than those obtained earlier by extrapolation (20). The modulus is five times greater and the strength is not less than three times (possibly five times) greater than previously estimated. If the further increase in strength for threads redrawn after water hydration is also observable at 100% relative humidity, then wall strength would be not less than 12 N/mm² (possibly 20 N/mm²). A turgor pressure of 20 atm (2,026 kPa) (15) would thus not, as previously suggested (20), lead to fracture of the wall. A difficulty which remains is that 100% relative humidity cannot properly represent the in vivo state. Attempts are being made to measure the properties of thread and macrofiber in the fully hydrated state.

Macrofibers, Bacterial Threads, and Conventional Rod-Shaped Cells

We believe that the properties studied in macrofibers concerning regulation of cell surface conformational states and in bacterial threads concerning mechanical properties of cell wall reflect basic aspects of the cell wall pertinent to wild-type *B. subtilis* cells as well. The macrofiber and thread systems provide amplified material which makes possible the unique approaches that we have developed. The fact that wild-type cells when cultured in ways that suppress autolysins will grow into long separation-suppressed cell filaments akin to those from which FJ7 threads are made, and also will produce macrofibers, though short-lived ones, makes it very unlikely that the presence of the *lyt* mutation in the strains that we have studied has altered the properties of the cell walls in any significant manner with respect to the parameters that we have studied.

ACKNOWLEDGMENTS. This work was supported by a Public Health Service grant from the National Institute of General Medical Sciences and by a Collaborative Research Grant from the North Atlantic Treaty Organization.

LITERATURE CITED

1. **Bondi, A., J. Kornblum, and C. Forte.** 1957. Inhibition of antibacterial activity of cycloserine by alpha-alanine. *Proc. Soc. Exp. Biol. Med.* **96:**270–272.
2. **Burman, L. G., and J. T. Park.** 1984. Molecular model for elongation of the murein sacculus of *Escherichia coli. Proc. Natl. Acad. Sci. USA* **81:**1844–1848.
3. **Clark, V. L., and F. E. Young.** 1978. D-Alanine incorporation into macromolecules and effects of D-alanine deprivation on active transport in *Bacillus subtilis. J. Bacteriol.* **133:**1339–1350.
4. **Favre, D., D. Karamata, and N. H. Mendelson.** 1985. Temperature-pulse-induced "memory" in *Bacillus subtilis* macrofibers and a role for protein(s) in the left-handed-twist state. *J. Bacteriol.* **164:**1141–1145.
5. **Favre, D., N. H. Mendelson, and J. J. Thwaites.** 1986. Relaxation motions induced in

Bacillus subtilis macrofibers by cleavage of peptidoglycan. *J. Gen. Microbiol.* **132:**2377–2385.

6. **Knaysi, G., J. Hillier, and C. Fabricant.** 1950. The cytology of an avian strain of Mycobacterium tuberculosis studied with electron and light microscopes. *J. Bacteriol.* **60:**423–447.

7. **Marquis, R. E.** 1968. Salt-induced contraction of bacterial cell walls. *J. Bacteriol.* **95:**775–781.

8. **Mendelson, N. H.** 1976. Helical growth of *Bacillus subtilis*: a new model of cell growth. *Proc. Natl. Acad. Sci. USA* **73:**1740–1744.

9. **Mendelson, N. H.** 1978. Helical *Bacillus subtilis* macrofibers: morphogenesis of a bacterial multicellular macroorganism. *Proc. Natl. Acad. Sci. USA* **75:**2478–2482.

10. **Mendelson, N. H.** 1982. Dynamics of *Bacillus subtilis* helical macrofiber morphogenesis: writhing, folding, close-packing, and contraction. *J. Bacteriol.* **151:**438–449.

11. **Mendelson, N. H., and D. Favre.** 1987. Regulation of *Bacillus subtilis* macrofiber twist development by ions: effects of magnesium and ammonium. *J. Bacteriol.* **169:**519–525.

12. **Mendelson, N. H., D. Favre, and J. J. Thwaites.** 1984. Twisted states of *Bacillus subtilis* macrofibers reflect structural states of the cell wall. *Proc. Natl. Acad. Sci. USA* **81:**3562–3566.

13. **Mendelson, N. H., and D. Karamata.** 1982. Inversion of helix orientation in *Bacillus subtilis* macrofibers. *J. Bacteriol.* **151:**450–454.

14. **Mendelson, N. H., J. J. Thwaites, D. Favre, U. Surana, M. M. Briehl, and A. Wolfe.** 1985. Factors contributing to helical shape determination and maintenance in *Bacillus subtilis* macrofibers. *Ann. Inst. Pasteur Microbiol.* **136A:**99–103.

15. **Mitchell, P., and J. Moyle.** 1956. Osmotic function and structure of bacteria. *Symp. Soc. Gen. Microbiol.* **6:**150–180.

16. **Ou, L., and R. E. Marquis.** 1970. Electrochemical interactions in cell walls of gram-positive cocci. *J. Bacteriol.* **101:**92–101.

17. **Scheraga, H. A.** 1980. Protein folding: application to ribonuclease, p. 261–268. *In* R. Juenicke (ed.), *Protein Folding.* Elsevier/North-Holland Publishing Co., Amsterdam.

18. **Shockman, G. D.** 1959. Reversal of cycloserine inhibition by D-alanine. *Proc. Soc. Exp. Biol. Med.* **101:**693–695.

19. **Thor, C. J. B., and W. F. Henderson.** 1940. The preparation of alkali chitin. *Am. Dyest. Rep.* **29:**461–464.

20. **Thwaites, J. J., and N. H. Mendelson.** 1985. Biomechanics of bacterial walls: studies of bacterial thread made from *Bacillus subtilis*. *Proc. Natl. Acad. Sci. USA* **82:**2163–2167.

21. **Tilby, M. J.** 1977. Helical shape and wall synthesis in a bacterium. *Nature* (London) **266:**450–452.

22. **Weis-Fogh, T.** 1960. A rubber-like protein in insect cuticle. *J. Exp. Biol.* **37:**889–906.

23. **Zaritsky, A., and N. H. Mendelson.** 1984. Helical macrofiber formation in *Bacillus subtilis*: inhibition by penicillin G. *J. Bacteriol.* **158:**1182–1187.

III.   WALL STRUCTURE AND BIOSYNTHESIS

Chapter 11

Mass Spectrometry of Peptidoglycans

Stephen A. Martin

During the past 8 years, mass spectrometry (MS) has experienced a series of advances in instrumentation and experimental methodologies which have resulted in the capability to determine the primary structure of a wide range of biological molecules. These current and developing techniques have pushed MS to the forefront of primary structural characterization, providing complementary information to the more established techniques such as gas-phase Edman degradation and two-dimensional nuclear magnetic resonance.

One of the most important advances in methodology was the introduction of fast atom bombardment (FAB) MS by Barber et al. in 1980 (1, 2). FAB MS is one of a series of "soft ionization" techniques in which large, thermally labile samples can be ionized without the extensive chemical derivatization required in more classical ionization modes, such as electron and chemical ionization. The most abundant sample-related ions in FAB MS are associated with the molecular weight of the component(s) of interest, typically $(M + H)^+$; in addition, there are a series of very abundant ions characteristic of the liquid matrix in which the sample is dissolved. In FAB MS it is possible to observe ions below the mass of the protonated molecular ion which are structure specific. However, in these examples a large amount of material is required, ~5 to 15 nmol/μl of matrix, and the sample must consist of only a single component. If the sample is a mixture or only a small amount of material is available, or both, only molecular weight information can be obtained. The com-

Stephen A. Martin • Department of Chemistry, Massachusetts Institute of Technology, Cambridge, Massachusetts 02139 (on leave of absence from the Department of Cell and Molecular Pharmacology and Experimental Therapeutics, Medical University of South Carolina, Charleston, SC 29425).

bination of matrix background and predominantly molecular weight information from the sample limits the extent to which fragment ions associated with the primary structure of an unknown compound can be assigned. This is the major limitation of FAB MS. To overcome this limitation, tandem MS (MS/MS) has been coupled with FAB MS. In MS/MS the sample ions of interest are selected in the first mass spectrometer and transmitted into the collision cell containing helium gas, and the products of this collision are detected with the second mass spectrometer. The products of the collisions are fragment ions associated with the primary structure of the compound, enabling its sequence to be deduced.

During the past 2 years, the application of high-performance MS/MS for the structural characterization of biological molecules has been refined primarily with the sequence determination of peptides and proteins (3, 5, 8a). A series of observations concerning the manner in which the primary structure can be deduced from the tandem mass spectra has resulted from these investigations. This information is applicable to a wide range of compound classes including peptidoglycan, glycoconjugates, and oligonucleotides.

Integrated Approach to Structural Analysis of Peptidoglycan Monomers

Recently, FAB MS (6, 10) and MS/MS (11), in conjunction with high-performance liquid chromatography (HPLC), have been applied to the determination of the structures of a series of peptidoglycan monomers from a variety of bacteria. This approach offers several advantages over classical techniques: it allows structural characterization from mixtures and location and identification of any structural modification, and it is unaffected by N-terminal blocking groups and requires only a few days for the partial separation by HPLC and full characterization of the intact peptidoglycan monomer by MS and MS/MS. Employing these procedures, a series of structure-function studies related to the somnogenicity have been carried out (6, 8). The integrated approach of HPLC, MS, and MS/MS offers a powerful means to identify all compounds of a particular preparation and to isolate and characterize those components which are responsible for the observed response in the system under study.

Application with Selected Samples

Materials. The *Chalaropsis* and anhydro monomers from *Neisseria gonorrhoeae* RD5 and FA19 were provided by R. S. Rosenthal of the Department of Microbiology and Immunology, Indiana University School of Medicine, Indianapolis, as were the *Chalaropsis* dimers from *N. gonorrhoeae*. Peptidoglycan dimers from *Actinomadura* sp. strain R39 were provided by M. Guinand of the Department of Biochemistry and Microbiology, University Claude Bernard, Villeurbanne, France. The human insulin was supplied by Eli Lilly & Co., Indianapolis, Ind., and the molecular ion region of the thioredoxin of *Chromatium vinosum* is from the work of Johnson and Biemann (5).

HPLC. HPLC separation of the unreduced peptidoglycan monomers and dimers was accomplished with a Waters (Millipore Corp., Bedford, Mass.) binary pumping system consisting of two model 510 pumps, a gradient controller, a model 441 single-wavelength detector monitoring 214 nm, and a model 740 integrator. The samples were separated with either C_4 or C_{18} (4 by 250 mm) reversed-phase Vydac columns (The Separations Group, Hesperia, Calif.). The elution program was linear, typically ranging from 100% H_2O (containing 0.05% CF_3COOH) to 25% CH_3CN (containing 0.035% CF_3COOH) over a period of 30 min at a flow rate of 1.0 ml/min.

Sample preparation. The success of the analysis of a sample by FAB MS is very dependent on the manner in which the sample is prepared (9). The sample should be free from all buffers and salts as these components can suppress or mask the sample-related ions. Typically, these types of impurities are removed via column chromatography, i.e., Sep-Pak, molecular sieves, or reversed-phase HPLC using volatile buffers such as CF_3COOH. Once the buffers have been removed, the sample is dissolved in water-miscible solvents to which a liquid matrix is added. The liquid matrix is required to observe long-lasting secondary ion signals associated with the species of interest. The compounds under investigation must dissolve in the matrix, and the matrix should have a vapor pressure low enough that it remains a liquid under high vacuum (133 × 10^{-6} Pa) for several minutes. Glycerol, which is the most widely used matrix for the analysis of biological molecules, was employed in these studies of peptidoglycan monomers. The sample is placed in 1.0-ml conical vials (Sarstedt Inc, Princeton, N.J.) with glycerol–5% aqueous acetic acid (5:1, vol/vol). Sample concentrations range from 0.01 to 15 nmol/μl of matrix with a total volume of 3 to 5 μl. Although the samples must be free from all salts and buffers, mixtures of several components of similar structure can be analyzed and all components can be identified. The sample preparation and analysis methodologies are summarized in Fig. 1.

FAB MS. A JEOL HX110/HX110 high-performance tandem mass spectrometer was used in the present studies (Fig. 2) (12). This instrument consists of two identical high-resolution double-focusing mass spectrometers containing an electric field for energy focusing and a magnetic field for momentum analysis. Both the first mass spectrometer, MS-1, and the second, MS-2, have a mass range of 14,500 daltons at an accelerating potential of 10 kV. Either

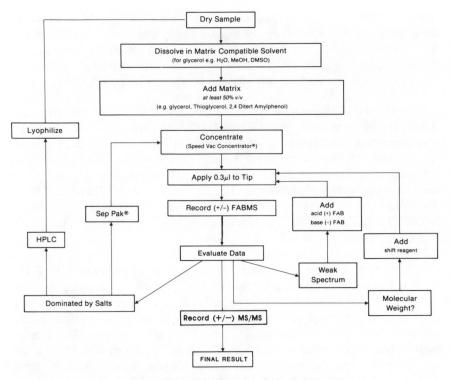

Figure 1. Sample preparation and analysis methodologies.

instrument can be operated as a single double-focusing mass spectrometer or they can be operated in the tandem mode with an EBEB geometry. The interface region between the two mass spectrometers consists of a series of lenses which focus the precursor ion beam into the collision cell. The collision cell is also located in the interface region, and the type employed in this work is an integral part of ion source two and is, therefore, operated at ground potential.

In the normal two-sector mode of operation, approximately 0.5 μl of the sample, dissolved in a liquid matrix (see above), is deposited via a syringe to the tip of a strainless-steel sample stage affixed to the end of a vacuum-sealed rod. The probe tip is placed under vacuum and inserted into the center of the ionization region where a primary beam of xenon atoms from a JEOL ion/ neutral beam source (10 mA, 6 kV) strikes the liquid matrix containing the sample. The momentum transferred in the interaction of the primary xenon beam with the sample results in the ejection from the probe surface of a wide range of particles including positive and negative ions. Either the positive or negative ions can then be selected for analysis, depending on the accelerating potential of the mass spectrometer. The ions sputtered off the probe tip are

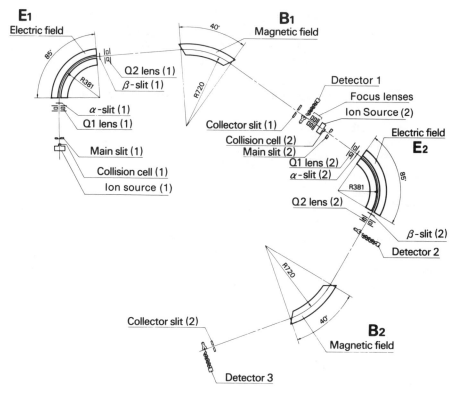

Figure 2. Ion optics of the JEOL HX110/HX110 high-performance tandem mass spectrometer.

accelerated to 10-kV translational energy, focused, energy and mass analyzed, and detected with a 16-stage secondary electron multiplier.

As discussed above, the major species detected in a normal two-sector mass spectrum are ions corresponding to the molecular weight of the samples and provide very little of the primary structure information which is necessary to determine the nature of the compound. Furthermore, the matrix contributes significantly to the mass spectrum, thereby complicating the identification of any ions which may be associated with the structure of the molecule. To increase the amount of primary structure information, MS/MS has been combined with FAB MS. In this mode the ion for which structural information is required, the precursor, is selected in MS-1 and transmitted into the collision region with unit resolution. The precursor ions which have 10 kV of translational energy collide with helium gas in the chamber, and a portion of the energy from the collision is converted to vibrational modes of freedom, resulting in the molecule's breaking apart into smaller structural components, product ions. The product ions exit the collision region and are detected by

MS-2. Since the precursor ion is selected with unit resolution, i.e., only ions of a single mass entered the collision region, all product ions analyzed by MS-2 must be related to the precursor and thus to its structure.

Molecular Weight Determination by Mass Spectrometry

The JEOL HX110/HX110 tandem mass spectrometer (Fig. 2) can be operated both as two double-focusing mass spectrometers or in the MS/MS mode as described above. The mass range, resolution, sensitivity, and mass assignment accuracy of its two-sector mode of operation are demonstrated in Fig. 3. In Fig. 3a the fully resolved molecular ion region from a few nanomoles of human insulin, $(M + H)^+$ m/z 5804.6, is shown. Each peak is 1 dalton in width, and the relative distribution of the peaks corresponds to the isotopic distribution predicted for the elemental composition $C_{257}H_{384}N_{65}O_{77}S_6$. The mass assignment accuracy is ± 0.5 dalton in this mass range. The high-mass-range capability of this instrument is demonstrated in Fig. 3b, in which the unresolved molecular ion region of the thioredoxin of *C. vinosum* (5) is plotted. The observed mass, based on the data system calibration, is within 2 daltons

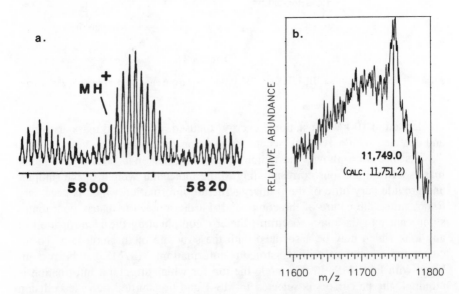

Figure 3. Mass range and resolution capabilities of the JEOL HX110/HX110. (a) FAB mass spectra of a single unit-resolved scan of the molecular ion region of human insulin. The $(M + H)^+$ is 5804.6 and the relative distribution of peaks separated by one mass unit represents the isotope distribution for a compound of elemental composition $C_{257}H_{384}N_{65}O_{77}S_6$. (b) Unresolved isotopic multiplet of the thioredoxin $(M + H)^+$ ion of *C. vinosum*. Mass value corresponds to the center of the peak (5).

of the calculated mass for this protein, providing confirmation that the correct thioredoxin has been characterized. Finally, the specificity for identifying the $(M + H)^+$ ions from two different peptidoglycan dimers which differ by 2 daltons, because of the mass difference resulting from the substitution of two COOH groups by two $CONH_2$ moieties, is shown in Fig. 4.

Although the high-mass and high-resolution capability of a single double-focusing mass spectrometer is noteworthy and by itself solves a number of important problems, the unique aspect of the JEOL HX110/HX110 is its MS/MS capabilities. To highlight some of the limitations of a normal two-sector mass spectrum and the advantages of MS/MS for primary structure characterization or verification, an MS mass spectrum and a series of tandem mass spectra from *N. gonorrhoeae* RD5 are discussed below.

Primary Structure Characterization by Tandem Mass Spectrometry

The first step in acquiring information on a particular compound or mixture of compounds is the determination of their molecular weights by recording a two-sector mass spectrum such as the one shown in Fig. 5 for a mixture from *N. gonorrhoeae* RD5. The mass spectrum is dominated over most of the mass range by a series of abundant ions (labeled with asterisks) which are due to the glycerol matrix, as well as a constant chemical background which results in ions of low intensity at every mass, indicated by the dark, continuous background. At the high-mass end a series of $(M + H)^+$ ions characteristic of peptidoglycan monomers from strain RD5 (Table 1) are observed. It would be impossible to relate any fragment ions observed below the molecular ion region

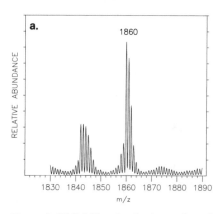
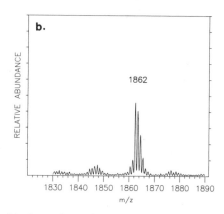

Figure 4. FAB MS molecular ion region of peptidoglycan dimers from (a) *Actinomadura* sp. strain R39 and (b) *N. gonorrhoeae* in which the structural difference responsible for the 2-dalton mass shift of the $(M + H)^+$ ion is the replacement of two $CONH_2$ groups in panel a by two COOH moieties in panel b.

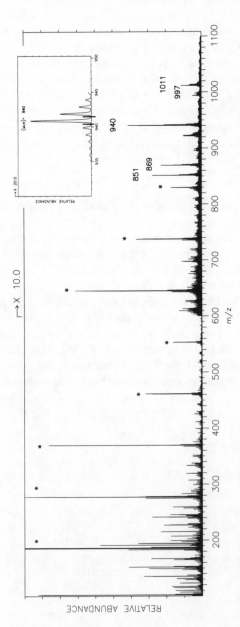

Figure 5. Normal FAB mass spectrum of an unseparated preparation of *Chalaropsis* monomers from *N. gonorrhoeae* RD5 (Table 1). The glycerol cluster ions are labeled with asterisks. The insert is a plot of the expanded region around $(M + H)^+$ m/z 940 to demonstrate the resolution of this mass spectrum. The vertical lines bracket the 1-dalton window containing $(M + H)^+$ m/z 940 which has been selected for MS/MS (Fig. 6 and 7).

Table 1. Structures of *Chalaropsis* and anhydro monomers from
N. gonorrhoeae RD5 determined by FAB MS/MS

$(M + H)^+$ observed[a]	Primary structure[b]
Chalaropsis[c]	
697	GlcNAc-MurNAc-Ala-Glu
851[d]	GlcNAc-(1,6-anhydro)MurNAc-Ala-Glu-A$_2$pm
869	GlcNAc-MurNAc-Ala-Glu-A$_2$pm
926	GlcNAc-MurNAc-Ala-Glu-A$_2$pm-Gly
940[d]	GlcNAc-MurNAc-Ala-Glu-A$_2$pm-Ala
997	GlcNAc-MurNAc-Ala-Glu-A$_2$pm-Ala-Gly
1,011	GlcNAc-MurNAc-Ala-Glu-A$_2$pm-Ala-Ala
Anhydro[c]	
679	GlcNAc-(1,6-anhydro)MurNAc-Ala-Glu
719	(1,6-anhydro)MurNAc-Ala-Glu-A$_2$pm-Ala
851[d]	GlcNAc-(1,6-anhydro)MurNAc-Ala-Glu-A$_2$pm
908	GlcNAc-(1,6-anhydro)MurNAc-Ala-Glu-A$_2$pm-Gly
922[d]	GlcNAc-(1,6-anhydro)MurNAc-Ala-Glu-A$_2$pm-Ala
979	GlcNAc-(1,6-anhydro)MurNAc-Ala-Glu-A$_2$pm-Ala-Gly
993	GlcNAc-(1,6-anhydro)MurNAc-Ala-Glu-A$_2$pm-Ala-Ala

[a]$(M + H)^+$ determined by FAB MS.
[b]GlcNAc, *N*-Acetylglucosamine; MurNAc, *N*-acetylmuramic acid; A$_2$pm, diaminopimelic acid.
[c]Separated by HPLC into single-component fractions for structure-function studies (6, 8).
[d]Verified by high-resolution exact mass measurement (11).

to any one of these $(M + H)^+$ ions because of the glycerol background and the presence of several components. When only a single compound is present in the FAB mass spectrum and its primary sequence is known or postulated on the basis of homology, it is possible to assign low-mass ions to various structural components. However, it is not possible to do this with samples of unknown structure.

Once the normal mass spectrum has been recorded and the masses of the peaks of interest have been calculated by the data system, the precursor ion of interest is selected with unit resolution and transmitted into the collision cell. In the insert of Fig. 5, the one-mass-unit region bracketed by the two vertical lines corresponding to $(M + H)^+$ *m/z* 940, GlcNAc-MurNAc-Ala-Glu-A$_2$pm-Ala, has been transmitted into the collision region. The resulting product ion spectrum of $(M + H)^+$ *m/z* 940 is shown in Fig. 6, along with its corresponding structure and mass assignments based on the interpretation of the tandem mass spectrum. There are several differences between this mass spectrum and the normal two-sector mass spectrum (Fig. 5), the most obvious being the absence of ions associated with the glycerol matrix, especially at low mass. In Fig. 5, only glycerol ions are observed at low mass. In addition, the signal-to-noise ratio is increased in the tandem mass spectrum, and most important, because only a single mass was transmitted into the collision cell, all the prod-

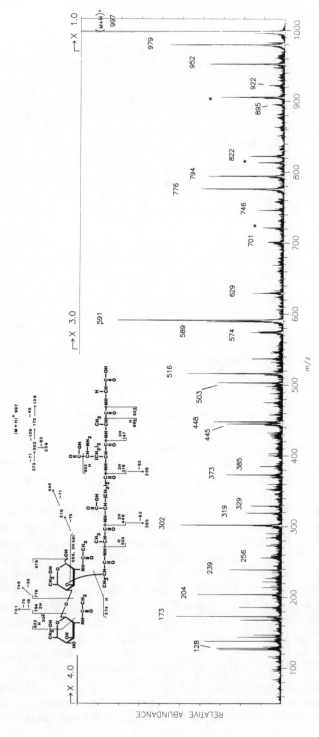

Figure 6. FAB MS/MS mass spectrum of GlcNAc-MurNAc-Ala-Glu-A₂pm-Ala, (M + H)⁺ *m/z* 940, and its structure and associated sequence ions observed in the MS/MS spectrum. The numerical values refer to the mass of the fragments produced by cleavage along the bonds indicated plus 1 or 2 if the notation H or 2H indicates hydrogens transferred to the species which is observed in the mass spectrum.

138

uct ions in the mass spectrum are related to the structure of the precursor. As a result of this high degree of selectivity and the mass assignment accuracy, ±0.3 dalton, reliable structural assignments can be made in the interpretation of the tandem mass spectra. In Fig. 7, the tandem mass spectrum of Fig. 6 is replotted so that it shows the relationship among the product ions to demonstrate the manner in which these spectra are interpreted. The most abundant

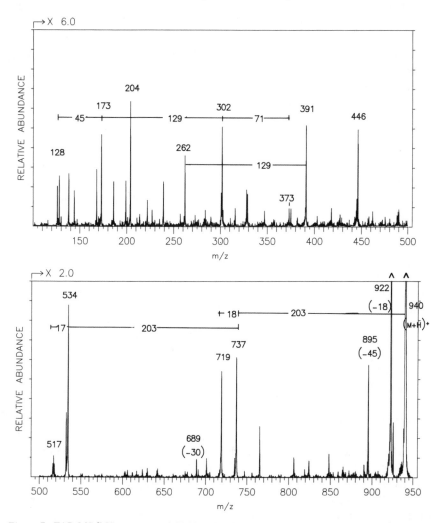

Figure 7. FAB MS/MS spectrum of GlcNAc-MurNAc-Ala-Glu-A₂pm-Ala, (M + H)⁺ *m/z* 940, from Fig. 6 expanded to illustrate the manner in which these mass spectra are interpreted. The numerical values bracketed by vertical lines indicate the mass difference between the two peaks connected by the vertical line, i.e., (M + H)⁺ 940 − 203 = *m/z* 737. Values in parentheses correspond to the mass difference between the labeled peak and a peak at higher mass.

ion corresponds to the precursor ion remaining after the collision process, i.e., $(M + H)^+$. The peak at m/z 922 corresponds to the loss of H_2O from $(M + H)^+$ and indicates the presence of a free C_1–OH group. The loss of 203 daltons resulting in m/z 737 and the subsequent loss of 18 mass units (m/z 719) corresponds to the cleavage on either side of the glycosidic oxygen and the loss of GlcNAc. Furthermore, the C_6 group of MurNAc is identified by the loss of 30 daltons from m/z 719 (i.e., m/z 689). The subsequent loss of 203 mass units from m/z 737, yielding m/z 534 and m/z 517 (m/z 737 − 220), characterizes the loss of the second sugar, MurNAc, resulting in an ion species containing only the peptide portion, m/z 517. Since each amino acid has a unique mass except Leu-Ile and Lys-Gln, the peptide portion can be characterized in the same manner as the carbohydrate moieties. For example, the mass difference from m/z 391 to m/z 262, 129 daltons, corresponds to the loss of Glu, and the fragment ion series m/z 373, 302, 173, 128 corresponds to the loss of 71, 129, and 45 daltons, identifying an internal peptide of the sequence -Ala-Glu-A_2pm-.

In this manner, the remainder of the ions in the tandem mass spectrum are assigned to structural components. As diagrammed in the structure in Figure 6, the vertical lines indicate cleavage of that bond with charge retention on either the carbohydrate or C-terminal peptide portion of the molecule, as indicated by the direction of the horizontal portion of the line below the mass value. The numerical values refer to the mass of the fragments produced by cleavage along the bonds indicated, plus 1 or 2 if the notation H or 2H indicates hydrogens transferred to the species which is observed in the mass spectrum. For example, the ion at m/z 391 in the structure in Fig. 6 results from cleavage of the amide linkage and the addition of two hydrogens to yield a C-terminal fragment ion of mass 391. This fragment, as well as m/z 262, loses an additional 63 mass units corresponding to the loss of H_2O (−18) from the C-terminal amino acid and the epsilon COOH group (−45) from A_2pm. If the epsilon carbon of A_2pm is an amide, then a corresponding loss of 44 daltons is observed for a total loss of 62 daltons, identifying the site of amidation.

On the basis of the interpretation of a large number of peptidoglycan monomers, it is possible to rapidly characterize slight structural differences, such as three peptidoglycan monomers which differ only in modifications to the MurNAc moiety. Three compounds which illustrate this point are GlcNAc-(1,6-anhydro)MurNAc-Ala-Glu-A_2pm-Ala, $(M + H)^+$ m/z 922; GlcNAc-MurNAc-Ala-Glu-A_2pm-Ala, $(M + H)^+$ m/z 940; and GlcNAc-(O-acetyl)MurNAc-Ala-Glu-A_2pm-Ala, $(M + H)^+$ m/z 982. The high-mass portion of their tandem mass spectra is shown in Fig. 8a–c, respectively. Comparison of Fig. 8a and b indicates that these two compounds differ by 18 daltons (H_2O), and the location of this modification is assigned to the C_1 of MurNAc on the basis of the low abundance of the $(M + H)^+ - 18$ ion (i.e., 904) in Fig. 8a and the very high abundance of $(M + H)^+ - 18$ (m/z 922) in Fig. 8b. Fur-

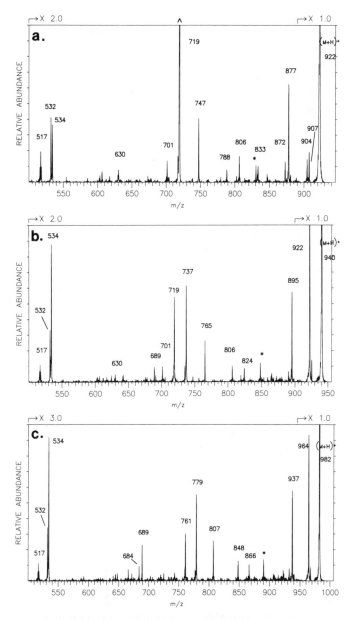

Figure 8. Comparison of the high-mass end of the FAB MS/MS mass spectra from the (M + H)⁺ of (a) GlcNAc-(1,6-anhydro)MurNAc-Ala-Glu-A₂pm-Ala, *m/z* 922, (b) GlcNAc-MurNAc-Ala-Glu-A₂pm-Ala, *m/z* 940, and (c) GlcNAc-(*O*-acetyl)MurNAc-Ala-Glu-A₂pm-Ala, *m/z* 982. In all three examples the peptide portion of the MS/MS spectra represented by *m/z* 534, 532, and 517 is the same. The structural modifications are associated with the MurNAc moiety.

thermore, all ions associated with the carbohydrate portion of the molecule in Fig. 8a are shifted down in mass by 18 daltons while those ions which contain only the peptide portion, i.e., m/z 534, 532, and 517, are the same in Fig. 8a and b, indicating that the peptide portions are identical. A comparision of Fig. 8b and c indicates a mass difference of 42 daltons between $(M + H)^+$ m/z 940 and 982. As in Fig. 8a and b there are abundant ions at m/z 534, 532, and 517, indicating that the peptide portion of the molecule is identical in all three components. The abundant ion due to the loss of 18 daltons from $(M + H)^+$ m/z 982, i.e., 964, indicates, as in Fig. 8b, that the C_1 group is $-OH$. The location of the modification is based on the ion at m/z 689, which is the same in both Fig. 8b and c and corresponds to the loss of the N-terminal sugar and the R group of the C_6 position of MurNAc. In Fig. 8b the loss from m/z 719 to m/z 689 is 30 daltons (CH_2O), and in Fig. 8c the loss from m/z 761 to m/z 689 is 72 daltons ($CHOCOCH_3$). Therefore, the modification is an O-acetyl group on the C_6 position of MurNAc.

The low-mass portion of the tandem mass spectrum displays several ions which are important in characterizing both the carbohydrate and peptide portions of the molecule. For example, the ion at m/z 173 (Fig. 5) corresponds to A_2pm which cyclizes to form the ion at m/z 128. If the molecule contained

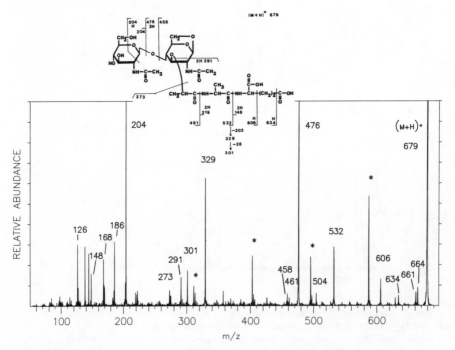

Figure 9. FAB MS/MS spectrum of GlcNAc-(1,6 anhydro)MurNAc-Ala-Glu, $(M + H)^+$ m/z 679, and its structure and associated sequence ions observed in the MS/MS spectrum.

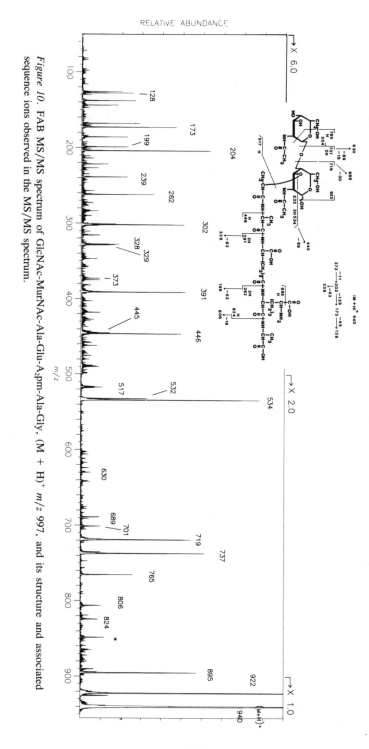

Figure 10. FAB MS/MS spectrum of GlcNAc-MurNAc-Ala-Glu-A₂pm-Ala-Gly, (M + H)⁺ *m/z* 997, and its structure and associated sequence ions observed in the MS/MS spectrum.

143

A_2pm-NH$_2$, the ions would be m/z 172 and m/z 127, respectively. In this manner, rapid verification of the components of the molecule's structure is possible. The tandem mass spectrum of another peptidoglycan monomer, shown in Fig. 9, lacks ions at m/z 173/172 and m/z 128/127 and thus indicates that there is no A_2pm/A_2pm-NH$_2$ in this compound. Furthermore, the ions at m/z 203, 186, and 168 indicate that the molecule contains carbohydrate moieties. As in Fig. 8a, the absence of a very abundant ion at m/z 661, $(M + H)^+ -$ 18, indicates that this compound contains a 1,6-anhydro sugar. Therefore, on the basis of these observations and the assignment of the remaining ions in the MS/MS spectrum, the compound with $(M + H)^+$ m/z 679 is identified as GlcNAc-(1,6-anhydro)MurNAc-Ala-Glu (Fig. 9).

When this technique is employed, minor components containing modified structures, each representing less than 1% of the total sample preparation, can be identified and fully characterized. One such modification is the addition of glycine. In strain RD5 the amino acid analysis consistently reported the presence of glycine, and the normal FAB mass spectrum confirmed molecular ions corresponding to a peptidoglycan containing an additional 57 daltons, the mass of a glycine residue. The tandem mass spectrum confirmed that it was glycine and assigned its position in the structure (Fig. 10).

Conclusion

The combination of FAB MS for molecular weight determination coupled with MS/MS provides a means to characterize or verify the primary structure of biological molecules such as peptidoglycan. Structural assignments including 1,6-anhydro sugars, sites of O acetylation, and location of amino acids such as glycine in the primary sequence can be made with a high degree of confidence. The selection of a single precursor ion, i.e., $(M + H)^+$, with one-mass-unit specificity in MS-1 and ± 0.3 dalton mass accuracy of the product ion mass spectrum ensures the ability to distinguish two compounds differing by COOH versus CONH$_2$ and accurately to assign all product ions. Furthermore, the interpretation of these modified peptidoglycans is based on the interpretation of more than 20 peptidoglycan compounds from several different strains of bacteria provided by a number of different laboratories, in addition to extensive experience in the characterization of pure peptides (4). With this technique, in conjunction with HPLC, it is possible to fully characterize nanomole quantities of a sample in a few days, providing a very valuable addition to the current methods for the characterization of low-molecular-weight peptidoglycan.

ACKNOWLEDGMENT. This work was supported by Public Health Service grant RR00317 from the National Institutes of Health Division of Research Resources (to Klaus Biemann).

LITERATURE CITED

1. **Barber, M., R. S. Bordoli, G. J. Elliot, R. D. Sedgwick, and A. N. Tyler.** 1982. Fast atom bombardment mass spectrometry. *Anal. Chem.* **54:**645a–657a.
2. **Barber, M., R. S. Bordoli, R. D. Sedgwick, and A. N. Tyler.** 1981. Fast atom bombardment of solids (F.A.B): a new ion source for mass spectrometry. *J. Chem. Soc. Chem. Commun.* **7:**325–327.
3. **Biemann, K.** 1986. Mass spectrometric methods for protein sequencing. *Anal. Chem.* **58:**1288a–1300a.
4. **Biemann, K., and S. A. Martin.** 1987. Mass spectrometric determination of the amino acid sequence of peptides and proteins. *Mass Spectrom. Rev.* **6:**1–76.
5. **Johnson, R. S., and K. Biemann.** 1987. The primary structure of thioredoxin from *Chromatium vinosum* determined by high performance tandem mass spectrometry. *Biochemistry* **27:**1209–1214.
6. **Krueger, J. M., D. Davenne, J. Walter, A. Shoham, S. L. Kubillus, R. S. Rosenthal, S. A. Martin, and K. Biemann.** 1987. Bacterial peptidoglycans as modulators of sleep. II. Effects of muramyl peptides on the structure of rabbit sleep. *Brain Res.* **403:**259–266.
7. **Krueger, J. M., M. L. Karnovsky, S. A. Martin, J. R. Pappenheimer, J. Walter, and K. Biemann.** 1984. Peptidoglycans as promoters of slow-wave sleep. II. Somnogenic and pyrogenic activities of some naturally occurring muramyl peptides; correlations with mass spectrometric structure determination. *J. Biol. Chem.* **259:**12659–12662.
8. **Krueger, J. M., R. S. Rosenthal, S. A. Martin, J. Walter, D. Davenne, A. Shoham, S. L. Kubillus, and K. Biemann.** 1987. Bacterial peptidoglycans as modulators of sleep. I. Anhydro forms of muramyl peptides enhance somnogenic potency. *Brain Res.* **403:**249–258.
8a. **Martin, S. A., and K. Biemann.** 1987. A comparison of fast atom bombardment mass spectra obtained with a two-sector mass spectrometer vs. a four-sector instrument. *Int. J. Mass Spectrom. Ion Processes* **78:**213–228.
9. **Martin, S. A., C. E. Costello, and K. Biemann.** 1982. Optimization of experimental procedures for fast atom bombardment mass spectrometry. *Anal. Chem.* **54:**2362–2368.
10. **Martin, S. A., M. L. Karnovsky, J. M. Krueger, J. R. Pappenheimer, and K. Biemann.** 1984. Peptidoglycans as promoters of sleep-promoting factor isolated from human urine. *J. Biol. Chem.* **259:**12652–12658.
11. **Martin, S. A., R. S. Rosenthal, and K. Biemann.** 1987. Fast atom bombardment mass spectrometry and tandem mass spectrometry of biologically active peptidoglycan monomers from *Neisseria gonorrhoeae*. *J. Biol. Chem.* **262:**7514–7522.
12. **Sato, K., T. Asada, M. Ishihara, F. Kunihiro, Y. Kammei, E. Kubota, C. E. Costello, S. A. Martin, H. A. Scoble, and K. Biemann.** 1987. High performance tandem mass spectrometry. Calibration and performance of linked scans of a four-sector instrument. *Anal. Chem.* **59:**1652–1659.

Chapter 12

Cell Wall Proteins in *Bacillus subtilis*

R. E. Studer
D. Karamata

It is widely recognized that the relatively thick cell wall of gram-positive rods is not an inert shell. However, little is known about what goes on between the moment of the insertion of wall polymers and their release into the medium as trichloroacetic acid-soluble products. Until now, with the exception of the autolysins, no enzyme or protein at the wall level has been shown to participate in the complex mechanism of cell wall extension. Although chemical analyses of purified walls have repeatedly revealed the presence of variable amounts of proteinaceous material (2, 5), no definite evidence regarding the nature of its association with the walls of *Bacillus subtilis* has been obtained.

We report analysis of purified native walls from *B. subtilis,* which have reproducibly revealed sodium dodecyl sulfate (SDS)-extractable proteins, and we propose to examine whether any of these proteins play a role in cell wall geometry. In particular, we have sought to assess whether they contain a "protease-sensitive factor," supposed to be involved in macrofiber orientation (3), by analysis of mutants with a fixed positive or negative orientation. Since macrofibers can be obtained only on autolysin-deficient strains, we first examined cell walls of the *flaD2* (*lyt-2*) parent strain (6), which is known to be 90 to 95% deficient in autolytic activity.

Preparation of Cell Walls and Identification of Cell Wall Proteins

Cells were grown at different temperatures (20 and 46°C) in tryptose broth (900 ml), with forced aeration. Late-exponential-phase cells were collected by

R. E. Studer and D. Karamata • Institut de Génétique et Biologie Microbiennes, 1005 Lausanne, Switzerland.

centrifugation (10 min, 13,000 × g), washed once with tryptose broth, and stored frozen at $-70°C$.

Cells were thawed, suspended in water (15 ml), and broken by shaking (2 to 3 min) with glass beads (45 g) under liquid CO_2 cooling. Beads were eliminated by filtration on sintered glass, and walls were separated from cytosol, unbroken cells, and other debris by repeated (usually three) cycles of centrifugation at low (5 min, 3,000 × g) and high (15 min, 27,000 × g) speed. "Native walls" were then washed with 0.9% NaCl (twice) and water (three times), freeze-dried, and stored at $-20°C$. All operations were carried out at 0 to 4°C, in the presence of 0.1 mM phenylmethylsulfonyl fluoride. Freeze-dried walls (1 to 2 mg) were extracted twice with sample buffer containing SDS (3 min, 100°C). Extracted proteins were separated on SDS-polyacrylamide gel electrophoresis (12% acrylamide) and detected by staining with Coomassie blue.

Characterization of Cell Wall Proteins

Gel electrophoresis of SDS extracts of native walls has consistently revealed a set of at least 12 well-defined bands (Fig. 1). Differences, essentially at the qualitative level, were found when cells were grown in different media

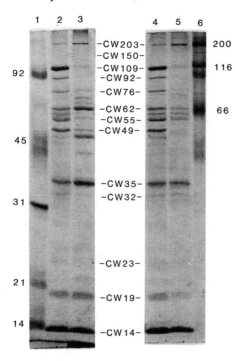

Figure 1. Characterization of cell wall proteins in *B. subtilis* and its *flaD2* (*lyt-2*) mutant. Purified native walls were extracted with SDS-containing buffer, and the proteins were separated on SDS-polyacrylamide gel electrophoresis (12% acrylamide). Lanes 1 and 6, Molecular size standards (kilodaltons); lanes 2 and 4, wild type at 20°C; lane 3, wild type at 46°C; lane 5, *flaD2* (*lyt-2*) at 20°C.

(not presented) as well as at different temperatures (see Fig. 1). Proteins in-
ventoried here correspond to those present on all electrophoretograms when
cultures were grown at lower temperature (20 to 34°C). The cell wall proteins
were designated according to their apparent molecular weight as CW203, 150,
109, 92, 76, 62, 55, 49, 35, 32, 23, 19, and 14. Minor bands, present in
particular conditions, were not reported. Visual inspection of one-dimensional
gels strongly suggests that cell wall proteins were absent from both the soluble
and the cytoplasmic membrane fractions (not presented). Examination of the
culture fluid (not presented) revealed five (or six) major bands, mainly in the
range of 40 to 70 kilodaltons. All but one of them (203 kilodaltons) appeared
clearly distinct from cell wall proteins on SDS-polyacrylamide gel electrophoresis, showing that cell wall proteins are not exocellular proteins trapped on
their way through the wall. Finally, comparison of cell wall proteins with penicillin-binding proteins of *B. subtilis,* on the basis of known molecular weights,
revealed no similarity (1).

Preliminary Search for Possible Functions of Cell Wall Proteins

In a preliminary search for possible functions of these proteins, we screened
certain mutants affected in cell wall metabolism or geometry. Since the helix
phenotype can be studied only on lytic-deficient strains, we first examined the
autolysin-deficient mutant *flaD2* (Fig. 1). While having a lower overall protein
content, *flaD2* walls appear to be devoid of bands corresponding to CW109,
92, 76, and 49. Absence of CW49, 76, and 92 is consistent with autolytic
deficiency, since molecular weights of these bands correspond to the amidase,
its activator, and the glucosaminidase, respectively (4, 7). However, the function of CW109, also absent from *flaD2* walls, remains unknown.

To examine whether any of the proteins identified above could be involved
in the orientation of helical macrofibers, we compared cell wall proteins of
mutants fixed in either positive or negative orientation.

Under the growth conditions used, the parent strain exhibits a thermosensitive helix phenotype; i.e., negative orientation occurs at high temperature
(46°C), whereas the positive one is characteristic of low temperature (20°C).
Negative orientation has been associated with the synthesis of a protease-sensitive factor (3), suggesting that fibers locked in this orientation are characterized, at both temperatures, by the presence of a given protein, whereas fibers
with the positive orientation should be devoid of that protein.

Electrophoretograms (Fig. 2) do not reveal any clear difference between
strains K5 and K8 (negative and positive orientation, respectively) at either
low or high temperature, suggesting that none of the above-identified cell wall
proteins corresponds to the hypothetical protease-sensitive factor. However, in
preliminary experiments (not presented) we had observed that protein CW55

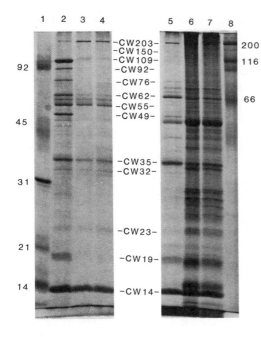

Figure 2. Cell wall proteins of fiber-producing strains. Fibers with fixed helix orientation were grown at high and low temperature. Cell wall proteins were obtained as described in the legend of Fig. 1. Lanes 1 and 8, Molecular size standards (kilodaltons); lanes 2 and 5, wild type at 20 and 46°C; lanes 3 and 6, strain K5 (negative orientation) at 20 and 46°C; lanes 4 and 7, strain K8 (positive orientation) at 20 and 46°C.

occasionally showed the behavior expected for the hypothetical factor involved in microfiber orientation. Unfortunately, the very sensitivity of CW55 to protease, which makes it an excellent candidate for a substance involved in helix orientation, complicated this study; i.e., its high instability led to poor reproducibility.

Preliminary observations reported here revealed the existence of a distinct set of cell wall proteins which include the previously characterized autolytic enzymes. No convincing evidence was obtained for the involvement in fiber geometry of any cell wall proteins. A more complete characterization of these proteins will be reported elsewhere.

LITERATURE CITED

1. **Blumberg, P. M., and J. L. Strominger.** 1972. Five penicillin-binding components occur in *Bacillus subtilis* membranes. *J. Biol. Chem.* **247:**8107–8113.
2. **Ellwood, D. C., and D. W. Tempest.** 1972. Effects of environment on bacterial wall content and composition. *Adv. Microb. Physiol.* **7:**83–117.
3. **Favre, D., D. Karamata, and N. H. Mendelson.** 1985. Temperature-pulse-induced "memory" in *Bacillus subtilis* macrofibers and a role for protein(s) in the left-handed-twist state. *J. Bacteriol.* **164:**1141–1145.
4. **Herbold, D., and L. Glaser.** 1975. *Bacillus subtilis* N-acetylmuramic acid L-alanine amidase. *J. Biol. Chem.* **250:**1676–1682.

5. **Mobley, H. T., J. Doyle, and L. K. Jolliffe.** 1983. Cell wall-polypeptide complexes in *Bacillus subtilis. Carbohydr. Res.* **116:**113–125.
6. **Pooley, H. M., and D. Karamata.** 1984. Genetic analysis of autolysin-deficient and flagellaless mutants of *Bacillus subtilis. J. Bacteriol.* **160:**1123–1129.
7. **Rogers, H. J., C. Taylor, S. Rayter, and J. B. Ward.** 1984. Purification and properties of autolytic endo-beta-N-acetylglucosaminidase and the N-acetylmuramyl-L-alaine amidase from *Bacillus subtilis* strain 168. *J. Gen. Microbiol.* **130:**2395–2402.

Chapter 13

Solubilization and Purification of the Glucosyltransferase Involved in the Biosynthesis of Teichuronic Acid by *Micrococcus luteus* Cell Membrane Fragments

Kim M. Hildebrandt
John S. Anderson

Teichuronic acid of the *Micrococcus luteus* cell wall is a polysaccharide consisting of alternating glucose and N-acetylmannosaminuronic acid residues (3). Biosynthesis of teichuronic acid has been studied by using enzymes present in cytoplasmic membrane fragments recovered from lysozyme digests of *M. luteus* cells (1, 5). These cytoplasmic membrane fragments contain glucosyltransferase which catalyzes the following reaction: UDP-glucose + acceptor → glucosyl-acceptor + UDP. P^1-N-Acetylmannosaminuronyl-(1,4)-N-acetylmannosaminuronyl-(1,3)-N-acetylglucosaminyl P^2-undecaprenyl diphosphate (component C) normally functions as the initial acceptor for the glucosyltransferase (1).

Assay of the Glucosyltransferase

Glucosyltransferase activity was determined with the following standard assay conditions: 0.7 mM UDP-[U-^{14}C]glucose (2 μCi/μmol), 50 mM HEPES (N-2-hydroxyethylpiperazine-N'-2-ethanesulfonic acid) (pH 8.2), 20 mM mag-

---

Kim M. Hildebrandt and John S. Anderson • Department of Biochemistry, University of Minnesota, St. Paul, Minnesota 55108.

151

nesium acetate, 5 mM 2-mercaptoethanol, 9 to 35 mg of acceptor per ml, and 0.15 to 15 mg of enzyme protein per ml with incubation at 37°C. Radioactive product was separated from residual substrate by descending paper chromatography in isobutyric acid–1 M NH_4OH (5:3, vol/vol). Product was quantitated by liquid scintillation counting of the radioactivity retained at origins (except when component C was the acceptor, in which case the product had a mobility of R_f 0.55).

Protein concentrations were determined by the A_{230}/A_{260} spectrophotometric method (2) as well as with bicinchoninic acid (4).

Alternative Acceptor for Glucosyltransferase

Although the formation of component C is relatively simple (incubation of cytoplasmic membrane fragments with UDP-*N*-acetylglucosamine and UDP-*N*-acetylmannosaminuronic acid), its isolation and use as a glucosyltransferase acceptor are fraught with difficulty. Not only is component C labile, but the amount needed for routine assays during glucosyltransferase purification is prohibitive. These considerations prompted the search for alternative acceptors.

Studies of teichuronic acid biosynthesis by wall-membrane preparations indicated that teichuronic acid chains linked to peptidoglycan were suitable acceptors for glucosyltransferase (6). These teichuronic acid-peptidoglycan complexes were recovered from lysozyme digests of *M. luteus* cells by fractionation on DEAE-cellulose with a linear salt gradient. Of the two peaks of anthrone-positive material eluted, the first peak (pool 1) was identified as mainly peptidoglycan, with relatively little teichuronic acid present; conversely, the second peak eluted (pool 2) contained predominantly teichuronic acid. Both preparations demonstrated glucosyltransferase acceptor capacity.

Purified teichuronic acid derived from cell walls proved to be the most effective acceptor. Cell walls were obtained from mechanically disrupted cells sequentially treated with DNase and RNase, sodium decyl sulfate, and sodium bicarbonate. Washed cell walls were lyophilized and subsequently hydrolyzed at 8 mg/ml in 50 mM HCl at 90°C. Teichuronic acid was released from walls predominantly during the first 60 min of treatment; as determined by ¹H-nuclear magnetic resonance spectroscopy, this teichuronic acid was devoid of peptidoglycan. With continued acid treatment of the cell wall residue, peptidoglycan fragments comprised an increased proportion of the solubilized material. The best glucosyltransferase acceptor capacity was recovered from early hydrolysis fractions (i.e., less than 60 min). After 90 min, acceptor capacity was reduced by one-half, and it was completely lacking after extensive hydrolysis (Fig. 1).

A comparison of acceptor capacities for several different acceptors is shown in Fig. 2. Teichuronic acid, devoid of peptidoglycan, clearly is the best acceptor available, showing 2-, 4- and 10-fold greater acceptor capacity relative

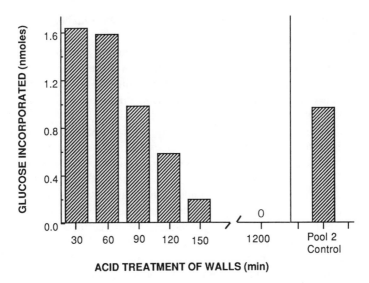

Figure 1. Teichuronic acid acceptor activity as released from walls by acid. Incorporation of glucose was measured with teichuronic acid used as acceptor at 9 mg/ml; in the control panel, pool 2 was used as acceptor, also at 9 mg/ml. All reactions were incubated for 60 min.

to that of pool 2, pool 1, and component C, respectively. In addition to demonstrated high acceptor capacity, teichuronic acid has the advantages of relative ease of preparation, stability during extended storage, and high solubility in the reaction mixtures used for glucosyltransferase assays.

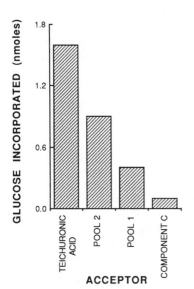

Figure 2. Alternative acceptors of glucosyltransferase. Incorporation of glucose was measured with acceptors used at 9 mg/ml (except component C with which conditions approached saturation). All reactions were incubated for 60 min.

Detergent Solubilization and Purification of the Glucosyltransferase

Purification of glucosyltransferase from crude cytoplasmic membrane fragments was initiated by detergent extraction. A variety of detergents were evaluated for efficacy of solubilization, maintenance of enzyme activity upon storage, absence of interference in protein determinations, and lack of sulfhydryl oxidizing contaminants. Of the detergents studied (Triton X-100, Nonidet P-40, Tween 20, Brij 58, and Thesit), only Thesit (dodecylalcohol polyoxyethylene ether) fell within our criteria for extraction parameters. Thesit was found to give optimal extraction of glucosyltransferase between 1 and 10%. A 3% solution was chosen as the working concentration. Optimal glucosyltransferase extraction appeared to be dependent on the reduction of magnesium ion concentration below 1 mM in cytoplasmic membrane fragments. After extraction with Thesit, 20 mM magnesium acetate, 15% glycerol, and 2 mM 2-mercaptoethanol were promptly added to stabilize enzymatic activity.

The second step of purification involved ammonium sulfate precipitation at 70% saturation. Glucosyltransferase activity could not be reproducibly fractionated away from contaminating proteins with small increases of ammonium sulfate concentration. Consequently, this precipitation step has been used primarily to concentrate the solubilized enzyme.

After dialysis, the resolubilized enzyme was applied to a DEAE-cellulose column and eluted with stepwise and linear NaCl gradients in column buffer (20 mM Tris [pH 7.4], 15% glycerol, 0.75% Thesit, 2 mM 2-mercaptoethanol, and 20 mM magnesium acetate) (Fig. 3). Several different preparations of the solubilized enzyme fractionated on DEAE-cellulose gave variable glucosyltransferase recoveries with a 50- to 200-fold increase in specific activity.

Analysis by sodium dodecyl sulfate-polyacrylamide gel electrophoresis has shown that protein homogeneity has not yet been achieved in the glucosyltransferase recovered from the DEAE-cellulose column. Additional purification attempts involving affinity chromatography have not yet eliminated all contaminating proteins or markedly increased the specific activity of the glucosyltransferase. Efforts to achieve a homogeneous preparation of the glucosyltransferase are continuing.

Conclusions

Teichuronic acid, released from purified cell walls by mild acid treatment, proved to be the most effective artificial acceptor in the in vitro assay system for glucosyltransferase. Glucosyltransferase was solubilized from cytoplasmic membrane fragments most effectively with Thesit. This detergent provided the highest recovery of enzymatic activity, allowed the longest retention of enzymatic activity during storage, and did not interfere with protein determina-

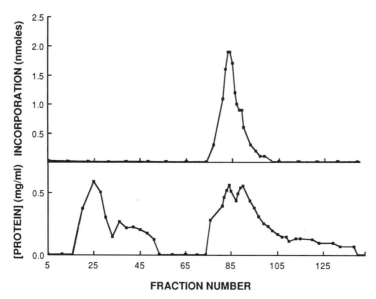

Figure 3. Fractionation of glucosyltransferase on DEAE-cellulose. Glucosyltransferase was prepared for fractionation by Thesit solubilization, ammonium sulfate precipitation, and dialysis. This material (900 mg of protein) was loaded onto a column (20 by 4.5 cm) of DEAE-cellulose equilibrated with column buffer and sequentially eluted with 1,000 ml of column buffer, 400 ml of 140 mM NaCl in column buffer, and an 800-ml linear gradient of column buffer from 175 to 625 mM NaCl at a flow rate of 72 ml/h at 4°C. Assay mixtures employed pool 2 (35 mg/ml) as acceptor and were incubated for 60 min.

tions. Purification of the solubilized glucosyltransferase included ammonium sulfate precipitation and DEAE-cellulose ion-exchange chromatography. The best preparations of glucosyltransferase obtained thus far have shown a 200-fold increase in specific activity.

ACKNOWLEDGMENT. This research was supported by Public Health Service grant 5 RO1-AI-08295 from the National Institutes of Health.

LITERATURE CITED

1. **Johnson, G. L., J. F. Hoger, J. H. Ratnayake, and J. S. Anderson.** 1984. Characterization of three intermediates in the biosynthesis of teichuronic acid of *Micrococcus luteus*. *Arch. Biochem. Biophys.* **235:**679–691.
2. **Kalb, V. F., and R. W. Bernlohr.** 1977. A new spectrophotometric assay for protein in cell extracts. *Anal. Biochem.* **82:**362–371.
3. **Perkins, H. R.** 1963. Polymer containing glucose and aminohexuronic acid isolated from the cell walls of *Micrococcus lysodeikticus*. *Biochem. J.* **86:**475–483.

4. **Smith, P. K., R. I. Krohn, G. T. Hermanson, A. K. Mallia, F. H. Gartner, M. D. Provenzano, E. K. Fujimoto, N. M. Goeke, B. J. Olson, and D. C. Klenk.** 1985. Measurement of protein using bicinchoninic acid. *Anal. Biochem.* **150:**76–85.
5. **Stark, N. J., G. N. Levy, T. E. Rohr, and J. S. Anderson.** 1977. Reactions of second stage of biosynthesis of teichuronic acid of *Micrococcus lysodeikticus* cell walls. *J. Biol. Chem.* **252:**3466–3472.
6. **Traxler, C. I., A. S. Goustin, and J. S. Anderson.** 1982. Elongation of teichuronic acid chains by a wall-membrane preparation from *Micrococcus luteus*. *J. Bacteriol.* **150:**649–656.

Chapter 14

Biophysics of Ampicillin Action on a Gas-Vacuolated Gram-Negative Rod

M. F. Suzanne Pinette
Arthur L. Koch

The surface stress theory (1, 2) postulates that cell growth modulates cell wall synthesis via stress developed in the murein of the bacterial cell wall. Higher stress favors autolysis of the existing stress-bearing wall needed for enlargement. Normally, secretion and cross-linking of non-stress-bearing peptidoglycan precede the autolytic step. The stress increases when the cell's turgor pressure rises as a result of the accumulation of cytoplasmic constituents during cell growth. Antibiotics which inhibit cell wall synthesis without blocking other phases of metabolism provide tools for studying this dynamic relationship between cytoplasm and cell wall growth.

When murein transpeptidases are blocked by penicillins, the turgor pressure inside the cell could rise as cytoplasmic growth continues. Alternatively, if the blockage of polymerization leads to a progressively weaker fabric, turgor pressure might gradually decrease as the wall expands or develops leaks, possibly due to residual autolytic activity that may not be fully inhibited. In the first case, turgor would rise toward the bursting point; in the second, turgor pressure would fall toward zero. We tested these hypotheses with the gas-vacuolated (i.e., vesiculated) gram-negative heterotroph *Ancylobacter aquaticus*. We used a 90-degree nephelometer to follow the collapse of the gas vesicles as external pressure was applied. We had previously (3) refined a method first developed by Walsby (5, 6) for measuring cell turgor pressure in organisms with gas vesicles.

M. F. Suzanne Pinette and Arthur L. Koch • Department of Biology, Indiana University, Bloomington, Indiana 47405.

It was found that the mean turgor pressure of growing cells decreased as a function of time of treatment with ampicillin. However, after 20 min the population of cells in ampicillin-treated cultures was heterogeneous with respect to turgor pressure. A computer curve-fitting process was developed to analyze the collapse curves. After 20 min of treatment, this analysis showed that the majority of the gas vesicles were either free or in cells with no turgor pressure, but the turgor pressure on the remaining 25% of the vesicles had increased from the initial value of 193 kPa to 297 kPa. Thus, two distinct effects of ampicillin on this gram-negative cell were found, suggesting that the original population was heterogeneous in its response to ampicillin.

Collapse Pressure Measurements

A. aquaticus M158 was obtained from A. E. Konopka. All batch cultures were maintained in balanced growth by successive dilution for at least 10 doublings before being used for any experiment. The cultures were restreaked from single cells at monthly intervals. Cultures were grown in CAGV medium and monitored as previously described (3, 4).

The 50% collapse pressure of the vesicles, C_a, in a population of cells was determined from the recorded $X - Y$ plots of relative light intensity due to gas vesicles versus externally applied pressure. Increasing the applied pressure gradually collapsed all the vesicles. The collapse pressure of individual cells was determined by viewing a cell adhering to a glass surface by phase microscopy, as described (4).

It was assumed that the vesicles were in a turgor-free environment after cell lysis by ampicillin or after treatment of cells with high concentrations of sucrose. In the former case, the wall had been breached; in the latter, the cell had become plasmolyzed and then the turgor pressure decreased to essentially zero. The mean collapse pressure of the vesicles, $\bar{C}$, can be estimated in two ways. One method measures C_a of intact cells after sucrose treatment (3). The second is to equate $\bar{C}$ with the C_a of component II vesicles. The results of both methods were comparable and were in the range of 480 to 495 kPa. The turgor pressure of untreated and component I vesicles was calculated by subtracting the observed collapse pressure from the collapse pressure observed with turgor-free vesicles.

Effect of a High Concentration of Ampicillin

In an earlier study (3), we made the initial observation that the collapse curve was biphasic after 20 min of ampicillin treatment (500 µg/ml). This was taken to be a reflection of the presence of two populations of cells in the growing culture with different responses to ampicillin attack on their cell walls. The group of vesicles collapsing at high pressures was designated component II (Fig. 1). These vesicles had been present in cells which had succumbed with loss of turgor pressure, presumably as a result of autolysis spurred by the ampicillin action. The other group of vesicles, designated component I, whose collapse pressure was much lower than normal suggested that ampicillin had caused an increase in the turgor pressure of the cells containing them. Presumably, these cells were still accumulating nutrients and making proteins while

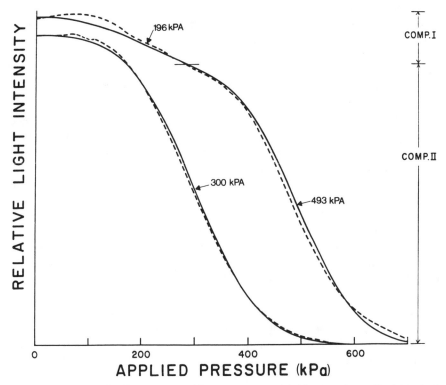

Figure 1. Analysis of collapse curves with a computer model. The experimentally determined collapse curve of an untreated culture of *A. aquaticus* is represented here as a sigmoidal curve (dotted line on left). The mid-point of the collapse curve gives a C_a of 300 ± 90 kPa, the pressure at which 50% of the vesicles in the sample collapsed. An experimental collapse curve obtained after 20 min of treatment with 500 μg of ampicillin per ml is shown by the dotted line on the right. The original data for this example have been published (3). The curve appears biphasic; i.e., the vesicles in the population are divided into two components by ampicillin activity. Component I has a lower than normal 50% collapse pressure of 196 kPa, while the collapse pressure of component II is higher, 493 ± 90 kPa, which is comparable to our previous measurements in turgor-free cells. The experimentally determined collapse curves were fitted by using a computer algorithm. The fitted curves (solid lines) were monophasic for untreated control samples and were composed of the sum of the integral of two normal distributions for ampicillin-treated samples.

their cell wall was prevented from growing by the antibiotic activity, but these cells did not develop leaks. Thus, the stress on the cell wall and the turgor increased to compensate for the increased osmotic pressure. Since the vesicles were under a higher cellular hydrostatic pressure than exists in normal growing cells, less applied external pressure was required to collapse them.

In the present studies we have confirmed our earlier results and have repeatedly observed a biphasic collapse curve with a component I fraction with a higher turgor pressure. This population appeared as a shoulder on the collapse curve of cultures treated with ampicillin. The biphasic character became obvious after a few minutes of ampicillin treatment but was most distinct after 15 min. At this time, the majority of cells had lost most of their phase contrast when examined by phase microscopy.

Computer Modeling of Biphasic Collapse Curves

To be sure that the resulting curves were actually the result of two distinct populations and to estimate the parameters characterizing these populations, we constructed a computer program which produced a biphasic curve by summing the integrals of two normal distributions. Model curves were produced by using only parameters for the means, standard deviations, and relative light intensities scattered by the two components. By trial and error, the six parameters were chosen so that the theoretical curves were superimposable with the observed experimental curves (see a typical example in Fig. 1). Using the same analysis on the untreated control, we found that a single cumulative normal distribution fit well. For the treated sample, accurate fits were possible only with the biphasic distribution model. This is strong evidence that cells under ampicillin treatment fall into two major distinct groups: one in which the turgor pressure increased above the turgor of untreated growing cells and one in which the turgor was reduced or completely removed even though the cell wall may or may not have yet disintegrated enough to liberate the vesicles into the medium.

Effect of Time and Drug Concentration

Changes over time in the relative proportions and parameters of the two components of the biphasic collapse curve of ampicillin-treated cultures were followed. Within the first few minutes only a single component of the collapse curve was discernible. Gradually, the curve became broader. After about 15 min, the biphasic character became evident. Then the breadth (standard deviation) of individual components returned to that of the original collapse curve. The amount of light refracted from component I vesicles increased gradually until about 20 min. After this time, the scattered light intensity did not change for several hours. Moreover, the 50% collapse pressure of this component did not change appreciably. Similar results were obtained when 10 μg of ampicillin per ml was used. Under these conditions this concentration is above that which stopped cell multiplication. The result (Table 1) is that again biphasic collapse curves develop with the same time course. Component II behaves in the same way, but component I is slightly different in that the turgor pressure rises more slowly and to a lesser extent.

Individual Collapse Pressures under Ampicillin Treatment

When the collapse of the vesicles in individual cells treated for about 20 min with 500 μg of ampicillin per ml was measured, the resulting collapse pressures also fell into two groups. For this kind of experiment, each cell was viewed in a phase microscope as the pressure was increased, and the pressure at which the light intensity just began to decrease was recorded. A histogram of these pressures formed a bimodal distribution. As expected, these pressures were lower than the 50% collapse pressures. There was a clear separation into two groups when 300 kPa was used as a dividing line. By dividing the histogram at this value, it was found that one group of cells lost refractility at a mean pressure of 176 ± 55 kPa ($n = 9$). The other group of cells lost refractility at a mean collapse pressure of 414 ± 80 kPa ($n = 15$). We assume that the former corresponds to component I and the latter corresponds to component II. From the number of cells involved, component I corresponds to 38% of the total. Because of the small number of cells examined, this is probably not different from the values measured nephelometrically, given in Table 1. Taken at its face value, the discrepancy between the two approaches could also be resolved if the two classes of cells have a different content of vesicles, with

Table 1. Turgor after ampicillin treatment

Ampicillin concn (μg/ml)	Component	Collapse pressure (kPa)	% of vesicles in component I
0 ($n = 4$)		283 ± 25[a]	
10 ($n = 8$)	I[b]	256 ± 11	25.4 ± 4.4
	II	481 ± 8	
500 ($n = 7$)	I	221 ± 12	24.0 ± 2.1
	II	490 ± 11	
		480[c]	

[a]Standard deviation.
[b]Measurement on samples taken between 15 and 35 min after ampicillin treatment. Data pooled from four independent cultures.
[c]Estimate of pressure to collapse turgor-free vesicles (from reference 3).

the cells under high turgor having more. Although the single-cell observations are independent confirmation of the bimodal population hypothesis, we were unable to discern a morphological basis for the difference between the two groups.

The Nature of the Two Components

With gas vesicles as a pressure probe, we have demonstrated that the hydrostatic pressure changes occur in different cells in two distinct and opposite ways. Since two components are sufficient to fit our data, it appears that both component I and component II are characterized by a uniform, but different, 50% collapse pressure. Both components are characterized by the same standard deviation, which is equal to that of the untreated control. Presumably, these standard deviation estimates all represent the inherent variability of the vesicles' resistance to collapse. Since the collapse pressure of component II approximates that of cultures of cells treated with high levels of sucrose (3), which eliminates the turgor, it can be concluded that the turgor pressure of the cells containing these vesicles is zero and that the loss of pressure is quite fast and complete. Except for the first 15 min of ampicillin treatment when the curve broadens, it has not been necessary either to fit the collapse curves to more than two components or to assume a standard deviation of the individual components larger than 90 kPa. Likewise, component I appears to be formed by a quite uniform process that allows the turgor pressure to increase to a maximum value in the first 20 min. At lower concentrations of ampicillin the turgor pressure reached the plateau as soon, but the increase in turgor pressure was not as great.

Initially, we tried a very high level of ampicillin in the expectation that the cell wall enlargement by insertion of new material would be totally halted, causing the pressure to increase in all the cells until they ruptured. Instead, the turgor pressure initially rose in one fraction and then remained constant, but in the majority of the cells it quickly fell to zero. The turbidity of the culture (measured after vesicles were collapsed) gradually declines as cell constituents leave the cell and are replaced by medium (data not shown). These two components were also evident when only 10 µg of ampicillin per ml was used. With this 50-fold smaller decrease in concentration, one might expect that a low level of wall enlargement should occur. This residual wall synthesis would have been expected, at least, to delay the response and alter the partition between the two components. But again, approximately 25% of the vesicles were in the component I class.

At this point, it is not clear how the two components are related, but we tend to think that they represent two physiologically different cellular responses: (i) component I, in which the cell does not become leaky and whose turgor pressure increases to a limit as the cells try to grow, and (ii) component II, in which the cell quickly becomes leaky, resulting in a loss of turgor pressure and of cell contents. The same biphasic character has been exhibited repeatedly after recloning. How this phenomenon relates to tolerance and resistance is also not clear. Perhaps this is a cell-cycle-dependent phenomenon. Perhaps the two responses are alternative strategies to overcome a common peril. In the latter case, perhaps this organism, even as a pure culture, has ways to respond so that different individuals employ quite opposite tactics (leaking or not leaking) in the hope that one or the other strategy will be successful.

ACKNOWLEDGMENT. This work was supported by Public Health Service grant GM-34222 from the National Institutes of Health.

LITERATURE CITED

1. **Koch, A. L.** 1983. The surface stress theory of microbial morphogenesis. *Adv. Microb. Physiol.* **24**:301–366.
2. **Koch, A. L.** 1985. How bacteria grow and divide in spite of internal hydrostatic pressure. *Can. J. Microbiol.* **31**:1071–1084.
3. **Koch, A. L., and M. F. S. Pinette.** 1987. Nephelometric determination of osmotic pressure in growing gram-negative bacteria. *J. Bacteriol.* **169**:3654–3663.
4. **Pinette, M. F. S., and A. L. Koch.** 1987. Variability of the turgor pressure of individual cells of the gram-negative heterotroph *Ancylobacter aquaticus*. *J. Bacteriol.* **169**:4737–4742.
5. **Walsby, A. E.** 1980. The water relations of gas-vacuolate prokaryotes. *Proc. R. Soc. London Ser. B* **208**:73–102.
6. **Walsby, A. E.** 1986. The pressure relationships of halophilic and non-halophilic prokaryotic cells determined by using gas vesicles as pressure probes. *FEMS Microbiol. Lett.* **39**:45–49.

Chapter 15

Processing of Nascent Peptidoglycan in *Gaffkya homari*

Francis C. Neuhaus
Paul W. Wrezel
Rabindra K. Sinha

In *Gaffkya homari* the in vitro processing of nascent peptidoglycan (PG) requires the regulated actions of at least three membrane-associated enzymes: DD-carboxypeptidase, LD-carboxypeptidase, and N^ϵ-(D-Ala)-Lys transpeptidase (8–10). The actions of the carboxypeptidases provide a population of tetrapeptide and tripeptide subunits in nascent PG which serve as required acceptors in the reaction catalyzed by the transpeptidase. In contrast to transpeptidases from other organisms, the enzyme from *G. homari* is essentially insensitive to inhibition by penicillin (8). Hammes (8) proposed that the target of penicillin action in this organism is the DD-carboxypeptidase.

Because of the unique target of penicillin action, several reviewers have placed *G. homari* in a class by itself (4, 17, 25). In this organism nascent PG is cross-linked to the preexisting PG with reverse polarity. This mechanism of cross-linking further distinguishes *G. homari* from other organisms. On the basis of our analyses of PG together with those from Nakel et al. (20), we suggest the schematic structure shown in Fig. 1. This PG matrix is 49% cross-linked by direct N^ϵ-(D-Ala)-Lys cross bridges (12). A unique feature of this PG, not previously recognized, is the presence of a significant fraction of unsubstituted *N*-acetylmuramyl (MurNAc) residues. Thus, in addition to the enzymes described above, we now report the participation of another in vitro processing enzyme, the MurNAc-L-alanine amidase.

Francis C. Neuhaus and Rabindra K. Sinha • Department of Biochemistry, Molecular Biology, and Cell Biology, Northwestern University, Evanston, Illinois 60208.   **Paul W. Wrezel** • Kraft, Inc., Glenview, Illinois 60025.

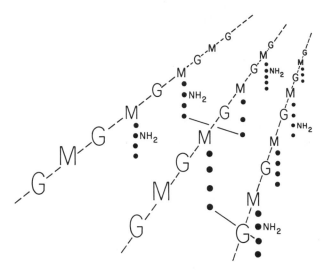

Figure 1. Schematic structure of PG from *G. homari.*

The objectives of these experiments were twofold: (i) to test the hypothesis proposed by Hammes (8) that the in vivo target of penicillin action in *G. homari* is DD-carboxypeptidase and (ii) to establish the role of MurNAc-L-alanine amidase in the in vitro processing of nascent PG.

In Vitro Target of Penicillin Action in G. homari

Benzylpenicillin acylates nine proteins (penicillin-binding proteins, PBPs) in membrane-walls from *G. homari* (Fig. 2). The molecular weights of these proteins range from 53,000 to 200,000. PBP 6 and PBP 9 account for 50% and 10% of the binding capacity, respectively. The identification of PBP 9 as a DD-carboxypeptidase was accomplished by direct enzyme assay (Fig. 2), β-lactam specificity studies (Table 1), and the correlation of concentrations of four β-lactams required to inhibit DD-carboxypeptidase activity by 50% (ID_{50}) with the concentrations required to bind 50% (B_{50}, ED_{50}) of PBP 9 (Table 1) (27).

Since PBP 9 is the DD-carboxypeptidase, it was possible to assess its participation in the synthesis of sodium dodecyl sulfate (SDS)-insoluble PG. The affinities of PBP 9 for benzylpenicillin, cefmenoxime, cephalothin, and cefoxitin were compared with the sensitivities of SDS-insoluble PG synthesis to these β-lactams (Table 1). These comparisons showed that the DD-carboxypeptidase is more sensitive to the action of each β-lactam than the synthesis of SDS-insoluble PG (Table 1). At concentrations which inhibited the DD-carboxypeptidase by 50%, the synthesis of SDS-insoluble PG was essentially un-

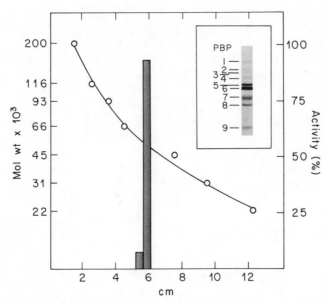

Figure 2. Histogram of DD-carboxypeptidase in sections of an SDS-polyacrylamide gel. The inset shows the nine PBPs (saturating conditions) in membrane-walls from *G. homari*. Reprinted with permission of the authors (27).

affected. Only when PBP 9 was bound by more than 50% with benzylpenicillin, cefmenoxime, cephalothin, or cefoxitin was inhibition of SDS-insoluble PG synthesis observed. At concentrations which inhibited the synthesis of SDS-insoluble PG by 50%, benzylpenicillin (2.0 μM), cefmenoxime (100 μM), and cephalothin (15 μM) acylated PBPs 1 through 8 by more than 95%. Therefore, PBPs 1 through 8 do not appear to play a role in the synthesis of SDS-insoluble PG. PBP 9, the DD-carboxypeptidase, appears to be the only PBP required for the synthesis of SDS-insoluble PG, and this enzyme functions in excess of the amount required for the synthesis of the insoluble polymer.

Table 1. Inhibition of DD-carboxypeptidase and SDS-insoluble PG synthesis by β-lactams in vitro[a]

β-Lactam	DD-Carboxypeptidase ID_{50}	PBP 9 B_{50} or ED_{50}[b] (μM)	SDS-insoluble PG synthesis ID_{50}
Cefoxitin	0.07	0.065	0.35
Benzylpenicillin	0.45	0.65	2.0
Cephalothin	3.5	2.5	15
Cefmenoxime	8.0	10	100

[a]Reprinted with permission of the authors (27).
[b]B_{50} is shown for benzylpenicillin, and ED_{50}s are shown for cefmenoxime, cephalothin, and cefoxitin.

In Vivo Target of Benzylpenicillin and Other β-Lactams in G. homari

The in vivo binding of [^{35}S]benzylpenicillin to PBPs was performed at low cell density by the method of Reynolds et al. (22). The B_{50}s were comparable to those established in the in vitro system. For benzylpenicillin, only PBPs 6 and 9 have B_{50}s which are similar to the minimal growth inhibitory concentration (MGIC) (Table 2). These results suggest that PBP 6, PBP 9, or both are candidates for the primary in vivo target of growth inhibition by benzylpenicillin.

Cefmenoxime, cephalothin, and cefoxitin were examined by the indirect competitive binding procedure (ED_{50}) to assess the relative importance of PBPs 6 and 9 as targets for growth inhibition by β-lactams. A comparison of the MGICs with the respective ED_{50}s for PBPs 6 and 9 indicated that PBP 9 is not a target of these antibiotics (Table 2). For example, the ED_{50}s of cefmenoxime and cephalothin for PBP 9 were 15- and 100-fold greater than their MGICs, respectively. On the other hand, the ED_{50} of cefoxitin for PBP 9 was one-hundredth of the MGIC. Thus, PBP 9 is eliminated as a primary target. In contrast, the ED_{50}s of the three β-lactams for PBP 6 correlated well with their respective MGICs. In addition, because of poor correlations, PBPs 1, 2, 4, 7, and 8 were also eliminated as primary candidates for targets of growth inhibition by these β-lactams. PBPs 3 and 5 could not be assessed in our in vivo binding studies (27). Thus, we suggest that PBP 6 is a primary target of growth inhibition by the cephamycin and the two cephalosporins.

In Vitro PG Processing as Measured by SDS Insolubility

Isolated membranes from *G. homari* synthesize SDS-insoluble PG poorly. By successive cycles of freezing and thawing (-196 versus $25°C$), these membranes can be reactivated for the synthesis of this polymer (13). In our exper-

Table 2. MGICs and affinities of the PBPs in *G. homari* for benzylpenicillin, cefmenoxime, cephalothin, and cefoxitin in vivo[a]

| β-Lactam | MGIC | B_{50} or ED_{50}[b] (μM) for: | | | | | | |
		PBP 1	PBP 2	PBP 4	PBP 6	PBP 7	PBP 8	PBP 9
Benzylpenicillin	0.25	<0.07	<0.07	<0.07	0.10	<0.07	<0.07	0.95
Cefmenoxime	0.25	0.10	0.10	0.07	0.30	0.05	0.20	>25
Cephalothin	0.50	0.04	0.03	0.05	0.25	0.09	0.04	7.5
Cefoxitin	5.0	0.40	0.30	0.40	5.0	0.20	0.03	0.05

[a]Reprinted with permission of the authors (27).
[b]B_{50}s are shown for benzylpenicillin, and ED_{50}s are shown for cefmenoxime, cephalothin, and cefoxitin.

iments with membranes, the time courses of cross-link formation did not correlate with those of SDS-insoluble PG synthesis (1). Insolubility of PG in 2% SDS has been used by several investigators to assess the cross-linkage of processed PG polymers (16, 24). Analyses of SDS-soluble PG synthesized by these membranes revealed significant cross-linking in this polymer, in some cases as high as that of SDS-insoluble PG. This comparison emphasized a lack of correlation between the concentration of cross-links in the PG and the amount of SDS-insoluble PG and suggested that additional features of processing may contribute to insolubility in 2% SDS. When the SDS-insoluble PG synthesized by reactivated membranes was labeled with [^{14}C]lysine and [^{3}H]*N*-acetylglucosamine ([^{3}H]GlcNAc), the muramidase digestion products contained less than the expected lysine-to-GlcNAc ratio of 1.0 (R. K. Sinha and F. C. Neuhaus, unpublished data). This observation indicated that a part of the glycan lacks peptide subunits containing [^{14}C]lysine. The action of either MurNAc-L-alanine amidase, γ-glutamyl-lysine endopeptidase, or both during the course of processing PG prior to its incorporation into SDS-insoluble PG is suggested. Determining whether the action of amidase provides the additional processing required for the formation of SDS-insoluble PG is the goal of the following experiments.

Action of MurNAc-L-Alanine Amidase in the Processing of Nascent PG

If MurNAc-L-alanine amidase is active in the processing and synthesis of SDS-insoluble PG by reactivated membranes, pseudodimers containing tetrasaccharide moieties as well as bis-disaccharide peptide dimer should be present in a muramidase (lysozyme) digest of this PG (Table 3). Tetrasaccharide pseudodimers are expected since lysozyme prefers to cleave at MurNAc residues bearing a peptide substituent (5). To detect the products of MurNAc-L-alanine amidase action, the high-performance liquid chromatography procedure devised by Glauner and Schwarz (6) was used to analyze the muramidase digestion products. This procedure was modified to include a prior Sephadex G-50–G-25 tandem separation of monomers and dimers (3). Figure 3 illustrates the separation of monomers isolated from the muramidase digest of [^{14}C]GlcNAc-labeled PG which had been synthesized by reactivated membranes (six cycles of freeze-thawing). The major radiolabeled products are the amidated and nonamidated disaccharide tetra- and pentapeptides. Relatively small amounts of radiolabeled amidated and nonamidated disaccharide tripeptides were detected. No evidence of amidase action was observed in the monomer fraction.

The chromatography of the [^{14}C]GlcNAc-labeled dimer fraction is illustrated in Fig. 4. Approximately 30 different peaks were detected in this separation. The tetrasaccharide and disaccharide moieties of these compounds were

Table 3. Summary of peptidoglycan cleavage products (muramidase)[a]

```
                                                          Dimers
     Monomers            Pseudodimers          True                        Truncated

  G-M      G-M      G-M-G-M   G-M-G-M      G-M       G-M-G-M        G-M
   |        |        |   |     |   |        |         |   |          |
   A        A        A   A     A   A        A         A   A          A
   |        |        |         |            |         |              |
   G      G-NH₂      G         G          G-NH₂     G-NH₂          G-NH₂
   |        |        |         |            |         |              |
   L        L        L       L-NH₂        L - A     L - A          L - A
   |        |        |         |          | |       | |            | |
   A        A        A         A         (A) L     (A) L          (A) L
   |        |        |         |               |         |              |
   A        A        A         A               G         G              G
                                               |         |              |
                                               A         A              A

  G-M      G-M      G-M-G-M   G-M-G-M      G-M        G-M
   |        |        |   |     |   |
   A        A        A   A     A   A
   |        |        |         |
   G      G-NH₂      G       G-NH₂
   |        |        |         |
   L        L        L         L      G-M        G-M-G-M       G-M-G-M
   |        |        |         |       |           |            |
   A        A        A         A       A           A            A
                                       |           |            |
                                     G-NH₂       G-NH₂        G-NH₂
                                       |           |            |
                                     L - A       L - A        L - A
                                     | |         | |          | |
  G-M      G-M                      (A) L       (A) L        (A) L
   |        |                        |           |            |
   A        A                      G-NH₂       G-NH₂          G
   |        |                        |           |            |
   G      G-NH₂                      A           A            A
   |        |                        |           |            |
   L        L                      G-M         G-M            G
```

[a]Only the amidated form is found in the intact cell (20). *G, N*-Acetylglucosaminyl; *M, N*-acetylmuramyl; A, L-alanyl or D-alanyl; G, D-isoglutamyl; G-NH₂, D-isoglutaminyl; L, L-lysyl.

identified after β-elimination by paper chromatography (23). The bis-disaccharide tetra-tetra dimer was eluted at 114 min, and its identification was based on a comparison with the standard compound isolated from the organism. In addition to the bis-disaccharide peptide, the dimer fraction contained four groups of dimer-related compounds. The first group of compounds (I), eluting between 35 and 52 min, was identified as amidated and nonamidated tetrasaccharide tetra- and pentapeptides (pseudodimers). The second (II, 78 to 88 min), third (III, 100 to 111 min), and fourth (IV, 112 to 132 min) groups were all characterized by the presence of tetrasaccharide peptides or complex combinations of tetrasaccharide and disaccharide peptides. On the basis of amino acid com-

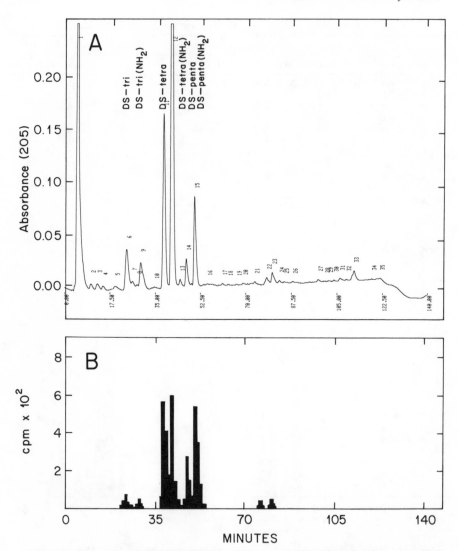

Figure 3. (A) Reverse-phase high-performance liquid chromatography of monomers from the muramidase digest of [^{14}C]GlcNAc-labeled PG synthesized by membranes freeze-thawed six cycles. (B) Profile of radiolabeled monomers.

position, it was proposed that groups II, III, and IV contain hexa-, hepta-, octa-, and nonapeptides. Such peptides would be consistent with those detected previously in this organism by Nakel et al. (20). The heterogeneity in each of these groups is based on the presence of both amidated and nonamidated compounds and tetrasaccharide-disaccharide peptide isomers. This unusually complex set of digestion products derived from the SDS-insoluble PG was also

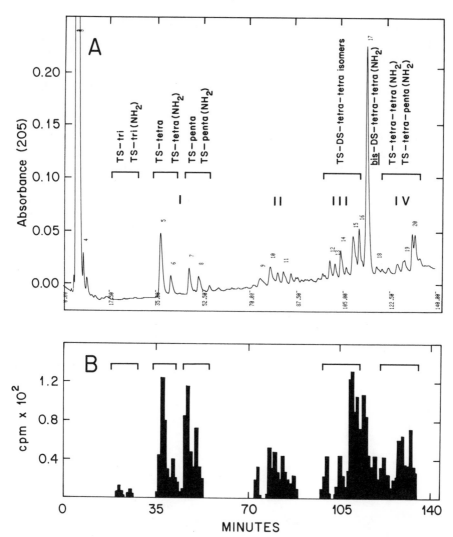

Figure 4. (A) Reverse-phase high-performance liquid chromatography of dimers from the muramidase digest of [^{14}C]GlcNAc-labeled PG synthesized by membranes freeze-thawed six cycles. (B) Profile of radiolabeled dimers.

observed in the peptidoglycan isolated from growing cells (Sinha and Neuhaus, unpublished data). Without prior separation of monomers and dimers by gel filtration on Sephadex G-50–G-25, it is likely that the pseudodimers (tetrasaccharide tetra- and pentapeptides) in the first group would have been missed.

In Table 3 the tentative structures of the fragments from the muramidase digest of SDS-insoluble PG are summarized. These products can be classified into four groups: (i) monomers (amidated and nonamidated disaccharide peptides), (ii) pseudodimers (amidated and nonamidated tetrasaccharide tetra- and pentapeptides), (iii) true dimers (bis-disaccharide tetra-tetra [tri] and tetrasaccharide disaccharide tetra-tetra [tri] isomers), and (iv) truncated dimers (tetrasaccharide tetra-tetra [tri]). Each of these modifications can be explained by the action of MurNAc-L-alanine amidase (Fig. 5).

The percentage composition of the muramidase digestion products of SDS-insoluble PG synthesized by membranes freeze-thawed for one cycle and six cycles is summarized in Table 4. Three important differences were observed: (i) a threefold decrease in the formation of tetrasaccharide tetrapeptide pseudodimers, (ii) the formation of a group of compounds containing a high proportion of tetrasaccharide hexapeptides, and (iii) a threefold increase in the amount of truncated dimers. From the changes in (ii) and (iii) it would appear that the process of freezing and thawing activates the MurNAc-L-alanine amidase in membranes and, thus, effects a significant change in the profile of digestion products.

Water-Soluble PG

One of the features of the in vitro processing system from *G. homari* is that penicillin and D-amino acids, which inhibit DD-carboxypeptidase and LD-carboxypeptidase, respectively, induce the release of water-soluble PG. The

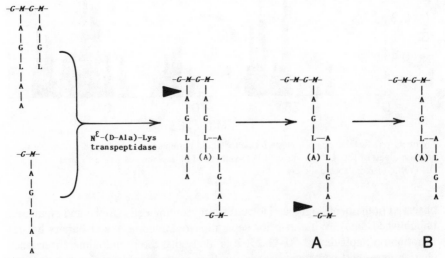

Figure 5. Proposed action of the MurNAc-L-alanine amidase in the formation of the tetrasaccharide disaccharide dimers (A) and truncated dimers (B). Arrowheads, MurNAc-L-alanine amidase.

Table 4. Composition of monomers and dimers in SDS-insoluble PG synthesized by membranes freeze-thawed one and six cycles

Product[a]	Composition (%)[b] from membranes freeze-thawed:	
	One cycle	Six cycles
Monomers (100%)		
DS-tri	11.8	2.8
DS-tri(NH$_2$)	7.2	3.1
DS-tetra	26.2	25.3
DS-tetra(NH$_2$)	29.2	29.6
DS-penta	11.7	12.0
DS-penta(NH$_2$)	13.9	27.2
Dimers (100%)		
TS-tri + TS-tri(NH$_2$)	7.9	4.6
TS-tetra + TS-tetra(NH$_2$)	35.2	13.7
TS-penta + TS-penta(NH$_2$)	19.8	16.0
TS-tetra-di[c]	0.0	11.0
TS-, DS-tetra-tetra(tri) isomers	20.7	16.7
bis-DS-tetra-tetra(NH$_2$)	8.7	12.5
TS-tetra-tetra(tri)(NH$_2$)	7.7	25.5

[a]DS-tri, Disaccharide-Ala-D-Glu-Lys, etc.; TS-tri, tetrasaccharide-Ala-D-Glu-Lys, etc.
[b]Percentages are based on radiolabeled products.
[c]Tentative identification

release of this PG has been studied in both membranes and membrane-walls of *G. homari* (21; Sinha and Neuhaus, unpublished data). With 10^{-4} M benzylpenicillin, as much as 60% of the SDS-insoluble PG synthesized by membrane-walls became water soluble (Fig. 6A). With freeze-thawed membranes 98% of the PG was released as water-soluble polymer at this concentration (data not shown). With 10^{-2} M D-methionine, all of the SDS-insoluble PG synthesized by membrane-walls became water soluble (Fig. 6B). The water-soluble PG in all cases contained only amidated and nonamidated monomers and, thus, was not cross-linked. No muramidase fragments containing tetrasaccharide were detected. This uncross-linked PG is essentially identical to that reported by others (2, 18, 26) which is secreted by intact cells in the presence of penicillin. The release of water-soluble PG in particulate enzyme fractions from *Escherichia coli* was first demonstrated by Izaki et al. (11). Whether the water-soluble PG contains undecaprenyl-phosphate or another lipophilic carrier involved in nascent PG synthesis has not been established (15). The inhibition of either LD-carboxypeptidase or DD-carboxypeptidase triggers a response leading to the release of nascent PG from the membrane. The mechanism for releasing this PG as a water-soluble fraction is not understood. The observation that the water-soluble glycan does not contain unsubstituted MurNAc residues, however, suggests that the action of the *G. homari* MurNAc-L-alanine amidase is only on the cross-linked polymer.

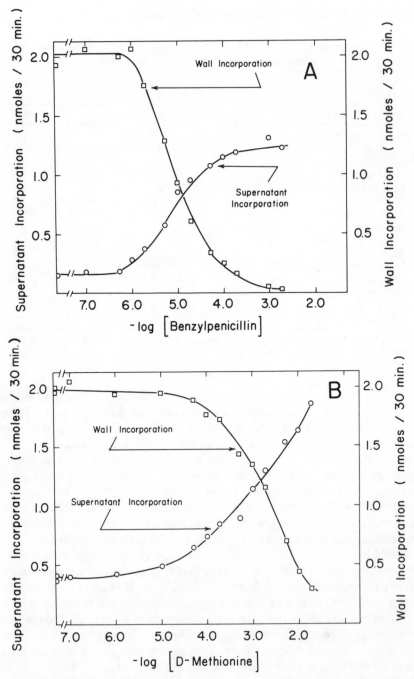

Figure 6. Formation of water-soluble PG by membrane-walls from *G. homari* in the presence of increasing concentrations of benzylpenicillin (A) and D-methionine (B) (C. E. Tobin and F. C. Neuhaus, unpublished data).

Conclusion

Our immediate goal is to understand the events in processing nascent PG catalyzed by the membrane-associated enzymes which function in the absence of the preexisting PG of the cell wall. Reactivated membranes from *G. homari* provide such a system. With an understanding developed at this level of organization, our ultimate goal is to address the more complex processing that exists in membrane-walls, toluene- and ether-permeabilized cells, and, finally, intact cells.

The enzymes in the flow diagram (Fig. 7) account for the synthesis of three classes of PG polymers: (i) SDS-insoluble PG, (ii) SDS-soluble PG, and (iii) water-soluble PG. Each class of polymers has its own unique physical features that reflect the action of these enzymes from *G. homari*. In earlier studies (1), we concluded that cross-linking is not the only feature required for SDS insolubility. Thus, we suggest that an additional feature(s) is necessary for the insolubility of PG in SDS.

It is proposed that the action of MurNAc-L-alanine amidase provides this additional feature required for the SDS insolubility of PG. Our results suggest that a prerequisite of amidase action is the cross-linking of the nascent polymer. Inhibition of cross-linking by either penicillin or D-amino acids renders the uncross-linked polymer insensitive to amidase action. Thus, to generate SDS-insoluble PG in an in vitro membrane system in the absence of walls requires both the synthesis of cross-links and the action of the amidase. It is our hypothesis that the action of this enzyme makes the cross-linked PG more chitinlike and, hence, insoluble in SDS. We estimate that 15 to 20% of the MurNAc residues in SDS-insoluble PG are unsubstituted.

These results may also provide a possible explanation why membrane-walls do not require freeze-thawing for activation (1, 13). With membrane-

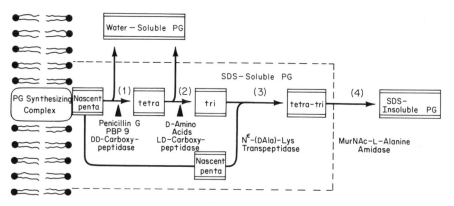

Figure 7. Proposed pathway of processing nascent PG in reactivated membranes from *G. homari*.

walls, covalent attachment of the SDS-soluble PG (cross-linked) to the preexisting PG of the wall renders the radiolabeled polymer SDS insoluble. Thus,
amidase action may not be necessary for the synthesis of SDS-insoluble PG by
membrane-walls.

The in vitro PG-synthesizing systems from *G. homari*, i.e., membranes
and membrane-walls, use a novel mechanism for designating the acceptor peptide in the cross-linking reaction. To accomplish this, DD-carboxypeptidase and
LD-carboxypeptidase form a population of tripeptide acceptors in the nascent
polymer. Utilizing these acceptors, the penicillin-insensitive transpeptidase cross-
links the glycan to either the donor pentapeptides of nascent PG (membranes)
or the donor pentapeptides of preexisting wall PG (membrane-walls). Inhibition
of either carboxypeptidase inhibits cross-linking. Thus, a population of pentapeptides must be converted to tripeptides to serve as acceptors in the reaction
catalyzed by the N^{ϵ}-(D-Ala)-Lys transpeptidase.

A variety of PG-synthesizing systems from *G. homari*, i.e., membranes,
membrane-walls, or toluene- and ether-treated cells, has been used to study
processing (unpublished data). Although these systems have retained many of
the features of PG processing found in the intact cell, a major question which
ultimately needs to be addressed is whether the processing events in each in
vitro system are identical to those of the cell (7, 14, 19). Two approaches to
address this question in the present work have used PBP analyses and muramidase digestion profiles of PG synthesized in vitro. In the in vitro system
the target of penicillin action is PBP 9, the DD-carboxypeptidase. In contrast,
the in vivo target is PBP 6. Cells with greater than 95% of PBP 9 in the bound
form show normal growth. The observation that PBP 9 is dispensable for growth
under laboratory conditions suggests that PG processing in intact cells is somewhat different from that observed in reactivated membranes. Our results indicate that MurNAc-L-alanine amidase functions in addition to DD-carboxypeptidase, LD-carboxypeptidase, and N^{ϵ}-(D-Ala)-Lys transpeptidase to process
nascent PG for the formation of SDS-insoluble PG in the absence of walls.
Thus, the further investigation of the roles of both PBP 9 and PBP 6 in the
intact cell is warranted.

In vitro systems which do not depend on preexisting wall PG, i.e., reactivated membranes, are essential for addressing the following questions. (i) Are
synthesis and elongation of nascent PG independent from cross-linking? (ii)
What are the intermediate PG polymers that are utilized for the synthesis of
SDS-insoluble PG? (iii) Is there a mechanism for designating acceptor peptides
in the reaction catalyzed by the N^{ϵ}-(D-Ala)-Lys transpeptidase? (iv) Is there a
system which monitors the degree of cross-linkage in nascent PG and triggers
its release as water-soluble PG in the presence of either penicillin or D-amino
acids? (v) In *G. homari* what are the functions of PBP 6 and PBP 9 in the
processing of PG? Reactivated membranes from *G. homari* may provide a system for answering these questions.

ACKNOWLEDGMENT. This study was supported by Public Health Service grant AI-04615 from the National Institute of Allergy and Infectious Diseases.

LITERATURE CITED

1. **Bardin, C., R. K. Sinha, E. Kalomiris, and F. C. Neuhaus.** 1984. Biosynthesis of peptidoglycan in *Gaffkya homari*: processing of nascent glycan by reactivated membranes. *J. Bacteriol.* **157**:398–404.

2. **Barrett, J. F., and G. D. Shockman.** 1984. Isolation and characterization of soluble peptidoglycan from several strains of *Streptococcus faecium*. *J. Bacteriol.* **159**:511–519.

3. **Dougherty, T. J.** 1985. Analysis of *Neisseria gonorrhoeae* peptidoglycan by reverse-phase, high-pressure liquid chromatography. *J. Bacteriol.* **163**:69–74.

4. **Frere, J. M., and B. Joris.** 1985. Penicillin-sensitive enzymes in peptidoglycan synthesis. *Crit. Rev. Microbiol.* **11**:299–396.

5. **Ghuysen, J.-M.** 1974. Substrate requirements of glycosidases for lytic activity on bacterial walls, p. 185–193. *In* E. F. Osserman, R. E. Canfield, and S. Beychok (ed.), *Lysozyme*. Academic Press, Inc., New York.

6. **Glauner, B., and U. Schwarz.** 1983. The analysis of murein composition with high-pressure-liquid chromatography, p. 29–34. *In* R. Hakenbeck, J.-V. Höltje, and H. Labischinski (ed.), *The Target of Penicillin: The Murein Sacculus of Bacterial Cell Walls, Architecture and Growth*. Walter de Gruyter & Co., Berlin.

7. **Goodell, E. W., Z. Markiewicz, and U. Schwarz.** 1983. Absence of oligomeric murein intermediates in *Escherichia coli*. *J. Bacteriol.* **156**:130–135.

8. **Hammes, W. P.** 1976. Biosynthesis of peptidoglycan in *Gaffkya homari:* the mode of action of penicillin G and mecillinam. *Eur. J. Biochem.* **70**:107–113.

9. **Hammes, W. P.** 1978. The LD-carboxypeptidase activity in *Gaffkya homari*: the target of the action of D-amino acids or glycine on the formation of wall-bound peptidoglycan. *Eur. J. Biochem.* **91**:501–507.

10. **Hammes, W. P., and H. Seidel.** 1978. The activities in vitro of DD-carboxypeptidase and LD-carboxypeptidase of *Gaffkya homari* during biosynthesis of peptidoglycan. *Eur. J. Biochem.* **84**:141–147.

11. **Izaki, K., M. Matsuhashi, and J. L. Strominger.** 1968. Biosynthesis of the peptidoglycan of bacterial cell walls (XIII). Peptidoglycan transpeptidase and D-alanine carboxypeptidase: penicillin-sensitive enzymatic reaction in strains of *Escherichia coli*. *J. Biol. Chem.* **243**:3180–3192.

12. **Jacob, G. S., J. Schaefer, and G. E. Wilson, Jr.** 1983. Direct measurement of peptidoglycan cross-linking in bacteria by ^{15}N nuclear magnetic resonance. *J. Biol. Chem.* **258**:10824–10826.

13. **Kalomiris, E., C. Bardin, and F. C. Neuhaus.** 1982. Biosynthesis of peptidoglycan in *Gaffkya homari*: reactivation of membranes by freeze-thawing in the presence and absence of walls. *J. Bacteriol.* **150**:535–544.

14. **Kraus, W., B. Glauner, and J.-V. Höltje.** 1985. UDP-*N*-acetylmuramylpentapeptide as acceptor in murein biosynthesis in *Escherichia coli* membranes and ether-permeabilized cells. *J. Bacteriol.* **162**:1000–1004.

15. **Mett, H., R. Bracha, and D. Mirelman.** 1980. Soluble nascent peptidoglycan in growing *Escherichia coli* cells. *J. Biol. Chem.* **255**:9884–9890.

16. **Mirelman, D.** 1979. Biosynthesis and assembly of cell wall peptidoglycan, p. 115–166. *In* M. Inouye (ed.), *Bacterial Outer Membranes: Biogenesis and Functions*. John Wiley & Sons, Inc., New York.

17. **Mirelman, D.** 1981. Assembly of wall peptidoglycan polymers, p. 67–86. *In* M. R. J. Salton and G. D. Shockman (ed.), β-*Lactam Antibiotics: Mode of Action, New Developments, and Future Prospects.* Academic Press, Inc., New York.

18. **Mirelman, D., R. Bracha, and N. Sharon.** 1974. Penicillin-induced secretion of a soluble, uncross-linked peptidoglycan by *Micrococcus luteus* cells. *Biochemistry* **13:**5045–5053.

19. **Mirelman, D., Y. Yashouv-Gan, and U. Schwarz.** 1976. Peptidoglycan biosynthesis in a thermosensitive division mutant of *Escherichia coli. Biochemistry* **15:**1781–1790.

20. **Nakel, M., J.-M. Ghuysen, and O. Kandler.** 1971. Wall peptidoglycan in *Aerococcus viridans* strains 201 Evans and ATCC 11563 and in *Gaffkya homari* strain ATCC 10400. *Biochemistry* **10:**2170–2175.

21. **Neuhaus, F. C., and W. P. Hammes.** 1981. Inhibition of cell wall biosynthesis by analogues of alanine. *Pharmacol. Ther.* **14:**265–319.

22. **Reynolds, P. E., S. T. Shephard, and H. A. Chase.** 1978. Identification of the binding protein which may be the target of penicillin action in *Bacillus megaterium. Nature* (London) **271:**568–570.

23. **Tipper, D. J.** 1968. Alkali-catalyzed elimination of a D-lactic acid from muramic acid and its derivatives and the determination of muramic acid. *Biochemistry* **7:**1441–1449.

24. **Tipper, D. J., and A. Wright.** 1979. The structure and biosynthesis of bacterial cell walls, p. 291–426. *In* I. C. Gunsalus, J. R. Sokatch, and L. N. Ornston (ed.), *The Bacteria: Mechanisms of Adaptation.* Academic Press, Inc., New York.

25. **Waxman, D. J., and J. L. Strominger.** 1983. Penicillin-binding proteins and the mechanism of action of β-lactam antibiotics. *Annu. Rev. Biochem.* **52:**825–869.

26. **Waxman, D. J., W. Yu, and J. L. Strominger.** 1980. Linear, uncross-linked peptidoglycan secreted by penicillin-treated *Bacillus subtilis*: isolation and characterization as a substrate for penicillin-sensitive D-alanine carboxypeptidase. *J. Biol. Chem.* **255:**11577–11587.

27. **Wrezel, P. W., L. F. Ellis, and F. C. Neuhaus.** 1986. In vivo target of benzylpenicillin in *Gaffkya homari. Antimicrob. Agents Chemother.* **29:**432–439.

IV. AUTOLYSIS AND PEPTIDOGLYCAN HYDROLASES

Chapter 16

Organization of the Major Autolysin in the Envelope of *Escherichia coli*

Joachim-Volker Höltje
Wolfgang Keck

Viability of *Escherichia coli* depends on the integrity of the rigid layer of the cell wall, the murein sacculus, with respect to mechanical strength and specific shape (10). Throughout growth and division, both have to be preserved. Numerous enzymes, murein hydrolases and synthetases, have to cooperate with one another in these processes. Only an efficient control of the enzymes in time and space guarantees the identical replication of the shape-maintaining giant macromolecule, murein. A specific distribution of the envelope-bound enzymes according to their functions among the cell wall substructures is, therefore, expected to be realized.

The cell wall of *E. coli* turns out to be a well-organized, complex structure. Outer and inner membranes define a specific compartment outside the cytoplasm, the periplasmic space, which may be subdivided further by periseptal annuli (17). Additional functional contacts between the two membranes seem to exist (1). The murein sacculus is embedded in the periplasm and is linked to the outer membrane via the lipoprotein anchor (24) and by strong associations with specific outer membrane proteins (23). Murein and outer membrane seem to constitute a joint structure withstanding the high osmotic pressure of the periplasm which is adjusted to the osmotic value of the cytoplasm by the synthesis of an osmoregulatory compound, the membrane-derived oligosaccharide (15).

Joachim-Volker Höltje • Max-Planck-Institut für Entwicklungsbiologie, Abteilung Biochemie, D-74 Tübingen, Federal Republic of Germany. **Wolfgang Keck** • Chemische Laboratoria der Rijksuniversiteit, 9747 AG-Groningen, The Netherlands.

We investigated in some detail the specific localization of the membrane-bound lytic transglycosylase (Mlt) (19) in the envelope of *E. coli*. This enzyme is unique in catalyzing an intramolecular muramyltransferase reaction concomitantly with the cleavage of the β-1,4-glycosidic bond between *N*-acetylmuramic acid and *N*-acetylglucosamine, yielding muropeptides with a 1,6-anhydromuramic acid (9). By this mechanism the enzyme is potentially able to totally degrade murein to low-molecular-weight products much the same as lysozyme does except that the released muropeptides are of the nonreducing 1,6-anhydro type. This lytic transglycosylase represents one of the most effective autolysins in *E. coli* and has to be carefully controlled by the cell. Two genetically distinct forms of the same specificity exist (9, 14, 19), one compartmentalized in the cytoplasm (9) and another present in the cell envelope (19). It is not known how the former enzyme is brought into action or what its specific function is. This communication deals with the membrane-bound lytic transglycosylase.

Several authors have reported on the localization of murein hydrolases in the envelope of *E. coli* (5, 6, 21, 25). An amidase which, in *E. coli,* seems not to function as an autolytic enzyme (21) has been found to reside in the outer membrane, and it has been speculated that this amidase is involved in the splitting process of the septum during cell separation (25). Another report claimed that the amidase is localized in the periplasm (21). Using a method for membrane separation based on trypsin-mediated detachment of the cytoplasmic from the outer membrane, the presence of murein lytic transglycosylases in both membrane fractions could be demonstrated (6). Interestingly, the activity could not be released completely, by Triton X-100 treatment, from the outer and intermediate membrane fraction but was easily released from the cytoplasmic membrane, indicating that this murein hydrolase is an integral component of the outer membrane.

A crude fractionation of the envelope complex is possible by isopycnic centrifugation of membrane vesicles that form after breakage of the cell (20). Instead of using the EDTA-lysozyme method of Osborn et al. (20), we opened the cells in a French press at a moderate pressure, as described (18). This enabled us to measure endogenous murein hydrolase activity in the separated membrane fractions, using an enzyme assay in which the release of soluble muropeptides from high-molecular-weight murein sacculi is determined (7). The protein profile of the sucrose gradient showed three peaks, designated H, M, and L (Fig. 1). All fractions were analyzed by sodium dodecyl sulfate-polyacrylamide gel electrophoresis (SDS-PAGE). The major outer membrane proteins were found exclusively in the fractions of protein peak H (Fig. 1). Determination of a protein specific for the cytoplasmic membrane, the D-lactate dehydrogenase, showed a specific distribution among the three membrane fractions, with most of the activity comigrating with the L fraction. A clean separation into outer and inner membrane and an additional membrane fraction of intermediate density (M) has, therefore, been accomplished.

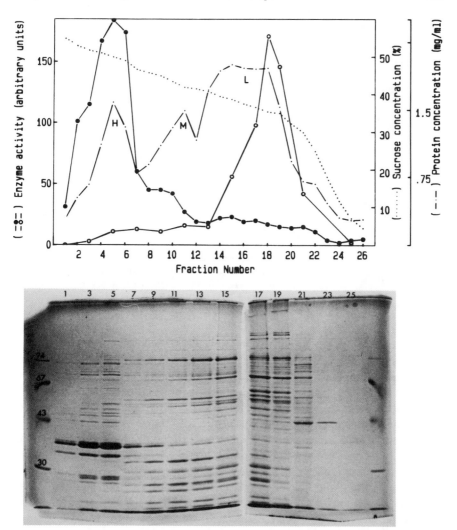

Figure 1. Separation of cytoplasmic and outer membrane by sucrose density centrifugation. (A) Protein concentration (–·–), D-lactate dehydrogenase activity (O), and lytic transglycosylase activity (●) are shown. (B) SDS-PAGE of the proteins in the various fractions.

To our surprise, lytic transglycosylase activity was predominantly found to be associated with the outer membrane fraction (Fig. 1). In contrast, the Mlt protein was found, by immunochemical staining on Western blots, to be present in equivalent amounts in the outer (H), intermediate (M), and inner (L) membrane fractions (Fig. 2). The difference in enzymatic activity of the Mlt protein, localized in either of the two membranes, is matched by a difference in the

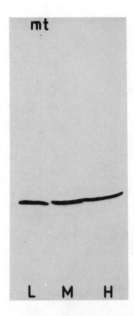

Figure 2. Distribution of the membrane-bound lytic transglyco-sylase protein among the membrane fractions obtained by sucrose density centrifugation as shown in Fig. 1. Membrane proteins (50 μg) were separated by SDS-PAGE. The Mlt protein was visualized on Western blots by immunochemical staining with the use of antiserum against purified Mlt protein.

isoelectric point (pI) of these two proteins. As shown by two-dimensional gel electrophoresis (Fig. 3), the protein present in the outer membrane displays a more acidic pI (pH 5.8) than the protein specific for the cytoplasmic membrane, which has a pI of pH 6.7. Thus, two forms of the Mlt exist, an OM form (in the outer membrane) and an IM form (in the inner membrane). Does this indicate a chemical modification of the lytic transglycosylase protein?

Attempts to demonstrate phosphorylation of the protein gave negative results. Also, neither of the two protein forms seems to be glycosylated, as is the case for one of the muramidases in *Streptococcus faecium* (13). It turned out, however, that the lytic transglycosylase can be released from the outer membrane very efficiently by a combination of Triton X-100 (0.4%) and KCl (2 M) and, at the same time, be transformed into the IM form, whereas solubilization of the Mlt protein by Triton X-100 alone releases the enzyme in the unchanged OM form (Fig. 4). Therefore, it seems unlikely that the transglycosylase protein is covalently modified. On the one hand, high salt concentrations convert one protein form into the other, whereas, on the other hand, the OM form is stable in the presence of 9.5 M urea during isoelectric focusing. This peculiar behavior is reminiscent of the features of a number of autolytic enzymes in gram-positive bacteria (3, 4, 22). It has been shown, in these cases, that the enzymes are strongly associated by multiple ionic interactions with murein. Chaotropic agents such as 8 M urea were unable to remove the enzymes from the walls (22), but high concentrations of CsCl, NH_4Cl, and LiCl did release the enzymes.

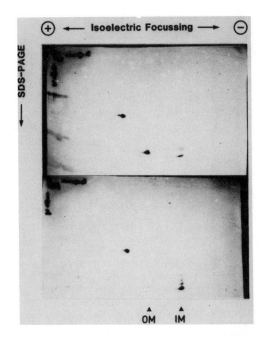

Figure 3. Analysis of the Mlt by two-dimensional gel electrophoresis of outer (top) and inner (bottom) membrane proteins (150 μm). Western blots were probed with antiserum against purified Mlt protein.

It is known that during separation of outer and inner membrane by sucrose gradient centrifugation the murein remains associated with the outer membrane. Thus, the lytic transglycosylase present in the outer membrane fraction may very well be bound to murein, and this would explain the necessity for high salt concentrations to efficiently solubilize the enzyme with Triton X-100. This indeed seems to be the case. In an experiment in which the unfractionated envelope from cells labeled with [^{3}H]diaminopimelic acid was analyzed by two-dimensional gel electrophoresis followed by fluorography, a radioactive spot was detected at the front line of the second dimension in the position of the pI of the OM form of the Mlt protein. Urea, used to dissociate the proteins for isoelectric focusing (first dimension), is likely to conserve an enzyme-[^{3}H]muropeptide complex that is focused at a more acidic pI than the free enzyme protein. During separation in the second dimension by SDS-PAGE this complex, however, dissociates and the [^{3}H]muropeptide migrates with the front. The chemical structure of the muropeptide has still to be analyzed, but we speculate that it is either a disaccharide or tetrasaccharide peptide. Being an exoenzyme, the lytic transglycosylase cleaves the glycan strands sequentially, starting at one chain end (2). The enzyme may stop at the other end with the ultimate disaccharide peptide structure still bound in its active site, which may stabilize the enzyme. This would be the OM form of the enzyme present in the outer membrane.

Overproduction of the Mlt protein in a strain harboring a runaway plasmid

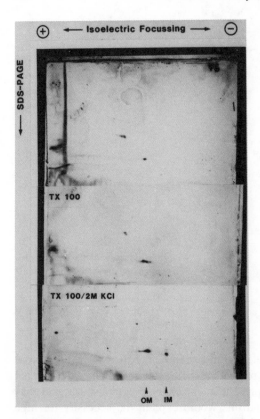

Figure 4. Solubilization of the Mlt from the outer membrane. Outer membrane proteins (150 μg; upper panel) were solubilized by 0.4% Triton X-100 (center panel) or by 0.4% Triton X-100 and 2 M KCl (lower panel) and were separated by two-dimensional gel electrophoresis. Mlt protein was visualized on Western blots by immunochemical staining with the use of antiserum against purified Mlt protein.

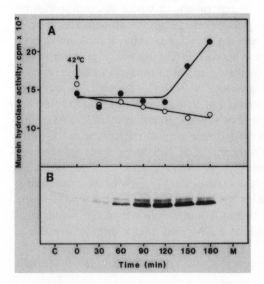

Figure 5. Overproduction of the Mlt in *E. coli* HB101(pWK2579) after temperature-induced plasmid amplification. (A) Murein hydrolase activity in crude membranes from strain HB101 (○) and the plasmid pWK2579-carrying strain (●). (B) Immunoblot of an SDS-PAGE of the membrane proteins of HB101 (pWK2579). Antiserum against Mlt protein was used. C, Noninduced control cells; M, 0.2 μg of Mlt protein.

pWK2579 (W. Keck, unpublished data) carrying the *mlt* structural gene did not result in a parallel increase in hydrolase activity with Mlt protein accumulation in the envelope (Fig. 5). This puzzling result is explained by the following finding. We could show that the extra protein is inserted into the cytoplasmic membrane only. Transport to the outer membrane and binding to the murein, which may activate or stabilize the enzyme, does not occur with the overproduced enzyme. Possibly, there are only a limited number of binding sites in the murein sacculus, as has been argued to be the case in *Lactobacillus acidophilus* (3).

Conclusion

Evidence is accumulating which suggests that the murein of *E. coli* is multilayered (8; see also Schwarz and Glauner, this volume) and expands according to the inside-to-outside growth mechanism (16). A spatial separation of the murein synthesis machinery from murein-degrading enzymes is, therefore, expected to be realized. In *E. coli* the outer membrane would be an ideal structure to harbor the murein hydrolases involved in the dissolution of the outer murein layer that allows wall expansion to take place. Murein synthesis, on the other hand, is known to be catalyzed in the cytoplasmic membrane (10). Our finding of a predominant localization of enzymatically active lytic transglycosylase in the outer membrane is consistent with this growth model. In addition, the strong association of the enzyme with the murein layer may serve as a kind of general topological control, as has been described before for autolytic enzymes in gram-positive bacteria (11, 12, 22).

LITERATURE CITED

1. **Bayer, M.** 1968. Areas of adhesion between wall and membrane of *Escherichia coli*. *J. Gen. Microbiol.* **53**:395–404.
2. **Beachey, E. H., W. Keck, M. A. de Pedro, and U. Schwarz.** 1981. Exoenzymatic activity of transglycosylase isolated from *Escherichia coli*. *Eur. J. Biochem.* **116**:355–358.
3. **Coyette, J., and J.-M. Ghuysen.** Wall autolysin of *Lactobacillus acidophilus* strain 63 AM Gasser. *Biochemistry* **9**:2952–2955.
4. **Fan, D. P.** 1970. Cell wall binding properties of the *Bacillus subtilis* autolysin(s). *J. Bacteriol.* **103**:488–493.
5. **Goodell, E. W., and U. Schwarz.** 1977. Enzymes synthesizing and hydrolyzing murein in *Escherichia coli*. Topographical distribution over the cell envelope. *Eur. J. Biochem.* **81**:205–210.
6. **Hakenbeck, R., E. W. Goodell, and U. Schwarz.** 1974. Compartmentalization of murein hydrolases in the envelope of *Escherichia coli*. *FEBS Lett.* **40**:261–264.
7. **Hartmann, R., J.-V. Höltje, and U. Schwarz.** 1972. Targets of penicillin action in *Escherichia coli*. *Nature* (London) **235**:426–429.

8. **Hobot, J. A., E. Carlemalm, W. Villinger, and E. Kellenberger.** 1984. Periplasmic gel: new concept resulting from the reinvestigation of bacterial cell envelope ultrastructure by new methods. *J. Bacteriol.* **160:**143–152.

9. **Höltje, J.-V., D. Mirelman, N. Sharon, and U. Schwarz.** 1975. Novel type of murein transglycosylase in *Escherichia coli. J. Bacteriol.* **124:**1067–1076.

10. **Höltje, J.-V., and U. Schwarz.** 1985. Biosynthesis and growth of the murein sacculus, p. 77–119. *In* N. Nanninga (ed.), *Molecular Cytology of Escherichia coli.* Academic Press, London.

11. **Höltje, J.-V., and A. Tomasz.** 1975. Specific recognition of choline residues in the cell wall teichoic acid by the N-acetyl-muramyl-L-alanine amidase of *Pneumococcus. J. Biol. Chem.* **250:**6072–6076.

12. **Joseph, R., and G. D. Shockman.** 1976. Autolytic formation of protoplasts (autoplasts) of *Streptococcus faecalis*: location of active and latent autolysin. *J. Bacteriol.* **127:**1482–1493.

13. **Kawamura, T., and G. D. Shockman.** 1983. Purification and some properties of the endogenous, autolytic N-acetylmuramoyl hydrolase of *Streptococcus faecium. J. Biol. Chem.* **258:**9514–9521.

14. **Keck, W., F. B. Wientjes, and U. Schwarz.** 1985. Comparison of two hydrolytic murein transglycosylases of *Escherichia coli. Eur. J. Biochem.* **148:**493–497.

15. **Kennedy, E. P.** 1982. Osmotic regulation and the biosynthesis of membrane-derived oligosaccharides in *Escherichia coli. Proc. Natl. Acad. Sci. USA* **79:**1092–1095.

16. **Koch, A. L.** 1985. Inside-to-outside growth and turnover of the wall of gram-positive rods. *J. Theor. Biol.* **117:**137–157.

17. **MacAlister, T. J., B. MacDonald, and L. I. Rothfield.** 1983. The periseptal annulus: an organelle associated with cell division in Gram-negative bacteria. *Proc. Natl. Acad. Sci. USA* **80:**1372–1376.

18. **MacGregor, C. H., C. W. Bishop, and J. E. Blech.** 1979. Localization of proteolytic activity in the outer membrane of *Escherichia coli. J. Bacteriol.* **137:**574–583.

19. **Mett, H., W. Keck, A. Funk, and U. Schwarz.** 1980. Two different species of murein transglycosylase in *Escherichia coli. J. Bacteriol.* **144:**45–52.

20. **Osborn, M. J., J. E. Gander, E. Parisi, and J. Carson.** 1972. Mechanism of assembly of the outer membrane of *Salmonella typhimurim. J. Biol. Chem.* **257:**3962–3972.

21. **Parquet, C., B. Flouret, M. Leduc, Y. Hirota, and J. van Heijenoort.** 1983. N-acetyl-muramoyl-L-alanine amidase of *Escherichia coli* K12. Possible physiological functions. *Eur. J. Biochem.* **133:**371–377.

22. **Pooley, H. M., J. M. Porres-Juan, and G. D. Shockman.** 1970. Dissociation of an autolytic enzyme-cell wall complex by treatment with unusually high concentrations of salt. *Biochem. Biophys. Res. Commun.* **38:**1134–1140.

23. **Rosenbusch, J. P.** 1974. Characterization of the major envelope protein from *Escherichia coli.* Regular arrangement on the peptidoglycan and unusual dodecylsulfate binding. *J. Biol. Chem.* **249:**8019–8029.

24. **Suzuki, H., Y. Nishimura, S. Yasuda, A. Nishimura, M. Yamada, and Y. Hirota.** 1978. Murein-lipoprotein of *Escherichia coli*: a protein involved in the stabilization of bacterial cell envelope. *Mol. Gen. Genet.* **167:**1–9.

25. **Wolf-Watz, H., and S. Normark.** 1976. Evidence for a role of N-acetylmuramyl-L-alanine amidase in septum separation in *Escherichia coli. J. Bacteriol.* **128:**580–586.

Chapter 17

Regulation of Peptidoglycan Biosynthesis and Antibiotic-Induced Autolysis in Nongrowing *Escherichia coli*: a Preliminary Model

E. E. Ishiguro
W. Kusser

A rapid accumulation of guanosine 5'-triphosphate 3'-diphosphate (pppGpp) and guanosine 5'-diphosphate 3'-diphosphate (ppGpp) occurs when the growth of *relA*[+] strains of *Escherichia coli* is arrested by amino acid deprivation. These compounds are proposed to represent signals for amino acid deficiency, and they apparently mediate the shutdown in the synthesis of key cellular macromolecules such as stable RNA species, phospholipids (PL), and cell wall peptidoglycan (PG). This regulatory mechanism has been termed the stringent response and apparently functions to immediately tone down macromolecular synthesis when bacteria are subjected to sudden nutritional stress (for reviews, see references 1 and 3). The synthesis of pppGpp and ppGpp is catalyzed by a ribosome-associated ATP:GTP 3'-pyrophosphotransferase (the product of the *relA* gene) which is activated during amino acid deprivation. Thus, amino acid-deprived *relA* mutants do not accumulate pppGpp and ppGpp, and as a consequence, they exhibit what is known as relaxed control; i.e., in this case, the syntheses of macromolecular species normally regulated by stringent control

E. E. Ishiguro and W. Kusser • Department of Biochemistry and Microbiology, University of Victoria, Victoria, British Columbia, Canada V8W 2Y2.

continue despite the inhibition of growth. It is also noteworthy that chloramphenicol and several other inhibitors of ribosome function antagonize the stringent response, apparently by blocking the function of the RelA enzyme and consequently preventing the accumulation of pppGpp and ppGpp by amino acid-deprived *relA*⁺ bacteria (2, 10). It is our purpose here to summarize the results of our studies on stringent control of PG metabolism to date in the form of a preliminary model.

The main steps in PG synthesis have been determined, although precise details of certain processes are lacking, e.g., the polymerization and maturation of macromolecular PG (6). The PG biosynthetic process occurs in all cellular compartments. A set of soluble enzymes carries out the synthesis of UDP-*N*-acetylglucosamine (UDP-GlcNac) and UDP-*N*-acetylmuramyl-pentapeptide (UDP-MurNac-pentapeptide). A series of membrane-bound reactions results in the formation of lipid-linked intermediates and in the polymerization of PG. The stringent response inhibits PG synthesis at two points: (i) an early step in the synthesis of UDP-MurNAc-pentapeptide (8, 11) and (ii) a late step in the polymerization of PG (11) which, on the basis of current information, presumably is catalyzed by one or more of the penicillin-binding proteins (PBPs). Stringent control apparently also inhibits an early step in UDP-GlcNAc synthesis (unpublished data). In our view, the combination of these events effectively prevents the unnecessary accumulation of PG and UDP-linked intermediates when growth is arrested by amino acid deprivation. The accumulation of lipid-linked intermediates appears to be restricted by the small amount of undecaprenol phosphate normally present in the bacteria.

The mechanisms of the stringent control of these processes are not yet known. However, studies on the membrane-bound reactions suggest that ppGpp does not simply act as an allosteric inhibitor of a PG polymerase. Instead, the PG polymerization reaction under stringent control is dependent on ongoing PL synthesis. It may be recalled that PL synthesis is also known to be under stringent control (3). Thus, PG synthesis and PL synthesis are inhibited in amino acid-deprived *relA*⁺ bacteria, and the syntheses of both can be relaxed by treatment with a stringent control antagonist such as chloramphenicol. Furthermore, the syntheses of both PG and PL are relaxed in amino acid-deprived *relA* mutants. If the relaxed synthesis of PL in these amino acid-deprived bacteria is interfered with, e.g., by the addition of cerulenin or by imposing glycerol deprivation on glycerol auxotrophs, the synthesis of PG is concomitantly inhibited (7). These results suggest that the inhibition of PG synthesis during the stringent response is a consequence of the inhibition of PL synthesis by ppGpp. It is also significant that the biosynthetic step which is dependent on PL synthesis is identical to the reaction previously identified as the site of stringent control, i.e., the membrane-associated polymerization of PG. The nature of the dependence of this reaction on PL synthesis is currently under investigation.

It has often been suggested that the synthesis and morphogenesis of PG require the coordinated activities of the PG biosynthetic enzymes and a set of PG hydrolases (see references 6 and 13 for reviews). This view has not been completely substantiated. Thus, the functions of the various PG hydrolases and the mechanism(s) by which their activities are regulated and coordinated with PG synthesis are largely speculative. However, it seems reasonably clear that in *E. coli* one or more of the PG hydrolases is responsible for the lysis induced by treatment with β-lactam antibiotics and other inhibitors of PG synthesis. Cellular autolysis occurs whenever PG synthesis is inhibited in growing *E. coli,* e.g., by treatment with specific inhibitors or by imposing mutational blocks in the biosynthetic pathway. Such conditions apparently deregulate the activities of the cellular PG hydrolases, but the details of the mechanism are at best sketchy.

Goodell and Tomasz (4) first presented evidence that the PG hydrolysis is yet another phenomenon regulated by stringent control. We have verified this finding and have extended it to the study of the antibiotic-induced lysis of nongrowing *E. coli* (9, 10). Nongrowing bacteria are generally tolerant to the lysis-inducing action of inhibitors of PG synthesis. In the case of amino acid-deprived *relA*[+] strains, this lysis tolerance can be abolished by simultaneously treating the bacteria with chloramphenicol (9) or several other inhibitors of ribosome function (10). All of the ribosome inhibitors active in destroying lysis tolerance are known antagonists of stringent control. These results suggest that the lysis mechanism is inactivated during the stringent response but can be reactivated without reinitiating growth by treatment with relaxing agents such as chloramphenicol. This interpretation is consistent with the observation that amino acid deprivation does not protect *relA* mutants from the lysis-inducing activities of various inhibitors of PG synthesis (4, 9).

Further support for the proposed involvement of the *relA* gene in the regulation of the cellular lysis mechanism comes from recent studies (10a) on temperature-sensitive, lysis-tolerant mutants. Previous studies on these mutants indicated that they fall into two classes: (i) some appear specifically tolerant to β-lactam antibiotics (12), and (ii) the others are more generally tolerant, i.e., not only to β-lactams but also to treatment with D-cycloserine or moenomycin and to diaminopimelate deprivation (5). The mutations conferring lysis tolerance have been assigned to two new loci designated *lytA* at 58 min on the *E. coli* linkage map and *lytB* at 0.5 min. The lysis-tolerant phenotypes of the *lytA* and *lytB* mutants are suppressed in *relA* genetic backgrounds. Furthermore, phenotypic suppression of the lysis tolerance can be achieved in *relA*[+] backgrounds by treatment of the mutants with chloramphenicol or other inhibitors of the stringent response. The *relA*[+] derivatives of the *lytA* and *lytB* mutants exhibit a temperature-dependent stringent response; i.e., upshift to the nonpermissive temperature results in the accumulation of ppGpp and in the concomitant inhibition of RNA and PG synthesis. We suggest that the *lytA* and *lytB* gene products may normally interact with the RelA enzyme to prevent its activation and that the thermoinactivation of the mutant LytA or LytB proteins at the nonpermissive temperature results in the activation of RelA and consequent synthesis of ppGpp. Furthermore, the inhibition of the autolysis mechanism by the stringent response at the nonpermissive temperature apparently accounts for the lysis-tolerant phenotypes of these mutants.

To summarize, we present in Fig. 1 one possible model to account for three phenomena related to the stringent response observed in nongrowing, amino acid-deprived bacteria: the inhibition of PG synthesis, the development of tolerance to antibiotic-induced autolysis, and the suppression of this lysis tolerance. In presenting this model, we realize that there are many unknowns, and our main intention is to provide a working hypothesis on which future experiments and discussions can be based. The highlights of the model are as follows.

(i) Amino acid deprivation of $relA^+$ bacteria results in the accumulation of ppGpp. Stringent control antagonists such as chloramphenicol prevent the accumulation of ppGpp.

(ii) One of several events mediated by ppGpp is the inhibition of PL synthesis.

(iii) As noted above, the terminal step in PG synthesis is inhibited during the stringent response, and on the basis of current information, this step is presumably catalyzed by a penicillin-sensitive PG polymerase (PBP). This enzyme can apparently undergo a reversible transition from active to inactive forms because the reaction can be reinstated in vivo without de novo protein

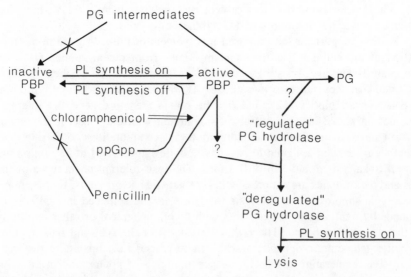

Figure 1. Model for regulation of PG synthesis and antibiotic-induced lysis in nongrowing *E. coli*. See text for details.

synthesis, e.g., by treatment of amino acid-deprived *relA*⁺ bacteria with strin-gent control antagonists such as chloramphenicol (9). The activity of this PBP is dependent on PL synthesis, and as indicated in the model, this forms the bases for the stringent control of this activity and for the reversible transition from active to inactive forms.

(iv) We go along with the notion that a "regulated" PG hydrolase system is involved in PG metabolism. However, whether it is coupled to the terminal steps in PG synthesis as depicted in the model is uncertain. Autolysis seems to be based on the "deregulation" of the hydrolase system.

(v) The interaction of penicillin with active PBPs appears necessary to initiate PG hydrolysis, but the molecular events involved are unknown. As-suming that penicillin is an analog of a PG intermediate, the inability of the antibiotic to interact with inactive PBPs may be sufficient to explain the pen-icillin tolerance observed during the stringent response.

(vi) Although the model focuses around the terminal step in PG synthesis and penicillin action, it should be recalled that the stringent response also ren-ders bacteria tolerant to blocks imposed earlier in the pathway which would normally cause lysis of growing bacteria, e.g., with D-cycloserine. The basis for this is uncertain, but it is possible that the "regulated" PG hydrolase system is coupled somehow to the terminal PG polymerase step. In this view, an early block in the pathway can be sensed only if the PG polymerase is in the active form. Alternatively, it is possible that the PG polymerase and PG hydrolase systems are independently regulated; i.e., both are concomitantly but indepen-dently inactivated by the stringent response.

(vii) Finally, it is noteworthy that lysis of nongrowing bacteria is also dependent on PL synthesis (studies in progress). It is not yet clear whether this requirement is for the PG hydrolase activity, for the PG polymerase activity, or for both activities. In any event, lysis tolerance during the stringent response may also be due to inactivation of the hydrolase system as a consequence of the inhibition of PL synthesis.

ACKNOWLEDGMENTS. Work in our laboratory was supported by grants from the Natural Sciences and Engineering Research Council of Canada. W.K. was supported in part by a postdoctoral fellowship from the Deutsche Forschungs-gemeinschaft.

LITERATURE CITED

1. **Cashel, M.** 1975. Regulation of bacterial ppGpp and pppGpp. *Annu. Rev. Microbiol.* **29**:301–318.
2. **Cortay, J. C., and A. J. Cozzone.** 1983. Effects of aminoglycoside antibiotics on the cou-

pling of protein and RNA syntheses in *Escherichia coli. Biochem. Biophys. Res. Commun.* **112**:801–808.

3. **Gallant, J. A.** 1979. Stringent control in *E. coli. Annu. Rev. Genet.* **13**:393–415.
4. **Goodell, W., and A. Tomasz.** 1980. Alteration of *Escherichia coli* murein during amino acid starvation. *J. Bacteriol.* **144**:1009–1016.
5. **Harkness, R. E., and E. E. Ishiguro.** 1983. Temperature-sensitive autolysis-defective mutants of *Escherichia coli. J. Bacteriol.* **155**:15–21.
6. **Höltje, J.-V., and U. Schwarz.** 1985. Biosynthesis and growth of the murein sacculus, p. 77–119. *In* N. Nanninga (ed.), *Molecular Cytology of Escherichia coli.* Academic Press, London.
7. **Ishiguro, E. E.** 1983. Mechanism of stringent control of peptidoglycan synthesis in *Escherichia coli*, p. 631–636. *In* R. Hakenbeck, J.-V. Höltje, and H. Labischinski (ed.), *The Target of Penicillin.* Walter de Gruyter & Co., Berlin.
8. **Ishiguro, E. E., and W. D. Ramey.** 1978. Involvement of the *relA* gene product and feedback inhibition in the regulation of UDP-*N*-acetylmuramyl-peptide synthesis in *Escherichia coli. J. Bacteriol.* **135**:766–774.
9. **Kusser, W., and E. E. Ishiguro.** 1985. Involvement of the *relA* gene in the autolysis of *Escherichia coli* induced by inhibitors of peptidoglycan biosynthesis. *J. Bacteriol.* **164**:861–865.
10. **Kusser, W., and E. E. Ishiguro.** 1986. Lysis of nongrowing *Escherichia coli* by combinations of β-lactam antibiotics and inhibitors of ribosome function. *Antimicrob. Agents Chemother.* **29**:451–455.
10a. **Kusser, W., and E. E. Ishiguro.** 1987. Suppression of mutations conferring penicillin tolerance by interference with the stringent control mechanism of *Escherichia coli. J. Bacteriol.* **169**:4396–4398.
11. **Ramey, W. D., and E. E. Ishiguro.** 1978. Site of inhibition of peptidoglycan biosynthesis during the stringent response in *Escherichia coli. J. Bacteriol.* **135**:71–77.
12. **Shimmin, L. C., D. Vanderwel, R. E. Harkness, B. R. Currie, A. Galloway, and E. E. Ishiguro.** 1984. Temperature-sensitive β-lactam-tolerant mutants of *Escherichia coli. J. Gen. Microbiol.* **130**:1315–1323.
13. **Tomasz, A.** 1979. The mechanism of the irreversible antimicrobial effects of penicillins: how the beta-lactam antibiotics kill and lyse bacteria. *Annu. Rev. Microbiol.* **33**:113–137.

Chapter 18

The Autolytic Peptidoglycan Hydrolases of *Streptococcus faecium:* Two Unusual Enzymes

G. D. Shockman
D. L. Dolinger
L. Daneo-Moore

Bacterial autolysis is a phenomenon well known to microbiologists. Not that long ago, autolysis was thought to be a complex phenomenon accompanied by a variety of enzymatic processes (12). The advent of the group of antibiotics that were later found to inhibit peptidoglycan assembly renewed interest in autolysis. It rapidly became clear that one mechanism by which penicillin and other "cell wall antibiotics," such as D-cycloserine and vancomycin, exerted their bactericidal effect was via the action of these endogenous enzyme systems that could hydrolyze bonds in the protective and essential bacterial cell wall peptidoglycan (36), thus leading to cellular lysis. Early on (34, 37, 41), bacterial autolysis was viewed as a manifestation of an "unbalance" in wall synthetic and hydrolytic systems, implying some sort of role for peptidoglycan hydrolases in proper cell wall assembly. Although indirect evidence for such a role has been obtained, to date a precise role remains to be defined.

A broad variety of bacterial species have been shown to possess peptidoglycan hydrolase activities (see references 10, 28, and 35 for reviews), and many of these organisms contain peptidoglycan hydrolases of more than one specificity. These activities include: (i) glycosidases such as N-acetylglucosaminidases and N-acetylmuramidases, (ii) N-acetylmuramyl-L-alanine amidases, and (iii) peptidases such as endopeptidases and carboxypeptidase-transpepti-

G. D. Shockman, D. L. Dolinger, and L. Daneo-Moore • Department of Microbiology and Immunology, Temple University School of Medicine, Philadelphia, Pennsylvania 19140.

dases. Hydrolysis of a sufficient number of bonds that contribute to the two- or three-dimensional meshwork of the wall peptidoglycan will result in a weakening of the wall and the bursting of the protoplast out of its "shell." Topological aspects of this process are of some interest. Hydrolysis of a number of bonds that are widely scattered over the peptidoglycan shell may not result in significant loss of wall strength and integrity, especially in the highly peptide-cross-linked peptidoglycans of some gram-positive bacteria, whereas hydrolysis of the same number of bonds in a topologically localized area of the wall could result in a relatively small, highly weakened area that no longer protects the protoplast membrane.

Hydrolysis of some of the bonds in the peptidoglycan is compatible with known biosynthetic processes (10, 28, 35). That is, hydrolysis will result in residues to which new precursors can be added. Such activities include the muramidases, which yield reducing N-acetylmuramic acid residues, and peptidases, which yield amino or carboxyl groups, or both, of amino acids that are substrates for transpeptidases. In contrast, the action of other peptidoglycan hydrolases (for example, amidases or muramidase-transglycosidases) that yield internal, 1,6-anhydro linkages (4, 18, 22–24) fails to yield substrates that are suitable for well-known biosynthetic processes. It is far easier to visualize a role for the former types of hydrolase in wall surface growth, division, or morphogenesis than it is for the latter types. Such a role could be in coordination with separate and distinct biosynthetic enzyme systems or could be due to the measurement of the hydrolytic rather than the biosynthetic direction of the enzymatic reaction (e.g., for transglycosidases or transpeptidases).

The Autolytic System of Streptococcus faecium ATCC 9790 (Enterococcus hirae)

A broad variety of data are completely consistent with the presence of only a single specificity of peptidoglycan hydrolase activity in Streptococcus faecium (11, 13, 29, 39). Only N-acetylmuramoylhydrolase (muramidase) activity has been detected, with no evidence of amidase or endopeptidase activity. (An active β-lactam-sensitive DD-carboxypeptidase activity is present [9], but its removal of C-terminal D-alanine residues does not affect the integrity of the wall peptidoglycan matrix.)

In addition, the following data are consistent with a role for autolysin activity in wall growth and division in this organism. (i) A series of observations demonstrated that, in bacteria from exponentially growing cultures, autolysin activity (both active and proteinase activatable, latent forms) is associated with newly assembled wall (19, 26) which, in this organism, is localized to nascent cross walls (15). This association was not observed in walls undergoing wall thickening, rather than surface expansion, after inhibition of protein synthesis due to chloramphenicol treatment or amino acid deprivation (26). Furthermore, ultrastructural observations (14, 19) of exponential-phase cells undergoing autolysis showed localized dissolution of walls at nascent cross walls. (ii) Rapidly growing exponential-phase bacteria autolyzed rapidly, whereas cells from cultures in which protein synthesis was inhibited were resistant to autolysis (26, 37). Such inhibitions occurred rapidly and were rapidly reversible (32). However, cells remained prone to autolysis after inhibition of DNA synthesis (32). (iii) Cellular autolysis rates were shown to decrease with decreasing growth rate of cultures (16). (iv) In synchronized cell populations the rate of autolysis was shown to vary with stage of division, with the slowest rates observed between 15 and 25 min before cell division (16, 17). (v) Autolytic activity appears to be regulated by several factors, including proteinase activation (26), its topological location in cells (26, 38), perhaps the activity of regulatory ligands such as lipoteichoic acid and certain lipids, notably cardiolipin (5–7), and substrate specificity (see below).

Purification and Some Properties of Muramidase-1 of S. faecium (21)

The latent form of muramidase-1 was rapidly extracted from isolated walls with 0.004 N NaOH at 0°C. Removal of walls and immediate neutralization was followed by precipitation of the enzyme with ammonium sulfate and chromatography on a column system consisting of a column of hemiglobin-Sepharose 4B followed by a column of concanavalin A (ConA)-Sepharose 4B. After removal of the hemiglobin-Sepharose column, the ConA-Sepharose column was extensively washed with a series of solutions, including 0.1% Triton X-100 and 1 M NaCl. Finally, latent enzyme activity was eluted with 50 mM α-methyl-D-mannopyranoside (or α-glucoside). Such eluates yielded latent muramidase-1 activity with an overall recovery of 70 to 100% that showed only a single Coomassie blue and periodate-Schiff (PAS) staining band at 130 kilodaltons (kDa) after sodium dodecyl sulfate (SDS)-gel electrophoresis (21). Trypsin treatment converted the 130-kDa protein to an 87-kDa form that retained PAS staining plus smaller peptides. Elution of gel slices, followed by unfolding in guanidine hydrochloride and renaturation, clearly showed that the 130-kDa band was the latent form requiring protease activation and that the 87-kDa form was the active form of the muramidase.

Strong evidence was obtained that binding of the enzyme to ConA-Sepharose and the positive reaction of the 87- and 130-kDa proteins with the PAS reagent were attributable to the presence of covalently linked glycan chains. The ability of these properties to survive high pH, high salt concentration, precipitation with trichloroacetic acid, detergents, and SDS-gel electrophoresis was consistent with that of covalently linked polysaccharides. In addition, we were able to demonstrate that CNBr cleavage and proteinase digestion each produced a family of polypeptides, several of which stained with the PAS reagent. Gas-liquid chromatographic analysis of the neutral sugars released from trichloroacetic acid-precipitated enzyme showed the presence of only glucose. β-Elimination and reduction in the presence of $[^3H]NaBH_4$ was used to show that about one-half of the glucose was present as oligosaccharides, with the remainder monosaccharides. The data were consistent with a glucoenzyme that is multiply substituted with glucose monomers and glucose oligosaccharides, probably O-linked to the polypeptide chain.

The purified enzyme was shown to be a muramidase (21), consistent with earlier data with crude enzyme preparations (39). The purified enzyme could hydrolyze re-*N*-acetylated peptidoglycans prepared from walls of several gram-positive bacterial species but had essentially no activity on intact walls of other bacterial species. It appears that the presence of covalently linked accessory wall polymers, other than those present in walls of *S. faecium,* inhibits this muramidase activity. Similarly, *N*-acetylation of the amino sugars was found to be required for enzyme activity.

Mechanism of Hydrolysis of Peptidoglycan by Muramidase-1

High-molecular-weight, soluble, un-cross-linked, linear peptidoglycan (sol-PG) chains produced and secreted by autolysis-deficient strains of *S. faecium* when incubated in a wall medium in the presence of benzylpenicillin were isolated, purified, and chemically characterized (3). These sol-PG chains were shown to consist of 42 to 47 fully peptide-substituted disaccharide-peptide (DSP_1) units (46 to 53 kDa). They were then used to study the mechanism of their hydrolysis by muramidase-1 (1, 2).

Comparison of the rate and extent of hydrolysis of the sol-PG of *Micrococcus luteus* by muramidase-1 and hen egg-white lysozyme (HEWL) showed that muramidase-1 hydrolysis resulted in a continuous increase in reducing groups at the same rate until about 90% of the available bonds were hydrolyzed (2). In contrast, HEWL hydrolyzed only about 60% of the potential susceptible bonds at a continuously decreasing rate of hydrolysis. Also with HEWL, but not with muramidase-1, the period of rapid increase in reducing groups was followed by a plateau which was then followed by a decrease and then a second increase in reducing groups (Fig. 1). The decrease in reducing groups was attributed to the known ability of HEWL to catalyze glycosylation (34). Although the ability of HEWL to carry out transglycosylation was easily demonstrated, no evidence was obtained that muramidase-1 could transglycosylate. Although we could easily demonstrate the production of shorter sol-PG chains by the action of HEWL, DSP monomers were nearly the sole product of muramidase-1 hydrolysis of *S. faecium* sol-PG. The apparent K_ms (in terms of DSP units) for muramidase-1 (73 ± 8 μM) and HEWL (81 ± 6 μM) were indis-

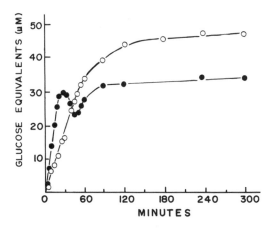

Figure 1. Comparison of the time course of hydrolysis of *M. luteus* sol-PG by muramidase-1 and by HEWL. The reaction mixtures contained 52 μM (○) or 60 μM (●) *M. luteus* sol-PG (as DSP_1) and either 10 U of trypsin-activated muramidase-1 per ml (○) or 10 U of HEWL per ml (●) in 1.8 ml of 10 mM sodium phosphate (pH 6.8).

tinguishable, whereas the V_{max} values were 230 and 380 nmol of DSP_1 released min^{-1} mg^{-1} for muramidase-1 and HEWL, respectively.

Further studies of hydrolysis of *S. faecium* sol-PG by muramidase-1 showed that this enzyme is a processive exodisaccharidase (1). A molecule of muramidase-1 binds tightly to a sol-PG chain so that it is not released and is not able to bind to other sol-PG chains present in the same incubation mixture until the enzyme finishes hydrolyzing all, or nearly all, susceptible bonds in that chain. Sol-PG chains were labeled at the nonreducing end with [^{14}C]galactose via the transfer of [^{14}C]galactose from UDP-galactose by the action of UDP-galactose transferase, and labeled with ^{3}H at the reducing end via the reduction of terminal muramic acid-reducing groups with [^{3}H]$NaBH_4$. Muramidase-1 hydrolysis of such end-labeled chains, at a high enzyme-to-substrate ratio, resulted in the very rapid release of ^{14}C, but a delay of about 1.5 min before ^{3}H at the reducing ends of the chains was released (Fig. 2). Hydrolysis of sol-PG chains at a low molar ratio of muramidase-1 to sol-PG chains (at 30°C) resulted in hydrolysis of chains to which the enzyme was bound, followed by a pause in hydrolysis of about 2.5 min and then a second round of hydrolysis (Fig. 3). These and other data were interpreted as consistent with the processive model of hydrolysis shown in Fig. 4. Muramidase-1 binds to nonreducing ends of sol-PG chains and processively hydrolyzes bonds in the chain, releasing DSP_1 units at a rate of about 91 bonds min^{-1}. As hydrolysis approaches the end of a sol-PG chain (e.g., DSP_2), it slows or stops, and the enzyme is very slowly released from the end of the chain.

At least some of the kinetic properties of muramidase-1 are admirably suited for an enzyme that hydrolyzes bonds in an insoluble substrate outside of the cellular permeability barrier. Examples are the high binding affinity (10^{-9} M for the dissociation constant) and lack of release during catalysis. Also, hydrolysis of many bonds in a localized area of a protective meshwork will result in a more pronounced but more localized weakening of the structure.

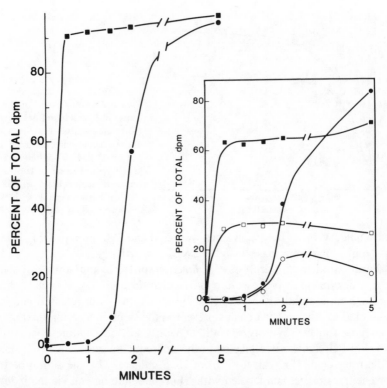

Figure 2. Direction of hydrolysis by muramidase-1. The reaction mixture, which contained dual radiolabeled sol-PG (0.4 nmol of chains), labeled with 6 and 51% efficiency by [^{14}C]galactose and ^{3}H, respectively, was mixed with 833 U of activated muramidase-1 (0.56 nmol) and 10 mM sodium phosphate (pH 6.8) in a total volume of 600 μl and incubated at 0°C. This represents an enzyme/substrate ratio of 1.4:1. The reaction mixture was quickly brought to 37°C, and samples of 100 μl were taken at timed intervals. After heat inactivation at 100°C for 5 min, the products were separated by descending paper chromatography and quantified by detection of radioactivity. Total [^{14}C]galactose-labeled products (■) precede release of ^{3}H-labeled products (●) by 1.5 min. The inset demonstrates the fraction of total counts which appeared as [^{14}C]galactose-DSP$_1$ (■), [^{14}C]galactose-DSP$_2$ (□), DSP$_2$ ([^{3}H]muramicitol) (●), and DSP$_1$ ([^{3}H]muramicitol) (○).

Additional Unusual Properties of Muramidase-1

To our surprise, the UV absorption spectrum of purified muramidase-1 showed a sharp maximum at about 260 nm rather than the expected maximum at about 280 nm (D. Dolinger, V. L. Schramm, and G. D. Shockman, manuscript in preparation). This observation, plus the presence of about 18 mol of phosphorus per mol of enzyme, led us to believe that the enzyme contained nucleotide. Phosphorus and the absorption maximum at 260 nm remained present after drastic treatments such as SDS-gel electrophoresis, trichloroacetic acid

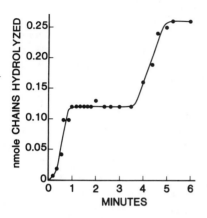

Figure 3. Time course of sol-PG hydrolysis. Latent muramidase-1 (0.14 nmol) was preincubated at 0°C for 5 min with 0.8 nmol of [^{14}C]glucose-DSP$_{46}$ and converted from the zymogen by 5 min of incubation at 0°C with 0.4 μg/ml of trypsin. The total reaction mixture (0.5 ml) was rapidly heated to 30°C, and at the indicated intervals, samples (20 μl) were taken and boiled to terminate the reaction. The amount of DSP$_1$ formed was established by chromatography and liquid scintillation counting.

precipitation, and unfolding in guanidine hydrochloride, consistent with a co-valently linked modifier. The presence of about 11 mol of SH groups per mol of enzyme in excess of those that could be accounted for as cysteine residues in the protein was demonstrated. Gas-liquid chromatography of trimethylsilane derivatives of β-eliminated saccharides (in this instance separated from the protein by gel filtration rather than by ion-exchange chromatography) revealed the

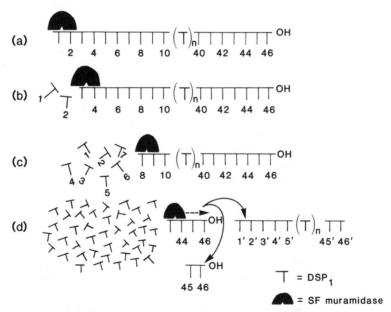

Figure 4. Model for hydrolysis of sol-PG by the muramidase-1. Shown are the stages of initial binding (a), initial hydrolyses (b), continued sequential hydrolysis (c), and hydrolysis of the last few bonds and release of DSP$_2$ and enzyme, before binding to another sol-PG strand (d).

presence of ribose plus a peak that could not be identified as a known saccharide. With several different solvents, thin-layer chromatography of the modifier, which had been treated with 70% formic acid for 30 min at 160°C, showed a single spot that did not cochromatograph with any of a variety of known purine or pyrimidine bases. However, the R_f values, the UV spectrum and spectral shifts, and the presence of SH groups suggested that the base could be 5-mercaptouracil. Therefore, 5-mercaptouracil was synthesized from uracil by use of chlorosulfonic acid. Thin-layer, paper, and high-pressure liquid chromatographic mobilities of the acid-treated modifier were identical to those of chemically synthesized 5-mercaptouracil. Finally, mass spectroscopy, using direct probe analysis, identified the isolated modifier as 5-mercaptouridine. The enzyme appears to be multiply (11 to 18) substituted with monomeric nucleotides, as determined by ion-exchange and paper chromatography of proteolytic hydrolysates. Phosphodiesterase treatment resulted in the release of about 50% of the phosphorus and nucleotides and reduction of about 50% in enzyme activity. These data suggest that the nucleotidylation may play a regulatory role. However, the exact role of the very unusual nucleotidylation of this enzyme remains to be elucidated.

Discovery of and Some Properties of the Second Peptidoglycan Hydrolase of *S. faecium* (Muramidase-2)

While purifying the glucoenzyme (muramidase-1), we observed the presence of an activity that could dissolve cell walls of *M. luteus* and that did not bind to the ConA-Sepharose column (20). This muramidase-2 activity was not observed previously because of its lack of ability to dissolve homologous, intact walls of *S. faecium*. (It can, however, hydrolyze susceptible bonds in this substrate.) This activity was partially purified from culture supernatants by taking advantage of its ability to bind to SDS-walls of *M. luteus*. Elution and renaturation of gel slices from SDS-gel electrophoresis showed that muramidase-2 activity was present in a protein band at about 75 kDa which did not stain with the PAS reagent.

Muramidase-2 has a number of properties that clearly distinguish it from muramidase-1 (Table 1). First, it differs in substrate specificity. For example, muramidase-2 is unable to dissolve intact, insoluble walls of *S. faecium*, although it does hydrolyze bonds in such walls. However, muramidase-2 can dissolve walls of *M. luteus*, the acid-insoluble peptidoglycan fraction of walls of *S. faecium*, or the peptidoglycan fraction of walls of *Lactobacillus acidophilus*. In contrast, purified muramidase-1 fails to dissolve walls of *M. luteus* or *L. acidophilus* or the peptidoglycan fraction of walls of *M. luteus*, *L. acidophilus*, or *S. faecium*. These differences in substrate specificity provide differential assays for both enzymes. Both muramidase-1 and muramidase-2 can

Table 1. Some properties of muramidase-1 and muramidase-2

Property	Muramidase-1 (E-1)	Muramidase-2 (E-2)
Molecular mass	130 kDa latent; 87 kDa (proteinase activated)	Two forms (probably related), 125 and 75 kDa
Glycosylated	Oligomeric and monomeric glucose	No
Nucleotidylation	5-Mercaptouridine ~10 monomers phosphodiester linked to polypeptide chain	No (absorption maximum is at 280 nm)
Hydrolysis mechanism	Processive, from nonreducing to reducing end of sol-PG chains; exodisaccharide	Probably random
Transglycosidase activity	Not detectable	+ (preliminary data)
pH and ionic optima	pH 6.8 to 7.0; 10 mM	pH 6 to 8; 20 mM
"True" K_m	0.17 μM	?
Apparent K_m	62 to 81 μM (dependent on chain length of sol-PG substrate)	?
Apparent V_{max}	173 to 233 nmol of disaccharide-peptide per min per mg (depending on sol-PG substrate)	?
K_d	1.5×10^{-9} M	?
Estimated no. of molecules per cell	1,000	?
Outside cyto-membrane	+	+
Found in wall fraction	Latent and active forms	Both forms
Extracellular (medium)	Probably unfolded; activated 87-kDa form	75-kDa form (active)
Binding affinities		
S. faecium SDS-walls	4+	±?
S. faecium re-*N*-acetylated PG[a]	4+	2+
S. faecium PG	−	4+
M. luteus walls	−	4+
Sol-PG of *S. faecium*	4+	?
Sol-PG of *M. luteus*	4+	?
ConA-Sepharose	4+	−
Hydrolytic specificities		
S. faecium SDS-cell walls	4+	± (dissolution)
Re-*N*-acetylated PG	3+	3+
PG (not re-*N*-acetylated)	−	3+
Sol-PG	4+	+ (preliminary data)
M. luteus SDS-cell walls	−	4+
Re-*N*-acetylated PG	±	4+
PG	±	4+
Sol-PG	4+	?
Staphylococcus aureus tar-1		
SDS-cell walls	−	±
Re-*N*-acetylated PG	−	2+

Table 1. Continued

Property	Muramidase-1 (E-1)	Muramidase-2 (E-2)
PG	±	1+
Extraction		
pH 12 (0.01 N NaOH)	Intact or disrupted bacteria	Better extraction from disrupted bacteria
Triton X-100	?	Yes
Renaturation with guanidine hydrochloride	100%	100%
Inhibition of hydrolytic activity:		
Lipoteichoic acid	–	±?; depends on substrate used
Penicillin G	–	–
Detergents		
Triton X-100	–	–
Zwittergent TM314	–	+ (0.02%)
EDTA	– at 0.1 mM	50% at 20 mM
Iodoacetate	– at 0.1 mM	24% at 1 mM
Reacts with monoclonal antibodies to:		
Muramidase-1	+	–
Muramidase-2	–	+
Binds [^{14}C]penicillin G	–	+ (both 125- and 75-kDa forms)
Sensitivity to proteolysis	+ (87 kDa more stable)	4+
Tendency to aggregate and precipitate	No	Strong
Hydrophobic	No	Strong
Hydrophobic	No	Highly

[a]PG, Peptidoglycan.

dissolve re-N-acetylated peptidoglycan fraction of walls of S. faecium, providing a common substrate for both enzyme activities. Muramidase-2 activity was not inhibited by 0.2% Triton X-100 but was inhibited by 0.02% Zwittergent TM-314. Muramidase-2 activity is present in the culture supernatant as well as in the cell wall fraction. In contrast, muramidase-1 activity is cell wall bound. Although muramidase-1 is also found in culture supernatants, it is present in an inactive (denatured) form that can be unfolded and renatured by using guanidine hydrochloride.

Recently we isolated and purified muramidase-2 from broken cell preparations (D. Dolinger, L. Daneo-Moore, and G. D. Shockman, manuscript in preparation). Muramidase-2 was present in the insoluble, crude wall-membrane complex of disrupted cells, from which it could be extracted by 0.01 to 0.02 N NaOH along with muramidase-1 and a complete assortment of the penicillin-binding proteins (PBPs) normally found in membrane preparations of S. faecium (I. Said, D. Dolinger, G. D. Shockman, and L. Daneo-Moore, manuscript in preparation). Extractions at pH 10 were used to remove most of the PBPs. The subsequent pH 12 extract was then mixed with SDS-walls of *M. luteus* or the peptidoglycan fraction of *S. faecium* walls. Both of these insoluble matrices bound muramidase-2 but not muramidase-1 (which could then be removed by binding to SDS-walls of *S. faecium*). The peptidoglycan fraction of *S. faecium* walls was then sequentially extracted at pH 8, at pH 10, and then at pH 12. The neutralized pH 12 extract contained high-specific-activity muramidase-2 and upon SDS-gel electrophoresis showed the presence of Coomassie blue-staining bands at 125 and 75 kDa. Elution and renaturation of gel slices showed that both the 125- and 75-kDa bands contained muramidase-2 activity. Incubation of this muramidase-2 preparation with [^{14}C]penicillin G, followed by SDS-gel electrophoresis and radioautography, showed the presence of three radioactive bands at 125, 75, and 55 kDa.

Evidence that the two proteins possessing muramidase-2 activity and PBPs 1 and 5 found in membranes of *S. faecium* (125 and 75 kDa, respectively) are identical was obtained as follows. A monoclonal antibody specific for muramidase-2 was covalently linked to CH-Sepharose 4B. A high-specific-activity preparation of muramidase-2 containing the 125- and 75-kDa proteins was incubated with [^{14}C]penicillin G and then exposed to the immunoaffinity column. Muramidase-2 activity, the 125- and 75-kDa proteins, and much of the ^{14}C was bound to the column. The column was then washed with buffer containing Triton X-100. This eluate contained little muramidase-2 activity. Muramidase-2 activity and considerable amounts of ^{14}C were then eluted from the immunoaffinity column with 0.2 N NH$_4$OH. SDS-gel electrophoresis showed the presence of three silver-staining bands at 125, 75, and 55 kDa, all of which contained ^{14}C. Chromatography of the 0.2 N NH$_4$OH eluate on a Mono Q column showed the presence of a single peak of muramidase-2 activity that also contained ^{14}C. Thus, it seems clear that the 125- and 75-kDa PBPs (PBPs 1 and 5) of *S. faecium* also possess muramidase-2 activity.

Evidence that Muramidase-1 and Muramidase-2 are Separate Gene Products

We (8, 27, 40) and other laboratories (summarized in reference 28) have attempted to obtain autolysis-negative mutants of *S. faecium* and of a variety of bacterial species. Although many autolysis-deficient strains have been obtained, to our knowledge no laboratory has succeeded in obtaining a truly autolysis-negative mutant. Perhaps the most striking example is that of *Streptococcus pneumoniae*, which has been reported to contain only *N*-acetylmuramyl-L-alanine amidase activity (25). Recently, Lopez and collaborators (31) obtained a mutant of *S. pneumoniae* in which the *lytA* gene, which codes for the amidase, was completely deleted. This mutant grew at a rate very similar to that of the isogenic parent strain. It differed from the wild type in that it grew as short chains of six to eight cells rather than as diplococci, failed to lyse when it reached the stationary growth phase, and showed a tolerant, rather than a lytic, response to penicillin G (31). However, the absence of the amidase enabled Lopez and collaborators to detect the presence of a second peptidoglycan hydrolase that appears to be a glycosidase (30).

Several years ago, using nitrosoguanidine mutagenesis, we isolated and partially characterized a conditional, temperature-sensitive, lytic-defective mutant of *S. faecium* which we called Lyt-14 (8). The phenotype of this strain is complex, and it seems likely that Lyt-14 contains more than one mutation. However, among the properties of this mutant was an inability of extracts to hydrolyze SDS-walls of *M. luteus,* which we were unable to explain at that time. However, discovery of muramidase-2, and its inability to hydrolyze walls of *M. luteus,* prompted a reexamination of Lyt-14 (J. Mays and G. D. Shockman, manuscript in preparation). Examination of alkaline extracts of Lyt-14 showed the presence of normal levels of both the latent and active forms of muramidase-1 and the presence of two proteins at 130 and 87 kDa that, in Western blots of SDS-gels, reacted with monoclonal antibodies to muramidase-1. However, the same alkaline extracts of Lyt-14 failed to show muramidase-2 activity in assays using either SDS-walls of *M. luteus* or the peptidoglycan of *S. faecium* as a substrate. Western blots using monoclonal antibodies to muramidase-2 showed reactivity with proteins at 125 and 75 kDa in SDS-gels of the parental strain, but failed to react with proteins of these molecular sizes in Lyt-14 extracts. However, the muramidase-2 monoclonal antibodies did react with polypeptides of other molecular sizes in Lyt-14 extracts. Silver staining of SDS-gel-electrophoresed samples showed the presence of normal, dark-staining bands at 125 and 75 kDa in extracts of the parent strain. However, in extracts of Lyt-14 grown at 42°C these two protein bands either excluded the stain or stained a blue-green color, depending on the silver staining procedure used. We interpret these results as indicating that Lyt-14 contains a mutation that results in a different protein(s) that no longer has muramidase-2 activity, no longer reacts with monoclonal antibodies to muramidase-2, and has unusual silver-staining properties after SDS-gel electrophoresis. Interestingly, when grown at 42°C, Lyt-14 appears to lack PBP 5 but to contain PBP 1 (I. Said, L. Marri, O. Massidda, and L. Daneo-Moore, manuscript in preparation). These results provide further confirmation that muramidase-2 covalently binds penicillin.

Redundancy of Peptidoglycan Hydrolases and Their Possible Functions in Surface Growth and Division

The presence of two large and complex peptidoglycan hydrolases in a bacterium of comparatively simple shape and mode of division raises many questions, including some that border on the philosophical. For example, to hydrolyze the same bond that HEWL, at a molecular weight of about 14,000, does rather efficiently, why does a bacterial cell synthesize two very large and very different proteins? How (and why) are these proteins processed (e.g., glycosylated, nucleotidylated) and secreted? The obvious thought is that each must be exquisitely regulated, and therefore contain regulatory sites, to prevent its unwanted and dangerous (to the producing cell) actions at the wrong time or place.

The presence of two independent systems and the observations that mutants can get along, grow, and divide in the absence of one system suggest that the presence of at least one is essential to the cell, with the second system present as a redundant, "backup" system. Previously (20) we have drawn attention to the presence of similar redundant systems in the cases of chromosome replication, multiple porin proteins in the outer membranes of enteric bacteria, and multiple PBPs.

This, of course, does not mean that the two muramidases do not have separate and distinct (or interrelated) functions in normal surface growth and division. For example, the presence of muramidase-2 activity in the culture medium at least provides a location in which it could facilitate the final stages of cell division and cell separation. The observation that Lyt-14 grows in long chains at 42°C (8) is consistent with such a function. Similarly, the processive mechanism of hydrolysis by muramidase-1 and its high affinity for peptidoglycan chains provide characteristics notable for deletion of specific chains, permitting changes in the shape of assembled wall (morphogenesis). The ability of muramidase-1 to hydrolyze bonds in a restricted location provides it with characteristics suitable for more efficient lysis of cells—which, as far as the bacterium is concerned, is a negative role. Their differences in substrate specificity and mechanism of hydrolysis could result in complementary and cooperative actions. Muramidase-2 hydrolysis in the middle of a glycan chain would provide additional nonreducing ends for the binding of muramidase-1. Muramidase-1 hydrolysis of chains carrying covalently linked nonpeptidoglycan wall polymers could provide muramidase-2 with access to susceptible bonds.

Additional data concerning these two enzymes at the molecular and cellular levels should provide new insights into both how bacteria assemble their cell walls and how we might be able to disrupt this process.

ACKNOWLEDGMENTS. This research was supported by Public Health Service Research grant AI05044 from the National Institute of Allergy and Infectious Diseases.

We thank V. L. Schramm, J. Mays, and R. Kariyama for their help and advice. We thank Smith Kline & French Laboratories and E. I. Du Pont de Nemours & Co. for providing us with large quantities of bacterial cells.

LITERATURE CITED

1. **Barrett, J. F., D. L. Dolinger, V. L. Schramm, and G. D. Shockman.** 1984. The mechanism of soluble peptidoglycan hydrolysis by an autolytic muramidase—a processive exodisaccharidase. *J. Biol. Chem.* **259**:11818–11827.
2. **Barrett, J. F., V. L. Schramm, and G. D. Shockman.** 1984. Hydrolysis of soluble, linear, un-cross-linked peptidoglycans by endogenous bacterial *N*-acetylmuramoylhydrolases. *J. Bacteriol.* **159**:520–526.
3. **Barrett, J. F., and G. D. Shockman.** 1984. Isolation and characterization of soluble peptidoglycan from several strains of *Streptococcus faecium. J. Bacteriol.* **159**:511–519.
4. **Beachey, E. H., W. Keck, M. A. de Pedro, and U. Schwarz.** 1981. Exoenzymatic activity of transglycosylase isolated from *Escherichia coli. Eur. J. Biochem.* **116**:355–358.
5. **Cleveland, R. F., L. Daneo-Moore, A. J. Wicken, and G. D. Shockman.** 1976. Effect of lipoteichoic acid and lipids on lysis of intact cells of *Streptococcus faecalis. J. Bacteriol.* **127**:1582–1584.
6. **Cleveland, R. F., J.-V. Holtje, A. J. Wicken, A. Tomasz, L. Daneo-Moore, and G. D. Shockman.** 1975. Inhibition of bacterial wall lysins by lipoteichoic acids and related compounds. *Biochem. Biophys. Res. Commun.* **67**:1128–1135.
7. **Cleveland, R. F., A. J. Wicken, L. Daneo-Moore, and G. D. Shockman.** 1976. Inhibition of wall autolysis in *Streptococcus faecalis* by lipoteichoic acid and lipids. *J. Bacteriol.* **126**:192–197.
8. **Cornett, J. B., B. E. Redman, and G. D. Shockman.** 1978. Autolytic defective mutant of *Streptococcus faecalis. J. Bacteriol.* **133**:631–640.
9. **Coyette, J., H. R. Perkins, I. Polacheck, G. D. Shockman, and J.-M. Ghuysen.** 1974. Membrane-bound DD-carboxypeptidase and LD-transpeptidase of *Streptococcus faecalis* ATCC 9790. *Eur. J. Biochem.* **44**:459–468.
10. **Daneo-Moore, L., and G. D. Shockman.** 1977. The bacterial cell surface in growth and division, p. 597–715. *In* G. Poste and G. L. Nicolson (ed.), *The Synthesis, Assembly and Turnover of Cell Surface Components.* Elsevier/North-Holland Biomedical Press, Amsterdam.
11. **Dezelee, P., and G. D. Shockman.** 1975. Studies of the formation of peptide cross-links in the cell wall peptidoglycan of *Streptococcus faecalis. J. Biol. Chem.* **250**:6806–6816.
12. **Dubos, R.** 1945. *The Bacterial Cell,* p. 94–99. Harvard University Press, Cambridge, Mass.
13. **Ghuysen, J.-M., E. Bricas, M. Leyh-Bouille, M. Lache, and G. D. Shockman.** 1967. The peptide *N*-(L-alanyl-D-isoglutaminyl)-*N*-(D-isoasparaginyl)-L-acetylmuramic acid in cell wall peptidoglycan of *Streptococcus faecalis* strain ATCC 9790. *Biochemistry* **6**:2607–2619.
14. **Higgins, M. L., H. M. Pooley, and G. D. Shockman.** 1970. Site of initiation of cellular autolysis in *Streptococcus faecalis* as seen by electron microscopy. *J. Bacteriol.* **103**:504–512.
15. **Higgins, M. L., and G. D. Shockman.** 1970. Model for cell wall growth of *Streptococcus faecalis. J. Bacteriol.* **101**:643–648.
16. **Hinks, R. P., L. Daneo-Moore, and G. D. Shockman.** 1978. Cellular autolytic activity in synchronized populations of *Streptococcus faecium. J. Bacteriol.* **133**:822–829.
17. **Hinks, R. P., L. Daneo-Moore, and G. D. Shockman.** 1978. Relationship between cellular autolytic activity, peptidoglycan synthesis, septation, and the cell cycle in synchronized populations of *Streptococcus faecium. J. Bacteriol.* **134**:1074–1080.
18. **Höltje, J. V., D. Mirelman, N. Sharon, and U. Schwarz.** 1975. Novel type of murein transglycosylase in *Escherichia coli. J. Bacteriol.* **124**:1067–1076.
19. **Joseph, R., and G. D. Shockman.** 1976. Autolytic formation of protoplasts (autoplasts) of

Streptococcus faecalis: location of active and latent autolysin. *J. Bacteriol.* **127**:1482–1493.

20. **Kawamura, T., and G. D. Shockman.** 1983. Evidence for the presence of a second peptidoglycan hydrolase in *Streptococcus faecium. FEMS Microbiol. Lett.* **19**:65–69.

21. **Kawamura, T., and G. D. Shockman.** 1983. Purification and some properties of the endogenous, autolytic *N*-acetylmuramoylhydrolase of *Streptococcus faecium,* a bacterial glycoenzyme. *J. Biol. Chem.* **258**:9514–9521.

22. **Keck, W., F. B. Wientjes, and U. Schwarz.** 1985. Comparison of two hydrolytic murein transglycosylases of *Escherichia coli. Eur. J. Biochem.* **148**:493–497.

23. **Kusser, W., and U. Schwarz.** 1980. *Escherichia coli* murein transglycosylase—purification by affinity chromatography and interaction with polynucleotides. *Eur. J. Biochem.* **103**:277–281.

24. **Mett, H., W. Keck, A. Funk, and U. Schwarz.** 1980. Two different species of murein transglycosylase in *Escherichia coli. J. Bacteriol.* **144**:45–52.

25. **Mosser, J. L., and A. Tomasz.** 1970. Choline-containing teichoic acid as a structural component of pneumococcal cell wall and its role in sensitivity to lysis by an autolytic enzyme. *J. Biol. Chem.* **245**:287–298.

26. **Pooley, H. M., and G. D. Shockman.** 1969. Relationship between the latent form and the active form of the autolytic enzyme of *Streptococcus faecalis. J. Bacteriol.* **100**:617–624.

27. **Pooley, H. M., G. D. Shockman, M. L. Higgins, and J. Porres-Juan.** 1972. Some properties of two autolytic-defective mutants of *Streptococcus faecalis* ATCC 9790. *J. Bacteriol.* **109**:423–431.

28. **Rogers, H. J., H. R. Perkins, and J. B. Ward.** 1980. *Microbial Cell Walls and Membranes,* p. 437–460. Chapman & Hall, Ltd., London.

29. **Rosenthal, R. S., and G. D. Shockman.** 1975. Characterization of the presumed peptide cross-links in the soluble peptidoglycan synthesized by protoplasts of *Streptococcus faecalis* ATCC 9790. *J. Bacteriol.* **124**:410–418.

30. **Sanchez-Puelles, J. M., C. Ronda, E. Garcia, E. Mendez, J. L. Garcia, and R. Lopez.** 1986. A new peptidoglycan hydrolase in *Streptococcus pneumoniae. FEMS Microbiol. Lett.* **35**:163–166.

31. **Sanchez-Puelles, J. M., C. Ronda, J. L. Garcia, P. Garcia, R. Lopez, and E. Garcia.** 1986. Searching for autolysin functions—characterization of a pneumococcal mutant deleted in the *lyt A* gene. *Eur. J. Biochem.* **158**:289–293.

32. **Sayare, M., L. Daneo-Moore, and G. D. Shockman.** 1972. Influence of macromolecular biosynthesis on cellular autolysis in *Streptococcus faecalis. J. Bacteriol.* **112**:337–344.

33. **Sharon, N., and S. Seifter.** 1964. A transglycosylation reaction catalyzed by lysozyme. *J. Biol. Chem.* **239**:PC2398–PC2399.

34. **Shockman, G. D.** 1965. Symposium on the fine structure and replication of bacteria and their parts. IV. Unbalanced cell-wall synthesis: autolysis and cell-wall thickening. *Bacteriol. Rev.* **29**:345–358.

35. **Shockman, G. D., and J. F. Barrett.** 1983. Structure, function, and assembly of cell walls of gram-positive bacteria. *Annu. Rev. Microbiol.* **3**:501–527.

36. **Shockman, G. D., L. Daneo-Moore, T. D. McDowell, and W. Wong.** 1981. Function and structure of the cell wall—its importance in the life and death of bacteria, p. 31–65. *In* M. R. J. Salton and G. D. Shockman (ed.), *β-Lactam Antibiotics.* Academic Press, Inc., New York.

37. **Shockman, G. D., J. J. Kolb, and G. Toennies.** 1958. Relations between bacterial cell wall synthesis, growth phase, and autolysis. *J. Biol. Chem.* **230**:961–977.

38. **Shockman, G. D., H. M. Pooley, and J. S. Thompson.** 1967. Autolytic enzyme system of *Streptococcus faecalis.* III. Localization of the autolysin at the sites of cell wall synthesis. *J. Bacteriol.* **94**:1525–1530.

39. **Shockman, G. D., J. S. Thompson, and M. J. Conover.** 1967. The autolytic enzyme

system of *Streptococcus faecalis*. II. Partial characterization of the autolysin and its substrate. *Biochemistry* **6**:1054–1065.

40. **Shungu, D. L., J. B. Cornett, and G. D. Shockman.** 1979. Morphological and physiological study of autolytic-defective *Streptococcus faecium* strains. *J. Bacteriol.* **138**:598–608.

41. **Weidel, W., and H. Pelzer.** 1964. Bag-shaped macromolecules—a new outlook on bacterial cell walls. *Adv. Enzymol.* **26**:193–232.

Chapter 19

Expression of a Muramidase Encoded by the Pneumococcal Bacteriophage Cp-1 in *Escherichia coli*

José L. García
Ernesto García
Enrique Méndez
Pedro García
Concepción Ronda
Rubens López

The study of bacteriophages infecting *Streptococcus pneumoniae* has been a matter of great interest in our laboratory for a long time (6). We have been particularly interested in enzymes postulated to participate in the lysis of the host cells because better understanding of these enzymes is fundamental for elucidation of the mechanisms of liberation of the phage progeny. Cp-1, a phage containing a terminal protein covalently linked to the 5′ end of each strand, has been well studied at the molecular level, and a partial restriction cleavage map of Cp-1 DNA has been constructed (7). This information makes Cp-1 an appropriate candidate for use in studying the characteristics of the lysin involved in the liberation of the phage progeny. We report here the cloning and expression, in *Escherichia coli*, of a 2.9-kilobase (kb) *Acc*I fragment of Cp-1

José L. García, Ernesto García, Pedro García, Concepción Ronda, and Rubens López • Centro de Investigaciones Biológicas, Consejo Superior de Investigaciones Científicas, Velázquez 144, 28006 Madrid, Spain.   Enrique Méndez • Departamento de Endocrinología, Centro Ramón y Cajal, Carretera de Colmenar, 28034 Madrid, Spain.

1 DNA containing the lysin gene (*cpl*, for Cp-1 lysin) as well as the purification and biochemical characterization of the cloned lytic enzyme (CPL).

In Vitro Transcription-Translation of Cp-1 DNA

To test whether proteinase K-treated Cp-1 DNA contains a gene encoding its own lysin, transcription-translation assays were carried out. The mixture containing the polypeptides synthesized in vitro (1.5 μg of Cp-1 DNA) was incubated at 37°C for 45 min in a final volume of 30 μl. Afterwards, 10 μl of [^{3}H]choline-labeled cell walls (total cpm, 7,500) and 10 μl of [^{14}C]ethanolamine-labeled cell walls (total cpm, 6,000) were added, incubation was continued at 37°C for 135 min, and the radioactivity released was determined as previously described (4). The preparation containing [^{3}H]choline-labeled walls released 1,932 cpm, whereas the preparation containing [^{14}C]ethanolamine-labeled walls released no radioactivity. This finding of choline-specific cell wall lytic activity indicated that Cp-1 DNA does code for a lytic enzyme.

Cloning of Cp-1 DNA in E. coli

Cp-1 DNA is cleaved once by *Pvu*II, *Hae*III, *Sph*I, and *Sau*96I and twice by *Hin*dIII and *Hpa*I (7), but is resistant to many other restriction endonucleases (8). On the other hand, it has been pointed out that the terminal protein-DNA bond is extremely resistant and is only partially hydrolyzed by treatment with strong alkali (3). This makes the cloning of the terminal fragments of Cp-1 DNA very difficult. The 8.6-kb *Hin*dIII fragment (i.e., the internal one) of Cp-1 DNA was cloned in pBR325, but no expression of the phage lysin was obtained. Furthermore, attempts to clone and express the lysin gene of Cp-1 by using the restriction endonucleases *Taq*I and *Dra*I were also unsuccessful. As these enzymes cleave Cp-1 DNA many times, producing rather small fragments, we chose *Acc*I, which generates only eight fragments ranging from 0.5 kb to about 8 kb (Fig. 1). We cloned the 2.9-kb *Acc*I fragment of Cp-1 DNA, first into the replicative form of M13 mp11 DNA and then into the high-copy-number plasmid pBR325. The complete strategy of this approach is also depicted in Fig. 1. The new recombinant plasmid, named pCIP50, expresses the CPL in *E. coli* (Table 1). The results reported in Table 1 also confirm that the CPL specifically requires the presence of choline in the teichoic acids of the cell walls; replacement of choline by its analog ethanolamine rendered the walls resistant to degradation by the enzyme. It is important to emphasize that this property is shared with the two lytic enzymes (an amidase and a glycosidase) already found in *S. pneumoniae* (9, 10).

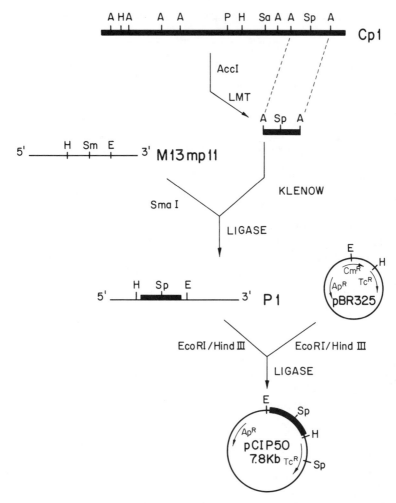

Figure 1. Construction of a plasmid containing the *cpl* gene. Plasmids are drawn as circles with the relevant elements and restriction sites indicated. The sizes of the recombinant plasmids are given in kilobases. Cp-1 and M13 DNAs are represented by a heavy and a light line, respectively. Only a limited number of restriction enzymes are shown. Abbreviations: A, *Acc*I; E. *Eco*RI; H, *Hind*III; P, *Pvu*II; Sa, *Sau*96I; Sm, *Sma*I; Sp, *Sph*I; Ap, ampicillin; Cm, chloramphenicol; Tc, tetracycline; LMT, low-melting-temperature agarose.

Purification of the CPL

By taking advantage of the affinity of the CPL for choline, this enzyme was purified by chromatography on a choline-Sepharose 6B column (1). Sonicated extracts obtained from *E. coli* HB101(pCIP50), containing 14 mg of

Table 1. Degradation of pneumococcal cell walls by extracts
obtained from different *E. coli* strains[a]

Strain	Substrate[b]	Activity (U/mg of protein)
HB101(pBR325)	C-cw	0
HB101(pCIP50)	C-cw	6,000
	EA-cw	0

[a]Lysates were obtained by sonication as previously described (2).
[b]C-cw, Choline-containing cell walls; EA-cw, ethanolamine-containing cell walls.

protein per ml and 8.6×10^4 U of CPL per ml, were applied on a choline-Sepharose 6B column (1.5 by 30 cm). The adsorbed CPL was eluted with 50 mM Tris-maleate buffer (pH 6.9) containing 2% choline and 0.1% Brij-58, and the active fractions were concentrated with polyethylene glycol 20,000. The specific activity of the purified CPL was 4.9×10^5 U/mg, and the recovery was about 70%. This preparation was examined for homogeneity by sodium dodecyl sulfate-polyacrylamide gel electrophoresis (Fig. 2). A single band with an apparent M_r of about 39,000 was found.

Biochemical Characterization of the CPL

Preliminary data showed that hydrolysis of pneumococcal walls was accompanied by a fivefold increase in the reducing power of the suspension, but with no concomitant increase in the number of free amino groups. To determine the nature of this glycosidase, samples of pneumococcal cell walls (6 mg/ml)

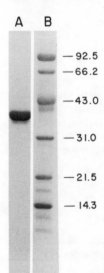

Figure 2. Sodium dodecyl sulfate-polyacrylamide gel electrophoresis of the purified CPL. Lane A, Purified CPL; lane B, protein standards (phosphorylase B, bovine serum albumin, ovalbumin, carbonic anhydrase, soybean trypsin inhibitor, and lysozyme). The M_rs are indicated in the margin in thousands.

were treated with the purified CPL (4.6 × 10^5 U/ml) at 37°C for 18 h and reduced with NaB^3H_4 (0.5 Ci/mmol), as described by Ward (11), to label the free reducing groups present. The reduced samples were hydrolyzed and applied on a Dowex 50 WX4 (200 to 400 mesh) column (0.9 by 20 cm) equilibrated in 0.1 M pyridine-acetate buffer (pH 2.8). Fractions (2 ml) were eluted with the same buffer (total, 70 ml) and then with 0.133 M pyridine-acetate buffer (pH 3.85) (arrow in Fig. 3). When samples of the fractions were analyzed for radioactivity, the radioactivity was found mainly as [^3H]muramitol (Fig. 3), demonstrating that the CPL is a muramidase.

N-Terminal Amino Acid Sequence of the CPL

The N-terminal amino acid sequence of the CPL was determined by Edman degradation with the use of an automatic sequenator (5). The defined sequence Val-Lys-Lys-Asn-Asp-Leu-Phe-Val-Asp-Val was obtained. This result correlates with the N-terminal amino acid composition derived from the partial nucleotide sequence of the *cpl* gene recently determined (E. García et al., this volume).

Concluding Remarks

The in vitro transcription-translation experiments using Cp-1 DNA as template showed that this DNA synthesizes an enzyme that specifically lyses pneumococcal cell walls containing choline in their teichoic acids. In addition, the

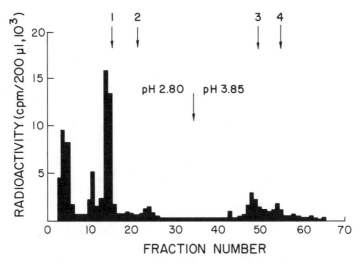

Figure 3. Analysis of free reducing groups in hydrolyzed cell wall of *S. pneumoniae*. 1, Muramitol; 2, muramic acid; 3, glucosaminol; 4, glucosamine.

results presented here demonstrated that a fragment of Cp-1 DNA cloned in plasmid pBR325 is capable of coding and expressing in *E. coli* a lytic activity that specifically recognizes in vitro choline-containing pneumococcal cell walls. Attempts to clone and express pneumococcal genes in *E. coli* have been unsuccessful until recently (2). To the best of our knowledge, the results presented here are the first report of cloning and expression of a pneumococcal phage gene in *E. coli*.

Briese and Hakenbeck (1) recently described a rapid and efficient method to purify the pneumococcal amidase. They reported that the specific and strong attachment of the amidase to the choline residues led to an improved affinity chromatographic system for the purification of this enzyme on choline-Sepharose columns. This system has turned out to be very useful for the purification of CPL. The results presented above demonstrate that the CPL is a muramidase that shares interesting properties with the pneumococcal and phage lysins studied so far (i.e., requirement of cell wall containing choline in the teichoic acid for activity and noncompetitive inhibition by the pneumococcal Forssman antigen and by free choline [data not shown]).

Preliminary Southern blot analyses (García et al., this volume) have indicated the existence of a certain degree of homology between M13 and Cp-1 DNAs and between the *lytA* gene and Cp-1 DNA. These results suggest the possibility of a homology between the muramidase encoded by Cp-1 DNA and the two lytic enzymes present in the host cell. These findings will facilitate the cloning of the host glycosidase, an enzyme that might play an important role in the synthesis or division, or both, of cell walls.

LITERATURE CITED

1. **Briese, T., and R. Hakenbeck.** 1985. Interaction of the pneumococcal amidase with lipoteichoic acid and choline. *Eur. J. Biochem.* **146:**417–427.
2. **García, E., J. L. García, C. Ronda, P. García, and R. López.** 1985. Cloning and expression of the pneumococcal autolysin gene in *Escherichia coli. Mol. Gen. Genet.* **201:**225–230.
3. **García, E., A. Gómez, C. Ronda, C. Escarmís, and R. López.** 1983. Pneumococcal bacteriophage Cp-1 contains a protein bound to the 5' termini of its DNA. *Virology* **128:**92–104.
4. **Höltje, J. V., and A. Tomasz.** 1976. Purification of the pneumococcal N-acetylmuramyl-L-alanine amidase to biochemical homogeneity. *J. Biol. Chem.* **251:**4199–4207.
5. **López, E., A. Grubb, and E. Méndez.** 1982. Human protein HC display variability in its carboxyl-terminal amino acid sequence. *FEBS Lett.* **144:**349–353.
6. **López, R., P. García, C. Ronda, and E. García.** 1982. Genetics and biology of pneumococcal phages, p. 141–144. *In* D. Schlessinger (ed.), *Microbiology—1982.* American Society for Microbiology, Washington, D.C.
7. **López, R., C. Ronda, P. García, C. Escarmís, and E. García.** 1984. Restriction cleavage maps of the DNAs of *Streptococcus pneumoniae* bacteriophages containing protein covalently bound to their 5' ends. *Mol. Gen. Genet.* **197:**67–74.

8. **Ronda, C., R. López, and E. García.** 1981. Isolation and characterization of a new bacteriophage, Cp-1, infecting *Streptococcus pneumoniae*. *J. Virol.* **40:**551–559.
9. **Sánchez-Puelles, J. M., C. Ronda, E. García, E. Méndez, J. L. García, and R. López.** 1986. A new peptidoglycan hydrolase in *Streptococcus pneumoniae*. *FEMS Microbiol. Lett.* **35:**163–166.
10. **Tomasz, A., M. Westphal, E. B. Briles, and P. Fletcher.** 1975. On the physiological functions of teichoic acids. *J. Supramol. Struct.* **3:**1–16.
11. **Ward, J. B.** 1973. The chain length of glycans in bacterial walls. *Biochem. J.* **133:**395–398.

Chapter 20

Functional Domains of the Lytic Enzymes Involved in the Recognition and Degradation of the Cell Wall of *Streptococcus pneumoniae*

Ernesto García
José L. García
Pedro García
José M. Sánchez-Puelles
Rubens López

Streptococcus pneumoniae contains two peptidoglycan hydrolases, a powerful *N*-acetylmuramic acid-L-alanine amidase (EC 3.1.1.3) (9) and a glycosidase (14). The *lytA* gene encoding the pneumococcal amidase has been cloned and expressed in *Escherichia coli* (3), and its nucleotide and deduced amino acid sequences have been determined (5). This enzyme appears to be involved in the liberation of the phage progeny at the end of the infective process (13), although other phage-encoded lysins are also implied, namely, the amidase from Dp-1 (4, 6) and the CPL muramidase encoded by the *cpl* gene of Cp-1 (J. L. García et al., this volume).

All the lytic enzymes of *S. pneumoniae* and its bacteriophages described so far require the presence of choline in the teichoic acids for activity. The

Ernesto García, José L. García, Pedro García, José M. Sánchez-Puelles, and Rubens López • Centro de Investigaciones Biologicas, Consejo Superior de Investigaciones Científicas, Velázquez 144, 28006 Madrid, Spain.

biosynthetic substitution for this amino alcohol of its analog ethanolamine renders the cell walls of this bacterium refractory to the degradative action of the lytic enzymes. In addition, these enzymes are inhibited by the pneumococcal lipoteichoic acid (8) (a lipopolysaccharide that also contains choline) as well as by free choline (1, 7). Therefore, the unique presence of choline in the cell wall of *S. pneumoniae* (15) appears to behave as a factor of strong selective pressure, suggesting that the evolutionary divergence between host and parasite lysins would be smaller than in other species. To test this hypothesis, we have partially determined the nucleotide sequence of the *cpl* gene encoding the muramidase of bacteriophage Cp-1.

Homology between the cpl and the lytA Genes

The biochemical similarities found between the host amidase and the CPL muramidase encoded by Cp-1 DNA prompted us to test whether the *lytA* and the *cpl* genes also showed homology at the nucleotide level. Figure 1 shows that pGL80 (*lytA*) hybridized intensely to a 2.9-kilobase *Acc*I fragment of Cp-1 DNA. This fragment was previously cloned in *E. coli* with pBR325 used as a vector, and the recombinant plasmid (pCIP50) allowed the expression of the phage-coded muramidase (CPL) in the heterologous system (García et al., this

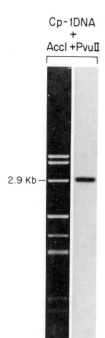

Cp-1DNA
+
AccI +PvuII

2.9 Kb—

Figure 1. Ethidium bromide-stained agarose gel and Southern blot analysis of Cp-1 DNA digested with *Acc*I and *Pvu*II. The gel was blotted to Hybond membranes and hybridized at 50°C for 5 h with ^{32}P-labeled pGL80 (5). The size of the restriction fragment of Cp-1 DNA that hybridizes with the probe is indicated. See Lopez et al. (11) and García et al. (this volume) for a restriction cleavage map of Cp-1 DNA.

volume). These results suggested that the *cpl* gene was responsible for the strong hybridization signal observed with the *lytA* gene. Other experiments not shown here indicated that the region of homology is located to the right of the unique *Sph*I site of the genome of Cp-1 (see Fig. 2).

Partial Nucleotide Sequence of the cpl Gene

Figure 2 shows the strategy used to partially determine the nucleotide sequence of the *cpl* gene encoding the CPL muramidase. Plasmid pCIP50 was digested with *Eco*RI and *Sph*I and cloned into M13 tg131 previously digested with the same enzymes. Similarly, pCIP50 was linearized with *Hind*III, digested with the *Bal*31 nuclease, and made blunt-ended with the Klenow fragment of DNA polymerase I in the presence of the four deoxynucleoside triphosphates. After digestion with *Eco*RI, the DNA fragment containing the *cpl* gene was inserted into M13 tg131 previously treated with *Eco*RI and *Hinc*II.

The DNA sequence and the amino acid translation appear in Fig. 3. The ATG initiation codon could be localized by comparison with the N-terminal amino acid sequence of the CPL muramidase previously determined (García et

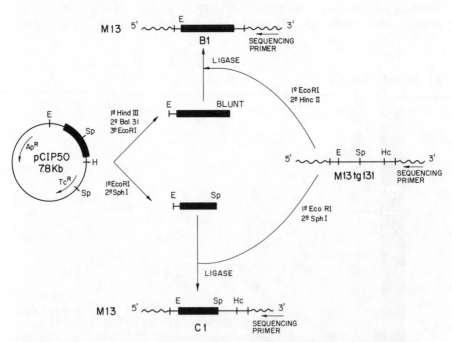

Figure 2. Partial restriction map of pCIP50 and the strategy for determining the nucleotide sequence of the *cpl* gene. The location of the restriction sites used for sequencing is indicated. Abbreviations: E, *Eco*RI; H, *Hind*III; Hc, *Hinc*II; Sp, *Sph*I. For details, see text.

N terminus

```
                              M   V   K   K   N   D   L   F   V   D   V   S   S   H   N   G   Y   D   I   T   G   I   L   E   Q   M   G
AAAAGACCTGCTAGAAGAATTTAAAGGAGAAAAGAAATAATGGTTAAAAAGAATGATTTATTTGTAGATGTTTCAAGTCACAACGGTTACGATATAACAGGTATCTTGGAGCAAATGGGA

  I   F   N   T   I   I   K   I   S   E   S   T   T   Y   L   N   P   C   L   S   A   Q   V   E   Q   S   N   P   I   G   F   Y   H   F   A   R   F   G   G   D
ACAACTAACACCATCATTAAAATTTCTGAAAGTACGACCTATTTAAACCCTTGCTTGTCTGCTCAAGTGGAGCAGTCAAACCCTATTGGCTTTTATCACTTCGCACGCTTTGGCGGAGAC

  V   A   E   A   E   R   E   A   Q   F   F   L   D   N   V   P   M   Q   V   K   Y   L   V   L   D   Y   E   D   G   P   S   G   D   A   Q   A   N   T   N   A
GTAGCAGAAGCCGAAAGAGAAGCGCAGTTTTTCCTTGACAACGTGCCTATGCAAGTTAAATACCTTGTATTGGACTACGAGGACGACCCAAGCGGAGACGCACAAGCGAACACTAACGCA

  C
TGC
```

C terminus

```
  M   V   T   G   W   V   K   Y   K   N   N   W   Y   Y   M   T   N   E   R   G   N   M   V   S   N   E   F   I   K   S   G   K   G   W   Y   F   M   N   T   N
ATGGTTACTGGTTGGGTCAAGTATAAGAATAACTGGTACTATATGACAAATGAACGTGGTAACATGGTTTCTAATGAATTTATTAAGTCTGGAAAAGGTTGGTATTTCATGAACACAAAC

  G   E   L   A   D   N   P   S   F   T   K   E   P   D   G   L   I   T   V   A end
GGAGAGCTTGCAGACAATCCAAGTTTCACGAAAGAACCAGACGGGCTTATAACCGTAGCATAAAAAAAGAAAAGCTAGTAGGATTTTCCTACTAGCTGTTTTTATAGTCTGCTATAATTTT

ATAAGCATCTTCGTCTGGATTATCCAGAGCGATGGAACAGATTGCA
```

Figure 3. Nucleotide sequence of the N- and C-terminal domains of the *cpl* gene. Only the non-coding strand of the DNA is shown. The sequence of the N terminus and that of the C terminus were obtained by sequencing the clones C1 and B1, respectively (see Fig. 2). The number of nucleotides that remain to be sequenced, as estimated by restriction mapping, is about 500.

al., this volume). Ten bases upstream from the initiator ATG codon, a ribo-some-binding sequence (AAGGAG) is found. The nucleotide sequence of the C terminus of the *cpl* gene is also shown. The correct open reading frame was determined by comparison with the sequence of the *lytA* gene previously reported (5). When both sequences were compared, it became evident that no statistically significant homology exists between their 5′ ends. However, about 50% identical nucleotides were found at the C termini of both genes. These results are in agreement with the hybridization analyses reported above which revealed a high degree of DNA-DNA homology between the *lytA* gene and the region of Cp-1 DNA corresponding to the right of the *Sph*I restriction site. As expected, similar results were found when we compared the amino acid sequences of both proteins; i.e., about 50% identical amino acids were observed in the C-terminal residues. Taking into account that in the homologous region many nonidentical amino acids are in fact conserved substitutions, the C-terminal domains of the CPL and of the amidase can be considered virtually identical.

Computer Searches for Amino Acid Homologies

Preliminary protein homology searches were conducted by using the National Biomedical Research Foundation Protein Sequence Data Bank with the computer programs of Lipman and Pearson (10). It was observed that the N termini of the CPL and of the *Chalaropsis* muramidase (2) were unexpectedly homologous (not shown). The latter enzyme is one of the few enzymes able to degrade either choline- or ethanolamine-containing pneumococcal cell walls. In addition, the two amino acid residues of the *Chalaropsis* muramidase involved in the active center of the enzyme (aspartic acid residue 6 and glutamic acid residue 33) are conserved in the sequence of CPL.

Concluding Remarks

The similar choline dependence for activity shown by the host amidase and the CPL muramidase suggested the existence of some structural similarities between the two enzymes. The availability of the *cpl* gene cloned into pBR325 (García et al., this volume) has allowed us to partially determine the sequence of this gene (Fig. 3) and to compare it with that of the *lytA* gene encoding the host amidase (5). The remarkably high homology found at the C termini of both enzymes, together with the lack of homology at their N ends, seems to indicate that the C-terminal domains of these lytic enzymes are involved in the recognition of the choline residues present in the cell wall of *S. pneumoniae*. On the other hand, the N-terminal amino acid sequence of CPL is strikingly similar to that of the muramidase of *Chalaropsis* sp., suggesting that the active center of CPL (and of the host amidase) might be located in this region. Taken together, all these results show that lysin genes of *S. pneumoniae* and its bacteriophages are composed of two different modules, i.e., one responsible for the attachment of the enzyme to its insoluble substrate and the other conferring specificity for the catalytic activity. Our results add experimental support to the hypothesis of modular evolution of phage DNAs (12).

LITERATURE CITED

1. **Briese, T., and R. Hakenbeck.** 1985. Interaction of the pneumococcal amidase with lipoteichoic acid and choline. *Eur. J. Biochem.* **146:**417–427.
2. **Fouche, P. B., and J. H. Hash.** 1978. The *N, O*-diacetylmuramidase of *Chalaropsis* species. Identification of aspartyl and glutamyl residues in the active site. *J. Biol. Chem.* **253:**6787–6793.
3. **García, E., J. L. García, C. Ronda, P. García, and R. López.** 1985. Cloning and expression of the pneumococcal autolysin gene in *Escherichia coli. Mol. Gen. Genet.* **201:**225–230.
4. **García, P., E. García, C. Ronda, R. López, and A. Tomasz.** 1983. A phage-associated murein hydrolase in *Streptococcus pneumoniae* infected with bacteriophage Dp-1. *J. Gen. Microbiol.* **129:**489–497.
5. **García, P., J. L. García, E. García, and R. López.** 1986. Nucleotide sequence of the pneumococcal autolysin gene from its own promoter in *Escherichia coli. Gene* **43:**265–272.
6. **García, P., E. Méndez, E. García, C. Ronda, and R. López.** 1984. Biochemical characterization of a murein hydrolase induced by bacteriophage Dp-1 in *Streptococcus pneu-*

moniae: comparative study between bacteriophage-associated lysin and the host amidase. *J. Bacteriol.* **159**:793–796.

7. **Giudicelli, S., and A. Tomasz.** 1984. Attachment of pneumococcal autolysin to wall teichoic acids, an essential step in enzymatic wall degradation. *J. Bacteriol.* **158**:1188–1190.

8. **Höltje, J. V., and A. Tomasz.** 1975. Lipoteichoic acid: a specific inhibitor of autolysin activity in pneumococcus. *Proc. Natl. Acad. Sci. USA* **72**:1690–1694.

9. **Howard, L. V., and H. Gooder.** 1974. Specificity of the autolysin of *Streptococcus (Diplococcus) pneumoniae*. *J. Bacteriol.* **117**:796–804.

10. **Lipman, D. J., and W. R. Pearson.** 1985. Rapid and sensitive protein similarity searches. *Science* **227**:1435–1441.

11. **López, R., C. Ronda, P. García, C. Escarmis, and E. García.** 1984. Restriction cleavage maps of the DNAs of *Streptococcus pneumoniae* bacteriophages containing protein covalently bound to their 5′ ends. *Mol. Gen. Genet.* **197**:67–74.

12. **Reanney, D. C., and H. W. Ackerman.** 1982. Comparative biology and evolution of bacteriophages. *Adv. Virus Res.* **27**:205–280.

13. **Ronda, C., R. López, A. Tapia, and A. Tomasz.** 1977. Role of the pneumococcal autolysin (murein hydrolase) in the release of progeny bacteriophage and in the phage-induced lysis of the host cells. *J. Virol.* **21**:366–374.

14. **Sánchez-Puelles, J. M., C. Ronda, E. García, E. Méndez, J. L. García, and R. López.** 1986. A new peptidoglycan hydrolase in *Streptococcus pneumoniae*. *FEMS Microbiol. Lett.* **35**:163–166.

15. **Tomasz, A.** 1967. Choline in the cell wall of a bacterium: novel type of polymer-linked choline in pneumococcus. *Science* **157**:694–697.

Chapter 21

Recovery of Gram-Negative Rods from Mecillinam: Role of Murein Hydrolases in Morphogenesis

E. W. Goodell
Anneleen Asmus

Salmonella typhimurium and *Escherichia coli* are both bacilli. In rich medium they have a mean diameter of 1 μm and an average length of 2.5 μm (11, 12). If, however, mecillinam is added to the culture, the cells gradually transform themselves into cocci; after two generations they are almost perfectly spherical, with a diameter of 2 μm (4). These spherical cells are unlike osmotically labile spheroplasts in two respects: (i) they have an intact sacculus (cell wall), and (ii) they quickly revert to rods after removal of the mecillinam (4). As they become rods, they also reduce their diameter and after three to four generations most have regained their original 1-μm diameter.

Since the bacteria retain an intact sacculus throughout the rod-to-sphere-to-rod transformations, they must be reshaping the sacculus as they change shape. The ability to mold the sacculus is not unique to mecillinam-treated cells; it is also evident under normal growth conditions. For example, during shift up from minimal to rich medium, the cells double in diameter, and when they are returned to minimal medium they again reduce their diameter by half (1). Thus, in situ, the sacculus appears to be a plastic structure which can assume a variety of shapes depending upon the culture conditions. In many organisms, such as plants and fungi, the cell wall is inherently plastic in that the glycan strands are held together only by hydrogen bonds and thus can be displaced relative to each other by osmotic pressure (9). In contrast, the sacculus of bacteria is a covalently interlinked network of polysaccharide chains

E. W. Goodell and Anneleen Asmus • Natural Science Department, State University of New York College of Technology, Utica, New York 13504.

224

and oligopeptides; any change in its overall shape would require an additional cutting and resynthesis of covalent bonds. Of the several possible mechanisms bacteria might use to adjust the shape of their cell walls, two appear most likely. (i) Murein with the incorrect morphology could simply be removed from the sacculus and discarded into the growth medium. Presumably, before this old murein can be removed, new murein with the correct shape would have to be added to the sacculus. (ii) A sufficient number of bonds in the old cell wall could be cut to allow displacement of parts of the sacculus relative to one another. These pieces would then have to be firmly reattached to one another, perhaps by a transpeptidation reaction involving newly synthesized murein.

In this chapter we describe some of the effects of sphere-rod morphogenesis on the metabolisms of murein in the sacculus. These shape changes increased the rate of loss of murein fragments from the sacculus, suggesting that *S. typhimurium* and *E. coli* adjust the shape of their sacculi by removing portions with the incorrect morphology.

Morphogenic Conversion of Bacteria

Bacteria and Culture Conditions

S. typhimurium CH44 and *E. coli* CH212 were obtained from Christopher Higgins. Both lack the oligopeptide permease (7, 8) and are unable to take up and reuse cell wall peptides (3).

The bacteria were grown in Minimal C medium (5) supplemented with glucose (2 mg/ml) and each of the amino acids (40 μg/ml) found in proteins, except cysteine, at 37°C with vigorous aeration. Turbidity of the cultures was measured at 578 nm; generation time under these conditions was 30 to 35 min.

Mecillinam at a concentration of 1 μg/ml was used to convert the cells to cocci. Cell division was inhibited by growing the bacteria in nalidixic acid (10 μg/ml). To label the murein, bacteria were grown in the presence of 2,6-diamino[G-^{3}H]pimelic acid (^{3}H-A$_2$pm; 1 Ci/mmol; Amersham Corp.); the final concentration of ^{3}H label in the culture was 1 μCi/ml.

The dimensions of bacteria were determined from photographs made with a light microscope; the grids of a Petroff-Hausser cell counter were used as an internal size standard.

Assay of Murein Synthesis and Degradation

Cells labeled with ^{3}H-A$_2$pm were separated from the culture medium by centrifugation in an Eppendorf microcentrifuge (10,000 × *g*, 1 min). Murein was isolated from the cells by resuspending the cell pellet in 1% sodium dodecyl sulfate, boiling the cells for 5 min, and centrifuging the produce for 15 min in the Eppendorf centrifuge to collect the sacculi. The sacculi were then washed twice with distilled water. Radiolabeled cell wall peptides in the culture fluid were separated and analyzed as described earlier (5).

Effect of Mecillinam on Cell Morphology

Figure 1 illustrates the change in cell shape induced by growing *E. coli* CH212 in 1 μg of mecillinam per ml. After two generations, the bacteria are essentially spherical and have a diameter twice that of the original rods. Removal of mecillinam allows the cells to revert to rods and to regain their original diameter. Inhibition of cell division (with 10 μg of nalidixic acid per ml) does not affect the reversion (Fig. 1). Thus, reduction in cell diameter and

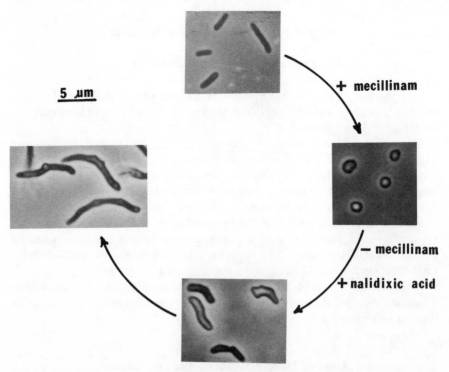

Figure 1. Morphogenesis of *E. coli* CH212. Mecillinam was added to an exponentially growing culture of rods (top photograph), and the cells were cultured for 60 min to convert them to spheres (right photograph). Then the mecillinam was removed by washing the bacteria, nalidixic acid was added to prevent cell division, and the bacteria were allowed to revert to rods (bottom and left photographs).

reformation of pole caps can take place during elongation of the sacculus and do not require the constriction of the cell sacculus associated with septum formation and cell division.

Loss of Material from the Sacculus During Morphogenesis of the Cells

During their normal cell cycle, *E. coli* and *S. typhimurium* (5; E. W. Goodell and N. O. Buswell, unpublished data) remove pieces of murein from the sacculus. The initial glycopeptide fragments are rapidly degraded to free peptides (L-Ala-D-Glu-A$_2$pm and LD-Glu-A$_2$pm-D-Ala) and saccharides (probably disaccharides). These molecules are small enough to pass through the porins of the outer membrane, and some do escape into the culture medium. The peptides, however, can be reused by the bacteria to synthesize new murein (2); they are returned to the cytoplasm by the oligopeptide permease (3) and are attached (without degradation) to the UDP-*N*-acetylmuramic acid to form UDP-*N*-acetylmuramyl-tripeptide (2). In wild-type bacteria with an active oligopeptide permease (opp^+), most of the tri- and tetrapeptides released from the sacculus appear to be recycled in this manner. Finally, the bacteria also lose a dipeptide, A$_2$pm-D-Ala, into the culture fluid. This compound appears to be cut directly from tetrapeptide side clains in the sacculus; it is not reutilized by the bacteria (2).

Bacteria which are recovering from mecillinam treatment might be expected to show increased loss of murein from their sacculi if they were removing portions of the wall which did not have the proper shape. To test this an *opp* strain of *S. typhimurium*, CH44, which cannot recycle cell wall peptides was converted to cocci by adding 1 µg of mecillinam to a 1-ml culture when the turbidity was $A_{578} = 0.07$. After two generations (60 min), when the cells were essentially spherical, their murein was labeled by growing them for an additional 30 min in ^{3}H-A$_2$pm. The ^{3}H-A$_2$pm and mecillinam were then removed by washing the cells and placing them in 16 ml of fresh medium. As a control, rod-shaped exponentially growing cells were also labeled for 30 min with ^{3}H-A$_2$pm. The loss of label from the murein sacculi of both samples was then followed by periodically removing 2-ml samples for 120 min until the cells undergoing sphere-rod morphogenesis had regained their original diameter.

Measurement of ^{3}H label in the murein sacculi as described above showed that the reverting spheres lost more label (44%; ca. 10% per generation) from their walls than did the control cells (10%) (Fig. 2). When we analyzed the spent medium from these cultures, we found about fivefold more ^{3}H-tripeptide and ^{3}H-tetrapeptide in the medium from reverting spheres than in the control culture. Thus, the bacteria undergoing sphere-rod morphogenesis appear to be cutting glycopeptide fragments from the cell walls at an increased rate. These

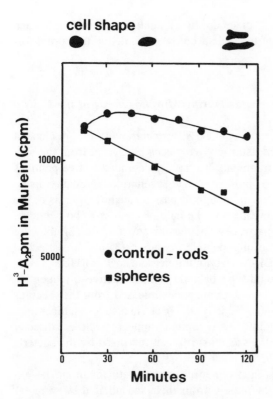

Figure 2. Effect of sphere-rod morphogenesis on murein metabolism.

cells also released the disaccharide-peptides *N*-acetyglucosamine (β-1,4)-1,6-anhydro-*N*-acetylmuramyl-L-Ala-D-Glu-A$_2$pm and *N*-acetylglucosamine (β-1,4)1,6-anhydro-*N*-acetylmuramyl-L-Ala-D-Glu-A$_2$pm-D-Ala into the culture medium (ca. 25% of the total label released), suggesting that the outer membrane had become somewhat leaky. We found equal amounts of the ^{3}H-dipeptide, A$_2$pm-D-Ala, in the two cultures.

Since cells undergoing sphere-rod morphogenesis have a higher rate of murein loss, one might expect the same of rods undergoing the reverse morphogenesis to spheres. To test this, we added ^{3}H-A$_2$pm to 1 ml of an *S. typhimurium* CH44 culture with a turbidity of 0.3. After 30 min, the label was removed by washing the cells. The cells were suspended in 16 ml of fresh medium containing mecillinam and allowed to grow for 2 h until they were round; 2-ml samples were periodically removed, and ^{3}H label in the murein sacculi was measured as described above. A control culture was treated in the same way except that it received no mecillinam. As expected, the reverting rods lost more label from their walls than did the control cells (Fig. 3).

The experiments depicted in Fig. 2 and 3 were repeated with an *E. coli opp* strain, CH212, with similar results.

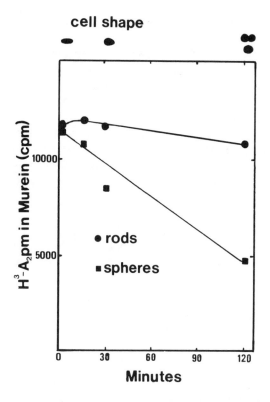

Figure 3. Effect of rod-sphere morphogenesis on murein metabolism.

Role of Murein Hydrolases in Morphogenesis of the Murein Sacculus

S. typhimurium and *E. coli* reverting from spheres to rods had an increased rate of loss of murein from their sacculi as did cells undergoing rod-to-sphere morphogenesis. The results suggest that a bacterium must remove murein from its sacculus when it reshapes it. It is, however, worth noting that approximately half of the ^{3}H label was still retained within the sacculus at the end of 2 h when morphogenesis was essentially complete. A perfect sphere with a diameter of 2 μm and a perfect rod with a length of 2.5 μm and a diameter of 1 μm (and hemispherical ends) have completely different surface geometries; that is, no part of the 2-μm-diameter sphere has the same radius of curvature as the rod that is 1 by 2.5 μm. Thus, if the above conclusion about removing murein with an incorrect shape was strictly true, bacteria reverting from spheres to rods would be expected to lose all their spherical murein. Figure 2 shows that this is clearly not the case. There are several possible explanations for this.

(i) The ^{3}H-murein which is still attached to the sacculus may no longer be shape determining. It may be partially degraded fragments of murein which are attached to the outer surface of the sacculus, as found in gram-positive baccili (10).

(ii) It could, however, just as well be an integral part of the sacculus since partial degradation of spherical murein may increase the flexibility to the point where it can become part of the rod-shaped sacculus. In this case morphogenesis of the sacculus would be a combination of removing portions of the old murein and reshaping the rest.

(iii) Finally, the resolution of a light microscope is not adequate to determine whether the morphogenesis is complete and the cells have become "perfect" rods. We are, therefore, planning to look at sacculi from reverting spheres in an electron microscope and to use autoradiography to determine where on the surface of the sacculus the ^3H label remains.

ACKNOWLEDGMENTS. This work was supported by Public Health Service grant 1-R15-AI231139-01 from the National Institute of Allergy and Infectious Diseases.

LITERATURE CITED

1. **Donachie, W. D.** 1981. The cell cycle of *Escherichia coli*, p. 63–83. *In* P. C. L. John (ed.), *The Cell Cycle.* Cambridge University Press, Cambridge.
2. **Goodell, E. W.** 1985. Recycling of murein by *Escherichia coli. J. Bacteriol.* **163**:305–310.
3. **Goodell, E. W., and C. F. Higgins.** 1987. Uptake of cell wall peptides by *Salmonella typhimurium* and *Escherichia coli. J. Bacteriol.* **169**:3861–3865.
4. **Goodell E. W., and U. Schwarz.** 1975. Sphere-rod morphogenesis of *Escherichia coli. J. Gen. Microbiol.* **86**:201–209.
5. **Goodell, E. W., and U. Schwarz.** 1985. Release of cell wall peptides into culture medium by exponentially growing *Escherichia coli. J. Bacteriol.* **162**:391–397.
6. **Helmstetter, C. E.** 1967. Rate of DNA synthesis during the division cycle of *Escherichia coli* B/r. *J. Mol. Biol.* **24**:417–427.
7. **Higgins, C. F., and M. M. Hardie.** 1983. Periplasmic protein associated with the oligopeptide permeases of *Salmonella typhimurium* and *Escherichia coli. J. Bacteriol.* **155**:1434–1438.
8. **Higgins, C. F., M. M. Hardie, D. J. Jamieson, and L. M. Powell.** 1983. Genetic map of the *opp* (oligopeptide permease) locus of *Salmonella typhimurium. J. Bacteriol.* **153**:830–836.
9. **Koch, A. L.** 1983. The surface stress theory of microbial morphogenesis. *Adv. Microb. Physiol.* **24**:301–366.
10. **Pooley, H. M.** 1976. Layered distribution, according to age, within the cell wall of *Bacillus subtilis. J. Bacteriol.* **125**:1139–1147.
11. **Schaechter, M., O. Maaløe, and N. O. Kjeldgaard.** 1958. Dependency on medium and temperature of cell size and chemical compositon during balanced growth of *Salmonella typhimurium. J. Gen. Microbiol.* **19**:592–606.
12. **Trueba, F. J., and C. F. Woldringh.** 1980. Changes in cell diameter during the division cycle of *Escherichia coli. J. Bacteriol.* **142**:869–878.

Chapter 22

Hydrolysis of RNA during Penicillin-Induced Nonlytic Death in a Group A Streptococcus

Thomas D. McDowell
Kelynne E. Reed

Exposure of bacterial cultures to antibiotics which inhibit cell wall synthesis, assembly, or both may ultimately lead to one of four responses: cells may lyse, die and then lyse, die without lysis (nonlytic death), or survive (tolerance). The secondary effects of, for example, exposure to benzylpenicillin which result in immediate lysis of cells have been well documented (12, 14). Likewise, many aspects of nonlytic responses have been described. Clearly, an absent, well-regulated or weakly expressed autolytic system is one prerequisite of both tolerance and nonlytic death (12, 14). Activation of a global regulatory circuit which enables inhibitions of peptidoglycan (PG) assembly to be communicated back to cytoplasmic processes (talk-back regulation) has been proposed as the basis of tolerance in streptococci (8; T. D. McDowell et al., submitted for publication). In regard to nonlytic death, sublytic autolysin activity (PG nicking) has been suggested as a possible mechanism (2). Recent studies, however, which quantitated changes in the chain length of PG polymers in cell walls isolated from cultures of group A streptococci after exposure to penicillin G found no evidence for sublytic autolysin activity (C. L. Lemanski and T. D. McDowell, *Abstr. Annu. Meet. Am. Soc. Microbiol. 1987*, A115, p. 20).

In this report we present results of experiments which describe the penicillin G-dependent hydrolysis and the subsequent loss of RNA from apparently

Thomas D. McDowell and Kelynne E. Reed • University of New Mexico School of Medicine, Albuquerque, New Mexico 87131.

intact cells of a group A streptococcus. In addition, we present evidence which suggests that this response is more closely associated with the bactericidal action of penicillin G than with hydrolysis of PG.

Effects of Penicillin G Treatment on the Accumulation of Protein and RNA

In a previous report (8) we described the dose-dependent sequential inhibition of PG, RNA, and protein synthesis subsequent to exposure of exponentially growing cultures of tolerant strains of *Streptococcus mutans* to penicillin G, the talk-back response. To determine whether this response was a characteristic of nonlytic phenotypes, the net accumulation of RNA and protein during exposure of group A streptococci to penicillin G was determined by radiolabeling techniques described previously (8). These methods, which utilize cultures grown for at least six generations in the presence of [^{3}H]leucine (protein) or [^{14}C]uracil (RNA), were designed to ensure that the radioactivity measured in trichloroacetic acid (TCA) precipitates collected on glass fiber filters would reflect the relative amounts of the macromolecules in the cells (8). Group A streptococcal strain 1224 was a clinical isolate obtained from St. Christophers Hospital for Children, Philadelphia, Pa. *Streptococcus faecium* ATCC 9790 was obtained from our laboratory stock cultures. All cultures were stored in the lyophilized state. For routine use, inocula were freshly prepared from frozen glycerol stocks and monitored for purity by plating on blood agar plates. Cultures were grown aerobically (8) in chemically defined media (CDM for strain 1224 [15] and SFM for strain 9790 [10]). Growth was monitored turbidimetrically at 675 nm in a Bausch & Lomb Spectronic 21 colorimeter, and optical density readings were corrected as previously described (13); one adjusted optical density unit equals 0.45 µg of cellular dry weight per ml. In all cases, inocula were no more than two transfers away from the original lyophilized cultures. Since conventional methods for determining minimal inhibitory concentrations (MIC) were considered unsatisfactory for these studies the 50% growth inhibitory concentrations (GIC$_{50}$) were determined for each strain as described previously (7). (Potassium penicillin G [1,595 U/mg] used in these experiments was a gift from Wyeth Laboratories, Philadelphia, Pa., and chloramphenicol was obtained from Parke, Davis & Co., Detroit, Mich.)

The inhibition of growth and net accumulation after exposure of exponentially growing cultures to a GIC$_{50}$ and saturating (10 × GIC$_{50}$) dose of penicillin G are shown in Fig. 1. Reductions in turbidity were dose dependent and accompanied by a similar reduction of label in TCA-precipitable RNA. Accumulation of label into protein was inhibited to a lesser degree; interestingly, no reduction in protein was observed. These results provided the first indication that exposure to pencillin G was inducing or deregulating cytoplasmic RNase(s)

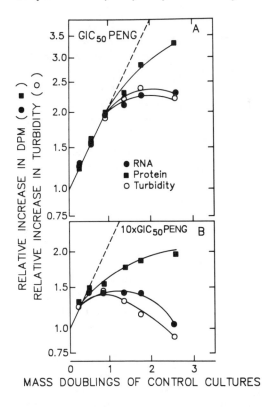

Figure 1. Comparison of the inhibition of growth (adjusted optical density units) and net accumulation of RNA and protein after exposure of exponentially growing cultures of group A streptococci to inhibitory (GIC_{50}, 0.1 μg/ml) and saturating (10 times the GIC_{50}) doses of penicillin G. (---) Untreated control culture. At a turbidity equivalent to 45 μg of cell dry weight per ml (time zero), cultures growing exponentially for at least six mass doublings in the presence of radiolabeled precursors ([^{14}C]uracil for RNA and [^{3}H]leucine for protein) were distributed into tubes containing penicillin G. At the intervals shown, samples were taken for determinations of the content of each macromolecule, as described in the text. At time zero, 6.8 × 10^3 and 2.4 × 10^4 dpm/ml were present in RNA and protein, respectively.

which hydrolyzed RNA and permitted the solubilized label to diffuse from cells. The results also demonstrated that inhibition of PG assembly could be communicated to (talk-back to) cytoplasmic processes in these sensitive streptococci in an apparently similar manner to that observed in tolerant strains (8).

Penicillin G-Induced Losses of Cellular Macromolecules and Viability

To further describe the effects of pencillin G on cellular macromolecules, the loss of RNA, protein, PG, and DNA from cells subsequent to exposure of exponentially growing cultures to penicillin G was determined by quantitating radiolabel, incorporated in macromolecules, which was retained on membrane filters (pore size, 0.45 μm). Cultures grown for at least six generations in the presence of [^{14}C]leucine (protein) and [^{14}C]uracil (RNA) or [^{3}H]lysine (PG) or [^{3}H]thymidine (DNA) were harvested and washed via filtration. Cells were suspended in unlabeled (chase) media containing concentrations of penicillin G indicated below. At intervals, 0.5-ml samples were collected on filters and washed with two 1-ml volumes of iced media. RNA and DNA levels were determined by counting the filters directly. Protein was determined after additional washes with iced TCA (10%, twice in 1 ml), 95°C TCA (10%, six times in 1 ml), iced H_2O (twice in 1 ml), and iced ethanol (95%, twice in 1 ml). Lysine levels present in PG were determined by the method of Boothby et. al. (1) by processing membrane filters directly. Label present in protein and PG in untreated control cells collected and processed on membrane filters was comparable to that present in parallel samples processed by traditional TCA precipitation techniques (1). All samples dried on filters were transferred to Filmware scintillation vials, and radioactivity was determined via liquid scintillation spectrophotometry. Corrections were made for quench and overlap of ^{14}C counts into the ^3H channel.

Comparison of the kinetics of loss of RNA, DNA, protein, and PG (Fig. 2) showed that the secondary effect of penicillin G on cellular RNA pools is specific and apparently is not the result of cellular lysis. Note that RNA ([^{14}C]uracil) was lost from cells at an exponential rate and in a dose-dependent manner. At the two highest concentrations (10 and 100 times the GIC_{50}), the loss was quite extensive (90% in 6 h for 10 mg of penicillin G per ml). This release of RNA was not a result of cellular lysis as is evident from the lack of a parallel loss of DNA, protein, or PG. Loss of label from DNA also showed a dose dependency; however, the timing of the loss lagged well behind RNA and may represent the degree of cytoplasmic disruption. Protein loss was incomplete (32%) and dose dependent. This partial loss of protein may reflect the loss of surface molecules, a limited proteolytic activity and/or loss of a low-molecular-weight subset of, for example, cytoplasmic proteins, or both. No evidence of penicillin G-induced PG hydrolysis was observed. The abrupt increase in label present in insoluble PG in control cultures and those exposed to a subinhibitory concentration of penicillin G, at a time equivalent to approximately three mass doublings, paralleled the transition from exponential- to stationary-phase growth. This increase may reflect cell wall thickening, which has been observed in other streptococci (5, 11), i.e., transfer of [^3H]lysine present in protein to PG.

Susceptibility to the lethal effects of penicillin G was determined by conventional agar overlay techniques. Exponentially growing cultures diluted to appropriate initial turbidities were transferred to tubes containing penicillin G at concentrations indicated below. At various time intervals, samples were collected and diluted in prewarmed media. Duplicate 0.1-ml samples were added to 3 ml of soft agar containing penicillinase (95 mU/ml) and plated on Todd Hewitt agar. Following incubation in a CO_2-enriched atmosphere at 37°C for 24 h, plates containing 30 to 300 colonies were counted. CFU were corrected for drug-induced changes in cell chain length as described previously (6).

Survival after exposure of exponentially growing cultures to penicillin G (10 times the GIC_{50}) for times equivalent to one, two, and three mass doublings

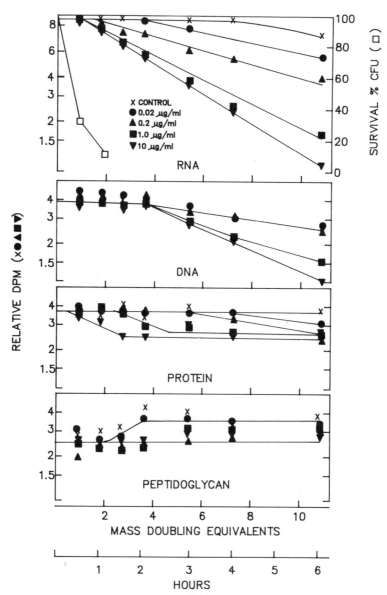

Figure 2. Comparison of the loss of viability and label from RNA, DNA, protein, and PG after exposure of exponentially growing cultures of group A streptococci to various concentrations of penicillin G. Cultures grown in the presence of appropriately labeled precursors ([^{14}C]uracil, RNA; [^{3}H]thymidine, DNA; [^{14}C]leucine, protein; and [^{3}H]lysine, PG) for six mass doublings in exponential growth were harvested and transferred to unlabeled medium containing penicillin G at an initial turbidity equivalent to 45 μg of cell dry weight per ml. At the intervals indicated, samples were collected on membrane filters and processed as described in the text. Viability (survival, percentage of time-zero control CFU) was determined in parallel experiments as described in the text. At time zero, 1.7×10^{4}, 7.8×10^{3}, 8×10^{4}, and 2.4×10^{4} dpm/ml were present in RNA, DNA, protein, and PG, respectively.

235

is also shown in Fig. 2. Together, these results domonstrate that in this group A streptococcus, reductions in viability subsequent to exposure to penicillin G are not associated with significant solubilization of PG or cellular lysis. Furthermore, the kinetics and extent of loss of RNA suggest that this feature of penicillin G treatment is significant enough to lead to losses in viability. In addition, results of preliminary studies with a tolerant strain, *Streptococcus rattus* FA-1 (data not presented), have shown that exposure to comparable concentrations of penicillin G does not trigger an extensive hydrolysis of RNA.

Effects of Penicillin G Treatment on a Lytic Phenotype

To confirm the validity of the experimental approach in distinguishing between lytic and nonlytic responses, experiments similar to those presented in Fig. 2 were performed with *Streptococcus faecium* 9790 (Fig. 3). The lytic nature of this species is clearly apparent (Fig. 3A). After exposure to a saturating dose of penicillin G, cultures rapidly lost turbidity which was accompanied by an extensive loss of PG and RNA (Fig. 3B). At a time equivalent to two mass doublings of the control, there was a 99.9% loss of viability (data not shown). Exposure to an inhibitory concentration of penicillin G, however, resulted in a quite different pattern. A rapid but partial reduction in label remaining in insoluble PG was observed. This degree of autolytic activity was not sufficient to lyse cells since neither a reduction in turbidity nor a loss of RNA was observed during the same time interval. At longer incubation times, the kinetics of RNA release observed after exposure to the GIC_{50} of penicillin G more closely resemble those observed after exposure of the nonlytic species to high concentrations of penicillin G (Fig. 2) than those of either lysis (10 times the GIC_{50}) or loss of PG (Fig. 3). Fifty percent of the culture retained viability after exposure to this concentration of penicillin G for a time equivalent to two mass doublings of the control (Fig. 3A). These results demonstrate the validity of the approach and suggest that exposure to penicillin G can also trigger the hydrolysis and release of RNA in this lytic bacterium. It should be noted that 30 years ago Prestidge and Pardee (9) reported similar findings in another lytic species, *Escherichia coli*.

Effects of Simultaneous Exposure to Chloramphenicol and Penicillin G

The antagonistic effects of, for example, chloramphenicol on penicillin G-induced killing have been attributed to the inhibition of expression of autolytic activity. In the light of the apparent lack of either cellular lysis or PG hydrolysis during periods of rapid reductions in viability subsequent to exposure of group

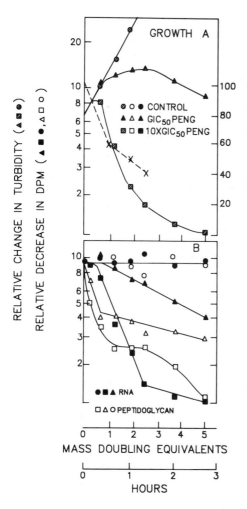

Figure 3. Comparison of growth inhibition, cell lysis (adjusted optical density units), loss of RNA ([^{14}C]uracil) and insoluble PG ([^{3}H]lysine), and viability (survival) after exposure of *S. faecium* 9790 to inhibitory (GIC$_{50}$, 2.5 μg/ml) and saturating (10 times the GIC$_{50}$) doses of penicillin G. Culturing, labeling, and processing conditions were as described in the legend for Fig. 2. Cells were suspended in unlabeled medium at an initial turbidity equivalent to 90 μg of cell dry weight per ml. (x---x, panel A) Percent survival after exposure to the GIC$_{50}$ of penicillin G. At time zero, 1.3 × 10^4 and 8 × 10^4 dpm/ml were present in PG and RNA, respectively.

A streptococci to penicillin G (Fig. 2), we examined the effects of simultaneous exposure to chloramphenicol and penicillin G on hydrolysis of RNA. The antagonistic effect of chloramphenicol on penicillin G-induced reduction of turbidity is shown in Fig. 4. Note again that reductions in turbidity after exposure to penicillin G agreed well with losses of label from RNA. Also note that simultaneous exposure to penicillin G and chloramphenicol resulted in an inhibition of the loss of RNA. In parallel experiments which quantitated penicillin G-induced killing, addition of chloramphenicol resulted in a greater than twofold increase in survival after an exposure interval equivalent to two mass doubling times (approximately 66 min). The effects of antibiotic treatment on label present in insoluble PG are also shown (Fig. 4C). Interestingly, chloramphen-

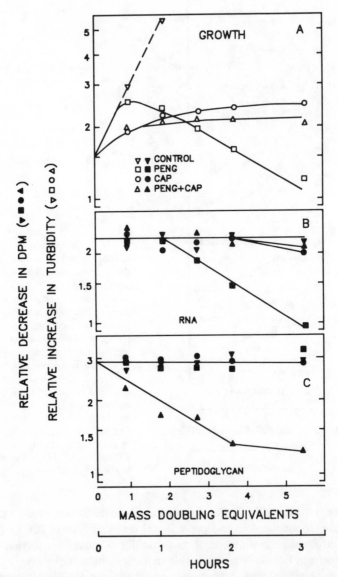

Figure 4. Effects of simultaneous exposure to chloramphenicol (10 times the GIC_{50}, 20 μg/ml) and penicillin G (10 times the GIC_{50}, 1 μg/ml) on growth (adjusted optical density units) and loss of RNA ([^{14}C]uracil) and insoluble PG ([^{3}H]lysine) in exponentially growing cultures of group A streptococci. For culturing, labeling, and processing details, see the legend for Fig. 2 and the text. At time zero, 1.6×10^{3} and 3×10^{4} dpm/ml were present in RNA and PG, respectively.

icol in combination with penicillin G induced an exponential reduction of label. This chloramphenicol-penicillin G induction of PG hydrolase(s), in combination with the chloramphenicol antagonism of penicillin G-induced killing and the absence of penicillin G-induced autolysis, clearly indicates that hydrolysis of the cell wall is unrelated to nonlytic death. Also, the chloramphenicol antagonism of both RNA release and killing suggests a relationship between these events.

The chloramphenicol-penicillin G-related expression of autolysis has been described previously in studies of *E. coli* (3). The mechanism is apparently related to the ability of chloramphenicol to uncouple the stringent regulation of PG synthesis. We previously reported the secondary inhibition of RNA and protein synthesis after exposure of tolerant streptococci to penicillin G (8). More recently, we described the similarity of the "talk-back" response to stringent control (submitted for publication). The results presented here provide additional indirect evidence for both the stringent control of PG synthesis in streptococci and the similarity of the mechanisms of talk-back and stringent regulation. The results presented in this report and elsewhere (Lemanski and McDowell, *Abstr. Annu. Meet. Am. Soc. Microbiol. 1987*) indicate that the mechanism of penicillin G-nonlytic death in this group A streptococcus is apparently not related to hydrolysis of the cell wall. Both the kinetics and extent of RNA loss suggest that this feature of penicillin G treatment is sufficient to account for losses in viability. The hypothesis is further supported by the observation that chloramphenicol antagonizes both penicillin G-induced killing and RNA hydrolysis. In addition, results obtained from studies with lytic streptococcus, *S. faecium* 9790, suggest that the penicillin G-induced hydrolysis of RNA is not restricted to nonlytic phenotypes. This observation is in good agreement with previously reported studies on *E. coli* (9) and *Proteus mirabilis* (4).

Summary

Studies of the effects of penicillin G on RNA and protein synthesis in a group A streptococcus revealed a dose-dependent inhibition of the net accumulation of both macromolecules with RNA synthesis being inhibited earlier than protein. Simultaneous treatment with both penicillin G and chloramphenicol resulted in a relaxation of the penicillin G-induced inhibition of RNA accumulation. At a saturating concentration of penicillin G, reductions in culture turbidity were observed which were accompanied by a parallel loss of RNA. Additional studies clearly demonstrated that the reductions of culture turbidity were not the result of cellular lysis but were apparently a reflection of the specific hydrolysis of RNA. Results of these and other experiments suggest that, in group A streptococci, penicillin G-induced inhibition of cell surface growth triggers a series of events which produce a specific inhibition of RNA synthesis and a subsequent deregulation/activation of RNase activity. Since neither turnover nor sublytic hydrolysis of the peptidoglycan can be detected in this strain, it is reasonable to expect that the hydrolysis of RNA may be the basis of penicillin G-induced death. In addition, the basis of chloramphenicol antagonism of penicillin G killing in this group A streptococcus is likely to be associated with the inhibition of expression of RNase rather than with the autolytic activity.

ACKNOWLEDGMENT. This work was supported by Public Health Service grant DE07077 from the National Institute of Dental Research.

LITERATURE CITED

1. **Boothby, D., L. Daneo-Moore, and G. D. Shockman.** 1971. A rapid, quantitative, and selective estimation of radioactively labeled peptidoglycan in Gram-positive bacteria. *Anal. Biochem.* **44:**645–653.
2. **Gutmann, L., and A. Tomasz.** 1982. Penicillin-resistant and penicillin-tolerant mutants of group A streptococci. *Antimicrob. Agents Chemother.* **22:**128–136.
3. **Kusser, W., and E. E. Ishiguro.** 1986. Lysis of nongrowing *Escherichia coli* by combinations of β-lactam antibiotics and inhibitors of ribosome function. *Antimicrob. Agents Chemother.* **29:**451–455.
4. **Lorian, U., L. D. Sabath, and M. Simionescu.** 1975. Decrease in ribosomal density of *Proteus mirabilis* exposed to subinhibitory concentrations of ampicillin or cephalothin (3888). *Proc. Soc. Exp. Biol. Med.* **149:**731–735.
5. **Mattingly, S. J., L. Daneo-Moore, and G. D. Shockman.** 1977. Factors regulating cell wall thickening and intracellular iodophilic polysaccharide storage in *Streptococcus mutans.* *Infect. Immun.* **16:**967–973.
6. **McDowell, T. D., C. E. Buchanan, J. Coyette, T. S. Swavely, and G. D. Shockman.** 1983. The effects of mecillinam and cefoxitin on growth, macromolecular synthesis, and penicillin binding proteins in a variety of gram-positive streptococci. *Antimicrob. Agents Chemother.* **23:**750–756.
7. **McDowell, T. D., T. S. Swavely, and G. D. Shockman.** 1982. A proposed method for quantifying and comparing the early inhibitory effects of antibiotics on growing bacterial cultures. *FEMS Microbiol. Lett.* **17:**59–62.
8. **Mychajlonka, M., T. D. McDowell, and G. D. Shockman.** 1980. Inhibition of peptidoglycan, ribonucleic acid, and protein synthesis in tolerant strains of *Streptococcus mutans.* *Antimicrob. Agents Chemother.* **17:**572–582.
9. **Prestidge, L. S., and A. B. Pardee.** 1957. Induction of bacterial lysis by penicillin. *J. Bacteriol.* **74:**48–59.
10. **Shockman, G. D.** 1963. Amino acids, p. 567–673. *In* F. Kavanagh (ed.), *Analytical Microbiology.* Academic Press, Inc., New York.
11. **Shockman, G. D.** 1965. Symposium on the fine structure and replication of bacteria and their parts. IV. Unbalanced cell-wall synthesis: autolysis and cell-wall thickening. *Bacteriol. Rev.* **29:**345–358.
12. **Shockman, G. D., L. Daneo-Moore, T. D. McDowell, and W. Wong.** 1981. Function and structure of the cell-wall—its importance in the life and death of bacteria, p. 31–35. *In*

M. J. Salton and G. D. Shockman (ed.), *β-Lactam Antibiotics. Mode of Action, New Developments, and Future Prospects.* Academic Press, Inc., New York.

13. **Toennies, G., and D. L. Gallant.** 1949. The relationship between photometric turbidity and bacterial concentration. *Growth* **13:**7–20.

14. **Tomasz, A.** 1979. The mechanism of the irreversible antimicrobial effects of penicillins: how the beta-lactam antibiotics kill and lyse bacteria. *Annu. Rev. Microbiol.* **33:**113–137.

15. **van de Rigin, I., and R. E. Kessler.** 1980. Growth characteristics of group A streptococci in a new chemically defined medium. *Infect. Immun.* **27:**444–448.

Chapter 23

Biochemical and Biophysical Investigations into the Cause of Penicillin-Induced Lytic Death of Staphylococci: Checking Predictions of the Murosome Model

Harald Labischinski
Heinrich Maidhof
Marita Franz
Dominique Krüger
Thomas Sidow
Peter Giesbrecht

Staphylococcus aureus SG511 "Berlin," like many other bacteria, responds to penicillin G treatment in a highly dose-dependent manner (Fig. 1). At very low drug concentrations (up to about 0.01 µg/ml), growth curves were only slightly affected. At an intermediate dose range (optimum at 0.1 µg/ml for our strain), bacteriolysis resulted, while high concentrations of the drug (≥ 1 µg/ml) did not lead to lysis; instead, the staphylococci suffered nonlytic death. From electron microscopy investigations of this strain, we recently proposed a new model

Harald Labischinski, Heinrich Maidhof, Marita Franz, Dominique Krüger, Thomas Sidow, and Peter Giesbrecht • Robert Koch Institute of the Federal Health Office, D-1000 Berlin (West) 65, Federal Republic of Germany.

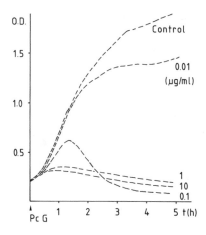

Figure 1. Growth curves of *S. aureus* SG511 Berlin for a control culture and cultures under the influence of sublytic (0.01 μg/ml), optimal lytic (0.1 μg/ml), and nonlytic (1 and 10 μg/ml) doses of penicillin G added at time zero (marked by the arrow). 7185

according to which penicillin-induced bacteriolysis in staphylococci involves the punching of a few minute holes into the peripheral cell wall (7). These perforations were thought to be the result of the lytic activity of extraplasmatic vesicular structures within the wall material, which we have named murosomes (6). Very briefly, this model states that at the relatively low bacteriolytic penicillin G concentration neither a breakdown of wall synthesis nor an overall activation of autolytic enzymes is a prerequisite of bacteriolysis (7, 11). The deadly incident seemed instead to be the wrong local distribution of wall material synthesized during penicillin treatment; since staphylococci are well known to create every new septum at right angles to the previous one, it was easy to demonstrate that wall material needed to construct the cross wall in the second division plane was incorrectly deposited further in the first division plane. This observation suggested that the minute holes in the peripheral wall (along the line marking the circumference of the second division plane after penicillin addition) were due to those murosomes responsible for cutting the peripheral wall to initiate cell separation. Thus, the normal initiation of cell separation (but after failure to assemble regular cross wall) was assumed to be the very reason for the penicillin-induced bacteriolysis.

This murosome model allows us to make several predictions: (i) the amount and the quality of cell wall material synthesized during penicillin action should be sufficient to protect the cell from being disrupted by mechanical forces, (ii) bacteriolysis should not occur before initiation of cell separation at the end of the second generation after penicillin addition, and (iii) in synchronized bacteria, the start of bacteriolysis inevitably should take place during the onset of cell separation, i.e., at the identical stage in the cell cycle, independent of the duration of the penicillin treatment. The present study was undertaken to check the validity of these predictions.

Treatment of Staphylococci with Penicillin G

Procedures for Sample Examination

Bacterial strains and growth media. S. *aureus* SG511 Berlin (from the culture collection of the Robert Koch Institute, Berlin) was used. The growth medium was peptone water (2.5% Bacto-Peptone [Difco Laboratories]) and 0.5% NaCl). Cells from an overnight culture at 37°C were transferred to fresh medium to give an initial optical density at 578 nm of 0.05. They were further grown under shaking, and during logarithmic growth penicillin G (potassium salt) was added. The generation time was approximately 30 min.

Labeling of cell walls, isolation of walls, and chemical analysis of the wall material. Growing staphylococci were specifically labeled in their cell walls by adding N-[^{14}C]acetylglucosamine (Amersham-Buchler, Braunschweig, Federal Republic of Germany) to the culture. The label was added together with the antibiotic to label exclusively the wall material synthesized during penicillin G action. Cell walls were isolated by established procedures (11). Standard techniques were used for the chemical analysis of the wall (12). Very briefly, the degree of cross-linking was either determined by analysis of free amino end groups of glycine using the dinitrofluorobenzene method or by gel chromatographic separation of walls digested with *Chalaropsis B*-muramidase into monomers, dimers, etc., of the peptidoglycan disaccharide-peptide units. *O*-Acetyl content of the walls was determined by gas chromatography as described by Fromme and Beilharz (5).

Synchronization of staphylococci. Staphylococci were synchronized by the elutriation method (4). The elutriator rotor used was a Beckman JE-6 model driven in a Beckman JB-5 centrifuge operated at 4°C to avoid growth of the cells. The fraction of cells to be used in the experiments was collected during a period of 1 h at a rotor speed of 5,820 rpm and a flow rate of 3.3 ml/min. The cells were suspended in fresh, prewarmed medium at 37°C for further use.

Cell counts, determination of cell size distributions, and microcalorimetry. Total bacterial cell numbers were determined by phase-contrast microscopy after fixation of the staphylococci. Viable counts were measured by using a spiral plater. Microcalorimetric determination of heat production by the bacterial cultures was performed continuously with an LKB 2107 flow instrument. Size distribution of the staphylococci was measured with the aid of a Coulter Counter (model ZBI) connected to a Channelyzer Z 1000. The counter was equipped with an orifice 30 μm wide. In some of the experiments Formalin-fixed cells were used. Measurements were done either in 0.15 M sodium chloride or directly in the growth medium, both thoroughly purified from other particles by filtration through membrane filters of 0.22-μm pore size (Millipore Corp.).

Release of [³H]uracil label. Staphylococci from the logarithmic phase of growth were suspended in 25 ml of fresh medium containing 150 μl of [5,6-³H]uracil (1 mCi/ml; Amersham-Buchler) to give an optical density of 0.1. After four generations of growth, bacteria were harvested by centrifugation and resuspended at an optical density of 0.2 in 50 ml of fresh medium containing penicillin G at the desired concentration. At intervals samples were withdrawn, and bacteria were removed by centrifugation. The supernatants were monitored for radioactive label by the usual procedures of scintillation counting.

Details of all experimental procedures and the equipment used will be published elsewhere.

Staphylococcal Reactions to Penicillin G

Amount and Quality of Cell Wall Material Synthesized during Penicillin G Action

To answer the question of whether penicillin would induce a shortage of cell wall material which might be causative for lysis, we compared the rate of synthesis of wall material by the culture at different penicillin concentrations with that of a control culture. Wall synthesis was inferred from the uptake of *N*-[¹⁴C]acetylglucosamine, since 95% of this label was shown to be incorporated into the amino sugars of the cell wall (1, 14). Results show wall label incorporated into the cells after 30 and 60 min of drug treatment expressed as a percentage of total label supplied to the culture (Fig. 2). At very low drug concentrations near the MIC (ca. 0.01 μg/ml), uptake of *N*-acetylglucosamine was actually enhanced. Most important, even the optimal lytic concentration of penicillin under these experimental conditions (see reference 11) led to only a slight decline in label incorporated into the wall (after 60 min, 76% of the total label supplied versus the 85% control value), while a serious reduction in wall synthesis took place only at rather high, nonlytic drug concentrations. It should be noted that even at a drug concentration of 6 μg/ml (approximately 300 times the MIC) a residual incorporation was detected representing 35% of that of the control culture. Thus, in quantitative terms, at lytic penicillin concentrations there was obviously no shortage of wall material sufficient to account for bacteriolysis.

Yet, the possibility existed that the wall material formed under the influence of penicillin differed from control walls in qualitative aspects, so that it might be regarded as useless for the mechanical protection of the cell. We therefore isolated the walls of bacteria pretreated for 60 min with penicillin and analyzed them for possible changes in chemical structure. We first looked for

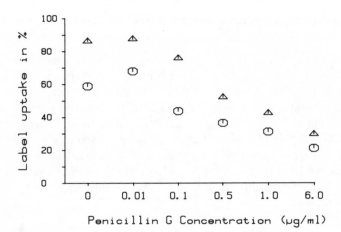

Figure 2. Ongoing cell wall synthesis of *S. aureus* SG511 Berlin after 30-min (△) and 60-min (○) treatment with penicillin at the concentrations given on the abscisssa, as measured by the incorporation of the cell wall-specific label, [^{14}C]*N*-acetylglucosamine. Values are given as a percentage of the total label supplied simultaneously with the drug to the culture at time zero.

alterations in overall amino acid or amino sugar composition by amino group analysis; however, we found no significant differences between the walls of penicillin-treated and untreated bacteria (data not shown). But as could in part be anticipated from earlier studies (for review, see reference 12), alterations in wall chemistry were, in fact, observed. Figure 3 shows the O acetylation of the muramic acid residues within the isolated peptidoglycan expressed as nano-

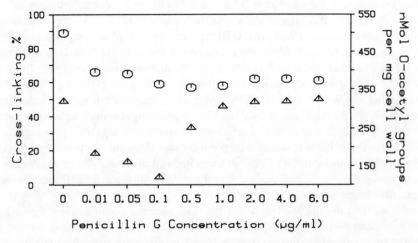

Figure 3. Alteration in the degree of O acetylation (△, expressed as nanomoles of *O*-acetyl groups per milligram of cell walls) and degree of cross-linking (○, in percent) of the wall after a 60-min treatment with penicillin at the doses indicated on the abscissa.

moles of *O*-acetyl groups per milligram of cell walls. As can be seen from Fig. 3, a drastic decrease in O acetylation occurred at very low penicillin concentrations, reaching a minimum at the optimal lytic concentration; this is similar to what has been observed with *Proteus mirabilis* and *Neisseria gonorrhoeae* (2, 10). Interestingly, at high, nonlytic doses a relative increase up to the control values in the degree of O acetylation was observed (Fig. 3). Although this finding is not completely understood, it is tempting to speculate that a low *O*-acetyl content might contribute to the low autolytic capacity of penicillin-treated staphylococci (11; Wong et al., unpublished results cited in reference 13), because it has been demonstrated that the activity of autolytic enzymes can be reduced by missing *O*-acetyl groups (3).

As far as wall stability is concerned, probably the most important chemical alteration in the fine structure of the wall is in the degree of cross-linking of the peptidoglycan peptide strands, which is well known to be reduced as a result of the blocking of transpeptidases by penicillin (12). Cross-linking of the wall material after a 60-min treatment with various doses of penicillin was therefore determined in our strain by measurement of free amino end groups of the glycine residues located in the interpeptide bridge of the staphylococcal peptidoglycan. The degree of cross-linking, expressed as bound N-terminal glycine as a percentage of total (free plus bound) N-terminal glycine, is also given in Fig. 3. As expected, at low penicillin concentrations the cross-linking was already significantly reduced. Unexpected, however, was the finding that nearly all of the cross-linking reduction achievable by the highest antibiotic doses tested (ca. 300 times the MIC, 6 μg/ml; cross-linking reduced from about 90% of the control value to about 60%) was already observed with the low, sublytic concentration of 0.01 μg/ml. This means that a 600-fold increase in inhibitor concentration did not further reduce the still very high degree of cross-linking of about 60% found in the isolated wall material. We obtained analogous results from an alternative cross-linking determination via gel-chromatographic separation of the wall fragments obtained by *Chalaropsis B*-muramidase treatment (data not shown). It should be stressed that a cross-linking of 60% is still higher than what has been found in most other bacteria, since in gram-positive organisms values around 50% and in gram-negative organisms values between 25 and 35% were usually determined (12).

Considering the complete data from the qualitative analysis of the staphylococcal cell wall synthesized during 60 min of penicillin treatment (a complete compilation of all data will be published elsewhere), it is hardly imaginable that the alterations observed would cause any weakness in wall structure sufficient to lead to a bursting of the wall. One might, however, argue that the comparison of the degree of cross-linking of the penicillin-treated staphylococcal wall with that of other bacterial species might be inadequate because of the relatively short sugar chains of staphylococcal peptidoglycan consisting of about 10 disaccharide units (12; unpublished data). One could reason that even if cross-linking of the wall was only slightly reduced by penicillin action, the cell might retain less mechanical stability than needed to balance its own osmotic pressure and, consequently, might swell and eventually burst. To check this possibility, we used Coulter Counter volume measurements to follow the alteration of size distribution (i.e., bacterial number versus volume) in a cell population induced by the optimal lytic dose of penicillin. Figure 4 shows the size distributions observed with bacteria synchronized by the elutriation method (4). The use of synchronously dividing bacteria is of great advantage, because their narrow size distribution makes any alterations more accurately observable than in the case of nonsynchronized cultures. Moreover, the results cannot be obscured by changes in the separation behavior of dividing cells as long as the observation period is restricted to the first generation time. Samples were withdrawn every 4 min and measured directly in fresh medium, starting 2 min before addition of the drug. As can be seen in Fig. 4, after 6 min of penicillin treatment the bacteria already appeared obviously swollen as compared with the untreated controls. A maximum increase in average volume of 20% relative to the controls was observed after 10 min of penicillin action. Unexpectedly, from that time until the end of the observation period, i.e., up to the start of the second generation time, no further volume increase occurred; in fact, compared with the control cells the size of the penicillin-treated cells was reduced

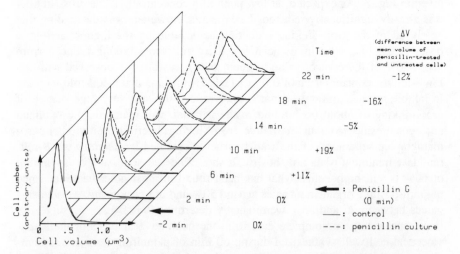

Figure 4. Alterations caused by penicillin (0.1 μg/ml) in cell size distribution during growth of synchronized *S. aureus* SG511 Berlin. The left-hand part of the figure shows the development of size distribution (bacterial cell numbers versus cell volume in cubic micrometers) with time in a control culture (———) and in a penicillin-treated culture (– – –). Zero time corresponds to addition of the drug (arrow). The right-hand part of the figure presents the differences, ΔV, between the mean volume of the penicillin-treated and untreated cells in percent: $\Delta V = 100(V_p - V_c)/V_c$, where V_p and V_c are the average volumes of the penicillin-treated and control cells, respectively.

(volume reduction by about 15% relative to the controls after 20 min of growth in the presence of the drug).

We will offer a more complete interpretation of this result at the end of this article. At the moment it is sufficient to state that although an alteration of elastic properties did occur under the influence of penicillin, the swelling observed was obviously not posing any problems to the mechanical stability of the wall fabric. It should be noted that we demonstrated earlier (9) that a much larger volume increase than the maximal value of 20% observed here for penicillin-treated cells was tolerated by the staphylococci without bursting. In summary, the first prediction of the murosome model of penicillin-induced bacteriolysis (that the amount and quality of cell wall material synthesized during penicillin action should be sufficient to protect the cell mechanically against lysis) could be experimentally verified.

Determination of the Number of Complete Cell Cycles that the Staphylococci Pass through under Lytic Doses of Penicillin

The murosome model of penicillin-induced bacteriolysis predicts, furthermore, that bacteriolysis will not occur before initiation of cell separation by the end of the generation consecutive to that during which penicillin was added. From optical density measurements, one would conclude that, during the first 30 min after addition of penicillin at its lytic concentration ($0.1 \, \mu g/ml$), growth of the staphylococci is hardly affected and that only between 30 and 50 min does a certain growth inhibition take place. An absolute decline of optical density was not observed before about 70 to 80 min of drug action (see Fig. 1). However, we must be well aware that optical density is a complex parameter dependent on various properties of the cell such as mass, size, and composition. We therefore tried to follow the fate of the bacteria more exactly by using not only optical density, but also measurements of heat production of the culture, determination of CFU (viable cells), and microscopic counting of the total number of bacteria (Fig. 5). At first view, the four graphs in Fig. 5 seem to belong to different time scales: optical density went down 70 min after addition of penicillin, reduction of total number of bacteria started only 90 min after drug application, and the number of CFU as well as heat production already showed a reduction after 50 to 60 min. This apparent puzzle can easily be resolved if we assume (as postulated) that bacteriolysis may only occur in the second division plane as a locally very restricted event. Obviously, it took the drug 50 to 60 min (i.e., two generations) to initiate lytic death of the bacteria, as clearly indicated by the simultaneous drop of CFU values and heat production. On the other hand, the overall number of growing cells already stagnated after 20 min of penicillin treatment (see Fig. 5), indicating that cell replication ceased and, hence, that the penicillin-treated staphylococci were not capable

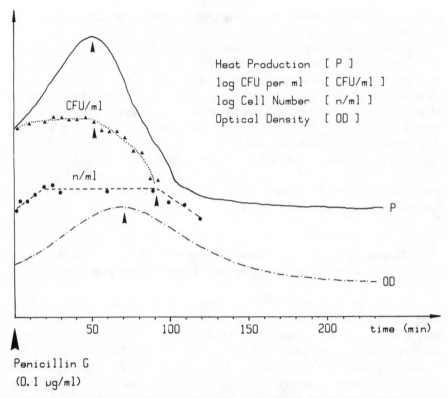

Penicillin G
(0.1 μg/ml)

Figure 5. Development of heat production (P, arbitrary units), logarithm of viable cell counts (CFU/ml), logarithm of total cell counts (dead or alive, n/ml), and optical density (OD) of a culture of *S. aureus* SG511 Berlin growing under the influence of the optimal lytic dose of penicillin (0.1 μg/ml). The large arrowhead marks the addition of penicillin (zero time); the four small arrowheads indicate the onset of decline of each parameter measured.

of performing more than a single division. In other words, the process of replication was in some way "frozen" during the second generation time, i.e., during formation of the second division plane; this means that bacteriolysis can take place only during the second generation time.

Of course, the question then arose as to why the total number of bacterial cells counted by light microscopy did not reflect the true onset of bacteriolysis 50 to 60 min after drug addition but remained constant for another 30 to 40 min. The convincing answer is that bacteriolysis was caused by the punching of tiny holes in the cell wall, tiny at least as compared with the dimensions of the whole cell. Since such holes cannot be detected under a light microscope because of its resolution, it takes a considerable period of time before these punched cells vanish, namely, until their complete, overall wall disintegration by postmortem activation of their autolytic system. Analogously, the optical density must also fail to indicate the true onset of bacteriolysis because a cell with tiny holes in its wall has indeed suffered lytic death but still contributes to optical density, and because it should be smaller than the nonlysed cell (as a result of the loss of cytoplasmic material), it could even bring about a slight increase in optical density (8).

In summary, the second prediction of the murosome model mentioned at the beginning of this chapter (that bacteriolysis should not occur before initiation of cell separation by the end of the next generation after penicillin addition) could also be experimentally verified.

Determination of the Cell Cycle Stage in which Bacteriolysis Takes Place

Again, staphylococci synchronized by the elutriation method were used for these experiments. The fraction of staphylococci collected by this technique showed synchronous growth for at least three generations (Fig. 6). This allowed us to perform the following experiment (Fig. 7). First, we followed the growth

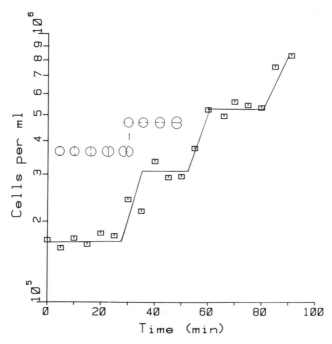

Figure 6. Growth of synchronized *S. aureus* SG511. The stepwise increase in cell number versus growth time (in minutes) indicates sufficient synchronization for at least three generation times. The schematic drawings of cells are meant only to provide a better visualization of the morphology of the cells at the corresponding cell cycle stage.

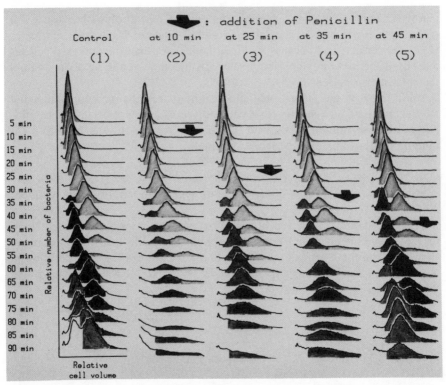

Figure 7. Effect of penicillin on the growth of *S. aureus* SG 511 at different stages of the cell cycle. Column 1, Control cells; columns 2 through 5, addition of 0.1 μg of penicillin per ml at the times indicated by the arrows. The time scale is given on the left axis in minutes after start of synchronous growth. Every 5 min, the size distribution of the culture (number of bacteria versus cell volume) was measured in a Coulter Counter. If penicillin was added at any time during the first cell cycle (columns 2 through 4), lysis occurred at the end of the second cycle. When penicillin was added after cell separation in the second generation, bacteria grew for one more complete generation before lysis occurred. In each column, the different generations are marked by different levels of gray. For details, see text.

of a synchronous control culture by measuring its bacterial size distribution versus growth time. Column 1 in Fig. 7 shows the development of the culture in 5-min intervals. The various graphs represent the number of bacteria (on the ordinate) and their size (on the abscissa). Since the cells are smallest at the beginning of each cell cycle and largest at its end, the successive generations can easily be recognized in these graphs (Fig. 7). Columns 2, 3, 4, and 5 in Fig. 7 represent the results of the analogous experiment in which the lytic penicillin dose had been added to the culture during the first generation of synchronous growth (at 10 min in column 2, at 25 min in column 3, and at 35 min in column 4, with zero time indicating the start of synchronous growth) or after completion of the first cell cycle (at 45 min in column 5). Whenever we added the penicillin during the first cycle, completion of the first and second generations was clearly detectable, and bacteriolysis started only thereafter. However, if we added penicillin after cell separation at the end of the first cycle (column 5 of Fig. 7), the staphylococci were capable of surviving one complete generation more. This means that, independent of a 25-min time variation in addition of the drug (columns 2, 3, and 4 in Fig. 7), lysis occurred at the same time, provided these time variations took place within one and the same cell cycle. On the other hand, a variation of only 10 min, but bridging two different generations (columns 4 and 5 in Fig. 7), allowed the bacteria to survive one more complete generation.

Thus, quite clearly, lysis starts at a very definite stage in a cell cycle, namely, when the bacteria initiate cell separation at the end of the generation consecutive to that in which penicillin had been added. It should be stressed that bacteriolysis occurred at the same moment as the second division was initiated regardless of whether penicillin had acted, by then, for only 40 to 45 min (column 4 in Fig. 7) or for 60 to 65 min (column 2 in Fig. 7). Consequently, it can be ruled out that penicillin-induced bacteriolysis might merely be the consequence of a continuously increasing damage of the wall. Therefore, bacteriolysis cannot be considered to be purely a chemical effect which should be postulated if penicillin was acting simply as an overall inhibitor of cell wall synthesis or an overall activator of enzymatic activities. On the contrary, our data show that penicillin-induced bacteriolysis must be regarded mainly as a morphogenetic effect.

Deadly Sequence Leading to Penicillin-Induced Lytic Death in Staphylococci

Before summarizing the sequence of events that lead to lytic death in staphylococci, we should report two more experiments that are of special importance in this context.

The first one underlines the great importance of not merely looking at penicillin as a pure "cell wall antibiotic" but also remembering the intimate connections between bacterial walls and membranes (12). Figure 8 shows in its upper part the course of the optical density of a staphylococcal culture labeled with [^{3}H]uracil as described above during treatment with various doses of penicillin. The lower part of the figure shows the release of labeled cytoplasmic material into the culture fluid. The release data (which turned out to be nearly the same when uracil was replaced by an amino acid or by thymidine as the label) clearly indicate a dose-dependent release of cytoplasmic material into the culture irrespective of bacteriolysis, which did not occur at all at a penicillin concentration of 10 μg/ml. The crossover of the graphs for 0.1 μg/ml and 10 μg/ml at about 60 to 70 min (Fig. 8) of penicillin treatment ob-

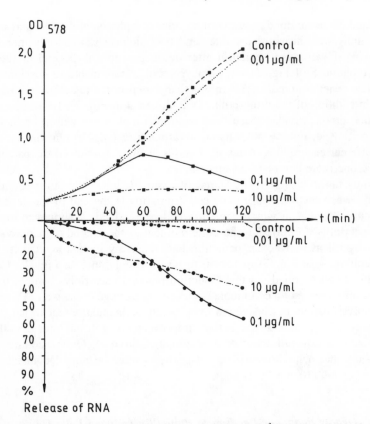

Figure 8. Development of optical density (OD_{578}) and release of [^{3}H]uracil ($\cong$ RNA) from pre-labeled *S. aureus* SG511 (as percentage of total label in the cells at time zero) into the supernatant during treatment with different concentrations of penicillin as indicated for each curve. Zero time corresponds to addition of the drug.

viously results from an additional release of the cytoplasmic material due to the onset of bacteriolysis at the lower drug concentration starting at about 50 to 60 min. No release could be detected with control cells (Fig. 8). Our findings point to dose-dependent membrane perturbations as a consequence of penicillin treatment, the more so since high-molecular-weight material also was released, as shown by trichloroacetic acid precipitation of the labeled material (data not shown). It should be recalled here that one of the earliest morphological alterations observed by electron microscopy of penicillin-treated staphylococci was the missing assembly of the membranelike-staining splitting system in the middle of the newly built cross wall of the first division plane (7).

The second experiment again concerns the size of the bacteria during penicillin treatment. We have already mentioned (see Fig. 4) that, after a short initial volume increase of the penicillin-treated bacteria, a size reduction as compared with the control cells was observed. The same, of course, should be true for the cytoplasmic membrane being pressed against the wall by the osmotic pressure within the cell, which in turn might lead to a functional disturbance and the onset of release of cytoplasmic material. Such a release of osmotically active material from the inside might be the reason for the relative shrinkage observed afterwards. However, when we looked at the volume of the penicillin-treated bacteria in a nonsynchronous culture for a longer period of time, we found that the volume alterations just described were again followed by a considerable increase in cell volume (Fig. 9). However, in this case we could unequivocally show that this volume increase was not due to a simple swelling, especially because at that time a considerable release of cytoplasmic material should already have lowered the internal pressure in the cell. Instead, electron microscopy data showed that this volume increase was due to special morphogenetic growth processes; the main factor in the volume increase could be shown to be a premature tearing up of the nascent cross wall along the "splitting line" (Fig. 10) during the cell separation process following the first division of the penicillin-treated cells.

From our present state of knowledge about these various and complex events, we propose the following "deadly sequence" that results in lytic death of staphylococci. (i) At 0 min, the optimal lytic dose of penicillin is added, which shortly afterwards leads to a blocking of the penicillin-binding proteins. (ii) After about 10 min, cell volume increases, release of cytoplasmic material

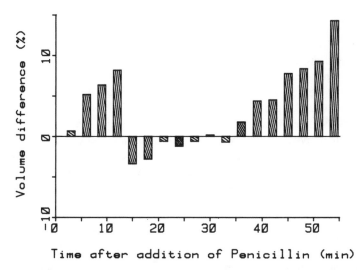

Figure 9. Development of the average volume versus time (in minutes) of cells of *S. aureus* SG511 during treatment with 0.1 μg of penicillin per ml (added at time zero) as measured with the aid of a Coulter Counter. Values are the volume difference expressed as a percentage of the average volume of an untreated control culture.

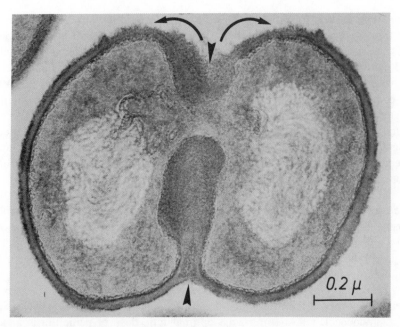

Figure 10. Electron micrograph of a thin section through a staphylococcal cell, revealing a morphogenetic increase of size by premature splitting up of the incomplete cross wall (arrows) under the influence of 0.1 µg of penicillin per ml.

becomes detectable, and assembly of the splitting system in the middle of the cross wall of the first division is impaired. (iii) After about 15 min, the size of the penicillin-treated cells lags behind that of the control cells; release of cytoplasmic material continues. (iv) After about 20 to 30 min, there is a morphogenetic increase in volume, blockage of cell separation first appears, the degree of cross-linking of the peptidoglycan is lowered from about 90 to 60%, O acetylation of the muramic acid residues is drastically reduced, wall turnover and autolytic activity are also reduced (11), and release of cytoplasmic material seems actually to become enhanced. (v) After about 50 to 60 min, while still in their second cell cycle, the bacteria suddenly lose huge additional amounts of their cytoplasmic material, which we ascribe to the murosome-induced perforation of their peripheral cell walls at the second division plane (= initiation of bacteriolysis); this substantial release of plasmatic components from the perforated cells causes the breakdown of the calorimetrically measurable biochemical processes within the staphylococci which leads to cell death. (vi) After about 80 min, postmortem processes begin, i.e., a multiple, overall disintegration of the peripheral cell wall of the staphylococci, still within the second cell cycle, and beginning of the continuous lysis of the whole bacterial cell (= overall cell dissociation).

In summary, all our experimental data are in agreement with the prediction, made by the murosome model, of a locally very restricted lysis at a certain site, namely, the site of initiation of the second cell separation after addition of penicillin.

ACKNOWLEDGMENTS. We are grateful to N. Nanninga and his staff for teaching us the application of the elutriation method. We thank H. Grob for the determination of CFU values.

LITERATURE CITED

1. **Blümel, P., W. Uecker, and P. Giesbrecht.** 1979. Zero order kinetics of cell wall turnover in *Staphylococcus aureus. Arch. Microbiol.* **121**:103–110.
2. **Blundell, J. K., and H. R. Perkins.** 1981. Effect of β-lactam antibiotics on peptidoglycan synthesis in growing *Neisseria gonorrhoeae* including changes in the degree of O-acetylation. *J. Bacteriol.* **147**:633–641.
3. **Blundell, J. K., and H. R. Perkins.** 1985. Selectivity for *O*-acetylated peptidoglycan during endopeptidase action by permeabilized *Neisseria gonorrhoeae. FEMS Microbiol. Lett.* **30**:67–69.
4. **Figdor, C. G., A. J. M. Olijhoek, S. Klencke, N. Nanninga, and W. S. Bont.** 1981. Isolation of small cells from an exponential growing culture of *Escherichia coli* by centrifugal elutriation. *FEMS Microbiol. Lett.* **10**:349–352.
5. **Fromme, I., and H. Beilharz.** 1978. Gas chromatic assay of total and O-acetyl groups in bacterial lipopolysaccharides. *Anal. Biochem.* **84**:347–353.
6. **Giesbrecht, P.** 1984. Novel bacterial wall organelles ("murosomes") in staphylococci: their involvement in wall assembly, p. 177–186. *In* C. Nombela (ed.), *Microbial Cell Wall Synthesis and Autolysis.* Elsevier Science Publishing, Inc., Amsterdam.
7. **Giesbrecht, P., H. Labischinski, and J. Wecke.** 1985. A special morphogenetic wall defect and the subsequent activity of "murosomes" as the very reason for penicillin-induced bacteriolysis in staphylococci. *Arch. Microbiol.* **141**:315–324.
8. **Koch, A.** 1984. Turbidity measurements in microbiology. *ASM News* **50**:473–477.
9. **Labischinski, H., G. Barnickel, and D. Naumann.** 1983. The state of order of bacterial peptidoglycan, p. 49–54. *In* R. Hakenbeck, J.-V. Höltje, and H. Labischinski (ed.), *The Target of Penicillin: the Murein Sacculus of Bacterial Cell Walls, Architecture and Growth.* Walter de Gruyter, Berlin.
10. **Martin, H. H., and J. Gmeiner.** 1979. Modification of peptidoglycan structure by penicillin action in cell walls of *Proteus mirabilis. Eur. J. Biochem.* **95**:487–495.
11. **Reinicke, B., P. Blümel, H. Labischinski, and P. Giesbrecht.** 1985. Neither an enhancement of autolytic wall degradation nor an inhibition of the incorporation of cell wall material are pre-requisites for penicillin-induced bacteriolysis in staphylococci. *Arch. Microbiol.* **141**:309–314.
12. **Rogers, H. J., H. R. Perkins, and J. B. Ward.** 1980. *Microbial Cell Walls and Membranes.* Chapman & Hall, Ltd., London.
13. **Shockman, G. D., L. Daneo-Moore, T. D. McDowell, and W. Wong.** 1981. Function and structure of the cell wall—its importance in the life and death of bacteria, p. 31–65. *In* M. R. J. Salton and G. D. Shockman (ed.), β-*Lactam Antibiotics: Mode of Action, New Developments, and Future Prospects.* Academic Press, Inc., New York.
14. **Wong, W., F. E. Young, and A. N. Chatterjee.** 1974. Regulation of bacterial cell walls: turnover of cell wall in *Staphylococcus aureus. J. Bacteriol.* **120**:837–843.

V. PENICILLIN-BINDING PROTEINS

Chapter 24

Molecular Graphics: Studying β-Lactam Inhibition in Three Dimensions

Judith A. Kelly
James R. Knox
Paul C. Moews
Jill Moring
Hai Ching Zhao

The molecules that make up living organisms are inherently inanimate. It is only when these molecules come together and interact that living systems develop and function. Taken individually, the components of bacterial cell walls are relatively simple building blocks, but once assembled, the cell wall provides the three-dimensional network essential for structural integrity. The effects of β-lactam antibiotics on cell wall assembly can be dramatic, leading to cell death in the extreme. Many approaches have been used to understand the interaction between β-lactams and the bacterial cell wall synthetic enzymes that are the targets of these drugs. These enzymes are the D-alanyl-D-alanine carboxypeptidases/transpeptidases (DD-peptidases) that form the peptide cross-links of the peptidoglycan in the final stages of cell wall synthesis (4, 13). While a great deal has been learned about β-lactam inhibition, no model has been developed that can explain why one antibiotic is effective against an enzyme while another, closely related antibiotic is not. Details of the inhibition at the molecular level have been missing. X-ray diffraction studies carried out on single crystals of a pure enzyme are the only method available that allows one

Judith A. Kelly, James R. Knox, Paul C. Moews, Jill Moring, and Hai Ching Zhao • Department of Molecular and Cell Biology and Institute of Materials Science, University of Connecticut, Storrs, Connecticut 06268.

to determine the exact three-dimensional structure of biological macromolecules and the complexes formed with small molecules. Diffraction studies, coupled with interactive computer graphics analyses, can provide the information necessary for developing an understanding of the effect of β-lactams on cell wall assembly. It is a long experimental road. One must have pure enzyme samples, crystals must be grown, hundreds of thousands of data points must be collected, the phase problem intrinsic to the diffraction technique must be overcome, and finally, one must interpret the electron density map that is calculated from the diffraction data. All this effort, however, is worthwhile because of the wealth of information that an accurate three-dimensional structure provides.

The specific enzyme inhibited by β-lactams that we have studied crystallographically is the DD-peptidase from *Streptomyces* sp. strain R61 (R61 enzyme). It is a water-soluble, extracellular enzyme isolated and purified by our colleagues Jean-Marie Ghuysen and Jean-Marie Frère at the University of Liège, Liège, Belgium (5). The primary structure of the 349-residue R61 enzyme was recently published (2, 10). The R61 enzyme is an active-site serine DD-peptidase that is able to catalyze either the carboxypeptidation or transpeptidation reaction, as shown in Fig. 1. The D-alanyl-D-alanine terminated donor peptide is cross-linked to a neighboring glycopeptide if there is an appropriate amino acceptor available, or the peptide is simply hydrolyzed if the acceptor is a water molecule (4). When the R61 enzyme is exposed to a penicillin or a cephalosporin, the enzyme mistakes the β-lactam for its natural peptide substrate and covalently binds the drug. However, in this case, the acyl-enzyme complex formed by the active-site serine is a stable one. The potency of β-lactams as DD-peptidase inhibitors is reflected in the half-lives of these complexes, which range from minutes to days (7).

Our X-ray diffraction studies of the R61 enzyme have shown that the protein has a compact, prolate spheroid shape with dimensions approximately 5.0 nm by 5.0 nm by 6.0 nm. There is an all-helical region and a region that contains a five-stranded β-sheet flanked by two helices on one face and a third

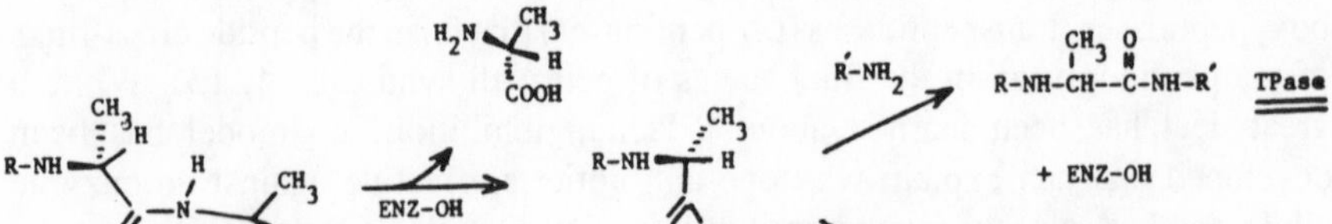

Figure 1. Reactions catalyzed by the R61 enzyme with R–D-alanyl-D-alanine as a substrate. R′–NH₂ represents an appropriate amino acceptor for the transpeptidation reaction in contrast to a water molecule for the carboxypeptidation reaction. TPase, Transpeptidation; CPase, carboxypeptidation.

helix on the other face. There has long been a question about the relationship between the cell wall synthetic DD-peptidases and the β-lactamases. These latter enzymes represent a principal line of defense for bacteria because the β-lactamases hydrolyze the β-lactam bond of the drug, producing products with no antibacterial effect. In work carried out with Otto Dideberg of the University of Liège, we used our Evans and Sutherland PS330 interactive computer graphics system to explore the structural relatedness of these two classes of enzymes. By simultaneously displaying the structures of the R61 enzyme and the β-lactamase from *Bacillus licheniformis,* we were able to rotate and translate the images of the two molecules until the five-stranded β-sheets of the enzymes were superimposed. When this was accomplished, seven of eight helices in the structures had common spatial positions (Fig. 2) (11). This analogy in the three-dimensional orientation of secondary structure elements is also observed between the R61 enzyme and the *B. cereus* β-lactamase (12). The striking similarity in the crystallographic structures supports the hypothesis of an evolutionary relationship between these two classes of penicillin-binding enzymes.

These secondary structure elements play a key role in the R61 enzyme. Not only do they represent the scaffolding of the proteins, but the active-site serine occurs at the N terminus of one of the helices in the all-helical region (Fig. 3). The regular, repeating pattern of residues in a helical configuration produces a dipole moment on the helix with a positive partial charge at the amino end of the helix (9). This delocalized positive field gradient may play a role in initial attraction of β-lactams to the active-site region of the enzyme. It has long been recognized that effective antibiotics must have a carboxylate in proximity to the β-lactam nitrogen. In addition, the position of the catalytic serine at the N terminus of the helix may activate this residue, making it more reactive than the 26 other serines found in the R61 enzyme. This is interesting in the context of the thiol protease papain, which has its reactive cysteine positioned at the N terminus of a helix. Having the reactive serine of the R61 enzyme at the end of a helix also explains the invariant spacing observed between this residue and lysine that occurs in all known penicillin-binding enzyme sequences. The pattern observed in both DD-peptidases and β-lactamases,

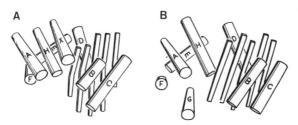

Figure 2. Distribution of secondary structure elements in the *B. licheniformis* β-lactamase (A) and the R61 DD-peptidase (B). Cylinders represent helices, and ribbons represent β strands.

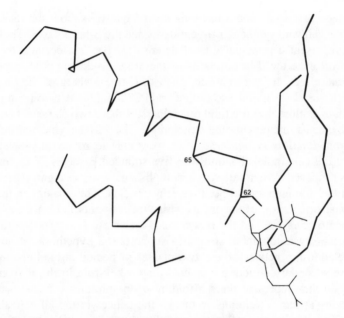

Figure 3. Closeup view of the active site region of the R61 enzyme as determined crystallo-graphically. The side chains of the reactive serine 62 and invariant lysine 65 are shown on the upper helix (designated helix H in Fig. 2). The configuration of cephalosporin C as determined by difference Fourier synthesis is also shown. Other secondary structure elements shown are the innermost two strands of the β sheet and helix A (as identified in Fig. 2).

with no known variations, is Ser-Xxx-Xxx-Lys (2). The periodicity of an α-helix is 3.6 residues per turn of helix. If this lysine is to participate in binding or catalysis, or both, as its highly conserved nature suggests, then it must be spaced two residues away from the serine to traverse the first turn of helix, bringing the basic side chain back into the active-site region.

One of the exciting aspects of macomolecular structural analysis by protein crystallography is that the structure need not be an endpoint in and of itself. The structure can provide the starting point for studies exploring the relation-ship between structure and function, substrate specificity, mechanisms of ac-tion, and inhibition. During the course of our structure determination of the native R61 enzyme, we have begun probing its active site by diffusing into the pregrown wet enzyme crystals a variety of β-lactam inhibitors ranging from penicillins through cephalosporins to monobactams. By a mathematical tech-nique called a difference Fourier synthesis, we are able to create an image of the drug molecule bound at the enzyme's active site. Since the difference Four-ier involves, in effect, subtracting the diffraction data for the native protein from the data for the protein-drug complex, the image calculated includes not only the drug which has been added, but also any differences in the protein conformation before and after binding. This provides the crystallographer with a powerful tool for studying interactions between macromolecules and small molecules.

The highest-resolution image we have for an inhibited R61 complex is with cephalosporin C. The half-life for this drug with the R61 enzyme is approximately 7 days (6). The stability of the complex is important in crystallographic studies because the time course for a diffraction experiment is on the order of days. The cephalosporin C difference Fourier map calculated at a resolution of 0.225 nm was displayed on the interactive computer graphics system along with the structure of the native enzyme. The binding site for the cephalosporin C was clearly shown to be at the active-site serine with the edge of the β-sheet flanking the drug molecule. When we displayed the structure of sodium cephalosporin C as determined by Hodgkin and Maslen (8), it became apparent that the intact drug molecule would not fit the experimentally observed electron density map. Several modifications had to be made to the model of the drug molecule to reflect the enzymatically induced changes that had occurred to the drug on binding to the R61 enzyme, as indicated by our map. It is clear that the β-lactam bond is open in the complex we isolated. This is confirmed by there being no density for the C3 acetate group of cephalosporin C. The ability of this substituent to act as a good leaving group has been related to the ease of opening the β-lactam bond and the acylation of the enzyme (1, 3). We also found that tetrahedral carbon atoms at the bridgehead (C_6 and C_7) fit the density much better than carbon atoms with the 90° angle imposed by the intact four-membered β-lactam ring. Finally, a rotation of approximately 50° about the C_6—C_7 bond is required to position the dihydrothiazene ring in the difference Fourier map (Fig. 4).

The fitting of the drug into the experimental data and the analysis of its position with respect to the enzyme model reveal interesting points. It can be seen that the C_7 side chain of the drug is in a hydrogen bonding position alongside the edge of the β-sheet (Fig. 3). If one remembers Tipper and Strominger's

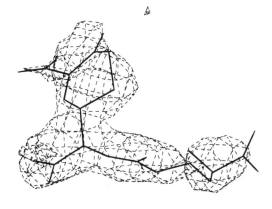

Figure 4. Fit of cephalosporin C into experimentally determined electron density (dashed lines) at 0.225 nm resolution. The β-lactam ring is open, and the carbonyl of the β-lactam bond is shown as the carboxylate reflecting its covalent attachment to the gamma oxygen of serine 62.

hypothesis of 1965 (14), this portion of the drug molecule is analogous to a peptide fragment. The position of the drug side chain we observed suggests that it may mimic a β strand, extending the antiparallel sheet. This interaction between the drug and the innermost strand of the β sheet may be an important, nonspecific interaction facilitating drug positioning for attack by the reactive serine. One also notes that the invariant lysine (three residues after the serine) is not near the cephalosporin C, as observed in the stable complex we isolated in our diffraction experiments. We should keep in mind, however, the rotation that occurred about the C_6—C_7 bond after β-lactam rupture. If one assumes that the position of the drug with respect to the reactive serine does not change dramatically and the interaction with the β sheet is stabilizing bonding, then that half of the drug molecule is fixed in position. If that is the case, one is left with the rotation observed being a rotation of the dihydrothiazene ring of the drug. Rotating that ring back to envision the intact drug molecule brings the dihydrothiazene ring and its carboxylate closer to the invariant lysine. The conserved nature of this lysine would then suggest that it plays a role in binding or catalysis.

These drug binding studies with the R61 enzyme underscore a critical point when considering interactions between small molecules and receptors in general. In the absence of sufficient information, one is often tempted to use the conformation of a small molecule to infer the shape of a receptor cavity to which it binds. We can see from our work that the surface of the receptor

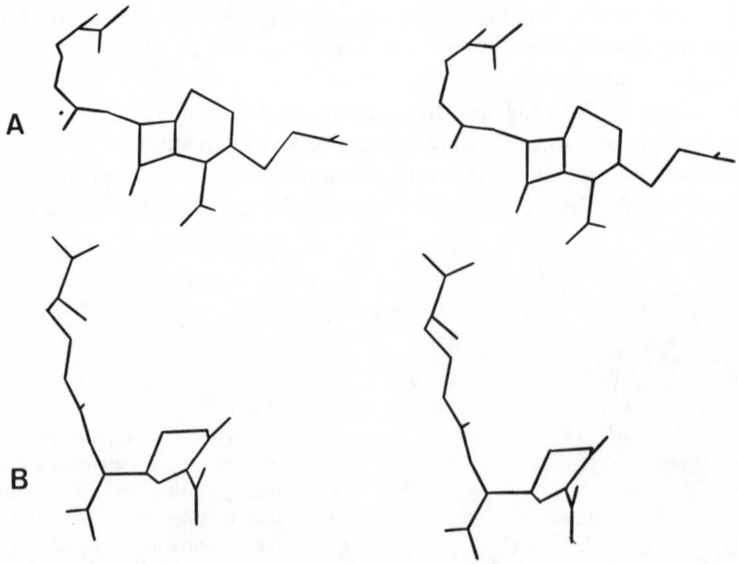

Figure 5. Stereoscopic views of cephalosporin C with the β-lactam intact (A) as opposed to open, as observed in crystallographic binding studies (B).

protein can greatly alter the shape of an effector molecule (Fig. 5), and caution must be exercised in inferring the conformation of a target protein in the absence of X-ray crystallographic data.

LITERATURE CITED

1. **Boyd, D. B.** 1985. Elucidating the leaving group effect in the β-lactam ring opening mechanism of cephalosporins. *J. Org. Chem.* **50**:886–888.
2. **Duez, C., C. Piron-Fraipont, B. Joris, J. Dusart, M. S. Urdea, J. A. Martial, J. M. Frere, and J. M. Ghuysen.** 1987. Primary structure of the *Streptomyces* R61 extracellular DD-peptidase. 1. Cloning into *Streptomyces lividans* and nucleotide sequence of the gene. *Eur. J. Biochem.* **162**:509–518.
3. **Faraci, W. S., and R. F. Pratt.** 1984. Elimination of a good leaving group from the 3'-position of a cephalosporin need not be concerted with β-lactam ring opening: TEM-2 β-lactamase-catalyzed hydrolysis of PADAC and cephaloridine. *J. Am. Chem. Soc.* **106**:1489–1490.
4. **Frere, J. M., and B. Joris.** 1985. Penicillin-sensitive enzymes in peptidoglycen biosynthesis. *Crit. Rev. Microbiol.* **11**:299–396.
5. **Frere, J. M., M. Leyh-Bouille, J. M. Ghuysen, M. Nieto, and H. R. Perkins.** 1976. Exocellular DD-carboxypeptidases-transpeptidases from *Streptomyces*. *Methods Enzymol.* **45B**:610–636.
6. **Ghuysen, J. M., J. M. Frere, M. Leyh-Bouille, J. Coyette, J. Dusart, and M. Nguyen-Distache.** 1979. Use of model enzymes in the determination of the mode of action of penicillins and Δ³-cephalosporins. *Annu. Rev. Biochem.* **48**:73–101.
7. **Ghuysen, J. M., J. M. Frere, M. Leyh-Bouille, M. Nguyen-Disteche, and J. Coyette.** 1986. Active-site-serine D-alanyl-D-alanine-cleaving-peptidase-catalyzed acyl-transfer reactions. *Biochem. J.* **235**:159–165.
8. **Hodgkin, D. C., and E. N. Maslen.** 1961. The X-ray analysis of the structure of cephalosporin C. *Biochem. J.* **79**:393–404.
9. **Hol, W. G. J., L. M. Halie, and C. Sander.** 1980. Dipoles of the α-helix and β-sheet: their role in protein folding. *Nature* (London) **294**:532–536.
10. **Joris, B., P. Jacques, J. M. Frere, J. M. Ghuysen, and J. Van Beeumen.** 1987. Primary structure of the *Streptomyces* R61 extracellular DD-peptidase. 2. Amino acid sequence data. *Eur. J. Biochem.* **162**:519–524.
11. **Kelly, J. A., O. Dideberg, P. Charlier, J. P. Wery, M. Libert, P. C. Moews, J. R. Knox, C. Duez, C. Fraipont, B. Joris, J. Dusart, J. M. Frere, and J. M. Ghuysen.** 1986. On the origin of bacterial resistance to penicillin: comparison of a β-lactamase and a penicillin target. *Science* **231**:1429–1431.
12. **Samraoui, B., B. J. Sutton, R. J. Todd, P. J. Artimyuk, S. G. Waley, and D. C. Phillips.** 1986. Tertiary structural similarity between a class A β-lactamase and a penicillin-sensitive D-alanyl carboxypeptidase-transpeptidase. *Nature* (London) **320**:378–380.
13. **Spratt, B. G.** 1983. Penicillin-binding proteins and the future of β-lactam antibiotics. *J. Gen. Microbiol.* **129**:1247–1260.
14. **Tipper, D. J., and J. L. Strominger.** 1965. Mechanism of action of penicillins: a proposal based on their structural similarity to acyl-D-alanyl-D-alanine. *Proc. Natl. Acad. Sci. USA* **54**:1131–1141.

Chapter 25

Evolution of DD-Peptidases and β-Lactamases

Jean-Marie Ghuysen

Peptide bonds adjacent to almost every kind of amino acid residue and in almost every position in the peptide are susceptible to being cleaved by one peptidase or another. Peptidase-catalyzed rupture of a peptide (amide) bond is made by transfer of the electrophilic group R—CO— of a carbonyl donor to a nucleophilic acceptor (Fig. 1). Positioning of the carbonyl donor is made by interaction between the R and R' substituents and various enzyme binding sites. Bond rupture is then carried out by several enzyme reagents that fulfill the required functions of an electrophile (or anion hole) and a proton abstractor-donor. The electrophile polarizes the C=O bond while the proton abstractor-donor achieves nucleophilic attack of the carbonyl carbon atom and protonation of the nitrogen atom. The reaction proceeds via a tetrahedral adduct with pyramidalization of the carbonyl carbon atom. Such a transition state has partially broken and partially made covalent bonds (Fig. 1).

There are many peptidase families, each with its characteristic set of functional amino acid residues arranged in a particular configuration to form the enzyme active site (19). The serine and metallo peptidases listed in Table 1 have been classified in terms of structure and mechanism. Members of a given family are believed to have evolved from a common ancestor. They are similar in the extent and distribution of secondary structures, and differences in the active sites are essentially limited to the substrate binding sites. Mechanistically related peptidases belonging to distinct families bear no relation in their amino acid sequences and three-dimensional structures but have similar active-site configurations.

The functions of the peptidases range from generalized protein degradation

Jean-Marie Ghuysen • Service de Microbiologie, Faculté de Médecine, Université de Liège, Institut de Chimie, B6, B-4000 Sart Tilman (Liège 1), Belgium.

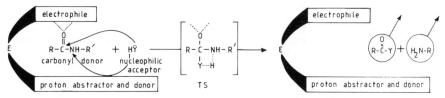

Figure 1. General mechanism of enzyme-catalyzed rupture of a peptide bond. TS, Transition state.

to specific processes. In humans, the zinc-containing angiotensin-converting enzyme (of unknown tertiary structure) catalyzes conversion of the decapeptide angiotensin I into the powerful vasoconstrictor octapeptide angiotensin II. Serine peptidases are involved in the activation of zymogens, the blood coagulation cascade, the complement system, the release of hormones and physiologically active peptides from precursors, etc. Metallo and serine peptidases are targets of great importance in the treatment of hypertension, emphysema, inflammation, and hemophilia and other dysfunctions of the hemostatic mechanism.

In this laboratory, we are concerned with defining the properties of bacterial serine and metallo peptidases (amidases) that characterize themselves by their ability to utilize as carbonyl donors D-alanyl-D-alanine-terminated peptides or β-lactam antibiotics (penicillins, cephalosporins, monobactams), or both.

Table 1. Known families of serine and metallo peptidases

Family and members	Endogenous nucleophile	Proton abstractor-donor	Electrophile	Exogenous nucleophile[a]
Serine peptidases I Trypsins Chymotrypsins Kallikreins Elastases	Ser-195	His-57-Asp-102	Hydrogen bonds from backbone NH of Ser-195 and Gly-193	HY
Serine peptidases II Subtilisins	Ser-221	His-64-Asp-32	Side chain of Asn-155 and hydrogen bond from backbone NH of Ser-221	HY
Zn peptidases I Carboxypeptidases A and B	His-69 Glu-72—Zn O⟨H H His-196	Glu-270		H_2O
Zn peptidase II Thermolysin	Glu-166 His-146—Zn O⟨H H His-142	Glu-143	Tyr-157 His-231	H_2O

[a]HY stands for H_2O, alcohol, amino compound.

They are related, directly or indirectly, to wall peptidoglycan metabolism and are important targets in antibacterial chemotherapy. The aim of the research is to understand the evolutionary relationships and catalytic mechanisms in structural terms.

Serine-DD-*Peptidases, Penicillin-Binding Proteins and* β-*Lactamases*

Functional and Mechanistic Homologies

Serine-assisted rupture of a peptide (amide) bond involves sequential transfer of the electrophilic portion R—CO— of the carbonyl donor to the enzyme active-site serine (which acts as an endogenous nucleophile) and then, from this, to an exogenous nucleophilic acceptor. Central to this mechanism is the formation and breakdown of a serine ester-linked acyl enzyme (Fig. 2).

The reaction can be represented by:

$$E + D \underset{}{\overset{K}{\rightleftharpoons}} E \cdot D \xrightarrow[P_1]{k_{+2}} E\text{-}D^* \xrightarrow[HY]{k_{+3}} E + P_2 \tag{1}$$

where E is the peptidase, D is the carbonyl donor, HY is the exogenous acceptor; $E \cdot D$ is the Michaelis complex, $E\text{-}D^*$ is the serine ester-linked acyl enzyme, K is the dissociation constant, k_{+2} and k_{+3} are first-order rate constants, P_1 is the leaving group, and P_2 is the second reaction product. It is assumed that $[D] >> [E]$, $E \cdot D$ is in rapid equilibrium with free enzyme and free donor, and HY is at a saturating concentration (11).

The proportion of total enzyme, $[E]_o$, which occurs as acyl enzyme, $[E\text{-}D^*]_{ss}$, at the steady state of the reaction is:

$$\frac{[E]_o}{[E\text{-}D^*]_{ss}} = 1 + \frac{k_{+3}}{k_{+2}} + \frac{Kk_{+3}}{k_{+2}[D]} \tag{2}$$

and the time necesary for the acyl enzyme to reach a certain percentage of its steady-state level is:

$$t = \frac{-\ln\left(1 - \dfrac{[E\text{-}D^*]}{[E\text{-}D^*]_{ss}}\right)}{k_{+3} + k_a} \tag{3}$$

where:

$$k_a = \frac{k_{+2}}{1 + (K/[D])}$$

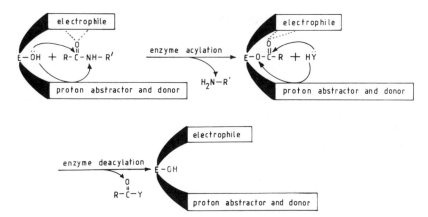

Figure 2. Active-site serine peptidase-catalyzed rupture of a peptide bond. E—OH, Active-site serine (endogenous nucleophile); HY stands for H_2O, R—OH, X—NH_2 (exogenous nucleophile); R'—NH_2, leaving group of the enzyme acylation step. Enzyme acylation and enzyme deacylation proceed via a transition state with pyramidalization of the carbonyl carbon (Fig. 1). The electrophile and proton abstractor-donor in the serine peptidases I and II are shown in Table 1.

is the pseudo-first-order rate constant of acyl enzyme formation at a given concentration of the carbonyl donor.

Enzyme inactivation requires a complementary geometry between the ligand and the cavity of the enzyme target. Inactivation of a serine peptidase can be achieved in different ways (25). Peptide halomethylketones inactivate various serine peptidases such as plasmin, thrombin, urokinase, factor Xa, and trypsin. These compounds whose —$COCH_2Cl$ functionality is sufficiently reactive to alkylate any nucleophile they may encounter in the enzyme active site, react, at least preferentially, with His-57 (while Ser-195 is favored in reactions with substrates). A second approach rests upon the transition state theory. Because the affinity of an enzyme for the transition state is very strong, even a crude transition state analog (in fact the only type which can be synthesized since the transition state itself has partially broken and partially made covalent bonds) binds to the enzyme active site very strongly. Diisopropyl-fluorophosphonate, phenylmethanesulfonyl fluoride, tosylsulfonyl fluoride, peptide aldehydes, and boron acids are good models for this theory; they are well-known inactivators of several serine peptidases I and II. A third approach is to use a substrate which gives rise to a stable acyl enzyme. One way to construct such a "suicide" substrate or mechanism-based inactivator is to incorporate the scissile peptide (amide) bond in a cyclic structure so that the leaving group remains part of the acyl enzyme and cannot leave the enzyme active site. The reaction is stoichiometric, and the enzyme is immobilized in the form of a stable acyl enzyme. This principle is best illustrated by the β-lactam compounds and is discussed below.

Serine DD-Peptidases

An important consequence of the double transfer mechanism shown in Fig. 2 is that the fate of the carbonyl donor varies, depending on the nature of the final acceptor. Attack of the acyl enzyme by water, an alcohol, or an amino compound causes hydrolysis, alcoholysis, or aminolysis (or transpeptidation), respectively, of the carbonyl donor. Though partitioning of the enzyme activity at the level of the acyl enzyme between several acceptors can be demonstrated in vitro, the serine peptidases I (from mammals) and II (from bacteria) function essentially as hydrolases.

The bacterial serine DD-peptidases behave differently. They catalyze two types of antagonistic reactions. Synthesis of new interpeptide linkages by transpeptidation of D-alanyl-D-alanine-terminated peptidoglycan precursors to the amino side chain of other peptidoglycan peptide units is a key reaction in wall peptidoglycan expansion and reticulation (reaction A in Fig. 3). Hydrolysis of the peptidoglycan precursors limits the number of carbonyl donors available for transpeptidation and presumably controls the extent of peptidoglycan cross-linking (reaction B in Fig. 3). The varying efficacy with which the serine DD-peptidases perform transpeptidation and hydrolysis is the result of a competition between various acceptors (i.e., the leaving group D-Ala, water, and an exogenous peptide) at the level of the acyl enzyme.

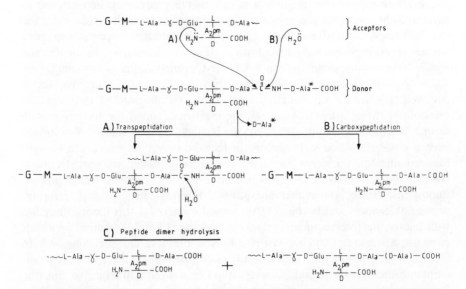

Figure 3. Main reactions catalyzed by the DD-peptidases. The active-site serine DD-peptidases catalyze reactions A and B on a competition basis. D-Ala*, Leaving group of the enzyme acylation step. The metallo (Zn) DD-peptidase is a strict hydrolase; it catalyzes reactions B and C without formation of an acyl enzyme intermediate.

With some serine DD-peptidases, the leaving group, D-Ala, as it is endogenously generated during enzyme acylation by the carbonyl donor, does not show, at this low concentration, any sign of acceptor activity. In water, hydrolysis of the donor proceeds to completion. However, millimolar concentrations of an exogenous, suitable peptide acceptor can compete with water. Part of the donor is then channeled to transpeptidated product. Both carboxypeptidation and transpeptidation occur on a competitive basis. With other serine DD-peptidases, the endogenously released D-Ala effectively suppresses the acceptor activity of water, performs attack of the acyl enzyme, and regenerates the original carbonyl donor. Though the enzyme turns over, little of the carbonyl donor is consumed, and hydrolysis is prevented. In turn, this situation is profoundly modified if a peptide (structurally related to wall peptidoglycan) of high acceptor activity is present. At low concentrations, such a peptide causes complete conversion of the donor into transpeptidated product. The enzyme functions as a strict transpeptidase (20).

β-Lactam compounds may have donor activity for every kind of serine peptidase. Reaction 1 and equations 2 and 3 apply, but given the endocyclic nature of the scissile amide bond, k_{+3} has a very low value. As a consequence, the serine peptidases behave as β-lactam compound-binding proteins (9, 11).

With the serine DD-peptidases, a negative charge must occur at $3'$ in the penicillins and $4'$ in the cephalosporins (this group is usually a carboxylate, but a sulfate is found in the monobactams), and the acyl side chain borne by the β-lactam ring must be on the β-face in a *cis* position to the thiazolidine (dihydrothiazine) ring in the penicillins (cephalosporins) (compounds I and II in Fig. 4). As a result of these substrate requirements, the β-lactam endocyclic amide bond is equivalent to a D-alanyl-D-alanine peptide bond. The serine peptidases I and II differ from the DD-peptidases in having endopeptidase versus carboxypeptidase activity and in preferring L versus D configuration. Given the different substrate requirements, cephalosporins have been modified accordingly and converted into effective inactivators of trypsin, chymotrypsin, elastase, plasmin, or thrombin (compounds III in Fig. 4) (6).

The efficacy of inactivation of a serine peptidase by a β-lactam compound depends on the values of the constants K, k_{+2}, and k_{+3} of reaction 1. Indeed, these constants determine both the proportion of enzyme which can be immobilized as acyl enzyme at the steady state of the reaction and the time which is necessary for the reaction to reach the steady state (11, 17). The higher the second-order rate constant of enzyme acylation (k_{+2}/K) and the smaller the first-order rate constant of acyl enzyme breakdown (k_{+3}), the lower the β-lactam compound concentration ($[D]$) for which the term $Kk_{+3}/k_{+2}[D]$ of equation 2 (which for $k_{+3} \ll k_{+2}$ simplifies to $[E]_o/[E\text{-}D^*]_{ss} = 1 + (Kk_{+3}/k_{+2}[D])$ becomes negligible. When this term is smaller than 0.01, virtually all the enzyme is immobilized as acyl enzyme at the steady state of the reaction. Moreover, assume four β-lactam compounds acting on a given serine peptidase.

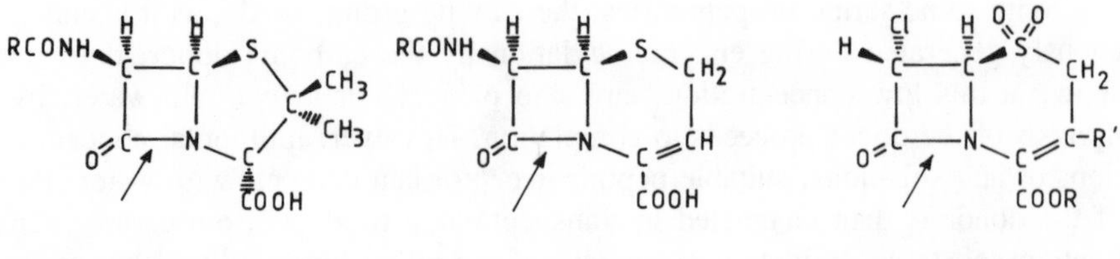

I   II   III

Figure 4. β-Lactam compounds. β-Lactam antibiotics I (penicillins) and II (cephalosporins) are mechanism-based inactivators of the serine DD-peptidases and substrates of the serine β-lactamases. β-Lactam compounds III (modified cephalosporins) are mechanism-based inactivators of the serine peptidases I (see Table 1). The scissile amide bond is shown by an arrow.

They are used at the same concentration (1 µM) and the k_{+3} value is the same (10^{-4} s^{-1}), but the relevant k_{+2}/K values are 10, 100, 1,000, and 10,000 M^{-1} s^{-1}, respectively. It follows from equations 2 and 3 that 9.1, 58, 90.9, and 99%, respectively, of the enzyme are acylated at the steady state and that it takes 698, 348, 170, and 7.6 min, respectively, for the reaction to reach 99% of the steady state.

Low (25,000 to 40,000)- and High (60,000 to 100,000)-M_r PBPs

All the bacteria possess, bound to the plasma membrane, multiple penicillin-binding proteins (PBPs). They are numbered in the order of decreasing molecular mass. Each bacterial species has its own assortment of PBPs, and the several PBPs present in a given bacterial cell show varying susceptibility to β-lactam antibiotics (varying k_{+2}/K and k_{+3} values), fulfill distinct cellular functions (related to peptidoglycan metabolism), may play compensatory roles, and do not show the same degree of essentiality. Some of them (PBPs 1A, 1B, 2, and 3 in *Escherichia coli*) have been assigned a role of prime importance in the killing of *E. coli* by the antibiotics (24), but this is a rather exceptional situation. In some cases, for example, the first symptom of growth inhibition by penicillin has been related to a characteristic degree of saturation of each of the PBPs present, suggesting that the antibacterial effect arises from a summation of the extent of inactivation of several PBPs or from the abnormal relative concentration of the PBPs left in a free form in the antibiotic-treated cells (26).

The low (25,000 to 50,000)-M_r PBPs are serine DD-peptidases. They effectively utilize, as carbonyl donor substrates, well-defined D-alanyl-D-alanine-terminated peptides of natural origin (Fig. 3, reactions A and B) or peptide and depsipeptide analogs (Fig. 5). These periplasmic proteins are attached to the underlying membrane via a short peptide stretch located at the carboxyl

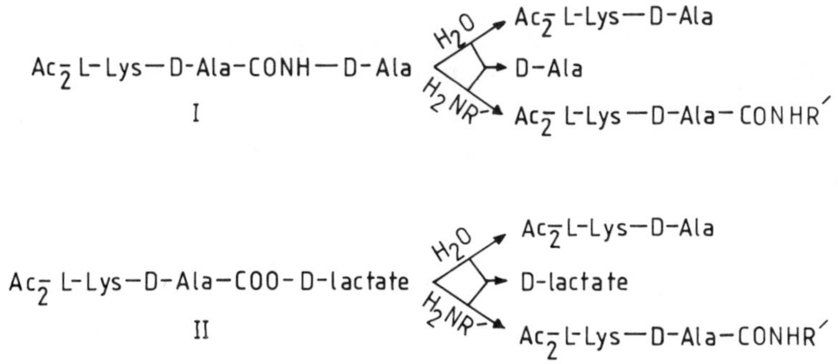

Figure 5. Serine DD-peptidase-catalyzed acyl transfer reactions with peptide (I) and ester (II) carbonyl donor analogs. The serine β-lactamases are devoid of peptidase activity on I, but some of them (of class C) have esterase activity on II.

end of the polypeptide chain. They can be detached from the membrane and converted into water-soluble proteins without alteration of the catalytic and penicillin-binding properties, by proteolytic treatment of isolated membranes (4) or by gene modifications resulting in the expression of a truncated poly-peptide that lacks the anchoring segment (L. C. S. Ferreira, U. Schwarz, W. Keck, P. Charlier, O. Dideberg, and J. M. Ghuysen, *Eur. J. Biochem.*, in press). In addition, bacteria, especially actinomycetes, excrete DD-peptidases during growth. The extracellular DD-peptidases/PBP of *Streptomyces* sp. strain R61 is synthesized in the form of a precursor which possesses both a cleavable N-terminal signal peptide and a cleavable carboxyl-terminal extension. Should it not be removed during maturation, this carboxyl-terminal segment might function as a halting signal through which the enzyme would become mem-brane bound (7).

The high (50,000 to 100,000)-M_r PBPs bind the β-lactam antibiotics by acylation of an active-site serine residue, but they lack detectable activity on short peptides or depsipeptides known to have donor activity for the low-M_r PBPs (Fig. 5). Some high-M_r PBPs (from *E. coli*) have been shown to catalyze, in vitro, peptide cross-linking by transpeptidation and glycan synthesis by transglycosylation. Their bifunctionality is revealed by the products that they generate upon incubation with the lipid-linked precursor *N*-acetylglucosaminyl-*N*-acetylmuramyl (pentapeptide)-diphosphoryl-undecaprenol (18). However, these reactions proceed with a low turnover number of 0.01 s^{-1} or less, suggesting that the conditions used poorly mimic the in vivo situation. PBP 3 of *E. coli* has received special attention. In all likelihood, fatty acids linked to an S-glyceryl cysteine at the amino-terminal region of the polypeptide chain anchor the protein to the plasma membrane (Y. Hirota, personal communication). A gene fusion that removes the amino-terminal 240-amino-acid region (which is probably responsible for the transglycosylase activity) and links the carboxyl-

terminal 349-amino-acid region to the amino terminus of β-galactosidase results in a truncated PBP 3 which still binds penicillin (14). In another construction, high-level expression of a PBP 3 that has been truncated downstream
of the cysteine residue results in the accumulation of large amounts of cytoplasmic granules that can be isolated from cell lysates by differential centrifugation. Solubilization of the granules (by guanidinium chloride unfolding and
refolding) yields a substantially purified, water-soluble PBP 3 that has virtually
the same size as the wild-type PBP 3. In spite of the fact that it is no longer
membrane bound, this water-soluble PBP 3 still lacks activity on short peptide
or depsipeptide analogs (J. Bartholomé-De Belder, M. Nguyen-Distèche, and
J. M. Ghuysen, unpublished data).

β-Lactamases

The β-lactamases are extracellular or periplasmic, water-soluble enzymes.
They degrade β-lactam antibiotics into biologically inactive metabolities by
opening the β-lactam ring through hydrolysis of the endocyclic amide bond.
The majority of the known β-lactamases are serine enzymes. Reaction 1 and
equations 2 and 3 apply. The serine β-lactamases have the same general requirements for β-lactam compounds as the serine DD-peptidases (Fig. 4; I, II),
but in contrast to these latter enzymes, they have adapted themselves to the
situation created by the presence of a nondiffusible leaving group in their active
sites. Water is an excellent attacking nucleophile of the acyl (penicilloyl, ce-
phalosporoyl, . . .) enzyme which, usually, is very short-lived (high k_{+3} values). The serine β-lactamases can hydrolyze β-lactam antibiotics with very
high k_{cat} values (up to 1,000 s^{-1}).

The distinction between the DD-peptidases and the β-lactamases in terms
of stability versus lability of the acyl enzymes formed by reaction with β-lactam
antibiotics is somewhat arbitrary since β-lactamases may also give rise to long-
lived acyl enzymes. Moreover, irrespective of the enzymes involved, long-
lived acyl enzymes may undergo intramolecular rearrangements. The first rearrangement reported in the literature describes the (slow) rupture of the C_5—C_6
bond in an acyl (benzylpenicilloyl) DD-peptidase (8). This rupture causes release of the leaving group and generates a new acyl enzyme which is very
labile. Another rearrangement, which is an integral part of the β-lactamase
inactivation mechanisms, involves rupture of the S_1—C_5 bond in the penam
sulfones and 6-β-bromo(iodo)penicillanate or the O_1—C_5 bond in clavulanate.
The established common feature is a branched pathway where β-elimination
of the acyl enzyme leads to a hydrolytically inert acyl enzyme or further modification of some residues of the enzyme active site (2), or both. A third type
of rearrangement, observed with those cephalosporins which have a good 3′
leaving group (for example, an acetoxy substituent), is elimination of this group

from the acyl enzyme, with formation of an exocyclic methylene. This elimination has important consequences with respect to cephalosporin turnover by the β-lactamases and inactivation of the DD-peptidases since the modified acyl enzymes are more hydrolytically inert than their parents (1).

Relatively little is known about serine β-lactamase-catalyzed acyl transfers involving noncyclic carbonyl donors. The β-lactamases appear to be devoid of peptidase activity, but at least some of them have esterase activity on these depsipeptides known to be substrates of the DD-peptidases (Fig. 5). Moreover, partitioning of the acyl enzyme between alternate nucleophiles occurs as expected for an acyl enzyme mechanism (22).

Structural Homologies

Gene sequencing has yielded the primary structures of eight β-lactamases (from *Bacillus licheniformis, B. cereus,* plasmid pBR322, *Staphylococcus aureus, Streptomyces albus* G, *E. coli* K-12, *Citrobacter freundii,* and *Salmonella typhimurium*), three low-M_r DD-peptidases/PBPs (the *Streptomyces* sp. strain R61 enzyme, the *E. coli* PBP 5, and the *B. subtilis* PBP 5), and four high-M_r, bifunctional PBPs (the *E. coli* PBPs 1A, 1B, 2 and 3). (For references, see B. Joris, J. M. Ghuysen, G. Dive, A. Renard, O. Dideberg, P. Charlier, J. A. Kelly, J. C. Boyington, P. C. Moews, and J. R. Knox [*Biochem J.*, in press].) In parallel to this, several β-lactamases and the (naturally occurring or genetically engineered) water-soluble DD-peptidases/PBPs of *Streptomyces* sp. strain R61 and *E. coli* (i.e., PBP 5) have been crystallized. X-ray crystallography has revealed details on the overall three-dimensional structure and active-site environment of some of these proteins (5a, 14a, 16, 23).

The β-lactamases and low-M_r DD-peptidases/PBPs possess two common "calibration marks." A conserved tetrad, Ser*-Xaa-Xaa-Lys, where Ser* is the active-site serine, occurs at the amino-terminal region of the proteins: at position 62 to 65 in the *Streptomyces* sp. strain R61 DD-peptidase or 48 to 51 in the *S. albus* G β-lactamase. In addition, a conserved triad, His-Thr-Gly, Lys-Thr-Gly, or Lys-Ser-Gly, occurs at the carboxyl-terminal region of the proteins: at position 288 to 300 in the 349-amino acid-containing *Streptomyces* sp. strain R61 DD-peptidase or 218 to 220 in the 273-amino acid-containing *S. albus* G β-lactamase. The high-M_r PBPs also possess these two calibration marks, and on this basis it has been assumed that the penicillin-binding domain of these bifunctional enzymes starts about 60 residues upstream of the tetrad Ser*-Xaa-Xaa-Lys and terminates about 60 residues downstream of the triad Lys-Ser-Gly or Lys-Thr-Gly (Joris et al., in press).

The significance of these two calibration marks is highlighted by the X-ray crystallography studies (5a, 14a, 16, 23). The *Streptomyces* sp. strain R61 DD-peptidase and the β-lactamases of class A (from *B. licheniformis, B. cereus,*

Staphylococcus aureus, and *S. albus* G) lack relatedness in their primary structure, yet they are very similar in the extent and distribution of their secondary structures. They are two-region proteins: one region has the "all-α" type of structure, and the other has a central core consisting of a five-stranded β-sheet protected by α-helices on both faces. Remarkably, the active-site serine of the tetrad Ser-48-Xaa-Xaa-Lys-51 in the *S. albus* G β-lactamase or Ser-62-Xaa-Xaa-Lys-65 in the *Streptomyces* sp. strain R61 DD-peptidase is at the amino terminus of one of the helices of the all-α region, so that after one turn of the helix, the side chain of the lysine residue is brought back within the active-site area (Fig. 6). In turn, the triad Lys-218-Thr-Gly-220 of strand S3 in the β-lactamase or His-298-Thr-Gly-300 in the DD-peptidase is on the other side of the pocket, with the lysine or imidazole side chain also pointing to the active site.

Search for homology between amino acid sequences (using the Goad-Kanehisa procedure [12a]) shows that all the penicillin-recognizing proteins or domains (except the *S. typhimurium* β-lactamase) form a family tree in which groups of closely related proteins are linked to each other through particular pairs of proteins (Joris et al., in press). Moreover, when aligned by reference to the *Streptomyces* sp. strain R61 DD-peptidase used as a template and in such a way that the two calibration marks defined above coincide, the β-lactamases, including that of *S. typhimurium*, the low-M_r DD-peptidases/PBPs of *E. coli* and *B. subtilis*, and the penicillin-binding domains of the *E. coli* high-M_r PBPs 1A, 1B, 2, and 3, all show significant homology with the *Streptomyces* DD-peptidase through major portions of their amino acid sequences. The cost of the editing, in terms of the percentage of amino acids eliminated by the alignments from the original sequences, ranges from 6 to 18% for the β-lactamases and does not exceed 24% for the low-M_r PBPs and the penicillin-binding domains of the high-M_r PBPs (in the case of PBP 1A, however, a 74-residue stretch 619 to 693 is excluded from the calculation).

Because homologous proteins have a similar three-dimensional structure, all the penicillin-recognizing proteins (domains) appear to be members of a single superfamily of active-site serine enzymes. They characterize themselves by a similar type of polypeptide scaffolding (Fig. 6) which is distinct from that of the trypsin and subtilisin families. Depending on the evolutionary distance, they may have very different sequences and distinct functionalities and specificities.

Refined three-dimensional structures and detailed knowledge of the active sites will highlight those amino acids possibly involved in substrate binding and catalysis. On the basis of the amino acid alignments discussed above, histidine is not a conserved amino acid residue among the penicillin-binding and penicillin-hydrolyzing enzymes (Joris et al., in press). This observation also contracts these enzymes with the peptidases of the trypsin and subtilisin families, in which histidine is an invariant element of the catalytic machinery. However, whatever the family to which a serine peptidase (amidase) belongs,

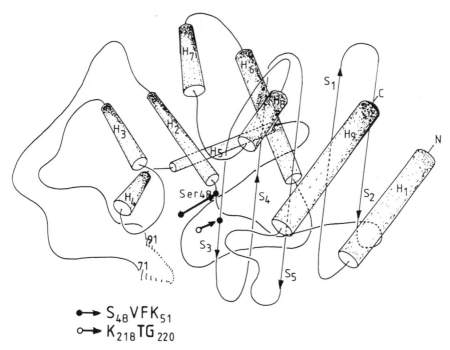

$\bullet\!\!\longrightarrow S_{48}VFK_{51}$
$\circ\!\!\longrightarrow K_{218}TG_{220}$

Figure 6. Positions of the Ser-48-Val-Phe-Lys-51 tetrad and the Lys-218-Thr-Gly-220 triad in the three-dimensional structure of the *S. albus* G β-lactamase. Adapted from Dideberg et al. (5a).

the mechanism of peptide (amide) bond ruptures is basically identical. Supporting this view and as already mentioned, cephalosporins, which for a long time were considered exclusively as antibacterial agents targeted against the PBPs, can be remodeled into mechanism-based inactivators of the endopeptidases of the trypsin family (6).

A possible explanation of the emergence of active-site serine β-lactamases in the bacterial world is that bacteria exposed to β-lactam antibiotics excrete one or several (nonessential) PBPs in the environment (gram-positive) or the periplasm (gram-negative). Further improvement of this detoxication mechanism resulted in the conversion of these water-soluble PBPs into β-lactam-hydrolyzing enzymes. Experiments designed to reproduce by protein engineering this proposed evolutionary transition from PBPs to β-lactamases are in progress.

Metallo (Zinc) DD-*Peptidase of S. albus*

Zinc-assisted rupture of a peptide bond involves, as acceptor of the transfer reaction, a water molecule that is prebound to the zinc cofactor (Fig. 7). Consequently, the metallopeptidases are strict hydrolases.

The metallo (Zn) DD-peptidase of *S. albus* G hydrolyzes D-alanyl-D-ala-nine-terminated peptidoglycan precursors. More specifically, it also effectively hydrolyzes the C-terminal D-alanyl-D-R interpeptide bonds in peptidoglycans of chemotypes I and IV (10), for example, the D-alanyl-(D)-*meso*-diaminopimelic acid linkages which cross-link the peptide units in the peptidoglycan of gram-negative bacteria (reaction C in Fig. 3). Its main or perhaps exclusive function is that of a bacteriolytic, peptidoglycan-hydrolyzing enzyme (12).

The metallo DD-peptidase of *S. albus* G is of known primary (15) and tertiary (5) structure. The amino-terminal region (76 residues) consisting of three α-helices is connected by a long loop to the carboxyl-terminal region (136 residues) consisting of three α-helices and a five-β-stranded sheet. The active site is located in the large region and is topologically defined by the β-sheet at the back and two loops, one on each side of the cavity (Fig. 8).

The metallo DD-peptidase of *S. albus* G lacks relatedness with the metallopeptidases I and II in amino acid sequence and tertiary structure. Yet, when the active site of the *Streptomyces* DD-peptidase (whose structure has been refined to 0.18 nm; J. P. Wéry, Ph.D. thesis, University of Liège, Liège Belgium, 1987) and that of the thermolysin (13) are similarly oriented with the zinc atoms and their ligands superimposed on each other (Asp-161, His-154, His-197 in the *Streptomyces* enzyme; Glu-166, His-146, His-142 in thermolysin), then the triad Tyr-189-His-192-His-195 in the *Streptomyces* DD-peptidase and the triad His-231-Tyr-157-Glu-143 in thermolysin are topologically equivalent. This observed correspondence at the level of the active sites suggests that Tyr-189 and His-192 in the *Streptomyces* DD-peptidase may play the role of the electrophile and His-195 (which is in interaction with Asp-194) may be involved in proton abstraction and donation (Fig. 9). The *Streptomyces* DD-peptidase differs from thermolysin in having carboxypeptidase versus endopeptidase activity. Another suggestion derived from the proposed model is that Arg 138 in the *Streptomyces* DD-peptidase may facilitate positioning of the carbonyl donor by charge pairing with the carboxylate grouping. Hence, in spite of differences in stereospecificity and polypeptide scaffolding, the *Streptomyces* DD-peptidase and thermolysin appear to have similar mechanistic prop-

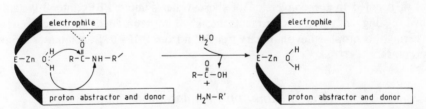

Figure 7. Metallo (Zn) peptidase-catalyzed rupture of a peptide bond. The exogenous nucleophile H_2O is bound to the zinc cofactor. The zinc ligands, electrophile, and proton abstractor-donor in the metallo peptides I and II are shown in Table 1.

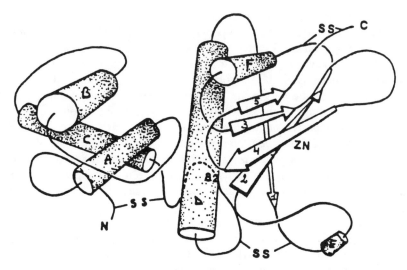

Figure 8. Three-dimensional structure of the *S. albus* G metallo (Zn) DD-peptidase. Zn, Position of the zinc cofactor. (Adapted from reference 5 and Wéry, thesis.)

erties. This may well be another example of convergent evolution; a similar mechanism dictated by common catalytic requirements has evolved from different starting tertiary structures.

The concept of "bi-product analogs" has led to the development of specific and potent inhibitors of the zinc-containing angiotensin-converting enzyme (21).

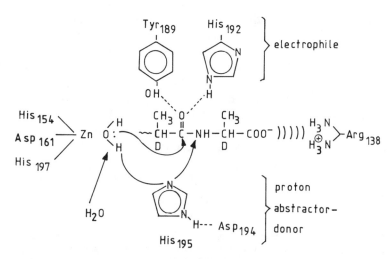

Figure 9. Putative catalytic mechanism of hydrolysis of a D-alanyl-D-alanine-terminated peptide by the *S. albus* G metallo (Zn) DD-peptidase (Wéry, Ph.D. thesis).

Some of these compounds are orally active antihypertensive drugs. By analogy with these studies, bifunctional compounds possessing, with the right spanning distance, both a C-terminal carboxylate and, at the other end of the molecule, a thiol, hydroxamate, or carboxylate function inhibit the metallo DD-peptidase of *S. albus* G, presumably by immobilization of the zinc cofactor (3). Classic β-lactam antibiotics are very weak inhibitors of the *Streptomyces* metallo DD-peptidase. Depending on the structure of the β-lactam compounds, enzyme inhibition is competitive or noncompetitive, and in this latter case, binding causes disruption of the protein crystal lattice. High concentrations of 6-β-iodopenicillanate inactivate the enzyme with a limit pseudo-first-order rate constant value of $7 \times 10^{-4} \text{ s}^{-1}$. It binds to the active site and superimposes His-192, suggesting that permanent inactivation is by reaction with this latter residue.

The metallo DD-peptidase of *S. albus* G may be the prototype of a family of bacteriolytic enzymes, autolysins and endolysins. The product of the gene pgR2 of phage λ and the so-called endopeptidase that is present in the periplasm of *E. coli* cells are DD-carboxypeptidases targeted against the peptidoglycan interpeptide D-alanyl-(D)-*meso*-diaminopimelic acid linkages. Autolysins are present in all bacteria. They may be involved in important processes related to wall peptidoglycan expansion, remodeling, and turnover, and they are subject to precise control. Inactivation of peptidoglycan-hydrolyzing metallo-DD-peptidase might cause dysfunctioning of the autolytic system, with harmful effects on bacterial growth.

Metallo (Zn) β-lactamases are known. They are not discussed here because of lack of information on their three-dimensional structure.

ACKNOWLEDGMENTS. I thank the colleagues who contributed to the work described here: P. Charlier, J. Coyette, O. Dideberg, G. Dive, C. Duez, J. Dusart, J. M. Frère, B. Joris, J. Lamotte-Brasseur, M. Leyh-Bouille, M. Nguyen-Distèche, and J. P. Wéry of the University of Liège, and J. Kelly and J. Knox of the University of Connecticut, Storrs.

The work in Liège is supported by the Fonds de Recherche de la Faculté de Médecine, the Fonds de la Recherche Scientifique Médicale, Brussels (contracts 3.4522.86 and 3.4507.83), the Gouvernement belge (action concertée 86/91-90), and the Région wallonne (C2/C16/conv. 246/20428).

LITERATURE CITED

1. **Boyd, D. B.** 1984. Electronic structure of cephalosporins and penicillins. Inductive effect of the 3-position side chain in cephalosporins. *J. Med. Chem.* **27**:63–66.
2. **Cartwright, S. J., and S. G. Waley.** 1983. β-Lactamase inhibitors. *Med. Res. Rev.* **3**:341–382.

3. **Charlier, P., O. Dideberg, J. C. Jamoulle, J. M. Frère, J. M. Ghuysen, G. Dive, and J. Lamotte-Brasseur.** 1984. Active-site directed inactivators of the Zn^{++} D-alanyl-D-alanine-cleaving carboxypeptidase of *Streptomyces albus* G. *Biochem. J.* **219**:763–772.

4. **Coyette, J., J. M. Ghuysen, and R. Fontana.** 1980. The penicillin-binding proteins in *Streptococcus faecalis* ATCC 9790. *Eur. J. Biochem.* **110**:445–456.

5. **Dideberg, O., P. Charlier, G. Dive, B. Joris, J. M. Frère, and J. M. Ghuysen.** 1982. Structure at 2.5 Å resolution of a Zn^{++}-containing D-alanyl-D-alanine-cleaving carboxypeptidase. *Nature* (London) **299**:469–470.

5a. **Dideberg, O., P. Charlier, J. P. Wéry, P. Dehottay, J. Dusart, T. Erpicum, J. M. Frère, and J. M. Ghuysen.** 1987. The crystal structure of the β-lactamase of *Streptomyces albus* G at 0.3 nm resolution. *Biochem. J.* **245**:911–913.

6. **Doherty, J. B., B. M. Ashe, L. W. Argenbright, P. L. Barker, R. J. Bonney, G. O. Chandler, M. E. Dahlgren, C. P. Dorn, Jr., P. E. Finke, R. A. Firestone, D. Fletcher, W. K. Hagmann, R. Mumford, L. O'Grady, A. L. Maycock, J. M. Pisano, S. K. Shah, K. R. Thompson, and M. Zimmerman.** 1986. Cephalosporin antibiotics can be modified to inhibit human leukocyte elastase. *Nature* (London) **322**:192–194.

7. **Duez, C., C. Piron-Fraipont, B. Joris, J. Dusart, M. S. Urdea, J. A. Martial, J. M. Frère, and J. M. Ghuysen.** 1987. Primary structure of the *Streptomyces* R61 extracellular DD-peptidase. 1. Cloning into *Streptomyces lividans* and nucleotide sequence of the gene. *Eur. J. Biochem.* **162**:509–518.

8. **Frère, J. M., J. M. Ghuysen, and J. De Graeve.** 1978. Fragmentation of penicillin catalysed by the exocellular DD-carboxypeptidase-transpeptidase of *Streptomyces* R61. Isotopic study of hydrogen fixation on carbon 6. *FEBS Lett.* **88**:147–150.

9. **Frère, J. M., and B. Joris.** 1985. Penicillin-sensitive enzymes in peptidoglycan biosynthesis. *Crit. Rev. Microbiol.* **11**:299–396.

10. **Ghuysen, J. M.** 1968. Use of bacteriolytic enzymes in determination of wall structure and their role in cell metabolism. *Bacteriol. Rev.* **32**:425–464.

11. **Ghuysen, J. M., J. M. Frère, M. Leyh-Bouille, M. Nguyen-Distèche, and J. Coyette.** 1986. Active-site serine D-alanyl-D-alanine-cleaving peptidases-catalysed acyl transfer reactions. Procedures for studying the penicillin-binding proteins of bacterial plasma membranes. *Biochem. J.* **235**:159–165.

12. **Ghuysen, J. M., M. Leyh-Bouille, R. Bonaly, M. Nieto, H. R. Perkins, K. H. Schleifer, and O. Kandler.** 1970. Isolation of DD-carboxypeptidase from *Streptomyces albus* G culture filtrates. *Biochemistry* **9**:2955–2961.

12a. **Goad, W. B., and H. I. Kanehisa.** 1982. Pattern recognition in nucleic acid sequences. I. A general method for finding local homologies and symmetries. *Nucleic Acids Res.* **10**:247–263.

13. **Hangauer, D. G., A. F. Monzengo, and B. W. Mathews.** 1984. An interactive computer graphics study of thermolysin-catalysed peptide cleavage and inhibition by N-carboxymethyl dipeptides. *Biochemistry* **23**:5730–5741.

14. **Hedge, P. J., and B. G. Spratt.** 1984. A gene fusion that localizes the penicillin-binding domain of penicillin-binding protein 3 of *Escherichia coli*. *FEBS Lett.* **176**:179–184.

14a. **Herzberg, O., and J. Moult.** 1987. Bacterial resistance to β-lactam antibiotics: crystal structures of β-lactamase from *Staphylococcus aureus* PC1 at 2.5 Å resolution. *Science* **236**:694–701.

15. **Joris, B., J. Van Beeumen, F. Casagrande, C. Gerday, J. M. Frère, and J. M. Ghuysen.** 1983. The complete amino acid sequence of the Zn^{++}-containing D-alanyl-D-alanine-cleaving carboxypeptidase of *Streptomyces albus* G. *Eur. J. Biochem.* **130**:53–69.

16. **Kelly, J. A., O. Dideberg, P. Charlier, J. P. Wéry, M. Libert, P. C. Moews, J. R. Knox, C. Duez, C. Fraipont, B. Joris, J. Dusart, J. M. Frère, and J. M. Ghuysen.** 1986. On the origin of bacterial resistance to penicillin. Comparison of a β-lactamase and a penicillin target. *Science* **231**:1429–1437.

17. **Leyh-Bouille, M., M. Nguyen-Distèche, S. Pirlot, A. Veithen, C. Bourguignon, and J. M. Ghuysen.** 1986. *Streptomyces* K15 DD-peptidase-catalysed reaction with suicide β-lactam carbonyl donors. *Biochem. J.* **235:**177–182.

18. **Matsuhashi, M., F. Ishino, S. Tamaki, S. Nakajima-Iijima, S. Tomioka, J. I. Nakagawa, A. Hirata, B. G. Spratt, T. Tsuruoka, S. Inouye, and Y. Yamada.** 1982. Mechanism of action of β-lactam antibiotics: inhibition of peptidoglycan transpeptidases and novel mechanisms of action, p. 99–114. *In* A. L. Demain, T. Hata, and C. R. Hutchinson (ed.), *Trends in Antibiotic Research.* Japan Antibiotics Research Association, Tokyo.

19. **Neurath, H.** 1984. Evolution of proteolytic enzymes. *Science* **224:**350–357.

20. **Nguyen-Distèche, M., M. Leyh-Bouille, S. Pirlot, J. M. Frère, and J. M. Ghuysen.** 1986. *Streptomyces* K15 DD-peptidase-catalysed reactions with ester and amide carbonyl donors. *Biochem. J.* **235:**167–176.

21. **Ondetti, M. A., D. W. Cushman, E. F. Sabo, S. Natarajari, J. Pluscec, and B. Rubin.** 1981. Angiotensin-converting enzyme inhibition: further explorations of an active-site model, p. 235–246. *In* T. P. Sanger and R. N. Ondarza (ed.), *Molecular Basis of Drug Action. Developments in Biochemistry*, vol. 19. Elsevier/North-Holland Publishing Co., Amsterdam.

22. **Pratt, R. F., and C. B. Govardhan.** 1984. β-Lactamase-catalysed hydrolysis of acyclic depsipeptides and acyl transfer to specific amino acid acceptors. *Proc. Natl. Acad. Sci. USA* **81:**1302–1306.

23. **Samraoui, B., B. J. Sutton, R. J. Todd, P. J. Artymiuk, S. G. Waley, and D. C. Phillips.** 1986. Tertiary structural similarity between a class A β-lactamase and a penicillin-sensitive D-alanyl carboxypeptidase-transpeptidase. *Nature* (London) **320:**378–380.

24. **Spratt, B. G.** 1983. Penicillin-binding proteins and the future of β-lactam antibiotics. *J. Gen. Microbiol.* **129:**1247–1260.

25. **Walsch, C.** 1979. *Enzymatic Reaction Mechanisms*, p. 56–94. W. H. Freedman and Co., San Francisco.

26. **Williamson, R., and A. Tomasz.** 1985. Inhibition of cell wall synthesis and acylation of the penicillin-binding proteins during prolonged exposure of growing *Streptococcus pneumoniae* to benzylpenicillin. *Eur. J. Biochem.* **151:**475–483.

Chapter 26

Crystallization and Preliminary Crystallographic Studies of the High-Molecular-Weight Penicillin-Binding Protein 1B-δ of *Escherichia coli*

Fumitoshi Ishino
Masaaki Wachi
Kuni-hiro Ueda
Yoshiyasu Ito
Robert A. Nicholas
Jack L. Strominger
Toshiya Senda
Kohki Ishikawa
Yukio Mitsui
Michio Matsuhashi

All of the higher-molecular-weight penicillin-binding proteins (PBPs) of *Escherichia coli*, PBPs 1A, 1Bs, 2, and 3, play an important role in duplication of the cell (8, 15). These proteins have been shown to be bifunctional enzyme

Fumitoshi Ishino, Masaaki Wachi, Kuni-hiro Ueda, Yoshiyasu Ito, and Michio Matsu-hashi • Institute of Applied Microbiology, University of Tokyo, Bunkyo-ku, Tokyo 113, Japan.   **Robert A. Nicholas and Jack L. Strominger** • Department of Biochemistry and Molecular Biology, Harvard University, Cambridge, Massachusetts 02138.   **Toshiya Senda, Kohki Ishikawa, and Yukio Mitsui** • Faculty of Pharmaceutical Sciences, University of Tokyo, Bunkyo-ku, Tokyo 113, Japan.

proteins, with activities of transglycosylation and transpeptidation, during the synthesis of the peptidoglycan network that covers the whole cell surface (3–5, 8, 10). Primary amino acid sequences have been deduced on the basis of the nucleotide sequences of their structural genes (1, 2, 11), and the amino acid sequences around the catalytic serine residue in the penicillin-binding site have also been elucidated for several PBPs (6, 12, 13). Many experiments, utilizing both biochemical and recombinant DNA techniques, have suggested that the transglycosylase domain is at the amino terminus, whereas the transpeptidase domain is located at the carboxyl terminus of these proteins (1, 10). These proteins are very hydrophobic, even though the hydrophobicity profiles, as estimated from their amino acid sequences, are rather moderate. It is possible that the molecules are embedded in the cytoplasmic membrane at a hydrophobic region of the folded chains and function as two enzymes at the more hydrophilic domains that are projected into the periplasm (10). Such an imaginary molecular shape of PBPs must be confirmed by actual elucidation of their three-dimensional structure. For this purpose, we tried to crystallize one of the higher-molecular-weight PBPs of *E. coli,* PBP 1B. Precise crystallographic studies will also provide important information on the configuration of the penicillin-binding site, which may certainly contribute to developing new and effective β-lactam or related antibiotics.

Isolation of PBP 1B-γ from E. coli

The gamma component (molecular weight, 88,800 [2]) of PPB 1B was extracted from *E. coli* JST975 (PBP 1Bs⁻) cells that contained plasmid pPR5, which carries the DNA fragment encoding the gamma component of PBP 1Bs and contains a mutation that results in overproduction of this protein (Y. Ito, F. Ishino, S. Ueno, S. Tamaki, and M. Matsuhashi, manuscript in preparation). Extraction was performed by differential solubilization with 1% Triton X-100 and then with 1.25% β-octylglucoside solution containing 0.4 M NaCl. The protein was then purified by ammonium sulfate fractionation (33% saturation) to a single protein on sodium dodecyl sulfate-polyacrylamide gel electrophoresis without detectable contamination of phospholipids or lipopolysaccharides.

Crystallization of PBP 1B-δ and Preliminary Crystallographic Study

Crystallization was carried out by a hanging drop-vapor diffusion method at a final concentration of 13 to 15% ammonium sulfate and 2.6 to 2.9 M NaCl.

Several crystal habits were obtained, as partly shown in Fig. 1. The largest

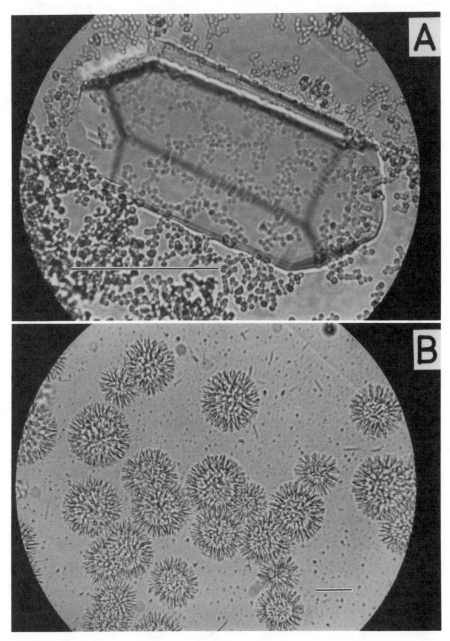

Figure 1. Crystals of *E. coli* PBP 1B-δ. A hexagonal pillar crystal (A) and smaller ones forming bur-shaped assemblies (B) are shown. Bar, 0.1 mm.

single crystals obtained were of hexagonal pillar shape, with dimensions of 1.0 (length) by 0.3 (width) mm. Under a polarizing microscope, the crystal exhibited moderate birefringence when viewed perpendicular to the pillar axis. Some crystals grew as irregular hexagonal disks. When this occurred, the birefringence was either very weak or absent, presumably because the viewing direction was close to the principal optical axis. Moreover, we have never observed the crystal morphology indicating the presence of a threefold inversion axis. We believe that these morphological and optical observations safely exclude all the crystal systems other than the hexagonal one. Unfortunately, the X-ray diffraction studies of these crystals were made considerably difficult by poor diffraction patterns. Although their mechanical strength seemed normal for protein crystals, the resolution of the diffraction patterns was in the range of 1 to 3 nm, depending on the crystal. From the best pattern shown in Fig. 2, however, we estimated the hexagonal lattice constants as being approximately $a = 34.0 \pm 2$ nm and $c = 16.5 \pm 1.5$ nm. According to the statistics of Matthews (9), the probable numbers of PBP 1B molecules (as one of the multiples of six) in the unit cell lie between the two extreme values 60 and 108 for which the V_M values as defined by Matthews (9) would be 3.10 and 1.72, respectively. This means that there are five to nine PBP 1B molecules

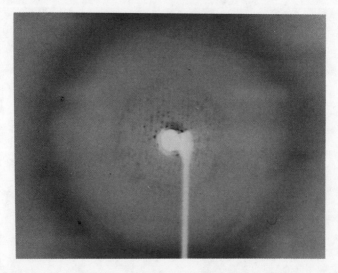

Figure 2. Precession photograph ($\mu = 0.5$ degrees), from a crystal similar to the one shown in Fig. 1A. The incident X-ray beam seems approximately perpendicular to the c axis, which is approximately horizontal. The circumscribing halo with an approximate spacing of 0.4 nm is due to the fused quartz capillary wall confining the crystal. The crystal size was 1.0 by 0.3 by 0.2 mm. An exposure time of 40 min was used with the graphite-monochromatized Cu K_α line from a Rigaku RU-200 rotating anode generator operating at 55 kV and 90 mA. The focus size was 3 by 0.3 mm, and a beam collimator of 0.2 mm diameter was used.

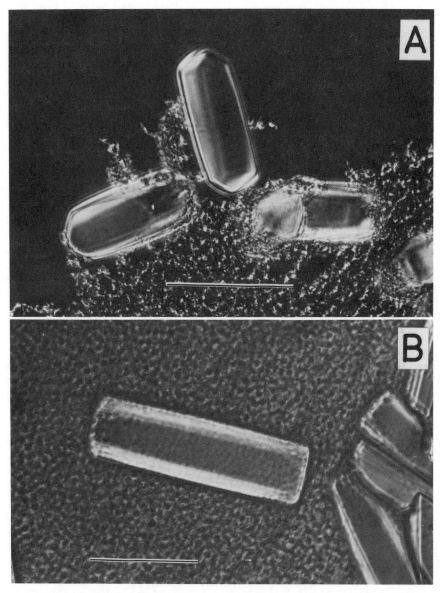

Figure 3. Crystals of benzylpenicillin-bound *E. coli* PBP 1B-δ. Two crystal habits are shown. Bar, 0.1 mm.

in an asymmetric unit, even if the point group *6 2 2* (rather than *6*), where there are 12 asymmetric units per unit cell, were assumed.

Sodium dodecyl sulfate-polyacrylamide gel electrophoresis showed that the crystals are the delta component of PBP 1B. The original gamma component synthesized by *E. coli* cells had probably been degraded by action of some peptidase during the process of crystallization. We are also trying to crystallize the undegraded gamma component. The protein that was bound to penicillins and cephalosporins could also be crystallized; two crystal habits of the benzylpenicillin-bound protein are shown in Fig. 3. The crystallized penicillin- or cephalosporin-bound protein also showed a mobility on sodium dodecyl sulfate-polyacrylamide gel electrophoresis similar to that of the delta component of PBP 1Bs. Thus, the putative protease activity that converted the gamma component into the delta component is not penicillin sensitive and may not be an autolytic activity of PBP 1B-γ.

With the success in crystallizing a higher-molecular-weight PBP of *E. coli,* we are still at the starting point in elucidating the three-dimensional structure of this important and interesting protein. Now that the precise crystallographic work on the lower-molecular-weight PBP and penicillinase has been achieved by other investigators (2a, 7, 14), our present success may have turned a decisive page in the investigation of membrane structure and function in the duplication of bacterial cells.

ACKNOWLEDGMENTS. This work was partially supported by Grant-in-Aid for Scientific Research 61470157 to M.M. from the Ministry of Education, Science and Culture of Japan; by grant 1985/6/7 for the U.S.-Japan Scientific Cooperation Program to M.M. and J.L.S. supported by the Japan Society for Promotion of Science and the National Science Foundation; and by Public Health Service grant AI09152 from the National Institutes of Health to J.L.S.

We thank Don C. Wiley, Michael Silver, and Boudjema Samroui of Harvard University for instructions on crystallization of membrane-bound proteins in the presence of detergent. We also thank Y. Iitaka for stimulating discussion on crystal morphology.

LITERATURE CITED

1. **Asoh, S., H. Matsuzawa, F. Ishino, J. L. Strominger, M. Matsuhashi, and T. Ohta.** 1986. Nucleotide sequence of the *pbpA* gene and characteristics of the deduced amino acid sequence of penicillin-binding protein 2 of *Escherichia coli* K12. *Eur. J. Biochem.* **160:**231– 238.

2. **Broome-Smith, J. K., A. Edelman, S. Yousif, and B. G. Spratt.** 1985. The nucleotide sequences of the *ponA* and *ponB* genes encoding penicillin-binding proteins 1A and 1B of *Escherichia coli* K12. *Eur. J. Biochem.* **147:**437–446.

2a. **Herzberg, O., and J. Moult.** 1987. Bacterial resistance to β-lactam antibiotics: crystal structure of β-lactamase from *Staphylococcus aureus* PC1 at 2.5 Å resolution. *Science* **236:**694– 701.

3. **Ishino, F., and M. Matsuhashi.** 1981. Peptidoglycan synthetic enzyme activities of highly

purified penicillin-binding protein 3 in *Escherichia coli:* a septum-forming reaction sequence. *Biochem. Biophys. Res. Commun.* **101:**905–911.

4. **Ishino, F., K. Mitsui, S. Tamaki, and M. Matsuhashi.** 1980. Dual enzyme activities of cell wall peptidoglycan synthesis, peptidoglycan transglycosylase and penicillin-sensitive transpeptidase, in purified preparations of *Escherichia coli* penicillin-binding protein 1A. *Biochem. Biophys. Res. Commun.* **97:**287–293.

5. **Ishino, F., W. Park, S. Tomioka, S. Tamaki, I. Takase, K. Kunugita, H. Matsuzawa, S. Asoh, T. Ohta, B. G. Spratt, and M. Matsuhashi.** 1986. Peptidoglycan synthetic activities in membranes of *Escherichia coli* caused by overproduction of penicillin-binding protein 2 and rodA protein. *J. Biol. Chem.* **261:**7024–7031.

6. **Keck, W., B. Glauner, U. Schwarz, J. K. Broome-Smith, and B. G. Spratt.** 1985. Sequences of the active-site peptides of three of the high-Mr penicillin-binding proteins of *Escherichia coli* K-12. *Proc. Natl. Acad. Sci. USA* **82:**1999–2003.

7. **Kelly, J. A., J. R. Knox, P. C. Moews, G. J. Hite, J. B. Bartolone, H. Zao, B. Joris, J.-M. Frere, and J.-M. Ghuysen.** 1985. 2.8-Å structure of penicillin-sensitive *D*-alanyl carboxypeptidase-transpeptidase from *Streptomyces* R61 and complexes with β-lactams. *J. Biol. Chem.* **260:**6449–6458.

8. **Matsuhashi, M., F. Ishino, S. Tamaki, S. Nakajima-Iijima, S. Tomioka, J. Nakagawa, A. Hirata, B. G. Spratt, T. Tsuruoka, S. Inouye, and Y. Yamada.** 1982. Mechanism of action of β-lactam antibiotics: inhibition of peptidoglycan transpeptidases and novel mechanisms of action, p. 99–114. *In* H. Umezawa, A. L. Demain, T. Hata, and C. R. Hutchinson (ed.), *Trends in Antibiotic Research.* Japan Antibiotics Research Association, Tokyo.

9. **Matthews, B. W.** 1968. Solvent content of protein crystals. *J. Mol. Biol.* **33:**491–497.

10. **Nakagawa, J., S. Tamaki, S. Tomioka, and M. Matsuhashi.** 1984. Functional biosynthesis of cell wall peptidoglycan by polymorphic bifunctional polypeptides. *J. Biol. Chem.* **259:**13937–13946.

11. **Nakamura, M., I. N. Maruyama, M. Soma, J. Kato, H. Suzuki, and Y. Hirota.** 1983. On the process of cellular division in *Escherichia coli:* nucleotide sequence of the gene for penicillin-binding protein 3. *Mol. Gen. Genet.* **191:**1–9.

12. **Nicholas, R. A., J. L. Strominger, H. Suzuki, and Y. Hirato.** 1985. Identification of the active site in penicillin-binding protein 3 of *Escherichia coli. J. Bacteriol.* **164:**456–460.

13. **Nicholas, R. A., H. Suzuki, Y. Hirota, and J. L. Strominger.** 1985. Purification and sequencing of the active site tryptic peptide from penicillin-binding protein 1b of *Escherichia coli. Biochemistry* **24:**3448–3453.

14. **Samraoui, B., B. J. Sutton, R. J. Todd, P. J. Artymiuk, S. G. Waley, and D. G. Phillips.** 1986. Tertiary structural similarity between a class A β-lactamase and a penicillin-sensitive *D*-alanyl carboxypeptidase-transpeptidase. *Nature* (London) **320:**378–380.

15. **Spratt, B. G.** 1975. Distinct penicillin binding proteins involved in the division, elongation, and shape of *Escherichia coli* K12. *Proc. Natl. Acad. Sci. USA* **72:**2999–3003.

Chapter 27

Membrane Topology of Penicillin-Binding Proteins 1b and 3 of *Escherichia coli* and the Production of Water-Soluble Forms of High-Molecular-Weight Penicillin-Binding Proteins

Brian G. Spratt
Lucas D. Bowler
Alex Edelman
Jenny K. Broome-Smith

The rational design of novel inhibitors of β-lactamases and penicillin-binding proteins (PBPs) depends on a detailed understanding of their three-dimensional structures. At present, X-ray crystallographic studies of PBPs have been limited to the DD-peptidase from *Streptomyces* sp. strain R61 since this enzyme, unlike other PBPs which are membrane bound, is secreted in a water-soluble form that can be readily purified and crystallized (5, 9). The three-dimensional structures of two class A β-lactamases were recently reported at low resolution (8, 14), and these studies have provided persuasive evidence for homology between the DD-peptidase of *Streptomyces* sp. strain R61 and the class A β-lactamases. It now seems likely that all β-lactamases and PBPs that utilize an

Brian G. Spratt, Lucas D. Bowler, and Jenny K. Broome-Smith • Microbial Genetics Group, School of Biological Sciences, University of Sussex, Falmer, Brighton BN1 9QG, United Kingdom. **Alex Edelman** • Unité de Biochimie Microbienne, Institut Pasteur, Paris 75724, Cedex 15, France.

active-site serine mechanism will form a superfamily of homologous enzymes that show substantial similarities in their three-dimensional structures but are highly diverged in their primary sequences (1, 2, 8, 11, 14).

The R61 DD-peptidase is probably analogous to the low-molecular-weight nonessential PBPs of other bacteria (5). In *Escherichia coli* the physiologically important PBPs have been identified as the high-molecular-weight PBPs 1a/ 1b, 2, and 3 (15, 17). These PBPs possess a carboxy-terminal domain that catalyzes the penicillin-sensitive peptidoglycan transpeptidase reaction and is probably homologous to low-molecular-weight PBPs and β-lactamases, but they also possess an amino-terminal domain that is absent from the latter classes of enzyme and is assumed to catalyze the penicillin-insensitive peptidoglycan transglycosylase reaction (2, 6, 10, 17). The determination of the three-dimensional structure of a high-molecular-weight PBP is an important aim, as it may provide a better model than the R61 DD-peptidase for the design of novel transpeptidase inhibitors and because it would provide the structure of a transglycosylase domain which is a potential target for antibacterial agents.

Since all PBPs other than the R61 DD-peptidase are membrane proteins (16), the determination of their three-dimensional structure requires either that they be crystallized from detergent solution or that soluble active forms can be obtained by partial proteolysis or through molecular genetics.

The feasibility of obtaining water-soluble, active forms of high-molecular-weight PBPs depends on their organization in the cytoplasmic membrane. At present we know that the low-molecular-weight PBPs 5 and 6 of *E. coli* and PBP 5 of *Bacillus subtilis* and *Bacillus stearothermophilus* have a simple mode of membrane insertion via a carboxy-terminal amphipathic membrane anchor, with the bulk of the proteins exposed on the periplasmic side of the membrane (12, 13, 19). If the high-molecular-weight PBPs also have a simple membrane topology and are inserted by a single amino- or carboxy-terminal membrane anchor, it should be possible to produce soluble forms for X-ray analysis. Alternatively, if high-molecular-weight PBPs have a complex membrane topology and span the membrane several times, the prospects of obtaining enzymatically active soluble forms are poor. At present, however, we know nothing about the organization of high-molecular-weight PBPs in the cytoplasmic membrane. We have therefore used a combination of molecular genetics with protease digestion and immunoblotting to analyze the organization of PBPs 1b and 3 in the cytoplasmic membrane of *E. coli*. The resulting information has been used to construct a water-soluble, enzymatically active, periplasmic form of PBP 3 that should be suitable for crystallization and X-ray analysis.

Organization of PBP 1b in the Cytoplasmic Membrane

PBP 1b exists in two forms which arise from the use of alternative translation starts in the transcript of the *ponB* gene (2, 7). PBP 1b is not synthesized as a preprotein, and the small form of the protein starts at Met-46 in the amino acid sequence of the large form (2, 18).

The β-lactamase fusion vector pJBS633 (3) was used to analyze the organization of PBP 1b in the *E. coli* cytoplasmic membrane. One hundred inframe fusions that joined the coding region for the mature form of TEM β-lactamase to random positions within the PBP 1b gene were obtained, and the nucleotide sequence across each fusion junction was determined by dideoxy sequencing (4). Several of the fusions were obtained more than once, and the series (the pAE30 series) contained constructs that expressed fusion proteins that had the mature form of TEM β-lactamase joined at 72 different positions within PBP 1b. In each of the pAE30 series the PBP 1b-β-lactamase fusion proteins should be expressed at the same level from the promoter of the tetracycline resistance gene of the vector.

The ability of a PBP 1b-β-lactamase fusion protein to protect single cells of *E. coli* from killing by ampicillin depends on whether the β-lactamase moiety is translocated across the cytoplasmic membrane (3). The plot of the "single cell" MIC against the position in PBP 1b at which the β-lactamase moiety is joined should therefore generate a map of those parts of PBP 1b which are on the cytoplasmic or periplasmic side of the membrane (see references 3 and 4 for further details). All fusion proteins that contained at least the amino-terminal 94 residues of PBP 1b provided single cells of *E. coli* with substantial levels of protection against ampicillin, suggesting that residues 94 to 844 of PBP 1b are translocated to the periplasmic side of the cytoplasmic membrane and that residues 1 to 94 contain a signal that acts to translocate the bulk of the protein across the membrane (Fig. 1).

PBP 1b is relatively hydrophilic and possesses only one region (residues 64 to 87) that is a candidate hydrophobic α-helical membrane-spanning segment (2). It seems likely that the amino-terminal region of PBP 1b acts as a noncleaved signallike peptide which serves both to translocate the bulk of the protein across the membrane and to anchor the protein in the membrane at its amino terminus.

The analysis of the organization of PBP 1b using β-lactamase fusions suggested a simple model in which residues 1 to 63 (or residues 46 to 63 of the small form of PBP 1b) are on the cytoplasmic side of the membrane, with residues 64 to 87 spanning the membrane and residues 88 to 844 on the periplasmic side of the membrane (4). If this model is correct, it predicts that PBP 1b should be completely degraded by protease digestion from the periplasmic side of the cytoplasmic membrane but should only be slightly reduced in size by protease digestion from the cytoplasmic side of the membrane. Protease K digestion of PBP 1b from the cytoplasmic side of the membrane, using inverted membrane vesicles, resulted in the conversion of the two forms of PBP 1b to a single form that was approximately the same size as the small form of PBP 1b (Fig. 2). The failure of protease K to digest PBP 1b further was due to

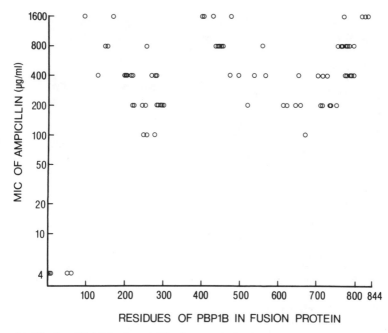

Figure 1. "Single cell" MIC of ampicillin for *E. coli* JM101, expressing PBP 1b-β-lactamase fusion proteins. Approximately 40 bacteria were spotted on Luria agar plates containing a range of concentrations of ampicillin, and the MIC was defined as the lowest concentration of ampicillin that prevented the growth of colonies. *E. coli* JM101 and *E. coli* JM101(pJBS633) required a "single cell" MIC of 4 µg/ml.

protection by the membrane since addition of protease K plus Triton X-100 resulted in complete digestion (Fig. 2). Since the two forms of PBP 1b differ at their amino termini, the conversion of the two forms into a single form implies that it is the amino terminus of PBP 1b that is exposed on the cytoplasmic side of the cytoplasmic membrane.

Protease digestion from the periplasmic side of the membrane, using spheroplasts, resulted in the complete digestion of PBP 1b (data not shown) and established that the bulk of the protein is exposed in the periplasm.

Organization of PBP 3 in the Cytoplasmic Membrane

PBP 3 has been shown to be synthesized as a preprotein with a signal peptide that results in the translocation of some or all of the protein across the cytoplasmic membrane (11). We have used protease digestion and immunoblotting to analyze the amount of PBP 3 that is exposed on the periplasmic side of the cytoplasmic membrane. PBP 3 was completely insensitive to protease

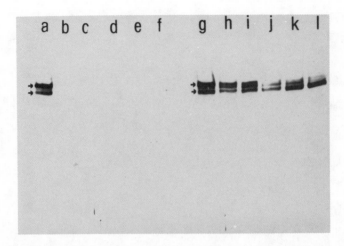

Figure 2. Protease digestion of PBP 1b in inverted membrane vesicles. Inverted membrane vesicles were prepared from *E. coli* C600 (12) and were incubated, in the presence (a–f) or absence (g–l) of 1% Triton X-100, with protease K at a final concentration of 0 (a and g), 4 (b and h), 17 (c and i), 75 (d and j), 313 (e and j), or 1,250 (f and l) μg/ml. After 30 min at room temperature, the protease was inhibited, and the samples were fractionated on a 10% sodium dodecyl sulfate-polyacrylamide gel (16) and immunoblotted using a polyclonal antiserum against PBP 1b. The arrows in lanes a and g indicate the positions of the large and small forms of PBP 1b.

K digestion from the cytoplasmic side of the membrane, using inverted membrane vesicles (Fig. 3). This suggests that very little, if any, of the protein is exposed on the cytoplasmic side of the membrane. The failure to digest PBP 3 was due to protection by the membrane since complete digestion was obtained when protease K was added in the presence of Triton X-100 (Fig. 3). Conversely, PBP 3 was completely digested by protease attack from the peri-

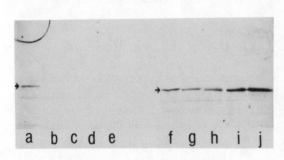

Figure 3. Protease digestion of PBP 3 in inverted membrane vesicles. Inverted membrane vesicles prepared from *E. coli* C600 were incubated, in the presence (a–e) or absence (f–j) of 1% Triton X-100, with protease K at a final concentration of 0 (a and f), 0.002 (b and g), 0.02 (c and h), 0.2 (d and i), or 2 (e and j) μg/ml and were treated as described for Fig. 2 except that immunoblotting was carried out with antiserum against PBP 3. The arrows in lanes a and f indicate the position of PBP 3.

plasmic face of the cytoplasmic membrane, using spheroplasts (data not shown).

Recently it has been proposed that PBP 3 is a lipoprotein in which fatty acyl groups are attached at Cys-30, which becomes the amino terminus of the mature protein, following the action of the lipoprotein signal peptidase (Y. Hirota, unpublished data). By analogy with other lipoproteins, it is likely that PBP 3 is embedded in the membrane by the fatty acyl chains attached to the amino-terminal cysteine residue. Our results are compatible with a model of PBP 3 in which the protein is attached to the cytoplasmic membrane only by these fatty acyl groups at the amino terminus, with all of the protein exposed on the periplasmic side of the membrane.

Attempts to Produce Soluble, Periplasmic Forms of PBPs 1b and 3

The results outlined above suggested that both PBP 1b and PBP 3 of *E. coli* are essentially periplasmic enzymes that are embedded in the membrane only at their amino termini. It should therefore be possible to obtain water-soluble, enzymatically active, periplasmic forms of both proteins by replacing the sequences encoding their amino termini with a sequence encoding an authentic signal peptide. We therefore constructed plasmids (pBS121 and PBS120) that expressed hybrid proteins consisting of the signal peptide and first 10 residues of mature PBP 5 fused, respectively, to residues 87 to 844 of PBP 1b and to residues 53 to 588 of pre-PBP 3 (i.e., residues 24 to 559 numbering from the amino terminus of the mature form of PBP 3).

Both of these constructs produced novel PBPs which, as expected, were slightly smaller than their parent proteins and which were expressed under the control of the *lac* promoter (Fig. 4). Only about 50 to 70% of the modified form of PBP 1b (PBP 1b*) was recovered in the periplasmic fraction, and expression of the protein, even at low levels, was highly deleterious to the growth of *E. coli*. It was unclear whether the nonquantitative recovery of PBP

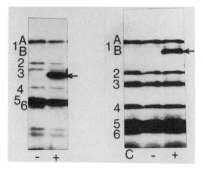

Figure 4. Expression of PBP 1b* and PBP 3* from pBS121 and pBS120. Left panel: PBPs were assayed as described (2) in freeze-thawed cells of *E. coli* JM101(pBS120) grown without (−) or with (+) isopropyl-β-D-thiogalactopyranoside. Right panel: PBPs of *E. coli* JM101 grown with 1 mM isopropyl-β-D-thiogalactopyranoside (C), and *E. coli* JM101(pBS121) grown without (−) or with (+) isopropyl-β-D-thiogalactopyranoside. The [³H]benzylpenicillin-labeled PBPs were fractionated on a 12% (left panel) or 10% (right panel) sodium dodecyl sulfate-polyacrylamide gel and were detected by fluorography (16). The arrows mark the position of PBP 3* (left panel) and PBP 1b* (right panel).

1b* in the periplasm occurred because a proportion of the protein was still associated with the membrane, e.g., by interaction with other proteins of the cytoplasmic membrane, or was the result of difficulties in the fractionation of the poorly growing cells that resulted from the expression of the toxic novel PBP. The protein was not studied further as it did not promise to provide a suitable soluble form of PBP 1b. Attempts to produce a cytoplasmic form of PBP 1b by fusing residues 87 to 844 to the start of β-galactosidase were also unsuccessful, as the resulting protein failed to bind penicillin, presumably because it failed to fold correctly in the cytoplasm.

The protein derived from PBP 3 (PBP 3*) was quantitatively recovered in the periplasmic fraction, whereas the normal PBP 3 was quantitatively recovered with the membrane fraction (Fig. 5). It therefore seems likely that PBP 3 is, as predicted, attached to the membrane only at its amino terminus and that the exchange of its lipoprotein signal peptide with a normal signal peptide results in the translocation of the modified form of PBP 3 across the cytoplasmic membrane and its processing by signal peptidase, to produce the soluble, periplasmic PBP 3*.

PBP 3* lacks only 23 residues from the amino terminus of normal membrane-bound PBP 3. This difference would not be expected to have any effect on the transpeptidase domain which is believed to extend from about residue 240 to the carboxy terminus (2, 6). In support of this, PBP 3* bound penicillin with the same affinity as normal membrane-bound PBP 3, and the stability of the penicillin-binding activity appeared to be unchanged from that of the normal enzyme. We would hope that the removal of the amino-terminal 23 residues would not eliminate the transglycosylase activity of PBP 3, but direct assay of this activity has not been attempted.

The periplasmic form of PBP 3 appears to be truly soluble since it remains in solution in the absence of detergents and fractionates with the predicted molecular weight on Sephadex G-100 in aqueous buffers.

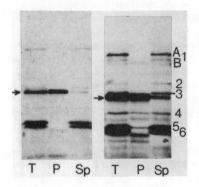

Figure 5. Periplasmic location of PBP 3* expressed from pBS120. *E. coli* JM101(pBS120) was grown in Luria broth to approximately 2×10^8 bacteria per ml, and 1 mM isopropyl-β-D-thiogalactopyranoside was added to induce expression of PBP 3*. After a further 60 min, the bacteria were converted to spheroplasts (3). The spheroplasts were centrifuged for 5 min in an Eppendorf centrifuge, and PBPs were assayed in the unfractionated spheroplasts (T) and in the supernatant (P = periplasm) and spheroplast pellet (Sp). The left panel shows an autoradiograph of the PBPs to illustrate the quantitative recovery of PBP 3* (arrow) in the periplasm, whereas the right panel shows a longer exposure of the same gel to show the quantitative recovery of the chromosomally expressed normal PBP 3 in the spheroplast pellet.

Conclusions

Our studies have shown that PBPs 1b and 3 of *E. coli* are essentially periplasmic enzymes that are held in the cytoplasmic membrane at their amino termini. The inability to produce a soluble, periplasmic form of PBP 1b by exchanging its amino-terminal noncleaved signal peptide with an authentic signal peptide may indicate that this protein is attached to the membrane at some position other than its amino terminus, but further experiments are required to eliminate other reasons for this result.

The situation with PBP 3 is more clear-cut since exchange of the lipoprotein signal peptide with an authentic signal peptide resulted in a protein that was quantitatively recovered in the periplasmic fraction. The properties of this protein make it a good candidate for crystallization and for the X-ray analysis of a high-molecular-weight PBP.

ACKNOWLEDGMENTS. This paper is dedicated to the memory of Yukinori Hirota. We are grateful to Wolfgang Keck for the generous gift of anti-PBP 1B and anti-PBP 3 serum, to Patrick Cassidy for providing tritiated benzylpenicillin, and to the Medical Research Council and Glaxo Group Research for supporting this work.

LITERATURE CITED

1. **Asoh, S., H. Matsuzawa, F. Ishino, J. L. Strominger, M. Matsuhashi, and T. Ohta.** 1986. Nucleotide sequence of the *pbpA* gene and characteristics of the deduced amino acid sequence of penicillin-binding protein 2 of *Escherichia coli*. *Eur. J. Biochem.* **160:**231–238.
2. **Broome-Smith, J. K., A. Edelman, S. Yousif, and B. G. Spratt.** 1985. The nucleotide sequences of the *ponA* and *ponB* genes encoding penicillin-binding proteins 1A and 1B of *Escherichia coli*. *Eur. J. Biochem.* **147:**437–446.
3. **Broome-Smith, J. K., and B. G. Spratt.** 1986. A vector for the construction of translational fusions to TEM β-lactamase and the analysis of protein export signals and membrane protein topology. *Gene* **49:**341–349.
4. **Edelman, A., L. Bowler, J. K. Broome-Smith, and B. G. Spratt.** 1987. Use of a β-lactamase fusion vector to investigate the organisation of penicillin-binding protein 1B in the cytoplasmic membrane of *Escherichia coli*. *Mol. Microbiol.* **1:**101–106.
5. **Frère, J.-M., and B. Joris.** 1985. Penicillin-sensitive enzymes in peptidoglycan biosynthesis. *Crit. Rev. Microbiol.* **11:**299–396.
6. **Hedge, P. J., and B. G. Spratt.** 1984. A gene fusion that localises the penicillin-binding domain of penicillin-binding protein 3 of *Escherichia coli*. *FEBS Lett.* **176:**179–184.
7. **Kato, J.-I., H. Suzuki, and Y. Hirota.** 1984. Overlapping of the coding regions for the α and γ components of penicillin-binding protein 1B in *Escherichia coli*. *Mol. Gen. Genet.* **196:**449–457.
8. **Kelly, J. A., O. Dideberg, P. Charlier, J. P. Wery, M. Libert, P. C. Moews, J. R. Knox, C. Duez, C. Fraipont, B. Joris, J. Dusart, J.-M. Frère, and J.-M. Ghuysen.**

1986. On the origin of bacterial resistance to penicillin: comparison of a β-lactamase and a penicillin target. *Science* **231**:1429–1431.

9. **Kelly, J. A., J. R. Knox, P. C. Moews, G. J. Hite, J. B. Bartolone, H. Zhao, B. Joris, J.-M. Frère, and J.-M. Ghuysen.** 1985. 2.8Å structure of penicillin-sensitive D-alanyl carboxypeptidase-transpeptidase from *Streptomyces* R61 and complexes with β-lactams. *J. Biol. Chem.* **260**:6449–6458.

10. **Nakagawa, J.-I., S. Tamaki, S. Tomioka, and M. Matsuhashi.** 1984. Functional biosynthesis of cell wall peptidoglycan by polymorphic bifunctional polypeptides. *J. Biol. Chem.* **259**:13937–13946.

11. **Nakamura, M., I. N. Maruyama, M. Soma, J.-I. Kato, H. Suzuki, and Y. Hirota.** 1983. On the process of cellular division in *Escherichia coli:* nucleotide sequence of the gene for penicillin-binding protein 3. *Mol. Gen. Genet.* **191**:1–9.

12. **Pratt, J. M., I. B. Holland, and B. G. Spratt.** 1981. Precursor forms of penicillin-binding proteins 5 and 6 of *E. coli* cytoplasmic membrane. *Nature* (London) **293**:307–309.

13. **Pratt, J. M., M. E. Jackson, and I. B. Holland.** 1986. The C-terminus of penicillin-binding protein 5 is essential for localisation to the *E. coli* inner membrane. *EMBO J.* **5**:2399–2405.

14. **Samraoui, B., B. J. Sutton, R. J. Todd, P. J. Artymiuk, S. G. Waley, and D. C. Phillips.** 1986. Tertiary structural similarity between a class A β-lactamase and a penicillin-sensitive D-alanyl carboxypeptidase-transpeptidase. *Nature* (London) **320**:378–380.

15. **Spratt, B. G.** 1975. Distinct penicillin binding proteins involved in the division, elongation, and shape of *Escherichia coli*. *Proc. Natl. Acad. Sci. USA* **72**:2999–3003.

16. **Spratt, B. G.** 1977. Properties of the penicillin-binding proteins of *Escherichia coli* K12. *Eur. J. Biochem.* **72**:341–352.

17. **Spratt, B. G.** 1983. Penicillin-binding proteins and the future of β-lactam antibiotics. *J. Gen. Microbiol.* **129**:1247–1260.

18. **Suzuki, H., J.-I. Kato, Y. Sakagami, M. Mori, A. Suzuki, and Y. Hirota.** 1987. Conversion of the α component of penicillin-binding protein 1b to the β component in *Escherichia coli*. *J. Bacteriol.* **169**:891–893.

19. **Waxman, D. J., and J. L. Strominger.** 1979. Cleavage of a COOH-terminal hydrophobic region from D-alanine carboxypeptidase, a penicillin-sensitive bacterial membrane enzyme. *J. Biol. Chem.* **254**:4863–4875.

Chapter 28

Alteration of Penicillin-Binding Protein 2 of *Escherichia coli* to a Water-Soluble Form with Penicillin-Binding Activity and Determination of Its Active Site

Hiroshi Matsuzawa
Hiroyuki Adachi
Akiko Takasuga
Takahisa Ohta

Penicillin-binding protein (PBP) 2 of *Escherichia coli* has been previously identified as a component of the cytoplasmic membrane. We recently reported the nucleotide sequence of the *pbpA* gene encoding PBP 2 and its deduced amino acid sequence (1). The N-terminal region has a signal-like sequence which is uncleaved. The hydropathy profile of the primary structure of PBP 2 suggests that the N-terminal hydrophobic region (a stretch of 25 nonionic amino acids; average hydropathy, +2.1) (Fig. 1) may anchor the protein to the cytoplasmic membrane. We have undertaken to delete the N-terminal hydrophobic region of PBP 2 by site-directed mutagenesis and have obtained a water-soluble form of PBP 2 called PBP 2*. PBP 2* was found to retain the penicillin-binding activity. Using this PBP 2*, we identified the active-site serine residue of PBP 2 which had previously been predicted on the basis of sequence homology (1).

Hiroshi Matsuzawa, Hiroyuki Adachi, Akiko Takasuga, and Takahisa Ohta • Department of Agricultural Chemistry, The University of Tokyo, Bunkyo-ku, Tokyo 113, Japan.

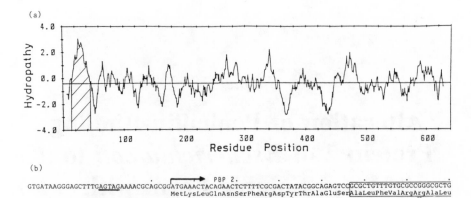

Figure 1. N-terminal region of PBP 2 containing the uncleaved signal-like sequence (1). (a) Hydropathy profile along the amino acid sequence of PBP 2. (b) Nucleotide sequence of the *pbpA* gene encoding the N-terminal amino acid sequence of PBP 2. The segment removed by site-directed mutagenesis is shown by cross-hatching (a) and the outlined block (b).

Alteration of PBP 2 to a Water-Soluble Form with Penicillin-Binding Activity

In the N-terminal amino acid sequence of PBP 2, there is a long stretch of 25 nonionic amino acids (residues 21 to 45) (1), as shown in Fig. 1b. The deletion of the 93 nucleotides coding for 31 amino acid residues (the 25 amino acid residues plus 6 amino acid residues preceding the long hydrophobic stretch) was introduced in the *pbpA* gene by site-directed mutagenesis using a synthetic oligonucleotide. The resulting *pbpA*(*) gene was unable to complement a *pbpA* mutant. The *pbpA*(*) gene was expressed by the *lpp-lac* or *lac* promoter of *E. coli*, and PBP 2* was expected to be localized in the cytoplasm, since the signal-like sequence had been removed. PBP 2* was localized in the soluble fraction and not detected in the membrane fraction (Fig. 2). In other experiments, no PBP 2* was detected in the periplasm (data not shown). These results indicate that PBP 2 was altered to a water-soluble form located in the cytoplasm. It is noteworthy that PBP 2* had penicillin-binding activity (Fig. 2) and that the overproduction of PBP 2* when the strong *lpp-lac* promoter was used (Fig. 2, e and g) was substantially greater than that of PBP 2 from the same promoter (Fig. 2, b and c).

In spite of the deletion of the 31 amino acids, the electrophoretic mobility of PBP 2* was indistinguishable from that of PBP 2 by sodium dodecyl sulfate-polyacrylamide gel electrophoresis (Fig. 2). However, as previously pointed out, the apparent molecular weight of PBP 2 is 66,000 by electrophoresis,

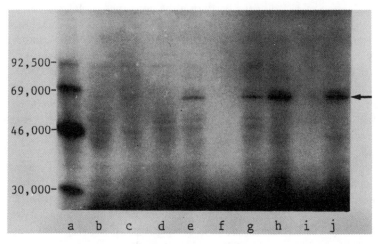

Figure 2. Identification of the localization of PBP 2*. *E. coli* cells, harboring the constructed plasmids carrying the *pbpA* gene (pHA102, b–d) or the *pbpA*(*) gene (pHA102Δ93, e–g; pHA104Δ93, h–j) preceded by the *lpp-lac* promoter, were cultured in the presence of isopropyl-β-D-thiogalactoside. The *ori* region of plasmid pHA104Δ93, being that of multicopy plasmid pUC19, differs from those of the other two plasmids. After the cells were harvested, disrupted cells were prepared by sonication, and then membrane and soluble fractions were separated by ultracentrifugation. The disrupted cells, membrane fractions, and soluble fractions were incubated with [³H]benzylpenicillin. [³H]benzylpenicillin-protein complexes were separated by sodium dodecyl sulfate-polyacrylamide gel electrophoresis and located by fluorography. (a) Molecular weight standards: (b, e, and h) disrupted cells; (c, f, and i) membrane fraction; (d, g, and j) soluble fraction. Arrow indicates the position of PBP 2 (b and c) and PBP 2* (e, g, h, and j).

whereas the calculated molecular weight is 70,867 (1). The molecular weight of PBP 2* is calculated to be about 67,500 on account of the deletion, being close to that determined by electrophoresis.

PBP 5 of *E. coli* is tightly bound to the cytoplasmic membrane by its C-terminal region, the removal of which results in the release of the truncated protein into the periplasm (7). By contrast, our results suggest that the N-terminal region of PBP 2 is not only an uncleaved signal-like sequence (1), but also an anchor of the protein to the cytoplasmic membrane.

Determination of Penicillin-Binding Site of PBP 2*

The penicillin-binding site of PBP 2 could not be easily determined, since the purification of the protein was difficult as a result of its low production and high hydrophobicity. PBP 2* could, however, be purified much more easily as a result of its high production (more than 5% of total soluble protein) and water solubility. PBP 2* was overproduced by isopropyl-β-D-thiogalactoside induction of *E. coli* cells harboring the constructed plasmid carrying the *pbpA*(∗) gene. PBP 2∗ was purified from the soluble fraction of disrupted cells by two purification steps, ammonium sulfate precipitation (40% saturation) and affinity chromatography on ampicillin-Sepharose 4B. Electrophoresis of purified PBP 2∗ revealed two bands (molecular weights, 63,000 and 58,000) besides a major band of PBP 2∗ (data not shown). The two bands seemed to be derivatives of PBP 2∗, since their penicillin-binding activities were inhibited by mecillinam but not by cefmetazole (data not shown). Therefore, this purified sample was used for the following experiment.

PBP 2∗ labeled with dansyl-penicillin was digested with lysyl-endopeptidase and trypsin. The tryptic fragments were separated by high-performance liquid chromatography on a C_8 column. One fragment showing fluorescence emission was obtained and was further digested with chymotrypsin. Finally, the fluorescence-labeled chymotryptic fragment was obtained and hydrolyzed in 6 N HCl at 110°C. The amino acid composition of the fragment was determined (Table 1) and was identical with that of the amino acid sequence Ala-Thr-Gln-Gly-Val-Tyr-Pro-Pro-Ala-Ser-Thr-Val-Lys-Pro (residues 321 to 334) of PBP 2, which had been deduced from the nucleotide sequence of the *pbpA* gene (1). One serine residue, Ser-330, was found in this segment, and penicillin seemed to bind to it. Thus, evidence showing that Ser-330 is the active-site residue was obtained, confirming the previous prediction based on sequence homology with other PBPs and β-lactamases (1). The penicillin-binding sites of PBPs 1A (3), 1B (3, 6), 3 (3, 5), and 5 (4) have also been identified.

We are going to crystallize PBP 2∗ to determine its three-dimensional structure by X-ray crystallographic analysis. This is essential for understanding

Table 1. Amino acid composition of chymotryptic dansyl-penicillin-bound peptide of PBP 2∗ and identification of its active site

Amino acid	Peptide analysis[a]	DNA sequence[b]
Ser	1.0	1
Glx	0.8	1
Gly	1.4	1
Thr	1.7	2
Ala	2.1	2
Pro	3.0	3
Tyr	0.7	1
Val	1.6	2
Lys	0.9	1

[a]Amino acid composition was analyzed after amino acids produced by acid hydrolysis of the peptide were treated with phenylisothiocyanate. Amounts of the amino acids are shown as values relative to Ser.
[b]The amino acid composition obtained on the basis of DNA sequence is that of the amino acid sequence, residues 321 to 334, of PBP 2. The complete primary structure of PBP 2 has been deduced from the nucleotide sequence of the *pbpA* gene (1).

the structure-function relationship of PBP 2. PBP 2 catalyzes the peptidoglycan transglycosylase and transpeptidase reactions in the presence of the RodA protein (2). Whether PBP 2* is able to catalyze the reactions should be investigated by in vitro reconstitution experiments with the RodA protein.

ACKNOWLEDGMENT. We thank Brian G. Spratt (University of Sussex, Brighton, U.K.) for the critical reading of the manuscript.

LITERATURE CITED

1. **Asoh, S., H. Matsuzawa, F. Ishino, J. L. Strominger, M. Matsuhashi, and T. Ohta.** 1986. Nucleotide sequence of the *pbpA* gene and characteristics of the deduced amino acid sequence of penicillin-binding protein 2 of *Escherichia coli* K12. *Eur. J. Biochem.* **160**:231–238.

2. **Ishino, F., W. Park, S. Tomioka, S. Tamaki, I. Takase, K. Kunugita, H. Matsuzawa, S. Asoh, T. Ohta, B. G. Spratt, and M. Matsuhashi.** 1986. Peptidoglycan synthetic activities in membranes of *Escherichia coli* caused by overproduction of penicillin-binding protein 2 and RodA protein. *J. Biol. Chem.* **261**:7024–7031.

3. **Keck, W., B. Glauner, U. Schwarz, J. K. Broome-Smith, and B. G. Spratt.** 1985. Sequence of the active-site peptides of three of the high-M_r penicillin-binding proteins of *Escherichia coli* K-12. *Proc. Natl. Acad. Sci. USA* **82**:1999–2003.

4. **Nicholas, R. A., F. Ishino, W. Park, M. Matsuhashi, and J. L. Strominger.** 1985. Purification and sequencing of the active site tryptic peptide from penicillin-binding protein 5 from the *dacA* mutant strain of *Escherichia coli* (TMRL 1222). *J. Biol. Chem.* **260**:6394–6397.

5. **Nicholas, R. A., J. L. Strominger, H. Suzuki, and Y. Hirota.** 1985. Identification of the active site in penicillin-binding protein 3 of *Escherichia coli*. *J. Bacteriol.* **164**:456–460.

6. **Nicholas, R. A., H. Suzuki, Y. Hirota, and J. L. Strominger.** 1985. Purification and sequencing of the active site tryptic peptide from penicillin-binding protein 1b of *Escherichia coli*. *Biochemistry* **24**:3448–3453.

7. **Pratt, J. M., M. E. Jackson, and I. B. Holland.** 1986. The C terminus of penicillin-binding protein 5 is essential for localisation to the *E. coli* inner membrane. *EMBO J.* **5**:2399–2405.

Chapter 29

Expression of *dacB*, the Structural Gene of Penicillin-Binding Protein 4, in *Escherichia coli*

Bogomira Korat
Wolfgang Keck

Penicillin-binding proteins (PBPs) involved in murein metabolism of bacteria are inhibited by β-lactam antibiotics through covalent binding of the drug which disturbs the delicate balance of the activities of the enzymes synthesizing and hydrolyzing murein. For a better understanding of the mechanism of murein growth and cell shape formation, the physiological function of the individual enzymes involved in murein metabolism has to be defined.

PBP 4 of *Escherichia coli* has been characterized in vitro as having carboxypeptidase IB and DD-endopeptidase activity (10). However, data exist suggesting that PBP 4 catalyzes the opposite reaction in vivo, i.e., the formation of a D-Ala-A$_2$pm peptide bond (3). To determine the physiological function of PBP 4, we used a straightforward approach: cloning of the structural gene of PBP 4, overexpression of the protein, and investigation of the effects of increased enzyme levels on cell growth and murein composition.

Investigation of Increased Levels of PBP4 in E. coli Levels

We used *E. coli* K-12 strains HB101 (*ara-14 galK2 hsdS20 proA2 rpsL20 xyl-5 mtl-1 supE44 recA13*) and JM83 [*ara* Δ(*lac-proAB*) *rpsL* Φ80 *lacZ* ΔM15] in our investigation. *E. coli* HB101 harboring pBK4 (DNA library [B. Korat and W. Keck, manuscript in preparation]) or pWK13 was cultured with aer-

Bogomira Korat • Max-Planck-Institut für Entwicklungsbiologie, 7400 Tübingen, Federal Republic of Germany. **Wolfgang Keck** • Biochemisch Laboratorium, Rijksuniversiteit Groningen, 9747 AG Groningen, The Netherlands.

ation in Luria broth (Difco Laboratories) supplemented with 100 μg of kanamycin per ml at 30°C and shifted to 42°C at the exponential growth phase
(optical density at 578 nm, 0.3) for 2 h. During this period of time these plasmids amplify and the cells accumulate plasmid-coded proteins, e.g., PBP 4
(pBK4). *E. coli* JM83 served as recipient of pUC19 and pBK19-1 (see below).
It was cultured in minimal medium (M9 plus 50 μg of ampicillin per ml, 0.2%
glucose) at 30°C and shifted to 37°C at exponential phase (optical density, 0.2).

Plasmids pWK13 and pBK4 are derivatives of the low-copy-number vector
pOU61 (5) harboring a temperature-sensitive replication control and a gene
encoding kanamycin resistance; pBK4 in addition codes for PBP 4 of *E. coli*
(Korat and Keck, in preparation). These plasmids are maintained at one copy
per cell at 30°C, but after a shift to 42°C replication becomes uncontrolled as
a result of the inactivation of the temperature-sensitive lambda repressor. At
this temperature the concentration of plasmid DNA increases up to more than
1,000 copies per cell; concomitantly, the encoded proteins accumulate. pUC19
(11), coding for ampicillin resistance and containing a multi-cloning site, was
used as a cloning vehicle for the structural gene of PBP 4. The copy number
of pUC19 and its derivatives is constant (ca. 30 copies per cell).

Preparation of the Membrane and Soluble Fraction of E. coli

E. coli HB101 harboring pWK13 or pBK4 was cultured in 100 ml of the
appropriate medium at 30°C. To get plasmid amplification and, in the case of
pBK4, PBP 4 accumulation, 50 ml of each culture was shifted to 42°C. Two
hours later, the cells were harvested by centrifugation (7,000 × *g*, 4°C), suspended in 2 ml of 10 mM sodium phosphate buffer (0.2 mM dithiothreitol, pH
7.0), and disrupted by sonication (three times, 60 W, 60 s). Membranes were
separated from the soluble fraction by centrifugation (100,000 × *g*, 4°C) and
solubilized in 50 mM sodium phosphate buffer (10 mM $MgCl_2$, 0.2 mM dithiothreitol, pH 7.0).

Restriction and Ligation of DNA

The restriction and ligation assays were done following the recommen-
dation by the supplier of the enzymes (Boehringer GmbH, Mannheim, Federal
Republic of Germany).

Overproduction of PBP 4 and Its Penicillin-Binding Activity in Different Compartments of E. coli HB101 Harboring pBK4 (dacB)

PBP 4-overproducing clones were screened from an *E. coli* DNA library
(HB101 [Korat and Keck, in preparation]) by applying specific antibodies and

using immunochemical staining by peroxidase reaction (8). Exponentially growing cultures of HB101 harboring pBK4 (*dacB*) or the control plasmid pWK13 were shifted from 30 to 42°C for 2 h. The cells were disrupted by sonication, and membrane and soluble fractions were separated by centrifugation. PBP 4 was shown to accumulate in both the soluble (Fig. 1, lane 3) and membrane (lane 7) fractions. In contrast to a previous publication (9), we were able to show that PBP 4 can be overproduced in *E. coli* at least 100-fold.

The activity of PBP 4 within the soluble and membrane fraction was determined by its ability to bind [^{14}C]penicillin G (58.9 mCi/mmol; Amersham Corp.) (7) (Fig. 2). Radioactively labeled PBPs were visualized by fluorography (2) after separation of the proteins by sodium dodecyl sulfate-polyacrylamide gel electrophoresis (6). Both the soluble and the membrane-bound PBP 4 bound penicillin covalently. Hence, incorporation of PBP 4 into the cytoplasmic membrane seems not to be required for its penicillin-binding activity.

Subcloning of the dacB Structural Gene of PBP 4 and Observation of the Growth Effects of PBP 4 Overproduction

Growth of the PBP 4-overproducing strain [HB101(pBK4)] at 42°C ceased after about 2 h as a result of the dramatic amplification of the plasmid. Therefore, the structural gene of PBP 4 was subcloned into pUC19 (Fig. 3), a vector with constant copy number (ca. 30), by using the restriction sites of *Eco*RI and *Sma*I. The resulting *dacB*-carrying plasmid, pBK19-1, was used to transform strain JM83 (4).

During exponential growth at 37°C in minimal medium, JM83(pBK19-1), which constitutively expresses PBP 4, started to lyse, whereas a control strain

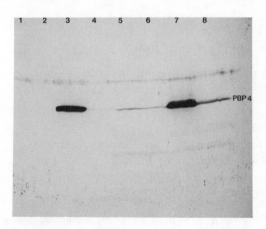

Figure 1. Detection of PBP 4 in soluble and membrane fractions of wild-type and PBP 4-overproducing strains. Amounts of 400 μg of membrane-bound (lanes 5–8) and soluble (lanes 1–4) proteins were separated by sodium dodecyl sulfate-polyacrylamide gel electrophoresis (6), blotted onto nitrocellulose filters (1), and immunochemically stained as described (8). Lanes 1, 2, 5, and 6, Fractions of wild-type cells [HB101(pWK13)] grown at 30°C (2 and 6) and at 42°C (1 and 5). Lanes 3, 4, 7, and 8, Fractions of HB101(pBK4) (overproducing PBP 4) grown at 30°C (4 and 8) and at 42°C (3 and 7).

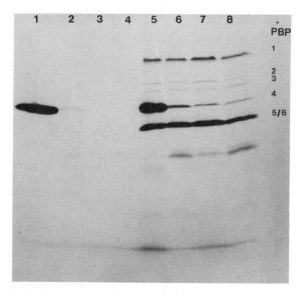

Figure 2. Fluorography of ^{14}C-labeled PBPs in soluble fractions (lanes 1–4) and membrane fractions (lanes 5–8). Lanes 3, 4, 7, and 8, Wild-type cells [HB101(pWK13)] grown at 30°C (4 and 8) and at 42°C (3 and 7). Lanes 1, 2, 5, and 6, HB101(pBK4) (*dacB*) grown at 30°C (2 and 6) and with plasmid amplification at 42°C (1 and 5).

[JM83(pUC19)] grew normally (Fig. 4). Lysis did not occur when the PBP 4-overproducing strain JM83(pBK19-1) was grown in rich medium. It seems that under certain growth conditions (as described below; Fig. 4) PBP 4 overproduction cannot be tolerated, ending in lysis of the cells.

Preliminary analysis of the murein structure (by high-performance liquid chromatography) revealed a lower degree of cross-linkage in all PBP 4-over-

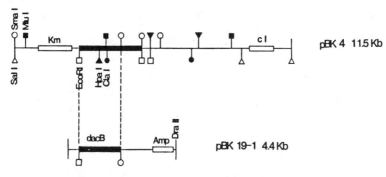

Figure 3. Restriction maps of pBK4 and pBK19-1. The *dacB*-harboring fragment was cut out from pBK4 with the restriction endonucleases *Eco*RI and *Sma*I and ligated into the *Eco*RI- and *Sma*I-digested pUC19 vector (pBK19-1).

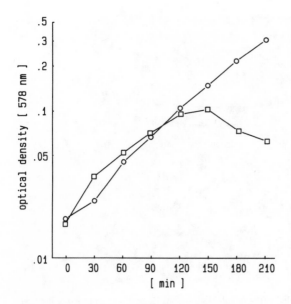

Figure 4. Growth curve of wild-type and of PBP 4-overproducing strains in minimal medium. At time zero, exponentially growing cultures of JM83(pBK19-1) (□) and JM83(pUC19) (○) were diluted to an optical density of 0.02 (578 nm) in prewarmed (37°C) minimal medium and incubated for another 210 min at 37°C under aeration.

producing strains as compared with wild-type cells. These results suggest that PBP 4 functions as an endopeptidase in vivo. Thus, by splitting the peptide bridges in the existing murein layer, PBP 4 stimulates the incorporation of new material into the murein sacculus, allowing the cell to enlarge.

LITERATURE CITED

1. **Bittner, M., P. Kupferer, and C. F. Morris.** 1980. Electrophoretic transfer of proteins and nucleic acids from slab gels to diazobenzyloxymethyl cellulose or nitro cellulose sheets. *Anal. Biochem.* **102:**459–471.
2. **Bonner, W. M., and R. A. Laskey.** 1974. A film detection method for tritium-labelled proteins and nucleic acids in polyacrylamide gels. *Eur. J. Biochem.* **46:**83–88.
3. **De Pedro, M. A., U. Schwarz, Y. Nishimura, and Y. Hirota.** 1980. On the biological role of penicillin-binding protein 4 and 5. *FEMS Microbiol. Lett.* **9:**219–221.
4. **Hanahan, D.** 1985. Techniques for transformation of *E. coli. In* D. M. Glover (ed.), *DNA Cloning, a Practical Approach,* vol. 1. IRL Press, Oxford.
5. **Larsen, J. E. L., K. Gerdes, J. Light, and S. Molin.** 1984. Low-copy-number plasmid-cloning vectors amplifiable by derepression of an inserted promotor. *Gene* **28:**45–54.
6. **Lugtenberg, B., J. Meijers, R. Peters, P. van der Hoek, and L. van Alphen.** 1975. An electrophoretic resolution of the major outer membrane protein of *E. coli* K12 into four bands. *FEBS Lett.* **58:**254–258.
7. **Spratt, B. G., and A. B. Pardee.** 1975. Penicillin-binding proteins and cell shape in *E. coli. Nature* (London) **254:**516–517.
8. **Sternberger, L. A., P. H. Hardy, Jr., J. J. Cuculis, and H. G. Meyer.** 1970. The unlabelled antibody enzyme method of immunochemistry. Preparation and properties of soluble antigen-antibody complex (horseradish peroxidase-antihorseradish peroxidase) and its use in identification of spirochetes. *J. Histochem. Cytochem.* **18:**314–333.

9. **Takeda, Y., Y. Nishimura, M. Yamada, S. Yasuda, H. Suzuki, and Y. Hirota.** 1981. Synthetic ColE1 plasmids carrying genes for penicillin-binding proteins in *E. coli. Plasmid* **6:**86–98.

10. **Tamura, T., Y. Imae, and J. L. Strominger.** 1976. Purification to homogenity and properties of two D-alanine carboxypeptidases I from *E. coli. J. Biol. Chem.* **251:**414–423.

11. **Vieira, J., and J. Messing.** 1982. The pUC plasmids, an M13mp7-derived system for insertion mutagenesis and sequencing with synthetic universal primers. *Gene* **19:**259–269.

Chapter 30

The Lone Cysteine Residue of Penicillin-Binding Protein 5 from *Escherichia coli* Is Not Required for Enzymatic Activity

Robert A. Nicholas
Jack L. Strominger

Penicillin-binding protein (PBP) 5 is one of the most abundant of the PBPs present in the cytoplasmic membrane of *Escherichia coli* (13) and is known to catalyze the major D-alanine carboxypeptidase activity in vivo (9). It is encoded by the *dacA* gene, which has been cloned and sequenced (3), and the amino acid nucleophile that forms the covalent bond with both substrate and penicillin has been identified as serine 44 (6, 10). A recent study on the interaction of PBP 5 with the cytoplasmic membrane has shown that the removal of as little as 10 amino acids from the carboxyl terminus is sufficient to result in the release of the protein into the periplasmic space (12). PBP 5 can thus be classified as an "ectoprotein" that is a normal globular protein attached to the membrane by a short stretch of amino acids.

β-Lactamase Activity of PBP 5

PBP 5 is one of the few PBPs characterized thus far that has been shown to have a weak β-lactamase activity; at pH 7 and 30°C, the half-life of the penicilloyl-PBP 5 complex is about 10 min (1). It had initially been reported by Tamura et al. (14) that sulfhydryl reagents severely inhibit the carboxypep-

---

Robert A. Nicholas and Jack L. Strominger • Department of Biochemistry and Molecular Biology, Harvard University, Cambridge, Massachusetts 02138.

tidase activity of PBP 5, but not the binding of penicillin to the enzyme. Further investigation by Curtis and Strominger (4) showed that the addition of sulfhydryl reagents only slightly affects the binding of penicillin to PBP 5 (as well as cell wall substrate), but greatly increases the half-life of the penicilloyl-PBP 5 complex. The amino acid sequence of PBP 5, deduced from the DNA sequence of the *dacA* gene, has shown that only one cysteine residue is present in mature (i.e., processed) PBP 5, at residue 115 of the mature protein (3). This lone cysteine residue was proposed to reside within the active site of PBP 5 (4, 10), and thereby catalyze the hydrolysis of the acyl-enzyme intermediate. The hypothesis of an intimate involvement of the sulfhydryl moiety in the catalytic mechanism was further supported by the properties of a naturally occurring mutant, termed PBP 5' (8). PBP 5' displays a near-normal acylation rate with respect to [^{14}C]penicillin G, but shows a drastic increase in the half-life of the [^{14}C]penicilloyl complex (2). It is also devoid of any detectable D-alanine carboxypeptidase activity (2, 8). This mutant thus mimics the effect of sulfhydryl alkylation on the activity of the wild type enzyme. The mutation represents a single base change that alters amino acid 105 from glycine to aspartic acid (3a), and this mutation is only 10 amino acids distant from the location of the cysteine residue. It can be argued, however, that structural constraints in this region, either the addition of a bulky alkylating group or the replacement of a small neutral amino acid with a large negatively charged amino acid, are responsible for the properties of both PBP 5' and alkylated PBP 5. Thus, the actual importance of the cysteine residue itself in the catalytic mechanism remains to be clarified. The present study examines the role of this cysteine residue in the hydrolysis of the acyl-enzyme intermediate, using oligonucleotide-directed mutagenesis.

Placement of the dacA Gene under Control of an Inducible Promoter

The construction of the plasmid pBS42-4 was recently reported (12). This plasmid encodes for a truncated, soluble form of PBP 5, designated sPBP 5. The ability to generate a soluble PBP 5 greatly simplifies the purification and manipulation of the protein. pBS42-4 is under strict copy number control, however, which makes it undesirable for use as an expression vector. Since the overproduction of PBP 5 is lethal, it was necessary to place the transcription of the *dacA* gene under the control of an inducible promoter. The expression vector that was chosen was pKK223-3 (Pharmacia), which contains a polylinker adjacent to the inducible promoter, p_{tac}. This promoter is a fusion between the *trp* and *lac* promoters containing the binding site for the *lacI* repressor, and can be induced by the addition of β-isopropylthiogalactoside. To construct a promoterless *dacA* gene, BAL 31 deletions were carried out on the *dacA* promoter region to remove most of the endogenous promoter sequence. One of these deletions, designated pBAL 10, contained 15 base pairs 5' to the ATG codon of the *dacA* gene. The truncated, promoterless, naturally occurring mutant *dacA* gene was then reconstructed and cloned into pKK223-3. Since this plasmid uses ampicillin resistance for its selection marker, which is detrimental to the detection and purification of PBPs, the ampicillin gene was disrupted and replaced with the gene encoding resistance to kanamycin. This plasmid was designated pKAN20.4. The final 15 base pairs were removed by using loop-out mutagenesis (11, 15).

Site-Directed Mutagenesis

A 400-base-pair *Bst*EII-*Sal*I fragment from the *dacA* gene encompassing the codon for the lone cysteine residue in PBP 5 was cloned into M13mp9, and site-directed mutagenesis was performed (see Fig. 1; 11, 15). The cysteine codon was mutated to either a serine codon or an alanine codon, and the positive bacteriophage were selected and sequenced. The mutated DNA fragments were subcloned into pUC9, and the entire *dacA* genes encoding for truncated wild-type PBP 5 (sPBP 5), cysteine-to-serine PBP 5 (sPBP 5^{c-s}), and cysteine-to-alanine PBP 5 (sPBP 5^{c-a} were then reconstructed and placed adjacent to the p_{tac} promoter to give the construction pATG. The final plasmid construction is shown in Fig. 2.

Expression and Purification of Wild-Type and Mutant Proteins

The expression of the sPBP 5 variants was acheived by inducing a culture of *E. coli* XA90 (*fhuA Δlac-proAB* F' *lacI*^{q} *proAB*), harboring a plasmid encoding for either sPBP 5, sPBP 5^{c-s}, or sPBP 5^{c-a}, with 1 mM β-isopropylthiogalactoside and allowing the induction to proceed for 4 h. Large-scale purification of these proteins was achieved by preparing osmotic shock fluid (7) from 2 liters of culture, followed by covalent ampicillin affinity chromatography (10). Sodium dodecyl sulfate-polyacrylamide gel electrophoresis confirmed the purity of the preparations (Fig. 3).

Assay of Mutant Proteins

To test the effects of the mutations, penicillin hydrolysis and carboxypeptidase assays were performed. To measure the half-life of the penicilloyl-PBP complex, the appropriate sPBP 5 was incubated with 50 μg of [^{14}C]penicillin G per ml for 20 min at 30°C, followed by the addition of a 100-fold excess

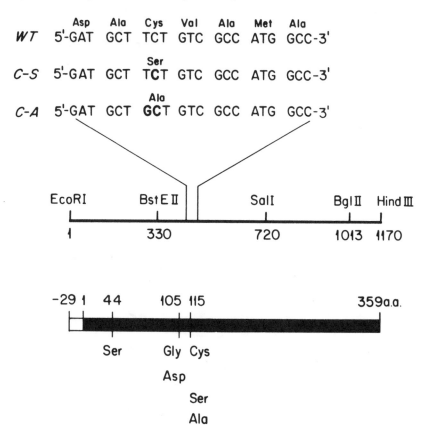

Figure 1. Relationship between the DNA and protein in the *dacA* gene encoding sPBP 5. The sites of glycine 105 and cysteine 115 are shown, along with the mutations that are described in the text. Serine 44 is the amino acid nucleophile that forms a covalent bond with β-lactam antibiotics.

of unlabeled penicillin at time zero. At various times, samples were removed and denatured in sodium dodecyl sulfate, and the radioactivity attached to the PBP was assayed by high-performance liquid chromatography. Figure 4 shows a semilog plot of the hydrolysis of penicillin from the sPBPs. It is evident that the mutations had a slight effect on the half-life of the complex, but certainly not the dramatic effect that is seen with sPBP 5'. It is therefore unlikely that the cysteine residue is directly involved in the hydrolysis of the acyl-enzyme intermediate. Mutation of cysteine 115 to serine or alanine had no affect on the affinity of the PBP for penicillin. To test the effect of the mutations upon its natural activity, D-alanine carboxypeptidase assays were performed as described by Frere et al. (5), using the peptide L-Ala-D-γ-Glu-L-Lys-D-Ala-D-Ala (Sigma Chemical Co.). These results are summarized in Table 1; again it is

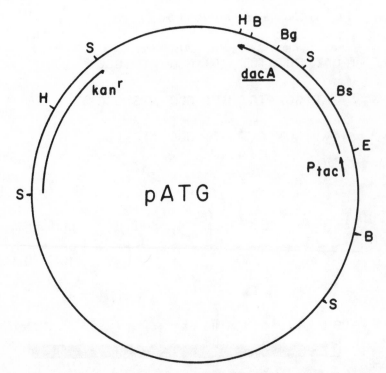

Figure 2. Plasmid construct used for the overproduction of PBP 5 variants. To construct the mutant and wild-type plasmids, the unique *Bst*EII-*Bgl*II fragment from pATG20.4 was replaced with the analogous fragment containing the appropriate sequence. The single letters indicate cleavage sites for restriction endonucleases: H, *Hin*dIII; B, *Bam*HI; Bg, *Bgl*II; S, *Sal*I; Bs, *Bst*EII; E, *Eco*RI.

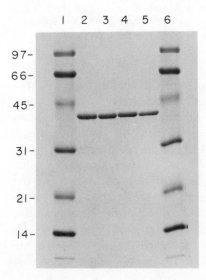

Figure 3. Sodium dodecyl sulfate-polyacrylamide gel electrophoresis of purified wild-type and mutant sPBP 5s. Lanes: 1 and 6, molecular weight standards; 2, sPBP 5′; 3, sPBP 5; 4, sPBP 5$^{c\text{-}s}$; 5, sPBP 5$^{c\text{-}a}$. Lanes 2 through 5 contain 2 μg of protein. Numbers for standards are in kilodaltons.

316

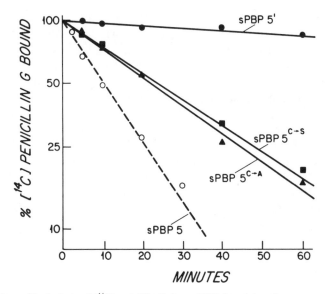

Figure 4. Rate of hydrolysis of [^{14}C]penicillin G from sPBP 5 proteins. Assays were performed as described in the text and are plotted on a semilog plot as percent [^{14}C]penicillin G remaining (relative to time zero) versus time.

evident that the mutations had little or no effect on the activity of the protein. In fact, the Cys-to-Ser and Cys-to-Ala mutations appear to have slightly increased the efficiency of the enzyme (see values for V_{max}/K_m), although the numbers are close enough to be considered essentially the same. As reported previously, sPBP 5' displayed no detectable carboxypeptidase activity (8).

Conclusion

The results presented above disprove the notion that the lone cysteine residue of PBP 5 is directly involved in the molecular mechanism of the hydrolysis of the acyl-enzyme intermediate. There is a minor discrepancy in the effects

Table 1. Kinetic constants of wild-type and mutant proteins

sPBP 5	k_3 (min^{-1})^a	K_m (mM)	V_{max} (μmol min^{-1} mg^{-1})	V_{max}/K_m
sPBP 5	0.066	5.02 ± 0.77 (5)	2.54 ± 0.39 (5)	0.51
sPBP 5^{c-s}	0.029	3.54 ± 0.81 (4)	3.30 ± 0.17 (4)	0.93
sPBP 5^{c-a}	0.032	7.75 ± 0.10 (2)	5.11 ± 0.30 (2)	0.66
sPBP 5'	0.003	ND	ND	ND

^{a}Values are for the rate of hydrolysis of [^{14}C]penicillin G. ND, Not determined. All other values were calculated from double reciprocal plots of initial rates. The numbers in parentheses indicate the number of determinations.

of the mutations between penicillin hydrolysis (twofold slower) and carboxy-peptidase activity (virtually no overall effect); the reason for this is not clear. It is known that reaction of PBP 5 with sulfhydryl reagents affects the same step in the similar reactions of substrate and penicillin, i.e., the hydrolysis of the acyl-enzyme intermediate, but not the binding. It is also not clear whether the differences in the K_m and V_{max} values seen with the two mutants are indicative of subtle changes in the active site of the enzyme or result from errors in the assays to determine these values. This will have to wait for the elucidation of the three-dimensional structure of PBP 5 by X-ray crystallography. It can be concluded, however, that the replacement of cysteine 115 by a structurally related serine or alanine residue has no damaging effect upon the function of PBP 5 and that alkylation of the cysteine residue or replacement of Gly-105 with Asp-105 (PBP 5') prevents the hydrolysis of the acyl-enzyme intermediate by sterically or spatially altering a mechanistically important amino acid.

ACKNOWLEDGMENTS. This work was supported by Public Health Service grant AT-09152 from the National Institutes of Health to J.L.S., and a National Institutes of Health postdoctoral grant to R.A.N.

We thank Jerry Boss, Dan Weeks, Robin Reed, Joan Gorga, and Carolyn Doyle for many helpful discussions and suggestions. We also thank Julie Pratt, who kindly supplied pBS42-4 before publication, and Michio Matsuhashi for supplying the *dacA* mutant gene.

LITERATURE CITED

1. **Amanuma, H., and J. L. Strominger.** 1980. Purification and properties of penicillin-binding proteins 5 and 6 from *Escherichia coli* membranes. *J. Biol. Chem.* **255**:11173–11180.
2. **Amanuma, H., and J. L. Strominger.** 1984. Purification and properties of penicillin-binding proteins 5 and 6 from the *dacA* mutant strain of *Escherichia coli* (JE 11191). *J. Biol. Chem.* **259**:1294–1298.
3. **Broome-Smith, J., A. Edelman, and B. G. Spratt.** 1983. Sequence of penicillin-binding protein 5 of *Escherichia coli*, p. 403–408. *In* R. Hakenbeck, J.-V. Holtje, and H. Labischinski (ed.), *The Target of Penicillin*. Walter de Gruyter and Co., Berlin.
3a. **Broome-Smith, J., and B. G. Spratt.** 1984. An amino acid substitution that blocks the deacylation step in the enzyme mechanism of penicillin-binding protein 5 of *Escherichia coli*. *FEBS Lett.* **165**:185–189.
4. **Curtis, S. J., and J. L. Strominger.** Effects of sulfhydryl reagents on the binding and release of penicillin G by D-alanine carboxypeptidase IA of *Escherichia coli*. *J. Biol. Chem.* **253**:2584–2588.
5. **Frere, J.-M., M. Leyh-Bouille, J.-M. Ghuysen, M. Nieto, and H. R. Perkins.** 1977. Exocellular DD-carboxypeptidases-transpeptidases from *Streptomyces*. *Methods Enzymol.* **45**:610–636.

6. **Glauner, B., W. Keck, and U. Schwarz.** 1984. Analysis of penicillin-binding sites with a micro method—the binding site of PBP 5 from *Escherichia coli. FEMS Microbiol. Lett.* **23:**151–155.

7. **Kustu, S. G., and G. F.-L. Ames.** 1974. The histidine-binding protein J, a histidine transport component, has two different functional sites. *J. Biol. Chem.* **249:**6976–6983.

8. **Matsuhashi, M., I. N. Maruyama, Y. Takagaki, S. Tamaki, Y. Nishimura, and Y. Hirota.** 1978. Isolation of a mutant of *Escherichia coli* lacking penicillin-sensitive D-alanine carboxypeptidase IA. *Proc. Natl. Acad. Sci. USA* **75:**2631–2635.

9. **Matsuhashi, M., S. Tamaki, S. J. Curtis, and J. S. Strominger.** 1979. Mutational evidence for identity of penicillin-binding protein 5 in *Escherichia coli* with the major D-alanine carboxypeptidase IA activity. *J. Bacteriol.* **137:**644–647.

10. **Nicholas, R. A., F. Ishino, W. Park, M. Matsuhashi, and J. L. Strominger.** 1985. Purification and sequencing of the active site tryptic peptide from penicillin-binding protein 5 from the *dacA* mutant strain of *Escherichia coli* (TMRL 1222). *J. Biol. Chem.* **260:**6394–6397.

11. **Nisbet, I. T., and M. W. Beilharz.** 1985. Simplified DNA manipulations based on in vitro mutagenesis. *Gene Anal. Techn.* **2:**23–29.

12. **Pratt, J. M., M. E. Jackson, and I. B. Holland.** 1986. The C terminus of penicillin-binding protein 5 is essential for localisation to the *E. coli* inner membrane. *EMBO J.* **5:**2399–2405.

13. **Spratt, B. G.** 1977. Properties of the penicillin-binding proteins of *Escherichia coli* K-12. *Eur. J. Biochem.* **72:**341–352.

14. **Tamura, T., Y. Iame, and J. L. Strominger.** 1976. Purification to homogeneity and properties of two D-alanine carboxypeptidases I from *Escherichia coli. J. Biol. Chem.* **251:**414–423.

15. **Zoller, M. J., and M. Smith.** 1983. Oligonucleotide-directed mutagenesis of DNA fragments cloned into M13 vectors. *Methods Enzymol.* **100:**468–500.

Chapter 31

Distribution of Membrane-Associated Proteins: Photo-Cross-Linking, Postembedding-Immunolabeling

M. H. Bayer

S. Haberer

M. E. Bayer

Understanding the functional organization of a cell requires the localization of its components. Difficulties arise with the localization of small molecules because of their capability of rapid diffusion. We searched for a method which would reduce or, optimally, prevent intracellular redistribution and diffusion processes. Ideally, an immobilization should work for cytoplasmic as well as for membrane-associated molecules.

It has been calculated that cytoplasmic molecules, diffusing within the cell in a three-dimensional path, may travel from one end to the other of an *Escherichia coli* cell within fractions of a second (6). Also, membrane-associated molecules may diffuse in two-dimensional paths, or may rotate and also flip-flop; furthermore, they may dissociate from the membrane in response to a variety of conditions, such as ionic changes or experimental manipulations during membrane isolation.

For localization of cell components at high resolution, electron microscopy

M. H. Bayer, S. Haberer, and M. E. Bayer • Fox Chase Cancer Center, Institute for Cancer Research, Philadelphia, Pennsylvania 19111.

320

is required. This method involves conventional chemical fixation procedures in which aldehydes and osmium tetroxide interact with the cell structures. These reactions are relatively slow and allow for redistribution of antigens as well as for extraction of material during fixation and dehydration. While the method of physical immobilization by cryofixation and cryosubstitution with subsequent low-temperature embedding (7) could probably circumvent these artifacts, we chose to explore another approach, namely, the rapid chemical cross-linking of small-molecular-weight proteins such as thioredoxin. The localization of this protein has been uncertain in the past, so that new technical approaches seemed to be necessary. Thioredoxin is a dithiol protein (molecular weight, 11,700) ubiquitous in bacteria, plants, and animal organs (for reviews, see references 8, 10, and 15). Its function has not been well established, although roles in protein transmembrane transport and in biosynthetic processes such as bacteriophage T7 replication (1, 17) have been suggested (20, 22); its sites of activity in the export of filamentous phage have been shown to be associated with membrane adhesion sites (3, 12). The protein has been reported to be localized in the cytoplasm, chromosomal area, periplasm, and envelope of bacteria (19). More recently, biochemical data indicated that thioredoxin is localized in a membrane fraction corresponding to the membrane adhesion sites (16), sites which represent domains of inner and outer membrane adherence (2). We exploited the fact that thioredoxin contains a highly reactive thiol-disulfide bridge, which could cross-link the protein to neighboring molecules and possibly stabilize its in vivo distribution. We used the heterobifunctional compound *p*-azidophenacylbromide (pAPA), which had previously been shown to concentrate thioredoxin in the inner membrane and in a hybrid membrane fraction of *E. coli* (16).

Our data show that rapid photo-cross-linking permits the localization of thioredoxin in wild-type *E. coli* and in thioredoxin-overproducing cell constructs. The data are statistically highly significant.

We also report here preliminary results of immunolabeling of both ultrathin sections and membrane fractions to localize penicillin-binding protein 1b (PBP 1b) in *E. coli* strains.

Photo-Cross-Linking as a Prefixation Method for Immunoelectron Microscopy

To demonstrate how we reached our goal of arresting macromolecular diffusion, a few technical details have been included in this chapter. In particular, the use of very fast cross-linking procedures, a novel approach to fixation in electron microscopy, is briefly mentioned, as well as the quantitation of label and the definition of a "membrane-associated area."

Photo-cross-linking was executed with pAPA. Protein A was obtained from Pharmacia (Piscataway, N.J.), and Lowicryl K4M was obtained from Polysciences (Warminster, Pa.). Rabbit antibodies to thioredoxin were obtained from the laboratory of V. Pigiet (14).

The bacterial strains used for these studies were *E. coli* SK3981, a thioredoxin overproducer, and SK3969, a thioredoxin nonproducer; both strains were F⁻ *argE4 his-4 ilvD188 lac mal mtl xyl rpsL supE44 trxA1*. Strain SK3969 carried plasmid pBR325, and SK3981 carried pBR325*trx*⁺. These and other thioredoxin mutants and wild-type *E. coli* W3110 have been described (13). For PBP studies, strains KN126 and KN126(pHK231) (which overproduces PBP 1b at 37°C [11]) were used; these strains have been characterized before (23). These two strains and purified antibody to PBP 1b were a gift of W. Keck (Max Planck Institut für Entwicklungsbiologie, Tübingen, Federal Republic of Germany).

The experiments were performed with a number of different growth media. *E. coli* W3110 was grown at 37°C either in Tris minimal medium (16) containing Casamino Acids and glucose (14) or in nutrient broth with tryptone, yeast extract, 0.1% glucose, and 0.5% NaCl (5). The construct strains SK3981 and SK3969 were grown to cell densities of 6×10^8 cells per ml in Tris minimal medium plus ampicillin (20 μg/ml). Stationary-phase cells (6×10^9 cells per ml) were derived from overnight aerated liquid cultures. *E. coli* KN126(pHK231) was grown at 28°C in the presence of 100 μg of kanamycin per ml. PBP 1b overproduction was induced at 37°C, and the cells were grown for 1.5 h before being prepared for either fractionation or electron microscopy.

For detection of the proteins, both immunotransfer and autoradiography were used. Extracts of whole cells were fractionated by 15% sodium dodecyl sulfate-polyacrylamide gel electrophoresis, and the proteins of the gel tracks were electrophoretically transferred to nitrocellulose (18). Immunoassays of PBP 1b were carried out on membrane fractions (French press) of the envelopes of *E. coli* KN126 and its derivatives (5). Fractions of the sucrose gradients were numbered 4 to 5 for the outer membrane and 10 through 12 for the inner membrane.

Our fast procedure of photo-cross-linking and subsequent preparation for microscopy is a unique preparative approach to stabilize the structural relation of thioredoxin to the inner membrane. To photo-cross-link cells, 12-ml cultures (SK strains and W3110) were suspended in Tris minimal medium containing 1 mM EDTA and then exposed to pAPA (20 μg/ml) and irradiated with a UV flash (4). After irradiation, cells were fixed for 30 min in 2% formaldehyde, centrifuged, and enrobed in agar. Dehydration (in ethyl alcohol) and the Lowicryl-embedding steps were performed at −20°C, and the Lowicryl was polymerized under UV light at −70°C. Ultrathin sections were reacted for 20 to 30 min with affinity-purified thioredoxin-specific antibody as described earlier (4), and protein A-gold labeling was subsequently performed. Prints were made from the micrographs of the labeled ultrathin sections and used for area measurements and gold particle counting. Since a cytoplasmic area next to the inner membrane was seen to be mostly labeled, a suitable quantitation method had to be developed.

Quantitative Evaluation of Immunolabeled Domains of Membranes and Cytoplasm

The entire area of cytoplasm plus membrane was measured with a Hipad electronic digitizer by following the contour of the inner membrane. A "membrane-associated" area of the cytoplasm was established by delineating an area corresponding to about 40 nm at ×50,000 magnification, bordering the inner contour of the inner membrane. Gold particles in both areas were counted (Table 1). With the aid of a computer program (developed by S. Litwin of the biostatistics group of the Fox Chase Cancer Center), the particle counts were used to test the hypothesis that the observed gold bead concentration in the membrane-associated area could have been the result of uniform random particle distribution throughout the cytoplasm. A statistical test based on the Poisson distribution of particle counts gave rise to a chi-squared test of the above hypothesis. In particular, a generalized likelihood ratio test was used to test

Table 1. Effect of cross-linking on distribution of thioredoxin-specific immunolabel

Strain (phenotype)[a]	No. of cells counted	Label density[b] (gold particles/unit area)	
		IM	Cyto
SK3981 (Trx⁺)[c]			
Untreated	32	0.32	0.48
Treated[d]	18	1.10	2.23
SK3981 (Trx⁺)			
Untreated	12	2.94	1.83
Treated	40	24.65	7.22
W3110 (Trx)			
Untreated	6	5.76	3.79
Treated	19	3.21	1.28
SK3969 (Trx⁻)			
Untreated	6	0.55	0.37
Treated	16	1.72	1.92

[a]Trx, Thioredoxin level of wild-type cells; Trx⁺, thioredoxin overproducer; Trx⁻, thioredoxin nonproducer.
[b]IM, Inner membrane and membrane-associated area; Cyto, cytoplasmic area (less membrane-associated area).
[c]Control. No immunoglobulin was used. All other specimens were treated with immunoglobulin.
[d]Treated cells were exposed to pAPA and UV irradiation.

the null hypothesis that the Poisson parameter lambda 1 (particles per unit area in membrane-associated area) equals lambda 2 (particles per unit area in bulk cytoplasm) versus the alternative that lambda 1 is not equal to lambda 2. The values of chi-squared have degrees of freedom equal to the number of sectioned cells observed. The chances of observing chi-squares larger than those reported are given as *P* values (Table 2).

Thioredoxin Distribution

The distribution of the low-molecular-weight thioredoxin was measured in both non-photo-cross-linked and photo-cross-linked cells.

Non-Cross-Linked Cells

Protein blotting. Immunotransfer executed on un-cross-linked *E. coli* W3110 (wild type) and the thioredoxin overproducer *E. coli* SK3981 (Trx$^+$) revealed a single (labeled) species corresponding to 11,700 to 12,000 daltons.

Electron microscopy. Since thioredoxin is a protein with weak membrane association (16), low-temperature embedding in Lowicryl K4M (with and without photo-cross-linking) appeared to be the method of choice. Unfortunately, aldehyde-fixed and Lowicryl-embedded cells exhibited poor membrane contrast, and attempts to enhance the contrast with prolonged staining were met

Table 2. Probability of nonrandom thioredoxin distribution[a]

Strain	UV and pAPA cross-linking	Immunoglobulin treatment	χ^2	No. of cells counted	*P*
SK3981	−	−	37.56	32	2.29×10^{-1}
	+	−	40.35	18	4.49×10^{-3}[b]
	−	+	36.56	12	2.67×10^{-4}
	+	+	285.98	40	$<6 \times 10^{-8}$
W3110	−	−	12.14	6	5.88×10^{-2}
	+	−	41.69	42	4.84×10^{-1}
	−	+	40.9	6	2.98×10^{-7}
	+	+	98.09	19	$<6 \times 10^{-8}$
SK3969	−	+	12.14	6	5.9×10^{-2}
	+	+	44.00	16	1.97×10^{-4}[b]

[a]The probability of nonrandom association of thioredoxin with the inner membrane was determined from quantitative measurements of immunoelectron micrographs by using a computer program. A *P* value of ≤0.05 is statistically significant and suggests an affinity of the label for the inner membrane.
[b]These results point toward a somewhat higher label density in the cytoplasmic area.

with little success. Furthermore, when conventional fixation in aldehyde-OsO$_4$ was employed, all strains used in this study showed an ultrastructure characteristic of *E. coli* (including typical membrane adhesion sites).

The localization of label for (un-cross-linked) thioredoxin found in *E. coli* W3110 (wild type), SK3969 (a thioredoxin nonproducer), and SK3981 (the overproducer) is shown in Table 2.

pAPA-Cross-Linked Cells

Protein blotting. Immunoelectrophoresis of samples of whole-cell extracts showed that the 12,000-dalton thioredoxin band increased significantly after pAPA treatment of both W3110 and SK3981 cells. Furthermore, cross-linked stationary-phase SK3981 cells revealed much more label than cross-linked exponentially growing cultures, whereas no label was observed in extracts from the thioredoxin-minus mutant SK3969 (Fig. 1). In addition to the 12,000-dalton band, extracts from both SK3981 and W3110 showed weaker bands at higher molecular masses; one of these bands appears to correspond to a dimer of thioredoxin.

Electron microscopy. Ultrathin sections showed a significantly increased number of gold particles in the pAPA-cross-linked strains W3110 (Fig. 2A) and SK3981 (Fig. 2C). Quantitation of the number of gold particles revealed, in both strains W3110 and SK3981, 2.5 to 3 times more particles associated with the inner membrane than in the cytoplasm, indicating a statistically significant uneven distribution (Table 1).

Membrane-associated thioredoxin was also observed in cells grown to stationary phase.

In spite of the poor contrast of the Lowicryl embedding, adhesion sites between the outer and inner bacterial membranes were observed. Some but not all of the sites were labeled, indicating that at least a portion of the cell's thioredoxin was associated with membrane adhesion sites. In the strains examined, the periplasmic space had not accumulated significant amounts of label.

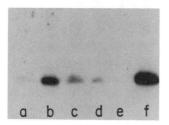

Figure 1. Protein transfer of cellular thioredoxin after pAPA treatment and UV cross-linking. *E. coli* cells in growing phase (3 × 10^8 cells per ml): (a) W3110, (b) SK3981, (c) B. Cells in stationary phase (6 × 10^9 cells per ml): (d) W3110, (e) SK3969, (f) SK3981. Molecular weight of thioredoxin, 12,000.

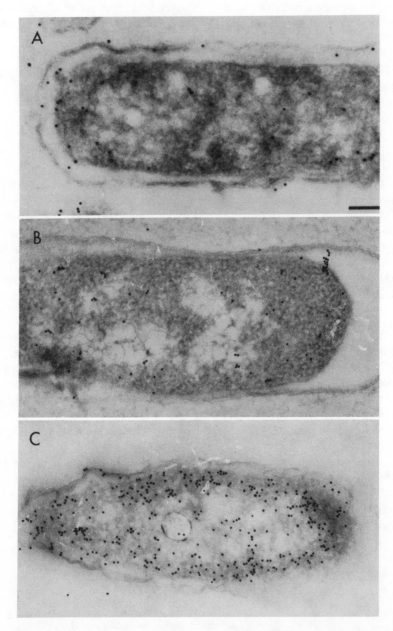

Figure 2. Thioredoxin-immunolabeled sections. (A) *E. coli* W3110 after pAPA-UV cross-linking (bar, 0.2 μm). (B) *E. coli* SK3981 (thioredoxin overproducer), no UV cross-linking. (C) *E. coli* SK3981 after UV cross-linking; note the density of label compared with panel B.

PBP 1b

Protein Blotting

Immunoassays of membrane fractions of the PBP 1b-overproducing strain KN126(pHK231) (grown at 28°C in the presence of kanamycin) showed an even distribution of PBP 1b in vesicle fractions 4 through 12 (Fig. 3). A similar distribution pattern for PBP 1b was detected in cells shifted up to 37°C. However, in *E. coli* KN126 cells lacking pHK231, only very weak labeling could be detected in the outer membrane fraction.

Electron Microscopy

The PBP 1b-overproducing cells (grown at 37°C) showed the highest label concentration at the inner membrane and in the membrane-associated area (a domain selected to include about 40 nm of the cytoplasm to the inner membrane) (Fig. 4). Comparatively little label was seen at the outer membrane and in the central regions of the cytoplasm. The periplasmic space did not show significant label. The label distribution in cells grown at 28°C (not overproducing PBP 1b) was indistinguishable from that in cells from temperature shift-up cultures (37°C). However, the PBP 1b label concentration appeared to be reduced in cell cultures grown at 28°C. A statistical evaluation, similar to that done for thioredoxin, has as yet not been completed.

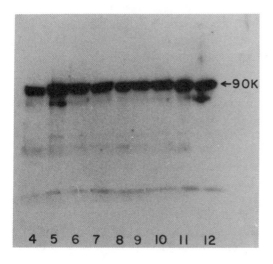

Figure 3. Protein transfer of PBP 1b of membrane fractions of *E. coli* KN126(pHK231). Cells were grown at 28°C to a density of 2×10^8 cells per ml in kanamycin medium before French press fractionation. Fractions: 4 and 5, outer membrane; 6 through 9, intermediate membrane region; 10 through 12, inner membrane. PBP 1b molecular weight, 90,000 (arrow); 25 μg of membrane protein per sample was applied.

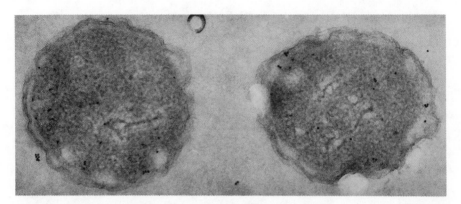

Figure 4. Cross section of *E. coli* KN126(pHK231) showing the binding of gold-labeled antibody to PBP 1b.

Conclusion

Although thioredoxin exhibits a broad in vitro specificity, viable mutants lacking thioredoxin have been reported (17). The known in vivo activities of this protein involve bacteriophage replication (thioredoxin is a subunit of the T7 DNA polymerase [1]), bacteriophage assembly, and export of filamentous phages (22). The newly developed flash cross-linking procedure, as well as the fixation and post-embedding methods described here, improve the likelihood for a realistic localization of thioredoxin.

Several data argue against the theoretical possibility that a thiol-specific cross-linker could create the observed localization of thioredoxin. The cross-linker appears to be efficient, as indicated by the Western blot (immunoblot) data. Thioredoxin, which behaves as a soluble protein in terms of classical cell fractionation techniques, has more recently been described as being associated with the cell membrane (16). The ability to observe cytoplasmically localized thioredoxin, especially in cells overproducing thioredoxin (Fig. 2), suggests that the reagent does not attach most or all of the available thioredoxin to the membrane. One may therefore assume that some of the thioredoxin exists in the cytoplasm. Previous biochemical studies with pAPA (16) also showed that the reagent does not generally cross-link other proteins to the cell membrane, suggesting that the cross-linking activity occurs at thioredoxin sites. Since the electronic flash produces a peak output in less than 1/500 s, we hypothesize that any directional diffusion of the excited thioredoxin toward the membrane will take considerably more time than the lifetime of an activated molecule, i.e., its binding to a neighboring molecule or its deactivation.

We observed in our Western blot studies that cell envelopes prepared by the French pressure cell procedure (5) showed, in addition to the major (11,700- to 12,000-dalton) thioredoxin band, some minor bands very similar to those found in whole-cell preparations. However, while whole-cell preparations revealed a relatively pronounced band in a thioredoxin dimer position (23,500 to 24,000 daltons), the isolated cell envelope fractions showed pronounced bands mainly of trimer and tetramer molecular sizes, suggesting that the envelope may contain higher multimers, possibly trimers and tetramers of thioredoxin.

Another laboratory (19) has reported that thioredoxin resides in the periplasm and that it is also associated with the cell nucleoid. We have been unable to confirm these observations (for discussion of this point, see reference 4). Instead of the relatively massive localization of thioredoxin in the nucleoid as described (19), we found that immunolabel may bind occasionally to very short stretches of the fibrous component within the bacterial nucleoid. Such fibers can be interpreted as DNA (9). However, Lowicryl sections have a relatively rough surface, as demonstrated by shadowcasting methods, and an entrapment or unspecific absorption of gold label at some of the (protruding?) DNA fibers of the nucleoid cannot be excluded.

We conclude from our thioredoxin studies that (i) the use of the photo-cross-linker pAPA reduces the loss of thioredoxin during the procedures of aldehyde fixation and embedding, (ii) thioredoxin is largely localized at the area of the inner membrane, (iii) overproducing cells also reveal high amounts of thioredoxin in the cytoplasm, (iv) thioredoxin is present in a number of membrane adhesion sites, and (v) the periplasmic space appears to lack the protein.

For the localization of PBP 1b, two approaches using binding-protein-specific antibody were chosen: protein transfer of sodium dodecyl sulfate-polyacrylamide gel electrophoresis of envelope fractions, and immunoelectron microscopy of aldehyde-fixed and Lowicryl-K4M-embedded cells. Antibody to the binding protein labeled all of the cell envelope fractions. However, the outer membrane often showed label of less intensity than did intermediate-density and inner membrane fractions. A good separation of other typical protein bands of the membrane fractions was routinely achieved in these gradients, as tested with Commassie blue and silver stain procedures. It should be noted that our procedure does not involve lysozyme-EDTA pretreatment. Our data are in partial agreement with penicillin-binding studies of other laboratories (21). The immunolabeling data from ultrathin sections of PBP 1b-producing whole cells show a slightly different aspect: most of the label is present in the membrane of the cells and in the adjacent cytoplasm, much less label is found in the central cytoplasm, and still smaller amounts are located in the outer membrane. The periplasm appears to be devoid of label.

In summary, the localization of a low-molecular-weight protein, thioredoxin, and that of a relatively large molecular-weight protein, PBP 1b, can be determined in *E. coli* by immunoelectron microscopy. A significant loss of protein, as observed during preparation of cells for electron microscopy, can be avoided by immobilizing the protein (such as thioredoxin) with the heterobifunctional photoactivatable cross-linker pAPA. Very fast cross-linking is achievable by use of a UV flash. The pAPA-treated cells, wild-type strain W3110 and strain SK3981, revealed in thin sections labeled with thioredoxin antiserum that thioredoxin was largely associated with the area of the inner membrane. The protein was also present in the cytoplasm and at membrane adhesion sites; the periplasm was not labeled. Statistical analysis strongly supported these results. Localization of PBP 1b showed this protein in various amounts in all of the membrane fractions of the envelope. Similar to the thioredoxin distribution, PBP 1b was concentrated in membrane-associated areas and, to a lesser extent, in the cytoplasm and the outer membrane. The membrane-associated area was found to concentrate both thioredoxin and PBP 1b. This area was not known to exist as an antigenically separable domain in the bacterial cytoplasm. Whether this domain is involved in the localization of still other membrane proteins will have to be studied by the use of new fixation and immobilization methods.

ACKNOWLEDGMENTS. The present work was supported by Public Health Service grant AI-10414 from the National Institutes of Health and National Science Foundation grant DCB-85-03684 (to M.E.B. and M.H.B.) and by Public Health Service grants CA-06927 and RR-05539 from the National Institutes of Health and an appropriation from the Commonwealth of Pennsylvania to the Institute for Cancer Research.

We thank J. Rech for data handling on the computer and A. Capriotti for typing of the manuscript.

LITERATURE CITED

1. **Adler, S., and P. Modrich.** 1983. T7-induced DNA polymerase: requirement for thioredoxin sulfhydryl groups. *J. Biol. Chem.* **258:**6956–6962.
2. **Bayer, M. E.** 1979. The fusion sites between outer membrane and cytoplasmic membrane of bacteria: their role in membrane assembly and virus infection, p. 167–202, *In* M. Inouye (ed.), *Bacterial Outer Membranes: Biogenesis and Functions.* John Wiley & Sons, Inc., New York.
3. **Bayer, M. E., and M. H. Bayer.** 1986. Effects of bacteriophage fd infection on *Escherichia coli* HB11 envelope: a morphological and biochemical study. *J. Virol.* **57:**258–266.
4. **Bayer, M. E., M. H. Bayer, C. A. Lunn, and V. Pigiet.** 1987. Association of thioredoxin with the inner membrane and adhesion sites in *Escherichia coli. J. Bacteriol.* **169:**2659–2666.
5. **Bayer, M. H., G. P. Costello, and M. E. Bayer.** 1982. Isolation and partial characterization of membrane vesicles carrying markers of the membrane adhesion sites. *J. Bacteriol.* **149:**758–767.
6. **Berg, H. C.** 1983. *Random Walks in Biology.* Princeton University Press, Princeton, N.J.
7. **Carlemalm, E., R. M. Garavito, and W. Villiger.** 1982. Resin development for electron microscopy and an analysis of embedding at low temperature. *J. Microsc.* (Oxford) **126:**123–143.

8. **Hansson, H. A., A. Holmgren, B. Rozel, and S. Stemme.** 1986. Localization of thioredoxin, thioredoxin reductase and ribonucleotide reductase in cells. Immunohistochemistry aspects, p. 237–248. *In* A. Holmgren, C.-I. Branden, H. Jornvall, and B.-M. Sjoberg (ed.), *Thioredoxin and Glutathione Systems*. 9th Karolinska Institute Nobel Conference, 27–31 May 1985. Raven Press, New York.

9. **Hobot, J. A., W. Villiger, J. Escaig, M. Maeder, A. Ryter, and E. Kellenberger.** 1985. Shape and fine structure of nucleoids observed on sections of ultrarapidly frozen and cryosubstituted bacteria. *J. Bacteriol.* **162:**960–971.

10. **Holmgren, A.** 1985. Thioredoxin. *Annu. Rev. Biochem.* **54:**237–271.

11. **Kraut, H., W. Keck, and Y. Hirota.** 1981. Cloning of pon B and fts I gene of *E. coli* and purification of penicillin binding proteins IB and 3. *Annu. Rep. Natl. Inst. Genet.* **31:**24–29.

12. **Lopez, J., and R. E. Webster.** 1985. Assembly site of the bacteriophage f1 corresponds to adhesion zones between the inner and outer membranes of the host cell. *J. Bacteriol.* **163:**1270–1274.

13. **Lunn, C. A., S. Kathju, B. J. Wallace, S. R. Kushner, and V. Pigiet.** 1984. Amplification and purification of plasmid-encoded thioredoxin from *Escherichia coli* K12. *J. Biol. Chem.* **259:**10469–10474.

14. **Lunn, C. A., and V. Pigiet.** 1982. Localization of thioredoxin from *Escherichia coli* in an osmotically sensitive compartment. *J. Biol. Chem.* **257:**11424–11430.

15. **Lunn, C. A., and V. Pigiet.** 1986. Localization of thioredoxin in *Escherichia coli*, p. 165–176. *In* A. Holmgren, C.-I. Branden, H. Jornvall, and B.-M. Sjoberg (ed.), *Thioredoxin and Glutathione Systems*. 9th Karolinska Institute Nobel Conference, 27–31 May 1985. Raven Press, New York.

16. **Lunn, C. A., and V. Pigiet.** 1986. Chemical cross-linking of thioredoxin to hybrid membrane fraction in *Escherichia coli*. *J. Biol. Chem.* **261:**832–838.

17. **Mark, D. F., and C. C. Richardson.** 1976. *Escherichia coli* thioredoxin: a subunit of bacteriophage T7 DNA polymerase. *Proc. Natl. Acad. Sci. USA* **73:**780–784.

18. **Nelson, W. J., and E. Lazarides.** 1984. Globin (ankyrin) in striated muscle: identification of the potential membrane receptor for erythroid spectrin in muscle cells. *Proc. Natl. Acad. Sci. USA* **81:**3292–3296.

19. **Nygren, H., B. Rozel, A. Holmgren, and H. A. Hansson.** 1981. Immunoelectron microscopic localization of glutaredoxin and thioredoxin in *Escherichia coli* cells. *FEBS Lett.* **133:**145–150.

20. **Pollitt, S., and H. Zalkin.** 1983. Role of primary structure and disulfide bond formation in β-lactamase secretion. *J. Bacteriol.* **153:**27–32.

21. **Rodriguez-Tebar, J., A. Barbas, and D. Vasquez.** 1985. Location of some proteins involved in pepidoglycan synthesis and cell division in the inner and outer membranes of *Escherichia coli*. *J. Bacteriol.* **161:**243–248.

22. **Russel, M., and P. Model.** 1985. Thioredoxin is required for filamentous phage assembly. *Proc. Natl. Acad. Sci. USA* **82:**29–33.

23. **Spratt, B. G.** 1977. Properties of the penicillin-binding proteins of *Escherichia coli* K12. *Eur. J. Biochem.* **72:**341–352.

Chapter 32

Variations in the Penicillin-Binding Proteins of *Bacillus subtilis*

Christine E. Buchanan

There are six penicillin-binding proteins (PBPs) located in the cytoplasmic membrane of *Bacillus subtilis*. At least one additional protein, PBP 5*, is synthesized only during sporulation (21, 29). Ever since their discovery, it has been the working hypothesis that these PBPs are enzymes active in cell wall metabolism; inhibition of a PBP or a combination of PBPs by penicillin is believed to be responsible for the well-known effects of this antibiotic on cellular morphology and viability (1, 2). Much of what is now known about the in vivo functions and the essential nature of the PBPs has come from work with *Escherichia coli*, for which PBP mutants, and more recently the purified PBPs, have been studied in detail (3, 17, 19, 23–25, 27, 32). In contrast, what little we know specifically about the functions of the *B. subtilis* PBPs has not come from examination of mutants, of which there are very few, but primarily from two other approaches, biochemical analyses of the partially purified PBPs and physiological studies on the wild-type organism. For example, the amounts of PBPs 1 and 4 that are detectable in *B. subtilis* membranes drop to a very low level early during sporulation, which leads us to believe that these two PBPs participate primarily, if not exclusively, in vegetative cell wall metabolism (6, 21). PBP 1 has transglycosylase activity in vitro (16, 33). No enzymatic activity has been confirmed for PBP 4 yet, but by analogy with the PBP 4 from *B. megaterium* (26), it too may be a transglycosylase. The suggestion that the activity of PBP 4 may duplicate that of another PBP is consistent with the conclusion (see below) that the normal role of PBP 4 is as a backup or compensatory enzyme under conditions in which another PBP is impaired.

Christine E. Buchanan • Department of Biology, Southern Methodist University, Dallas, Texas 75275.

PBP 2a is believed to have strictly a vegetative function, because its activity is undetectable in most sporulating cells. This PBP is restored in the germinating spore at a time that is consistent with a role for this protein in cell elongation (6, 20, 21).

Various observations indicate the PBP 2b is probably required for both vegetative and spore-specific septum formation. For example, there is enhanced synthesis of PBP 2b specifically during stage II of sporulation, which is when the asymmetric cell division occurs that forms the forespore (21). PBP 2b is the only vegetative PBP that is not detectable in the mature spore (5). Moreover, it is not detectable until more than 30 min after heat activation of the spores, which indicates that it can have no function during germination proper. PBP 2b appears and begins to increase exponentially shortly before the first cell division (20). It is obviously not involved in determining the site of septum formation, since it seems to participate in both symmetric and asymmetric septations.

PBP 3 is also likely to be involved in a reaction that is common to wall formation in both vegetative cells and spores. Synthesis of PBP 3 is enhanced during stage III of sporulation, which is the period when the enzymes needed for cortical peptidoglycan synthesis are made, and in addition, this PBP is rapidly restored to its vegetative level during the outgrowth phase of germination (20, 21).

PBP 5* first appears during stage III of sporulation and is primarily found in the outer forespore membrane (5, 6, 21). It has never been detected in vegetative cells or in any *spo0* mutants (21, 29). PBP 5* has D-alanine carboxypeptidase activity in vitro (28) and may be responsible for the relatively low level of cross-linked peptidoglycan found in the spore cortex.

PBP 5, which accounts for at least 70% of the penicillin-binding activity in vegetative cells, is also a D-alanine carboxypeptidase (1). Recently, Ellar and his colleagues succeeded in completely eliminating PBP 5 by insertional mutagenesis of its structural gene (30). Such mutant cells grow normally, but unexpectedly, they do not form normal spores. Thus, there seems to be a requirement for two carboxypeptidases during sporulation.

Despite the absence of a strong selection method for specific PBP mutants in *B. subtilis,* various phenotypically normal and mutant strains that have altered or missing PBPs have been identified. The preliminary characterization of some of these PBP variants is described here. The results suggest that highly different amounts of certain PBPs can be found among the so-called normal strains of this species and also reveal some unexpected complexities and possible interrelationships among these proteins.

The *B. subtilis* 168 *trp* strain that was used as the basis for all comparisons was obtained from J. H. Hageman, New Mexico State University, Las Cruces. The Porton strain and its derivatives have been described previously (7). All other strains were obtained from either the American Type Culture Collection,

Rockville Md., or the Bacillus Genetic Stock Center at The Ohio State University, Columbus. Details of most methods employed here have been published (4–7, 20–21); the remainder are described in the text.

Interstrain Variability of the B. subtilis PBPs

Strain 168 was chosen as the standard strain to which others were compared. This decision was based on several considerations; namely, there is more general knowledge about this strain than any other, its PBPs are the most thoroughly characterized under a variety of growth conditions, and the majority of B. subtilis mutants described in the literature are derived from this strain. The apparent molecular weights of its PBPs were determined on sodium dodecyl sulfate-polyacrylamide gels using molecular weight standards purchased from Bethesda Research Laboratories, Gaithersburg Md. The average molecular weight ± 1 standard deviation ($n = 25$) for each PBP was as follows: PBP 1, 100,500 $\pm$ 2,200; PBP 2a, 78,800 $\pm$ 1,100; PBP 2b, 76,500 $\pm$ 1,100; PBP 3, 70,700 $\pm$ 1,100; PBP 4, 67,000 $\pm$ 1,100; PBP 5, 46,300 $\pm$ 800; PBP 5*, 37,600 $\pm$ 900.

Other strains of B. subtilis had vegetative PBP profiles that were similar but not always identical to that of strain 168 (Fig. 1). For example, there was a consistent difference between the amounts of PBP 2b in the Porton and 168 strains. Relative to the amount of PBP 2b per milligram of membrane protein present in strain 168, Porton had a 3.3-fold higher level (compare lanes B and C in Fig. 1). In contrast, one derivative of the 168 strain, BR151 (35), consistently had a reduced amount of this PBP, approximately equivalent to 0.27 of the amount found in all other derivatives of this strain that were examined (compare lanes A and C). By varying the conditions of the binding assay, we determined that these differences most likely reflect different amounts of PBP 2b rather than different properties of the protein, such as a different sensitivity or rate of binding to penicillin (data not shown).

It is also apparent from Fig. 1 that PBP 5 from the Porton strain (lane B) had a slightly slower electrophoretic mobility on sodium dodecyl sufate gels compared with PBP 5 from BR151 or other 168 strains. This observation has been made previously (18). We calculated that the apparent molecular weight of PBP 5 in the Porton strain was approximately 50,000, which is the same value reported by Blumberg and Strominger (1).

Examination of the W23 strain (22) by our standard PBP assay (21) revealed that it had very little PBP 3 relative to the amount detected in the 168 strain (compare the B lanes in the two panels of Fig. 2). By increasing the amount of [^{3}H]penicillin in the binding assay, however, it was determined that PBP 3 of strain W23 was almost as abundant as in the other strain, but only half as sensitive to penicillin. In addition, PBP 3 from strain W23 clearly had

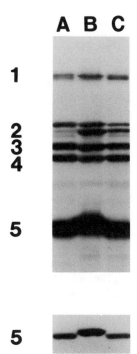

Figure 1. Fluorograph of [³H]penicillin-labeled PBPs in membranes from vegetative cells of *B. subtilis.* Lanes: A, BR151, a *lys-3 metB10 trpC2* derivative of strain 168; B, the Porton strain; C, a *trpC* derivative of strain 168. The PBPs are numbered in descending order of molecular weight. The three components of PBP 2 are designated PBPs 2a, 2b, and 2c. The upper panel is a 24-h exposure and the lower one is a 2-h exposure of the gel to X-ray film. Approximately the same amount of membrane protein was loaded into each lane of the gel.

a different electrophoretic mobility (compare the left and right panels of Fig. 2). There is a precedent in *B. subtilis* for an alteration of the molecular weight of a PBP occurring simultaneously with an increase in its penicillin resistance (7). Unfortunately, strain W23 is not well characterized, and it is not yet known whether there is a correlation between the increased resistance of PBP 3, its different molecular weight, and the MIC of benzylpenicillin for the organism.

PBP 4-Deficient Variants of Strain 168

A dramatic example of PBP variability within a single strain of *B. subtilis* is illustrated in Fig. 3. Two different samples of strain 168 were fortuitously discovered to have no detectable PBP 4. That these samples are derived from strain 168 was confirmed by phage typing (15). One of them (lane A, Fig. 3) was supposed to be completely wild type, except for a requirement for tryptophan. The other (lane C) has a known defect in sporulation, but there is no evidence to suggest any connection between the sporulation defect and the absence of PBP 4 (4a). As can be seen in Fig. 3, there was no significant increase in any one of the remaining PBPs in either of these variants, which suggests that none of the other PBPs was overproduced to compensate for the loss of

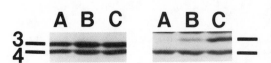

Figure 2. Fluorographs of PBPs 3 and 4 from *B. subtilis* 168 (left panel) and W23 (right panel). The standard PBP assay (21) was done on membranes from each strain with 0.04 (lanes A), 0.10 (lanes B), and 0.20 mM (lanes C) [3H]-labeled penicillin. Higher concentrations of penicillin gave results indistinguishable from those in the two C lanes.

PBP 4. PBP 4 was also undetectable when the length of the binding assay was increased from 10 to 60 min, which argues against the possibility that either mutant PBP 4 had slower binding kinetics (10). The concentration of [3H]penicillin used in the assay was several thousand-fold in excess of that required to saturate the wild-type PBP 4.

One of the PBP 4-deficient mutants, ATCC 23857 (lane A, Fig. 3), was extensively characterized in an effort to find some phenotypic effect of the missing PBP (4a). A summary of the results follows. Samples of strain 168 with supposedly the same genotype (i.e., *trpC*) were acquired from several sources and examined in parallel with ATCC 23857. All these samples had the normal amount of PBP 4, but by a number of other criteria the PBP 4-deficient mutant was indistinguishable from them. The generation time for ATCC 23857 was not significantly different from the times for the others in either Penassay broth (Difco Laboratories, Detroit, Mich.) or a minimal salts medium at 30, 37, or 45°C. By phase-contrast microscopy they all appeared to be normal motile rods. ATCC 23857 sporulated and germinated with both normal effi-

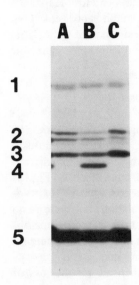

Figure 3. Fluorograph of PBPs in three derivatives of *B. subtilis* 168 acquired from different sources. Lanes: A, ATCC 23857, a *trpC* auxotroph; B, 168 *trpC* acquired from J. H. Hageman; C, BGSC 1S33, a *spoIID* mutant.

ciency and normal timing of developmental events. The cells were sensitive to lysozyme, and the activity of their principal autolysin, an *N*-acetylmuramoyl-L-alanine amidase, appeared to be within the normal range. Finally, the loss of one of its most penicillin-sensitive PBPs had no apparent effect on the susceptibility of ATCC 23857 to various β-lactam antibiotics such as cefoxitin, cephalothin, cloxacillin, penicillin G, and thienamycin.

Thus, the absence of PBP 4 does not measurably affect normal vegetative growth or sporulation under laboratory conditions. ATCC 23857 is not the first example of an organism that thrives without its full complement of PBPs (3, 7–9, 11–14, 17, 30–32, 36), but it is only the second example (loss of PBP la from *E. coli* being the first reported case; 17, 32) of the loss of a highly sensitive PBP with no obvious phenotypic effect. The tentative conclusion is that the normal role for PBP 4 is probably to serve as a backup or compensatory enzyme. The precedent for this suggestion has already been established in *E. coli* where duplication of enzyme activities and physiological functions among the PBPs has been described (3, 17, 19, 25, 27, 32, 34).

Further Characterization of a Complex PBP Mutant

The preceding examples are PBP variants that apparently arose with no selective pressure, which accounts, in part, for the absence of any measurable phenotypic changes in these strains. One complex PBP mutant that was isolated by a five-step selection for a highly cloxacillin-resistant mutant of *B. subtilis* Porton (7) is available. This mutant, named mutant 5 because it came from the fifth step in the selection protocol, has not been readily amenable to physiological analysis; it is defective in sporulation, as is the cloxacillin-sensitive parent, and it has acquired multiple changes in its PBP profile and phenotype. These changes, a few of which are obvious in Fig. 4 (compare lanes A and B), are that (i) mutant 5 is 180-fold more resistant than the wild type to cloxacillin, (ii) it has no detectable PBP 1, (iii) its PBP 2a has an altered electrophoretic mobility, (iv) its PBP 2a is highly resistant to cloxacillin, (v) it contains an increased amount of PBP 2a, and (vi) the cells have a reduced diameter. Several conclusions were made from an analysis of mutant 5 and the four mutants that preceded it in the selection scheme. First, it was concluded that PBP 2a was likely to be the lethal target of cloxacillin, because this PBP was the only one to change in its susceptibility to cloxacillin and because the stepwise changes in PBP 2a roughly paralleled the stepwise changes in resistance of the organism (7). Second, after examination of another mutant of *B. subtilis* that also had a narrow cell diameter, it was proposed that there was a correlation between this morphological characteristic and an altered or missing PBP 1 (18). Finally, it was proposed that an increase in the amount of PBP 2a occurred to compensate for the loss of PBP 1 (18). However, this last

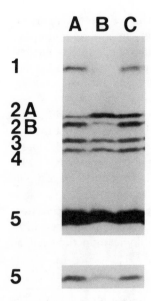

Figure 4. Fluorograph of PBPs from the cloxacillin-sensitive Porton strain (lane A) and two cloxacillin-resistant derivatives, mutant 5 (lane B) and mutant 4 (lane C).

proposal is inconsistent with the observation that PBP 2a increased in amount at a step preceding the disappearance of PBP 1 (compare lanes B and C with lane A in Fig. 4). In fact, upon reexamination of the PBP profiles from numerous membrane preparations, it became apparent that the selection of mutant 5 from mutant 4 was not only accompanied by the complete disappearance of PBP 1, but also by a significant decline in the membrane concentration of every one of the other PBPs, with the single exception of PBP 2a, which remained at a level twofold higher than the wild type. Although this result tends to focus attention on the essential nature of PBP 2a, there are no data to suggest that PBP 2a is the only essential PBP.

In an effort to further evaluate the effects of these PBP changes on the phenotype of the cell, the altered PBPs of mutant 5 were genetically transferred into a derivative of strain 168. Specifically, a saturating concentration of DNA from mutant 5 (cloxacillin MIC, 10 μg/ml) was used to transform a cloxacillin-sensitive strain (MIC, 0.1 μg/ml) to a low level of cloxacillin resistance. Attempts to select for higher levels of resistance failed, which is consistent with the notion that high-level resistance involves multiple genes.

Preliminary analysis of the cloxacillin-resistant transformants recovered from two separate transformations revealed that they had various combinations of the PBPs from the two parental strains (Fig. 5). One of the most unexpected results was that approximately 60% of the cloxacillin-resistant transformants had no detectable PBP 1 (lanes C and G, Fig. 5). This represents a significant linkage between the two characteristics, which should prove useful for mapping this PBP. In addition, it was observed that some of the transformants that lacked

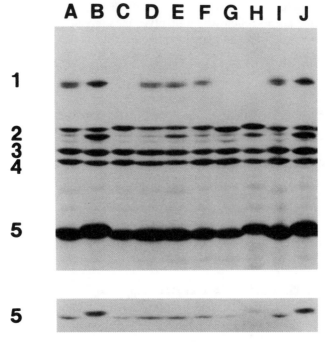

Figure 5. Flurograph of PBPs from parental strains and transformants. Lanes: A and I, duplicate samples of BR151, the cloxacillin-sensitive recipient; B and J, duplicate samples of Porton, the cloxacillin-sensitive parent of mutant 5; C through G, various low-level cloxacillin-resistant transformants of BR151; H, mutant 5, the cloxacillin-resistant donor.

PBP 1 had only an average amount of PBP 2a, which further weakens the proposal mentioned above that an overproduced PBP 2a may be needed to compensate for the loss of PBP 1.

All of the transformants that had no PBP 1 had a narrower-than-normal cell diameter, which was readily apparent by phase-contrast microscopy. This result indicates that PBP 1 probably has a unique role in cell wall synthesis that cannot be completely filled by any other available enzyme. However, whatever the activity of PBP 1 is, it is clearly dispensable under laboratory conditions. The transformants with no PBP 1 all had what appeared to be normal growth and sporulation properties.

Only a few of the transformants acquired the altered PBP 2a from the donor, mutant 5 (lanes C, D, E, and F, Fig. 5). Although it was determined that in every transformant in which PBP 2a appeared to have altered mobility on gels the PBP was also highly resistant to cloxacillin, no transformant was recovered for which the MIC of cloxacillin was correspondingly elevated. In other words, in no case was the presence of the highly cloxacillin-resistant PBP 2a sufficient to render the cells highly cloxacillin resistant. This tends to confirm the interpretation that a high level of resistance to β-lactam antibiotics is a multigenic phenomenon which involves the complicated interrelationships of at least several PBPs and perhaps other constituents as well.

Conclusions

A number of variations in the PBPs of *B. subtilis* are tolerated by the organism with few, if any, apparent effects on normal growth and sporulation properties. At least three of the PBPs can be lost or completely inactivated. The loss of PBP 5 leads to formation of abnormal spores but has no effect on vegetative growth (2, 30), the loss of PBP 1 has a nonlethal effect on cellular morphology (18), and the loss of PBP 4 has no known effect at all. While the individual PBPs may indeed play specific roles in normal cell wall metabolism, it is becoming increasingly apparent that at least some of their more essential functions are likely to be overlapping and interdependent.

ACKNOWLEDGMENT. This work was supported by Public Health Service grant AI-19829 from the National Institute of Allergy and Infectious Diseases.

LITERATURE CITED

1. **Blumberg, P. M., and J. L. Strominger.** 1972. Five penicillin-binding components occur in *Bacillus subtilis* membranes. *J. Biol. Chem.* **247**:8107–8113.
2. **Blumberg, P. M., and J. L. Strominger.** 1974. Interaction of penicillin with the bacterial cell: penicillin-binding proteins and penicillin-sensitive enzymes. *Bacteriol. Rev.* **38**:291–335.
3. **Broome-Smith, J. K.** 1985. Construction of a mutant of *Escherichia coli* that has deletions of both the penicillin-binding proteins 5 and 6 genes. *J. Gen. Microbiol.* **131**:2115–2118.
4. **Buchanan, C. E.** 1979. Altered membrane proteins in a minicell-producing mutant of *Bacillus subtilis*. *J. Bacteriol.* **139**:305–307.
4a. **Buchanan, C. E.** 1987. Absence of penicillin-binding protein 4 from an apparently normal strain of *Bacillus subtilis*. *J. Bacteriol.* **169**:5301–5303.
5. **Buchanan, C. E., and S. L. Neyman.** 1986. Correlation of penicillin-binding protein composition with different functions of two membranes in *Bacillus subtilis* forespores. *J. Bacteriol.* **165**:498–503.
6. **Buchanan, C. E., and M. O. Sowell.** 1983. Stability and synthesis of the penicillin-binding proteins during sporulation. *J. Bacteriol.* **156**:545–551.
7. **Buchanan, C. E., and J. L. Strominger.** 1976. Altered penicillin-binding components in penicillin-resistant mutants of *Bacillus subtilis*. *Proc. Natl. Acad. Sci. USA* **73**:1816–1820.
8. **Curtis, N. A., and M. V. Hayes.** 1981. A mutant of *Staphylococcus aureus* H deficient in penicillin-binding protein 1 is viable. *FEMS Microbiol. Lett.* **10**:227–229.
9. **Curtis, N. A. C., M. V. Hayes, A. W. Wyke, and J. B. Ward.** 1980. A mutant of *Staphylococcus aureus* H lacking penicillin-binding protein 4 and transpeptidase activity in vitro. *FEMS Microbiol. Lett.* **9**:263–266.

10. **Fontana, R., R. Cerini, P. Longoni, A. Grossato, and P. Canepari.** 1983. Identification of a streptococcal penicillin-binding protein that reacts very slowly with penicillin. *J. Bacteriol.* **155:**1343–1350.

11. **Fontana, R., A. Grossoto, L. Rossi, Y. R. Cheng, and G. Satta.** 1985. Transition from resistance to hypersusceptibility to β-lactam antibiotics associated with loss of a low-affinity penicillin-binding protein in a *Streptococcus faecium* mutant highly resistant to penicillin. *Antimicrob. Agents Chemother.* **28:**678–683.

12. **Godfrey, A. J., L. E. Bryan, and H. R. Rabin.** 1981. β-Lactam-resistant *Pseudomonas aeruginosa* with modified penicillin-binding proteins emerging during cystic fibrosis treatment. *Antimicrob. Agents Chemother.* **19:**705–711.

13. **Hakenbeck, R., H. Ellerbrok, T. Briese, S. Handwerger, and A. Tomasz.** 1986. Penicillin-binding proteins of penicillin-susceptible and -resistant pneumococci: immunological relatedness of altered proteins and changes in peptides carrying the β-lactam binding site. *Antimicrob. Agents Chemother.* **30:**553–558.

14. **Handwerger, S., and A. Tomasz.** 1986. Alterations in kinetic properties of penicillin-binding proteins of penicillin-resistant *Streptococcus pneumoniae*. *Antimicrob. Agents Chemother.* **30:**57–63.

15. **Hemphill, H. E., and H. R. Whiteley.** 1975. Bacteriophages of *Bacillus subtilis*. *Bacteriol. Rev.* **39:**257–315.

16. **Jackson, G. E. D., and J. L. Strominger.** 1984. Synthesis of peptidoglycan by high molecular weight penicillin-binding proteins of *Bacillus subtilis* and *Bacillus stearothermophilus*. *J. Biol. Chem.* **259:**1483–1490.

17. **Kato, J., H. Suzuki, and Y. Hirota.** 1985. Dispensability of either penicillin-binding protein-1a or -1b involved in the essential process for cell elongation in *Escherichia coli*. *Mol. Gen. Genet.* **200:**272–277.

18. **Kleppe, G., W. Yu, and J. L. Strominger.** 1982. Penicillin-binding proteins in *Bacillus subtilis* mutants. *Antimicrob. Agents Chemother.* **21:**979–983.

19. **Nakagawa, J., and M. Matsuhashi.** 1982. Molecular divergence of a major peptidoglycan synthetase with transglycosylase-transpeptidase activities in *Escherichia coli*—penicillin-binding protein 1Bs. *Biochem. Biophys. Res. Commun.* **105:**1546–1553.

20. **Neyman, S. L., and C. E. Buchanan.** 1985. Restoration of vegetative penicillin-binding proteins during germination and outgrowth of *Bacillus subtilis* spores: relationship of individual proteins to specific cell cycle events. *J. Bacteriol.* **161:**164–168.

21. **Sowell, M. O., and C. E. Buchanan.** 1983. Changes in penicillin-binding proteins during sporulation of *Bacillus subtilis*. *J. Bacteriol.* **153:**1331–1337.

22. **Spizizen, J.** 1958. Transformation of biochemically deficient strains of *Bacillus subtilis* by deoxyribonucleate. *Proc. Natl. Acad. Sci. USA* **44:**1072–1078.

23. **Spratt, B. G.** 1975. Distinct penicillin binding proteins involved in the division, elongation, and shape of *Escherichia coli* K12. *Proc. Natl. Acad. Sci. USA* **72:**2999–3003.

24. **Spratt, B. G.** 1977. Properties of the penicillin-binding proteins of *Escherichia coli* K12. *Eur. J. Biochem.* **72:**341–352.

25. **Spratt, B. G.** 1983. Penicillin-binding proteins and the future of β-lactam antibiotics. *J. Gen. Microbiol.* **129:**1247–1260.

26. **Taku, A., M. Stuckey, and D. P. Fan.** 1982. Purification of the peptidoglycan transglycosylase of *Bacillus megaterium*. *J. Biol. Chem.* **257:**5018–5022.

27. **Tamaki, S., J. Nakagawa, I. N. Maruyama, and M. Matsuhashi.** 1978. Supersensitivity to β-lactam antibiotics in *Escherchia coli* caused by D-alanine carboxypeptidase 1A mutation. *Agric. Biol. Chem.* **42:**2147–2150.

28. **Todd, J. A., E. T. Bone, and D. J. Ellar.** 1985. The sporulation-specific penicillin-binding protein 5a from *Bacillus subtilis* is a DD-carboxypeptidase in vitro. *Biochem. J.* **230:**825–828.

29. **Todd, J. A., E. J. Bone, P. J. Piggot, and D. J. Ellar.** 1983. Differential expression of penicillin-binding protein structural genes during *Bacillus subtilis* sporulation. *FEMS Microbiol. Lett.* **18:**197–202.

30. **Todd, J. A., A. N. Roberts, K. Johnstone, P. J. Piggot, G. Winter, and D. J. Ellar.** 1986. Reduced heat resistance of mutant spores after cloning and mutagenesis of the *Bacillus subtilis* gene encoding penicillin-binding protein 5. *J. Bacteriol.* **167:**257–264.

31. **Ubukata, K., N. Yamashita, and M. Konno.** 1985. Occurrence of a β-lactam-inductible penicillin-binding protein in methicillin-resistant staphylococci. *Antimicrob. Agents Chemother.* **27:**851–857.

32. **Yousif, S. F., J. K. Broome-Smith, and B. G. Spratt.** 1985. Lysis of *Escherichia coli* by β-lactam antibiotics: deletion analysis of the role of penicillin-binding proteins 1A and 1B. *J. Gen. Microbiol.* **131:**2839–2845.

33. **Waxman, D. J., D. M. Lindgren, and J. L. Strominger.** 1981. High-molecular-weight penicillin-binding proteins from membranes of bacilli. *J. Bacteriol.* **148:**950–955.

34. **Waxman, D. J., and J. L. Strominger.** 1983. Penicillin-binding proteins and the mechanism of action of β-lactam antibiotics. *Annu. Rev. Biochem.* **52:**825–869.

35. **Young, F. E., C. Smith, and B. E. Reilly.** 1969. Chromosomal location of genes regulating resistance to bacteriophage in *Bacillus subtilis. J. Bacteriol.* **98:**1087–1097.

36. **Zighelboim, S., and A. Tomasz.** 1980. Penicillin-binding proteins of multiply antibiotic-resistant South African strains of *Streptococcus pneumoniae. Antimicrob. Agents Chemother.* **17:**434–442.

Chapter 33

The Essential Nature of Staphylococcal Penicillin-Binding Proteins

Peter E. Reynolds

Most bacteria possess several membrane-bound proteins which bind β-lactam antibiotics covalently and are believed to function as transpeptidases and DD-carboxypeptidases in the final stages of wall peptidoglycan biosynthesis. The functions of penicillin-binding proteins (PBPs) and their essential nature have been investigated principally in *Escherichia coli* by means of genetic manipulation, observation of the morphological consequences of saturation and hence inactivation of a PBP, and biochemical investigations of the purified proteins (5, 6). Several of the high-molecular-mass PBPs of *E. coli* which function as bifunctional transglycosylases/transpeptidases (11) are lethal targets of β-lactam antibiotics, while the lower-molecular-mass PBPs appear to be dispensable when the organism is grown in the laboratory. Studies of the high-molecular-mass PBPs of gram-positive bacteria are less advanced, though PBP 1 of *Bacillus megaterium* appears to be a transpeptidase and a lethal target of β-lactams, but not the sole important target (12, 15).

In recent years much attention has been focused on *Staphylococcus aureus*, as the introduction of β-lactamase-stable penicillins and cephalosporins has resulted in the emergence of highly resistant strains in hospitals. In addition to the normal four PBPs, these methicillin-resistant *S. aureus* strains possess an additional PBP which has a very low affinity for β-lactam antibiotics (1, 9, 13, 14, 16, 18). When these strains are grown in the presence of β-lactams,

Peter E. Reynolds • Department of Biochemistry, University of Cambridge, Cambridge CB2 1QW, England.

the additional protein is produced in large amounts and is the only PBP not saturated with antibiotic; under these conditions it is presumably the only functional PBP (14). This finding prompted an investigation of the PBPs of laboratory strains of S. aureus that were sensitive to β-lactam antibiotics to ascertain how many of the PBPs were essential and hence the number of different functions for which the additional PBP of methicillin-resistant S. aureus strains might be able to substitute.

Previous studies showed that PBP 4 of S. aureus is dispensable; a mutant lacking it was viable, and specific saturation with low concentrations of cefoxitin had no effect on growth (3, 19). It was claimed that PBP 1 was also inessential since a mutant strain, H/28, that was slightly more sensitive to cefotaxime than the parent apparently lacked it, but nevertheless grew normally and had an unchanged degree of cross-linkage in the peptidoglycan (2). These observations ensured that attention was focused on PBPs 2 and 3 in the search for the lethal target of β-lactams in this organism, and it has been suggested, as a result of experiments in which binding affinities of β-lactams to PBPs have been measured, that at least one of them is essential but probably not both; the possibility exists that either can substitute for the function of the other. However, in some binding experiments PBPs 2 and 3 were apparently uncomplexed with antibiotic and therefore were still active at concentrations of β-lactams which inhibited bacterial growth, implying either that a different target was involved or that only a small reduction in the activity of one or both was necessary for the inhibition of bacterial growth. To resolve the question of which staphylococcal PBPs are important, we reexamined some of the original claims and carried out experiments on the consequences of the selective inhibition of PBPs, singly and in combination, on growth, on the saturation of PBPs in growing cultures, and on an attachment transpeptidase involved in cell wall thickening in nongrowing cultures.

Investigation of PBPs of S. aureus

The PBPs of S. aureus are difficult to investigate: the three of high molecular mass have similar mobilities on polyacrylamide gel electrophoresis, while PBP 1 is unstable and frequently appears as a very faint band on fluorographs when membrane preparations are labeled with β-lactams. For these reasons we have carried out all labeling studies with intact cells (to maximize visualization of PBP 1) and have improved the separation of PBPs 2 and 3 by continuing the electrophoretic separation for 150 V-h after the dye front reached the bottom of the separating gel. When it was important to visualize PBP 4, counterlabeling with [³H]benzylpenicillin was carried out after preparation of the membrane fraction, as the complex of this β-lactam with PBP 4 has a short half-life (10).

The Essential Nature of *S. aureus* PBPs

The evidence that PBP 4 is unnecessary for cell growth under laboratory conditions was incontrovertible, but a number of observations indicated that PBP 1, previously discarded as a target on the basis of the viability of mutant strain H/28 which lacked it, might be important for cell growth. Labeling studies with intact cells, using a saturating concentration of radiolabeled benzylpenicillin, revealed that strain H/28 possessed normal amounts of PBP 1; consequently, all three high-molecular-mass PBPs could be regarded as potential targets.

In most investigations in which attempts have been made to correlate MIC values with binding of β-lactams to PBPs, labeling studies are carried out for short periods with membrane preparations at high protein densities under conditions totally different from those used in the determination of MICs. In assessing the importance of individual PBPs for cellular economy, we have carried out binding studies in growing cultures under conditions identical to those used for the measurement of the effect of β-lactams on growth rate. Cultures were grown for not less than four generations with sub-MICs of different β-lactam antibiotics that effectively saturated a specific PBP. At these sub-MIC levels several β-lactams (cefotaxime, cefuroxime, clavulanic acid) bound specifically to PBP 2, methicillin or cephradine bound to PBP 3, and cefoxitin bound to PBP 4. Few β-lactams bind specifically to PBP 1, but at the MIC imipenem (and some monobactams) binds to this PBP without affecting PBPs 2 and 3. It is tempting to speculate that, individually, PBP 2, 3, or 4 in growing cultures can be saturated with no effect on growth while inhibition of the activity of PBP 1 alone is sufficient for growth inhibition, but imipenem also binds, at least partially, to PBP 4 at the concentration used.

Roles of Individual PBPs

In vivo, PBP 4 catalyzes a transpeptidase reaction involved in the secondary cross-linking of peptidoglycan, a role analogous to that proposed for PBP 4 of *E. coli* (4). In cultures treated with low concentrations of cefoxitin or in the mutant lacking PBP 4, the peptidoglycan is less cross-linked than in the control or parent strain (19). In the laboratory this reduced amount of cross-linking in the peptidoglycan can be tolerated with little effect on growth rate and with only a marginal increase in sensitivity to β-lactams (3).

It has not proved possible to demonstrate directly any enzymatic activity of the high-molecular-mass PBPs of *S. aureus*. However, a large amount of circumstantial evidence suggests that PBP 2 catalyzes an attachment transpeptidase reaction linking newly synthesized nascent peptidoglycan chains to the existing cell wall when cultures are incubated under nongrowing conditions, i.e., when only cell wall thickening rather than expansion can occur. This reaction was monitored by following the incorporation of [^{14}C]glycine (a wall amino acid) into sodium dodecyl sulfate-insoluble peptidoglycan, a process which can be completely inhibited by low concentrations of some β-lactam antibiotics. Two lines of evidence indicated that this reaction is catalyzed mainly by PBP 2: first, there was a firm correlation of the binding affinities of at least 20 β-lactam antibiotics for PBP 2 with inhibition of the attachment transpeptidase, and second, after total inhibition of the enzyme activity followed by destruction or removal of excess antibiotic not bound to PBPs, the rate of recovery of the attachment transpeptidase activity could be correlated with the rate of loss of a particular β-lactam from its complex with PBP 2. Activity was recovered rapidly when methicillin was used (50% in 12 min), very similar to the half-life of the methicillin-PBP 2 complex. Much longer half-lives of the complexes of cephalothin, cloxacillin, and benzylpenicillin with PBP 2 were observed, with the kinetics of recovery of transpeptidase activity being similar.

Consequences of Saturation of Multiple PBPs of S. aureus

Although a single PBP may be dispensable, it does not necessarily follow that its role is unimportant if the activity of another nonessential PBP is reduced as a result of inhibition by β-lactam antibiotics. To investigate this possibility, β-lactams which effectively saturated PBP 2, 3, or 4 were used in combination to study the consequences of inhibiting the activity of two or more PBPs simultaneously. The effects were studied of the antibiotic combinations on growth from small inocula similar to those used in classical MIC investigations, on growth rate over several generations, on the attachment transpeptidase, and on the PBP profile. As expected, the effects of β-lactams on the attachment transpeptidase were purely additive: β-lactams binding to PBP 1, 3, or 4 had only a marginal additive effect on the inhibition observed when PBP 2 was also complexed with a β-lactam.

Preliminary tests established that certain combinations of β-lactams were particularly effective in inhibiting growth from a small inoculum. When cefotaxime and methicillin (inhibitors of PBPs 2 and 3) or cefotaxime and cefoxitin (inhibitors of PBPs 2 and 4) were present together, the antibiotics apparently exerted a synergistic effect; growth was inhibited at low concentrations of each, very substantially below the MICs (Fig. 1). This response was not obtained when methicillin and cefoxitin were used (inhibitors of PBPs 3 and 4).

Similar results were obtained when the effect of combinations of β-lactams on growth rate was investigated. Incubation for several generations in the presence of β-lactams whose predicted binding was to PBPs 2 and 3 or PBPs 2 and 4 resulted in growth inhibition, but effective saturation of PBPs 3 and 4

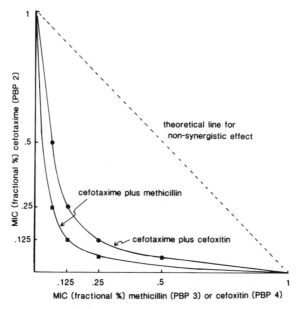

Figure 1. Effect of β-lactam antibiotics added in combination on the growth of *S. aureus* H. Organisms from a culture growing exponentially were inoculated into tubes of media containing tryptone, yeast extract, and glucose (pH 7.0) to give a density of 10^5 cells per ml. Antibiotic supplements were added singly or in combination at twofold dilutions of the MIC, and the tubes were incubated at 37°C for 12 h. The points on the curves represent the lowest concentrations of antibiotic in tubes in which no growth occurred.

could occur without serious effect (Fig. 2). Binding of the same β-lactams to the PBPs was carried out under identical incubation conditions to ascertain which PBPs were still present and active in the growing or nongrowing cultures. Therefore, in these experiments the binding patterns of β-lactams to PBPs represent the net effect of rate of synthesis of new PBPs and rate of binding and consequent inactivation. The amount of free PBP remaining in intact cells after growth in an unlabeled β-lactam was revealed by counterlabeling with an excess of [³H]benzylpenicillin followed by rapid lysis, recovery of the membrane fraction by centrifugation, and sodium dodecyl sulfate-gel electrophoresis of the solubilized labeled proteins. Surprisingly, the predicted binding patterns were not always obtained. In those instances in which growth was affected (inhibition of PBPs 2 and 3 or 2 and 4), in addition to the predicted effects, PBP 1 had also been affected and was present in severely limited amounts. This resulted either from binding of the β-lactams to PBP 1 in addition to their primary targets or from a reduced rate of synthesis of the PBP. Consequently, it is not necessarily the saturation of PBPs 2 and 3 or 2 and 4 in these experiments that is the important event, since binding of β-lactams to PBP 1 alone results in growth inhibition.

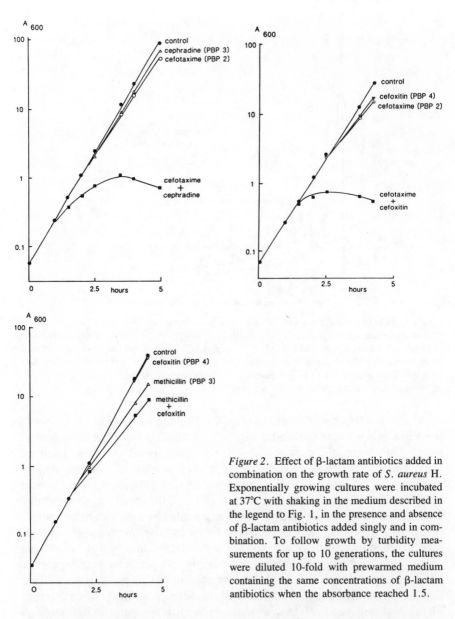

Figure 2. Effect of β-lactam antibiotics added in combination on the growth rate of *S. aureus* H. Exponentially growing cultures were incubated at 37°C with shaking in the medium described in the legend to Fig. 1, in the presence and absence of β-lactam antibiotics added singly and in combination. To follow growth by turbidity measurements for up to 10 generations, the cultures were diluted 10-fold with prewarmed medium containing the same concentrations of β-lactam antibiotics when the absorbance reached 1.5.

This finding was investigated further by reducing the concentrations of β-lactams used to bind specifically to PBPs 2, 3, and 4. When *S. aureus* was grown in the presence of clavulanic acid, cephradine, and cefoxitin in all possible combinations, growth was unaffected for at least five generations, and the binding of the β-lactams to PBPs was predictable. PBPs 2, 3, and 4 were effectively saturated (greater than 90%), while PBP 1 was present in virtually unchanged amounts compared with the control (Fig. 3). These results suggest that *S. aureus* can grow with PBP 1 as the sole functional PBP, though the necessity for small amounts of the remaining PBPs cannot be totally excluded.

Cells growing essentially in the absence of PBPs 2, 3, and 4 were considerably enlarged compared with the controls, suggesting that the peptidoglycan produced under these conditions is less cross-linked and consequently less rigid than that in cells possessing a full complement of PBPs or that septation has been impaired but not totally inhibited. It is therefore probable that PBPs 2, 3, and 4 (or at least one of them) have an important role in *S. aureus*, though they are less vital than PBP 1. The production of enlarged cells was also observed when *S. aureus* was grown with slightly sub-growth-inhibitory concentrations of imipenem and when methicillin-resistant strains were grown under conditions in which only PBP 2′ was functional.

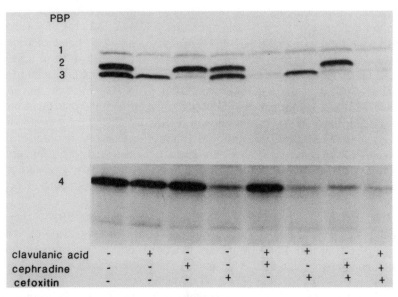

Figure 3. Binding of β-lactams to PBPs. Cultures of *S. aureus* H similar to those described in the legend to Fig. 2 were incubated in the presence and absence of clavulanic acid (3 mg/liter), cephradine (0.025 mg/liter), and cefoxitin (0.05 mg/liter), singly and in all possible combinations. The concentrations were selected as the highest which, when all three antibiotics were present, permitted an increase in turbidity similar to that of the control culture over five generations. Organisms were harvested and washed, and the PBPs not saturated with unlabeled antibiotic were revealed by counterlabeling with [^{3}H]benzylpenicillin. Separation of membrane proteins and fluorography were carried out as described (14). The fluorograph shown is a composite of two: PBPs 1, 2, and 3 were revealed by radiolabeling of intact cells prior to lysis and recovery of the membrane fraction, whereas the amount of free PBP 4 was detected by radiolabeling after membrane preparation.

The importance of PBP 1 is supported by the results of investigations with a cephradine-resistant strain of *S. aureus*. This strain is different from that examined by Georgopapadakou et al. (8): it is approximately 10-fold more resistant than sensitive strains and is specifically resistant to cephradine and a few other β-lactams. It has none of the characteristics of methicillin-resistant *S. aureus* strains and in particular does not possess the additional PBP 2′. The mutant possesses a PBP 1 with an altered affinity for cephradine and other β-lactams, and this strain is not as susceptible to growth inhibition by combinations of cefotaxime and methicillin or cefotaxime and cefoxitin which inhibit strain H and which result in diminished amounts of the normal PBP 1.

Conclusions

The experimental observations reported here emphasize the important role of PBP 1 of *S. aureus*. The probing of specific PBPs with β-lactams, used either singly or in combination at sub-MIC levels, indicates that PBP 1 is essential for cell growth. On the basis of predicted binding patterns of β-lactams to PBPs, the saturation of PBPs 2 and 3 or PBPs 2 and 4 was also considered to have deleterious effects, but the amounts of active PBP 1 were severely reduced under these conditions, and growth inhibition probably resulted from the indirect effect of the β-lactams on PBP 1. This protein is present in *S. aureus* H/28, in which it had been reported to be absent, and it is the primary target of imipenem in sensitive strains of *S. aureus*. PBP 1 of a cephradine-resistant strain of *S. aureus* has a lower affinity for β-lactam antibiotics than the same PBP of sensitive strains; this property is believed to render the organism less susceptible to those combinations of β-lactams whose action on sensitive strains is explicable in terms of a lowering of the amount of active PBP 1.

Although these experiments emphasize the importance of PBP 1, they do not completely rule out roles for the remaining PBPs as suggested by others (7). If PBPs 2, 3, and 4 have important functions, only a small proportion of the total population of the PBP is necessary. The question as to what proportion of a PBP is necessary for unimpaired cell growth has not been resolved, though a recent report suggested that only newly synthesized molecules were active (17). The current series of experiments does not distinguish between old and newly synthesized PBPs. A definitive answer to the roles of PBPs 2 and 3 will be obtained only in studies of mutants in which at least part of the genes encoding them have been deleted.

ACKNOWLEDGMENTS. The skillful technical assistance of Alison Jenner is gratefully acknowledged. I am grateful to P. J. Cassidy, Merck Sharp & Dohme Ltd., for generous gifts of [^{3}H]benzylpenicillin.

LITERATURE CITED

1. **Brown, D. F. J., and P. E. Reynolds.** 1980. Intrinsic resistance to β-lactam antibiotics in *Staphylococcus aureus. FEBS Lett.* **122:**275–278.

2. **Curtis, N. A. C., and M. V. Hayes.** 1981. A mutant of *Staphylococcus aureus* H deficient in penicillin-binding protein 1 is viable. *FEMS Microbiol. Lett.* **10:**227–229.

3. **Curtis, N. A. C., M. V. Hayes, A. W. Wyke, and J. B. Ward.** 1980. A mutant of *Staphylococcus aureus* H lacking penicillin-binding protein 4 and transpeptidase activity *in vitro. FEMS Microbiol. Lett.* **9:**263–266.

4. **De Pedro, M. A., U. Schwarz, Y. Nishimura, and Y. Hirota.** 1980. On the biological role of penicillin-binding proteins 4 and 5. *FEMS Microbiol. Lett.* **9:**219–221.

5. **Frère, J.-M., and B. Joris.** 1985. Penicillin-sensitive enzymes in peptidoglycan biosynthesis. *Crit. Rev. Microbiol.* **11:**299–396.

6. **Gale, E. F., E. Cundliffe, P. E. Reynolds, M. H. Richmond, and M. J. Waring.** 1981. *The Molecular Basis of Antibiotic Action,* p. 49–174. John Wiley & Sons, Inc., London.

7. **Georgopapadakou, N. H., B. A. Dix, and Y. R. Mauriz.** 1986. Possible physiological functions of penicillin-binding proteins in *Staphylococcus aureus. Antimicrob. Agents Chemother.* **29:**333–336.

8. **Georgopapadakou, N. H., S. A. Smith, and D. P. Bonner.** 1982. Penicillin-binding proteins in a *Staphylococcus aureus* strain resistant to specific β-lactam antibiotics. *Antimicrob. Agents Chemother.* **22:**172–175.

9. **Hartman, B. J., and A. Tomasz.** 1984. Low-affinity penicillin-binding protein associated with β-lactam resistance in *Staphylococcus aureus. J. Bacteriol.* **158:**513–516.

10. **Kozarich, J. W., and J. L. Strominger.** 1978. A membrane enzyme from *Staphylococcus aureus* which catalyses transpeptidase, carboxypeptidase and penicillinase activities. *J. Biol. Chem.* **253:**1272–1278.

11. **Nakagawa, J., S. Tamaki, S. Tomioka, and M. Matsuhashi.** 1984. Functional biosynthesis of cell wall peptidoglycan by polymorphic bifunctional polypeptides. *J. Biol. Chem.* **259:**13937–13946.

12. **Reynolds, P. E.** 1983. Penicillin-binding proteins and peptidoglycan biosynthesis in *Bacillus megaterium,* p. 517–522. *In* R. Hakenbeck, J.-V. Höltje, and H. Labischinski (ed.), *The Target of Penicillin.* Walter de Gruyter, Berlin.

13. **Reynolds, P. E.** 1984. Resistance of the antibiotic target site. *Br. Med. Bull.* **40:**3–10.

14. **Reynolds, P. E., and D. F. J. Brown.** 1985. Penicillin-binding proteins of β-lactam-resistant strains of *Staphylococcus aureus. FEBS Lett.* **192:**28–32.

15. **Reynolds, P. E., S. T. Shepherd, and H. A. Chase.** 1978. Identification of the binding protein which may be the target of penicillin action in *Bacillus megaterium. Nature* (London) **271:**568–570.

16. **Rossi, L., E. Tonin, Y. R. Cheng, and R. Fontana.** 1985. Regulation of penicillin-binding protein activity: description of a methicillin-inducible penicillin-binding protein in *Staphylococcus aureus. Antimicrob. Agents Chemother.* **27:**828–831.

17. **Tuomanen, E.** 1986. Newly made enzymes determine ongoing cell wall synthesis and the antibacterial effects of cell wall synthesis inhibitors. *J. Bacteriol.* **167:**535–543.

18. **Ubukata, K., N. Yamashita, and M. Konno.** 1985. Occurrence of a β-lactam-inducible penicillin-binding protein in methicillin-resistant staphylococci. *Antimicrob. Agents Chemother.* **27:**851–857.

19. **Wyke, A. W., J. B. Ward, M. V. Hayes, and N. A. C. Curtis.** 1981. A role *in vivo* for penicillin-binding protein 4 of *Staphylococcus aureus. Eur. J. Biochem.* **119:**389–393.

Chapter 34

Primary Structure and Origin of the Gene Encoding the β-Lactam-Inducible Penicillin-Binding Protein Responsible for Methicillin Resistance in *Staphylococcus aureus*

Min Dong Song
Shigefumi Maesaki
Masaaki Wachi
Tomohiro Takahashi
Masaki Doi
Fumitoshi Ishino
Yoshimi Maeda
Kenji Okonogi
Akira Imada
Michio Matsuhashi

The occurrence of highly β-lactam-resistant strains of *Staphylococcus aureus* (methicillin-resistant *S. aureus*) is becoming a very serious problem in clinical microbiology. These strains always contain a new penicillin-binding protein (PBP) called *S. aureus* PBP 2′, which is characterized by its low binding af-

Min Dong Song, Shigefumi Maesaki, Masaaki Wachi, Tomohiro Takahashi, Masaki Doi, Fumitoshi Ishino, Yoshimi Maeda, and Michio Matsuhashi • Institute of Applied Microbiology, University of Tokyo, Bunkyo-ku, Tokyo 113, Japan. Kenji Okonogi and Akira Imada • Biology Laboratories, Takeda Chemical Industries, Ltd., Osaka, Japan.

finity to β-lactams and its induction by β-lactams (2, 9, 11, 12). The high β-lactam resistance of the methicillin-resistant *S. aureus* strains has therefore been attributed to the presence of this peculiar PBP. Cloning of the gene that encodes for *S. aureus* PBP 2′ has been accomplished by taking advantage of its close linkage on the chromosome with resistance to the aminoglycoside tobramycin (6).

In this chapter we describe our knowledge of the structure of the *S. aureus* PBP 2′ gene and its origin obtained from DNA sequencing (9a) and DNA hybridization experiments.

The Nucleotide Sequence of the Regulatory Region

The nucleotide sequence of the putative regulatory region of the gene encoding the methicillin-resistant *S. aureus* PBP, cloned from strain TK784 (6), is shown in Fig. 1 and is compared with the nucleotide sequence of a classical penicillinase gene of *S. aureus* (7). This region contained promoter and ribosome-binding sequences on both DNA strands and had an additional open reading frame or frames on the opposite DNA strand.

A palindromic sequence was found in the genes of both *S. aureus* PBP 2′ and the staphylococcal penicillinase. This sequence is most likely the binding site of a common repressor (7). Like most β-lactamase genes that have been studied, the upstream region of the coding frame for PBP 2′ also contained promoter- and Shine-Dalgarno-like sequences on the opposite DNA strand, which is followed by another open reading frame. The presence of an open reading frame or frames on the opposite DNA strand has been found in many β-lactamase genes and is thought to be involved in the mechanism of induction of these enzymes (3, 7). A striking similarity was found in the sequence of about 100 nucleotides in the putative regulatory region between the genes encoding *S. aureus* PBP 2′ and the staphylococcal penicillinase (Fig. 1).

The Coding Frame and Deduced Amino Acid Sequence

The amino acid sequence of *S. aureus* PBP 2′ deduced from the DNA sequence (9a) is shown in Fig. 2. Two regions that contained the amino acid sequence S-X-X-K (in the single-letter notation, X being any amino acid) showed a high similarity with penicillin-binding peptide sequences of penicillinases and other PBPs. The first of such sequences is located around the serine 25 residue and was similar to the amino acid sequence around the penicillin-binding residue, serine 63, of the classical staphylococcal penicillinase (Y-A-S-T-S-K-A-I-N-S-A-I), and the other was around serine 405, which was similar to the penicillin-binding peptide sequences of *Escherichia coli* PBP 2 (P-A-S-T-V-

Figure 1. Nucleotide sequence around the proposed regulatory region of the *S. aureus* PBP 2' gene from *S. aureus* TK784 (6). A correlating sequence for a classical *S. aureus* penicillinase (7) is also shown. Palindrome sequence (—) and identical residues (*) are indicated. Open reading frames on both strands of the DNAs are indicated by shaded letters, and coded amino acid sequences are designated by the single-letter notation. Putative −35 and −10 promoter sequences and Shine-Dalgarno (SD) sequences are also shown in both strands of DNA.

Figure 2. Peptide sequence of *S. aureus* PBP 2' (strain TK784) deduced from the DNA sequence of the gene. Two serine residues with neighboring amino acids similar to the penicillin-binding site of staphylococcal penicillinase and *E. coli* PBPs are indicated by arrowheads.

K, containing serine 330) or *E. coli* PBP 3 (P-G-S-T-V-K, containing serine 307). It is possible that the minor amino-terminal portion of *S. aureus* PBP 2' encompassing serine 25 may be derived from a penicillinase, and the major carboxyl-terminal portion of the protein encompassing serine 405 may be from a PBP of another bacterium; thus, *S. aureus* PBP 2' itself may be a hybrid protein. It is also possible that the region encompassing serine 25 is part of the signal sequence of *S. aureus* PBP 2' since this region shows some homology to other signal sequences. We have not yet tested this possibility.

Similarities of Amino Acid Sequence with Those of Penicillinase and PBPs of E. coli

Similarities of the amino acid sequence were confirmed by use of a Harr plot program. Figure 3 shows the homology pattern plotted with the amino acid sequence of *S. aureus* PBP 2' on the *x* axis and the sequences of the classical staphylococcal penicillinase (7) (Fig. 3A) or *E. coli* PBPs 2, 3, 1A, and 1B (Fig. 3B–E; see reference 1) plotted on the *y* axis. In this graph, nine identical amino acids in any sequence of 25 amino acid residues are shown.

A sequence of amino acids at the N-terminal portion of the staphylococcal penicillinase appeared repeatedly along the polypeptide chain of *S. aureus* PBP 2'. It is unknown whether this repeated motif that appears in the homology patterns relates to the mechanism of evolution of the *S. aureus* PBP 2' gene or to some additional events.

The greatest similarities with *S. aureus* PBP 2', however, were seen in *E. coli* PBP 2, followed by PBP 3. *E. coli* PBPs 1A and 1B displayed very poor homology with *S. aureus* PBP 2'. These results may suggest that the major portion of the molecule of *S. aureus* PBP 2' has derived from the gene of a

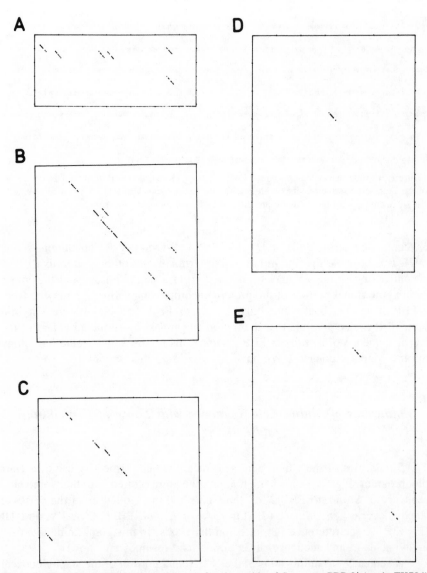

Figure 3. Harr plots indicating resemblance of amino acids of *S. aureus* PBP 2' (strain TK784) to those of staphylococcal penicillinase (A) and *E. coli* PBP 2 (B), PBP 3 (C), PBP 1A (D), and PBP 1B (E).

PBP that is related to *E. coli* PBP 2 or PBP 3. *E. coli* PBP 2 is thought to be a peptidoglycan synthetase in the formation of an initiation piece of peptido-glycan in cell elongation (5), and PBP 3 is a septum-forming peptidoglycan synthetase (4). It is likely that *S. aureus* PBP 2' may play a role in cell duplication similar to that of *E. coli* PBP 2 or 3. To determine from which organism the respective parts of *S. aureus* PBP 2′ may have been derived, we surveyed the homology of the PBP 2′ gene with DNA fragments from several β-lactam-sensitive and β-lactam-resistant strains of *S. aureus*, as well as other bacteria, by Southern blot hybridization (10).

S. aureus PBP 2′ from a Unique Source

The *S. aureus* PBP 2′ gene did not hybridize with DNA from *E. coli* or normal *S. aureus*. Thus, it is likely that the genes originated from other sources. We then compared DNA fragments from methicillin-resistant strains of *S. aureus* isolated in Japan and other countries (data not shown) by DNA hybridization (Southern blot).

As a probe, the 600-base-pair *Hinc*II-*Sau*3A1 fragment, containing a 250-base-pair portion of the inverted open reading frame, the promoter region, and a 243-base-pair portion of the coding frame from Met-1 to Asp-82, was used (Fig. 4). All of the 20 methicillin-resistant *S. aureus* strains tested (MIC of methicillin varying from 6.25 to 1,600 μg/ml) contained a 4-kilobase *Hind*III fragment that hybridized to the probe, but this fragment was absent from methicillin-sensitive *S. aureus* strains. The presence or absence of penicillinase activity had no effect on the appearance of this fragment. Also, β-lactam-resistant strains of *Enterococcus faecalis* (MIC of methicillin from 25 to 1,600 μg/ml) did not contain a DNA fragment that hybridized to the probe. Experiments with probes containing the major coding region of the PBP gene gave similar results.

These results suggest that the *S. aureus* PBP 2′ gene and its neighboring region are either identical or very similar in all of the methicillin-resistant *S. aureus* strains tested. More extensive work is required to determine whether the PBP 2′ gene from all methicillin-resistant *S. aureus* strains is unique and whether it has been formed from the same origin. Our recent results, however, indicate that, at least in Japan and in most other countries, the origin of *S. aureus* PBP 2′ is unique (unpublished data). These results are in accord with the previous observation of Reynolds and Fuller (8), who suggested the presence of an identical additional PBP in several strains of methicillin-resistant *S. aureus* from the United Kingdom and other European countries.

Conclusion

The nucleotide sequence of the gene and the deduced amino acid sequence of the β-lactam-inducible PBP of methicillin-resistant *S. aureus* (a PBP thought to cause the high β-lactam resistance in this bacterium) have been presented

kb

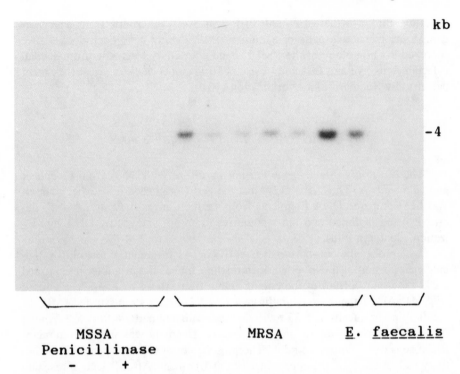

−4

MSSA   MRSA   E. faecalis
Penicillinase
    −   +

Figure 4. Southern blot hybridization indicating the presence of 4-kilobase *Hind*III fragments hybridizable with the PBP 2′ gene (strain TK784) in DNA of various methicillin-resistant *S. aureus* strains. For the probe, see the text. An autoradiogram is shown. The four methicillin-sensitive *S. aureus* (MSSA) strains tested were FDA209P (*bla*), N169 (*bla*), N122, and N284, and the seven methicillin-resistant *S. aureus* (MRSA) strains were N315, N56, SK189, SM147 (*bla*), SM230, N193, and N200, isolated in Japan. The two *E. faecalis* strains were TN1972 and TN1996.

and compared with those of staphylococcal penicillinase and *E. coli* PBPs. These comparisons and the results of Southern blot hybridization indicate that this PBP may have been derived by recombination of a penicillinase gene and a PBP gene. All of the β-lactam-resistant staphylococcal strains tested possessed a 4-kilobase *Hind*III DNA fragment that could hybridize with the unique PBP gene, suggesting their origin from a unique source.

ACKNOWLEDGMENTS. This work was partially supported by Grant-in-Aid for Scientific Research 61470157 to M.M. from the Ministry of Education, Science and Culture of Japan.

We thank Rob A. Nicholas, Department of Biochemistry and Molecular Biology at Harvard University, for critical reading of the manuscript.

LITERATURE CITED

1. **Asoh, S., H. Matsuzawa, F. Ishino, J. L. Strominger, M. Matsuhashi, and T. Ohta.** 1986. Nucleotide sequence of the *pbpA* gene and characteristics of the deduced amino acid sequence of penicillin-binding protein 2 of *Escherichia coli* K12. *Eur. J. Biochem.* **160:**231–238.
2. **Hartman, B. J., and A. Tomasz.** 1984. Low-affinity penicillin-binding protein associated with β-lactam resistance in *Staphylococcus aureus. J. Bacteriol.* **158:**513–516.
3. **Himeno, T., T. Imanaka, and S. Aiba.** 1986. Nucleotide sequence of the penicillinase repressor gene *penI* of *Bacillus licheniformis* and regulation of *penP* and *penI* by the repressor. *J. Bacteriol.* **168:**1128–1132.
4. **Ishino, F., and M. Matsuhashi.** 1981. Peptidoglycan synthetic enzyme activities of highly purified penicillin-binding protein 3 in *Escherichia coli*: a septum-forming reaction sequence. *Biochem. Biophys. Res. Commun.* **101:**905–911.
5. **Ishino, F., W. Park, S. Tomioka, S. Tamaki, I. Takase, K. Kunugita, H. Matsuzawa, S. Asoh, T. Ohta, B. G. Spratt, and M. Matsuhashi.** 1986. Peptidoglycan synthetic activities in membranes of *Escherichia coli* caused by overproduction of penicillin-binding protein 2 and RodA protein. *J. Biol. Chem.* **261:**7024–7031.
6. **Matsuhashi, M., M. D. Song, F. Ishino, M. Wachi, M. Doi, M. Inoue, K. Ubukata, N. Yamashita, and M. Konno.** 1986. Molecular cloning of the gene of a penicillin-binding protein supposed to cause high resistance to β-lactam antibiotics in *Staphylococcus aureus. J. Bacteriol.* **167:**975–980.
7. **McLaughlin, J. R., C. L. Murray, and J. C. Rabinowitz.** 1981. Unique features in the ribosome binding site sequence of the gram-positive *Staphylococcus aureus* β-lactamase gene. *J. Biol. Chem.* **256:**11283–11291.
8. **Reynolds, P. E., and C. Fuller.** 1986. Methicillin-resistant strains of *Staphylococcus aureus*; presence of identical additional penicillin-binding protein in all strains examined. *FEMS Microbiol. Lett.* **33:**251–254.
9. **Rossi, L., E. Tonin, Y. R. Cheng, and R. Fontana.** 1985. Regulation of penicillin-binding protein activity: description of a methicillin-inducible penicillin-binding protein in *Staphylococcus aureus. Antimicrob. Agents Chemother.* **27:**828–831.
9a. **Song, M. D., M. Wachi, M. Doi, F. Ishino, and M. Matsuhashi.** 1987. Evolution of an inducible penicillin target protein in methicillin-resistant *Staphylococcus aureus* by gene fusion. *FEBS Lett.* **221:**167–171.
10. **Southern, E. M.** 1975. Detection of specific sequences among DNA fragments separated by gel electrophoresis. *J. Mol. Biol.* **98:**503–517.
11. **Ubukata, K., N. Yamashita, and M. Konno.** 1985. Occurrence of a β-lactam-inducible penicillin-binding protein in methicillin-resistant staphylococci. *Antimicrob. Agents Chemother.* **27:**851–857.
12. **Utsui, Y., and T. Yokota.** 1985. Role of an altered penicillin-binding protein in methicillin- and cephem-resistant *Staphylococcus aureus. Antimicrob. Agents Chemother.* **28:**397–403.

Chapter 35

Role of the Penicillin-Binding Proteins of *Staphylococcus aureus* in the Induction of Bacteriolysis by β-Lactam Antibiotics

Frank Beise
Harald Labischinski
Peter Giesbrecht

Little is known about the physiological role of the high-molecular-weight penicillin-binding proteins (PBPs) of *Staphylococcus aureus,* although they are considered to be involved in the bacteriolytic effects of β-lactam antibiotics (2, 9, 10). Unfortunately, the various PBPs are usually inhibited at about the same concentration of any given antibiotic. We report here on a set of β-lactam antibiotics which exert preferential binding to each single PBP. Effects on growth were monitored by continuously measuring both the optical density and the heat production (catabolic activities, determined by thermography) of the staphylococci during drug treatment. The concentration of each selectively acting antibiotic needed to block an individual PBP was well below its MIC. This is in accordance with the assumption that at least two staphylococcal PBPs have to be blocked to induce a bacteriolytic effect. But only a limited number of combinations of the drugs caused bacteriolysis which did not, however, involve the previously proposed lethal combination of an inhibition of PBPs 2 and 3. Instead PBP 1 had to be blocked together with PBP 2, PBP 3, or both to induce bacteriolysis.

To examine the role of PBPs, we used *S. aureus* SG511 "Berlin" from

Frank Beise, Harald Labischinski, and Peter Giesbrecht • Robert Koch Institute of the Federal Health Office, D-1000 Berlin 65, Federal Republic of Germany.

the culture collection of the Robert Koch Institute. Bacteria grown overnight in a culture medium containing 2.5% peptone (Difco Laboratories, Detroit, Mich.) and 0.5% NaCl were added to prewarmed, fresh medium to establish an optical density of 0.15 to 0.2. This culture was allowed to grow at 37°C under aeration, up to an optical density of 1.0, when it served to inoculate the main culture.

Detection of PBPs

To test binding of antibiotics to staphylococcal PBPs, 5 ml of the main culture (optical density, 0.2) was pipetted into centrifugation tubes containing the appropriate amount of the drug(s). After 60 min of growth at 37°C, cell walls were removed by the addition of 20 μg of lysostaphin (Sigma, St. Louis, Mo.) per ml. Additionally, DNase (Serva, Heidelberg, Federal Republic of Germany) and phenylmethanesulfonyl fluoride (to prevent degradation of proteins; Fluka, Buchs, Switzerland) were applied, to give concentrations of 5 μg/ml and 1 mM, respectively. The then weakly opaque solution was rapidly cooled down and subjected to ultracentrifugation in an L2-65B ultracentrifuge (Beckman Instruments Inc., Stanford, Calif.) with a Ti 50 rotor (90 min, 140,000 × g, 4°C). The resulting membrane pellet was suspended by means of a Braunsonic sonicator (Braun, Melsungen, Federal Republic of Germany) in 20 μl of buffer consisting of 20 mM Tris hydrochloride (pH 7.4) (Tris base from Fluka), 20 mM NaCl, 5 mM 2-mercaptoethanol, 1 mM phenylmethanesulfonyl fluoride, 0.2% Triton X-100 (Serva), and 10% glycerol. Ten microliters of the resulting membrane suspension was then incubated with 10 μl of [*phenyl*-4-³H]benzylpenicillin (37 kBq/μl, 999 GBq/mmol; Amersham-Buchler, Braunschweig, Federal Republic of Germany) for 15 min. The reaction was terminated by the addition of 25 μl of a 0.1 M solution of cold penicillin G. For purification the proteins were precipitated by addition of a fourfold excess of ice-cold acetone and sedimented by centrifugation (30 min, 4°C, 15,000 rpm).

The protein pellet was redissolved in a sodium dodecyl sulfate (SDS)-containing sample buffer and subjected to SDS-polyacrylamide gel electrophoresis as described previously (7). The resolving gel (10% acrylamide, 0.13% bisacrylamide [see reference 6]) was about 18 cm in length. The PBPs were detected by fluorography as previously described (1). Quantification of the band intensity was performed by gravimetry of the peaks resulting from densitometry of the fluorograms with a Joyce-Loebl Mk III microdensitometer (Joyce-Loebl Co., Gateshead, United Kingdom).

Three β-lactam antibiotics which expressed selective binding to individual PBPs of *S. aureus* at subinhibitory doses (Table 1) have been found. Mecillinam, at a concentration of 2 μg/ml, bound almost specifically to PBP 3, whereas cefotaxime showed a strong preference for PBP 2. At 0.08 μg of this

Table 1. Selectively acting β-lactam antibiotics

Affected PBP	Antibiotic	Optimal concn for selective binding (μg/ml)	MIC (μg/ml)
PBP 1	Imipenem	0.003 or 0.006	0.01
PBP 2	Cefotaxime	0.08	0.5
PBP 3	Mecillinam	2.00	10.0
PBP 4	Cefoxitin[a]	0.05	1.0

[a]The conditions for the inhibition of PBP 4 were as described by Wyke et al. (11).

drug per ml, PBP 2 was blocked by about 80%, whereas PBP 1 was inhibited only to a minor degree (about one-third [Fig. 1]). Imipenem was found to bind preferentially to PBP 1. With 0.003 μg of imipenem per ml, only PBP 1 was affected by about 80%. At a concentration of 0.006 μg of imipenem per ml, this particular PBP was completely inhibited, but PBP 3 was blocked by about 50% too.

Growth Behavior of Staphylococci with Selectively Inhibited PBPs

Optical Density (Discontinuous)

Figure 2 shows the resulting graphs for the optical density of *S. aureus* treated with the antibiotics discussed above. Experiments were performed by cultivating the main culture until an optical density of 0.2 was reached. At this point the drug treatment started. The bacterial suspension was diluted six times before the turbidity at 578 nm was determined, to make good use of the optimal range (optical density, <1.0) of the photometer. Figure 2a represents the treatment with a single drug, aiming at an inhibition of one individual PBP. Cefoxitin at 0.05 μg/ml was used to inhibit PBP 4, as described previously (11).

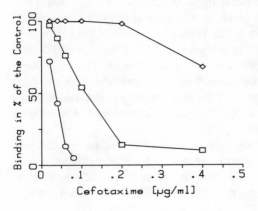

Figure 1. Saturation of staphylococcal PBPs by cefotaxime. Quantification was achieved by scanning the fluorograms obtained from gel electrophoresis of [³H]benzylpenicillin-labeled PBPs by means of a densitometer. Peak intensity was estimated by gravimetry and is expressed as a percentage of the particular peak intensity of the control. At 0.08 μg/ml, PBP 2 (○) was almost saturated with this drug, whereas PBP 1 (□) was affected to only a minor extent. PBP 3 (◇) retained full binding capacity.

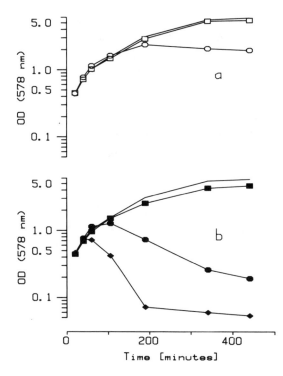

Figure 2. Growth curves of staphylococci with PBPs selectively inhibited by a single drug (a) or by drug combinations (b). Curves without symbols represent controls. (a) Use of 0.08 μg of cefotaxime per ml (□) to inhibit PBP 2 did not result in any significant deviation in growth, as revealed by optical density (OD), in contrast to the result with 0.006 μg of imipenem per ml (○). (b) Combined action of the selective β-lactam antibiotics. (●) Inhibition of PBPs 1 and 2 by 0.003 μg of imipenem and 0.08 μg of cefotaxime per ml; (■) inhibition of PBPs 2 and 3 by 0.08 μg of cefotaxime and 2 μg of mecillinam per ml; (◆) inhibition of all three high-molecular-weight PBPs with 0.006 μg of imipenem, 0.08 μg of cefotaxime, and 2 μg of mecillinam per ml.

With the exception of growth in 0.006 μg of imipenem per ml, all cultures grew apparently like the untreated control culture. The consequences of the application of drug combinations aimed at inhibiting two or three of the high-molecular-weight PBPs simultaneously are shown in Fig. 2b. Only combinations of antibiotics which inhibited PBP 1 along with PBP 2 or 3, or both, induced bacteriolysis. The combination of 0.003 μg of imipenem per ml plus 0.08 μg of cefotaxime per ml (i.e., inhibition of PBPs 1 and 2) induced the same bacteriolytic effect as 0.006 μg of imipenem per ml plus 2 μg of mecillinam per ml (inhibition of PBPs 1 and 3; data not shown). The combination (per ml) of 0.006 μg of imipenem, 0.08 μg of cefotaxime, and 2 μg of mecillinam, saturating all three high-molecular-weight PBPs, resulted in growth curves identical to those obtained by treatment with penicillin G at the optimal bacteriolytic dose (0.1 μg/ml) for this staphylococcal species (Fig. 2b).

Heat Production

The above-mentioned experiments were supplemented by a microcalorimetric examination of the heat production of the bacterial cultures. This method allows a certain insight into the bacterial physiology in terms of the heat (i.e., enthalpy) produced by catabolic processes. Continuous measurements of the optical density were performed with a Zeiss PM6 spectrophotometer (Zeiss, Oberkochen, Federal Republic of Germany) equipped with a flowthrough cell; the flow rate was 40 ml/min. With an optical density of 0.1 of the main culture, the drug treatment started.

Heat production of the bacteria was determined with an LKB 2107 microcalorimetry system (LKB, Bromma, Sweden) simultaneously with the measurement of the optical density; the flow rate was about 40 ml/min. Figure 3a shows the heat production and optical density of untreated staphylococci (control), monitored continuously and simultaneously. The same result was obtained after the application of 0.08 μg of cefotaxime per ml affecting PBP 2 (Fig. 3b). Staphylococci treated with 0.003 μg of imipenem per ml or 2 μg of mecillinam per ml reacted in the same way (data not shown). Again, 0.006 μg of imipenem per ml showed significant deviations, compared with the con-

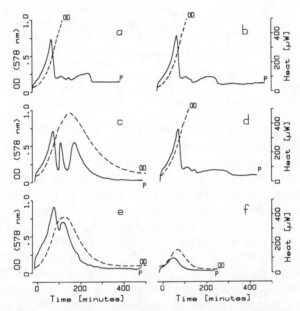

Figure 3. Growth of *S. aureus* in the presence of selective β-lactam antibiotics. Growth was monitored by measurement of the heat production (P in microwatts, ———) and the optical density (OD at 578 nm, – – –) of the cultures. Drugs were added at an OD of 0.1 (0 min). (a) A typical graph for an untreated control culture. (b) Effect of inhibition of a single PBP (PBP 2 by cefotaxime) on bacterial catabolism. (c) Effect of 0.006 μg of imipenem per ml. (d) Inhibition of PBPs 2 and 3 by 0.08 μg of cefotaxime and 2 μg of mecillinam per ml. (e) Combined inhibition of PBPs 1 and 2 by 0.003 μg of imipenem and 0.08 μg of cefotaxime; an analogous experiment with 0.006 μg of imipenem plus 2 μg of mecillinam per ml (inhibition of PBPs 1 and 3) gave an almost identical result (data not shown). (f) Inhibition of all three high-molecular-weight PBPs with 0.006 μg of imipenem, 0.08 μg of cefotaxime, and 2 μg of mecillinam.

trol (Fig. 3c). Combined inhibition of PBPs 2 and 3, induced by the application of 0.08 μg of cefotaxime per ml along with 2 μg of mecillinam per ml, did not result in any significant aberration (Fig. 3d). Bacteriolysis as indicated by a decline in heat production could, however, be induced by the addition of 0.003 μg of imipenem per ml plus 0.08 μg of cefotaxime per ml, inhibiting PBPs 1 and 2 (Fig. 3e), or 0.006 μg of imipenem per ml plus 2 μg of mecillinam per ml, inhibiting PBPs 1 and 3 (data not shown). Inhibition of all high-molecular-weight PBPs (see above) resulted once more in a typical penicillin G-like bacteriolysis (Fig. 3f).

Conclusions

This chapter shows that staphylococcal PBPs can be inhibited individually by using certain β-lactam antibiotics. The drugs used in the experiments were imipenem (0.003 or 0.006 μg/ml), cefotaxime (0.08 μg/ml), and mecillinam (2 μg/ml) to inhibit PBPs 1, 2, and 3, respectively. Furthermore, it could be shown that the inhibition of a single PBP did not affect bacterial growth. The seemingly bacteriostatic or weakly bacteriolytic activity of 0.006 μg of imipenem per ml (see Fig. 2a, 3c) is possibly caused by the additional saturation of PBP 3 by about 50%.

As has been concluded from various studies (4, 9, 10), only a combined inhibition of two high-molecular-weight PBPs of *S. aureus* may induce bacteriolysis. We could not, however, confirm the proposed lethal action of an inhibition of PBPs 2 and 3 (4, 9, 10). Only combinations involving PBP 1, hitherto considered to be nonessential (3, 11), induced bacteriolysis. Interestingly, preliminary data on the cross-linking of the staphylococcal cell wall after treatment with these selective β-lactam antibiotics revealed no differences between the simultaneous inhibition of PBPs 1 and 2 (or 3) and that of PBPs 2 and 3. These results indicate a more local event leading to bacteriolysis (5, 8; Labischinski et al., this volume) and not necessarily an event affecting the overall wall composition.

LITERATURE CITED

1. **Bonner, W. M., and R. A. Laskey.** 1974. A film detection method for tritium-labelled proteins and nucleic acids in polyacrylamide gels. *Eur. J. Biochem.* **46:**83–88.
2. **Brown, D. F. J., and P. E. Reynolds.** 1983. Transpeptidation in *Staphylococcus aureus* with intrinsic resistance to β-lactam antibiotics ("methicillin resistance"), p. 537–542. *In* R. Hakenbeck, J. V. Höltje, and H. Labischinski (ed.), *The Target of Penicillin*. Walter de Gruyter, Berlin.
3. **Curtis, N. A. C., and M. V. Hayes.** 1981. A mutant of *Staphylococcus aureus* H deficient in penicillin-binding protein 1 is viable. *FEMS Microbiol. Lett.* **10:**227–229.

4. **Georgopapadakou, N. H., B. A. Dix, and Y. R. Mauriz.** 1986. Possible physiological functions of penicillin-binding proteins in *Staphylococcus aureus*. *Antimicrob. Agents Chemother.* **29:**333–336.
5. **Giesbrecht, P., H. Labischinski, and J. Wecke.** 1985. A special morphogenetic wall defect and the subsequent activity of "murosomes" as the very reason for penicillin-induced bacteriolysis in staphylococci. *Arch. Microbiol.* **141:**315–324.
6. **Hartman, B. J., and A. Tomasz.** 1986. Expression of methicillin resistance in heterogenic strains of *Staphylococcus aureus*. *Antimicrob. Agents Chemother.* **29:**85–92.
7. **Laemmli, U. K., and M. Favre.** 1973. Mutation of the head of bacteriophage T4. I. DNA packing events. *J. Mol. Biol.* **80:**575–599.
8. **Reinicke, B., P. Blümel, H. Labischinski, and P. Giesbrecht.** 1985. Neither an enhancement of autolytic wall degradation nor an inhibition of the incorporation of cell wall material are pre-requisites for penicillin-induced bacteriolysis in staphylococci. *Arch. Microbiol.* **141:**309–314.
9. **Waxman, D. J., and J. L. Strominger.** 1983. Penicillin-binding proteins and the mechanism of action of β-lactam antibiotics. *Annu. Rev. Biochem.* **52:**825–869.
10. **Wyke, A. W., J. B. Ward, and M. V. Hayes.** 1983. Penicillin-sensitive enzymes in *Staphylococcus aureus*, p. 543–548. *In* R. Hakenbeck, J. V. Höltje, and H. Labischinski (ed.), *The Target of Penicillin.* Walter de Gruyter, Berlin.
11. **Wyke, A. W., J. B. Ward, M. V. Hayes, and N. A. C. Curtis.** 1981. A role *in vivo* for penicillin-binding protein 4 of *Staphylococcus aureus*. *Eur. J. Biochem.* **119:**389–393.

Chapter 36

Characterization of the Trypsin-Solubilized Penicillin-Binding Proteins of *Enterococcus hirae* (*Streptococcus faecium*)

Aboubaker El Kharroubi
Philippe Jacques
Graziella Piras
Jacques Coyette
Jean-Marie Ghuysen

Most penicillin-binding proteins (PBPs) behave as membrane-bound proteins which can be solubilized only by detergents or chaotropic agents, or both, and by proteolytic enzymes. When trypsin is used, several PBPs are cleaved off the membranes and lose a small terminal peptide mediating the anchorage of the proteins in the membranes. Usually, trypsin degrades the large soluble polypeptides released into smaller fragments, designated below as tPBPs, some of which are relatively stable and still active (i.e., able to hydrolyze a substrate or to bind β-lactams, or both) (2, 10).

Enterococcus hirae (*Streptococcus faecium*) strains usually possess seven PBPs ranging in size from 140 to 43 kilodaltons (kDa) (4). PBPs 2 and 3 are involved in cellular division (5). PBP 5 is a low-affinity PBP implicated in the β-lactam resistance of the *E. hirae* strains (7). PBP 6 is the only DD-carboxypeptidase activity measurable in vitro in the membranes (3).

Aboubaker El Kharroubi, Philippe Jacques, Graziella Piras, Jacques Coyette, and Jean-Marie Ghuysen • Service de Microbiologie, Université de Liège, Institut de Chimie, B6, B-4000 Sart Tilman (Liège 1), Belgium

In the work presented below, we determined precise experimental conditions for solubilizing, as much as possible specifically, the different PBPs from *E. hirae* membranes, and we identified the different tPBPs released.

PBP Degradation

PBP 4

PBP 4 was shown to be spontaneously released, probably by an endogenous protease, when *E. hirae* ATCC 9790 membranes were incubated at 37°C. The soluble active polypeptide isolated, PBP 4*, was 7 kDa smaller than the membrane protein (4).

Low concentrations of trypsin (given as percentage [wt/wt]) promoted the release of tPBP 4a, which was indistinguishable from PBP 4* (identical β-lactam affinities and behavior on sodium dodecyl sulfate-polyacrylamide gel electrophoresis [SDS-PAGE]) (Fig. 1 and 2). More than 90% of the membrane PBP 4 was solubilized as tPBP 4a, which remained stable in the presence of trypsin as long as the protease concentration and the incubation time did not exceed 0.5 to 1% and 30 min, respectively. Above these limits, tPBP 4a was degraded into smaller, yet unidentified cleavage products.

PBP 5

Two batches of membranes, isolated from the resistant mutant strain R40, which contains large amounts of PBP 5 (7), were labeled differently. Under one set of conditions all the PBPs, including PBP 5, were labeled; under the other, only PBP 5 and a small fraction of PBP 6 were labeled. Comparative analysis of these differently labeled membranes allowed the identification of PBP 5 trypsin degradation products (Fig. 1).

In the presence of trypsin, PBP 5 was cleaved into four different polypeptides designated as tPBPs 5a, 5b, 5c, and 5d (Fig. 1). All were soluble and still able to bind β-lactams. Cleavage kinetics showed that tPBP 5a was the first product released. tPBP 5a was about 7 kDa smaller than PBP 5 and was then degraded into PBPs 5b and 5c. Polypeptide 5d, appearing last, was the smallest "active" product. It was stable under the experimental conditions used, but if the trypsin treatment was increased too much, it was degraded into much smaller, yet unidentified peptides.

The optimal conditions for a maximal yield of tPBP 5a depended not only on the protease concentration and the incubation duration but also on the presence of glycerol at a concentration above 1% and on the membrane concen-

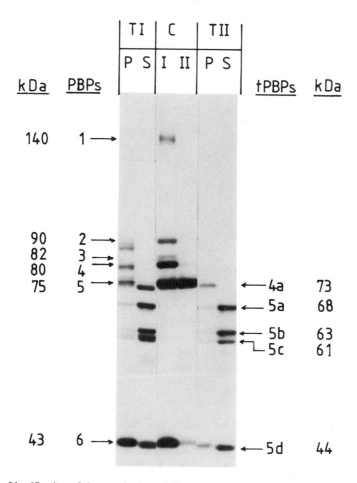

Figure 1. Identification of the trypsin degradation products of PBPs 4 and 5. Two batches of prelabeled R40 resistant mutant membranes were treated for 30 min at 37°C with 2% trypsin. Prelabeling was realized as follows: batch TI was labeled with 10^{-4} M [^{3}H]benzylpenicillin for 1 h; batch TII was preincubated with 1.5×10^{-5} M nonradioactive penicillin for 10 min and then saturated with 10^{-4} M [^{3}H]benzylpenicillin for 1 h. Control samples from each batch (CI and CII) were taken just before trypsin treatment. After proteolysis, supernatant (S) and pellet (P) fractions from a centrifugation (40,000 × *g*, 30 min) were collected and analyzed by SDS-PAGE followed by fluorography. The separating gel was 7.2% acrylamide.

tration, which should not exceed 10 mg/ml. On these bases, about 75, 3, and 4.5% of tPBPs 5a, 5b, and 5c, respectively, were released in the presence of 0.25% trypsin after 5 min of incubation at 37°C. About 18% of PBP 5 remained in the membranes. After 10 min of incubation, about 38% of total PBP 5 was already cleaved into the three smaller products.

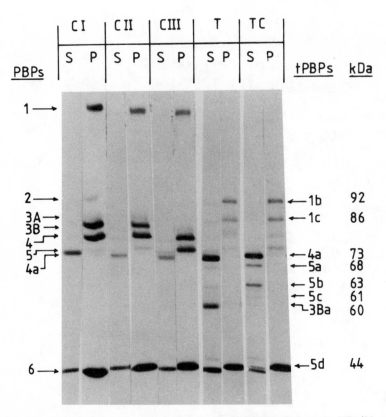

Figure 2. Identification of the trypsin degradation products of PBP 3B from the NT1/20 mutant membranes. Two membrane samples (150 µg/15 µl) were incubated with 2×10^{-5} M benzyl[^{35}S]penicillin for 20 min. One was kept as a control (CII). The other (T) was treated with 1% trypsin for 10 min. Proteolysis was stopped by the addition of a $10 \times$ excess of trypsin soybean inhibitor. Two other membrane samples were incubated with 5×10^{-6} M cefotaxime for 10 min and then with 10^{-4} M benzyl[^{35}S]penicillin for 30 min. One was kept as a control (CIII). The other (TC) was treated with trypsin as described for sample T. Membranes from *E. hirae* ATCC 9790 were used as a control (CI) and labeled as samples CII and T. After incubation in the presence or absence of trypsin, all the samples were centrifuged and separated into supernatant (S) and pellet (P) fractions which were submitted to SDS-PAGE and fluorography. The separating gel was 7.2% acrylamide.

PBP 3B

The degradation of PBP 3B was studied on membranes isolated from the thermosensitive division mutant NT1/20 of *E. hirae* ATCC 9790, which is constitutively devoid of PBP 2 (1). Labeling of NT1/20 membranes with benzyl[^{35}S]penicillin, with or without previous incubation in the presence of a low cefotaxime concentration, allowed us to find the PBP 3B derivatives among the different PBP degradation products released after trypsin treatment (Fig. 2).

The main cleavage product, tPBP 3Ba, still able to bind β-lactams, had a size of 60 kDa. No intermediate-size degradation product was identified. Kinetics of release showed that the optimal yields reached 95% when the proteolysis was done under the following conditions: 1% trypsin for 10 min at 37°C in 1 mM MgCl$_2$–5% glycerol–40 mM phosphate buffer (pH 7). After 15 min of incubation, degradation of PBP 3Ba began. The half-life of the polypeptide under these conditions was about 40 min. Thus tPBP 3Ba was relatively less stable in the presence of trypsin than tPBPs 4a and 5d.

PBP 6 (DD-Carboxypeptidase)

PBP 6 solubilization by trypsin was already described in 1980 (4). The experimental conditions for solubilization were reexamined and improved as described here. On the basis of the results presented above and to avoid the loss of different soluble tPBPs, membranes were treated with 1% trypsin for 10 min at 37°C prior to the trypsin treatment required for DD-carboxypeptidase solubilization. PBP 3B, 4, and 5 derivatives were found in the supernatant of a centrifugation done after the trypsin pretreatment (Fig. 3). They represented, respectively, about 60, 90, and 85% of the untreated membrane proteins. Usually, a small, apparently soluble fraction of intact PBP 6 was also present in this supernatant. Experimental results (not shown here) indicated that this fraction corresponded to micellar or microparticular PBP 6.

High trypsin concentrations induced the formation of two PBP 6 cleavage derivatives. Depending on the experimental conditions used, the largest polypeptide, tPBP 6a (40 kDa), was found after centrifugation either in supernatants or in pellets, in this latter case thus being still associated in some way with the treated membranes. It appeared to be an intermediate in the DD-carboxypeptidase degradation as it never exceeded 10% of the total PBP 6. The second derivative, tPBP 6b (formerly designated as PBP 6*) accumulated during the solubilization as a 30-kDa soluble polypeptide still possessing the enzymatic properties of the membrane PBP 6.

Optimal yields of solubilization of this 30-kDa DD-carboxypeptidase were obtained by using a two-step trypsin treatment (Fig. 3). DD-Carboxypeptidase activity or β-lactam binding capacity was measured in the two supernatants and the residual membrane suspension obtained.

Maximal solubilization reached 90 to 97% of the activity present in the pretreated membranes in analytical as well as preparative experiments.

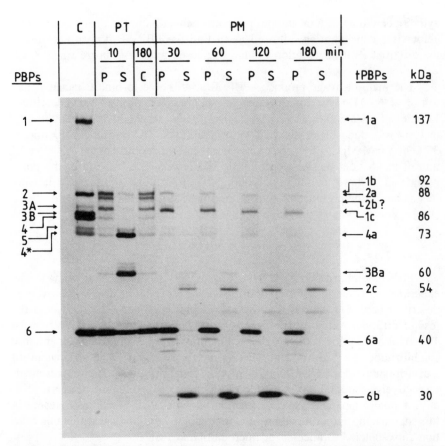

Figure 3. Conversion by trypsin of membrane-bound PBP 6 into water-soluble tPBP 6b. Samples of membranes (130 μg/15 μl), prelabeled with 5×10^{-5} M benzyl[^{35}S]penicillin for 30 min at 37°C, were incubated for 10 min at 37°C with 1% trypsin, centrifuged (40,000 × g, 30 min), and separated in pellets (P) and supernatants (S). The pellets, suspended in 100 mM NH₄HCO₃– 0.1 mM CaCl₂ solution (pH 7.8), were treated at 37°C in the presence of 10% trypsin for increasing times. After the second incubation, the soluble products were collected in supernatants (S) after centrifugation, and the residual membranes (P) were resuspended in the original volume (27 μl). The different fractions were submitted to SDS-PAGE (7.2% acrylamide) and fluorography. C, Control membranes; PT, pretreated membranes; PM, pretreated membranes submitted to the second trypsin treatment; 180C, control pretreated membranes incubated for 180 min in the absence of trypsin.

PBP 2

Present in trypsic hydrolysates of strain ATCC 9790 membranes and absent in similar hydrolysates of strain NT1/20 membranes, two PBP 2 derivatives were identified with certainty. Other minor cleavage products probably derived from PBP 2, but their identifications were not pursued (Fig. 3). One derivative of 88 kDa (2 kDa smaller than the native PBP) remained associated with the membranes and seemed to be an intermediate in the hydrolysis of PBP 2 (0.85 to 1% trypsin, 15 to 30 min). The second tPBP 2, 54 kDa, was solubilized with 5% trypsin after 2 h of incubation (Fig. 3). Optimal conditions for release were not determined, nor was its binding capacity further examined.

PBP 1

No soluble PBP 1 derivatives have been identified yet, but three membrane tPBPs 1 were found. One, tPBP 1a (sometimes seen in untreated membranes; Fig. 3, C), appeared in the presence of low trypsin concentrations (0.1% for 5 to 10 min). It was very unstable and degraded into tPBPs 1b and 1c (92 and 86 kDa, respectively). These membrane tPBPs 1b and 1c were quite stable, as they were still present after 30 min of treatment with 10% trypsin (Fig. 3, PM, pellets).

Both PBPs 1b and 1c were easily identified in NT1/20 trypsin-treated membranes (Fig. 2, T and TC). Optimal conditions for degradation or solubilization, as well as binding capacity, were not further examined.

PBP 3R

Membranes of *E. hirae* S185, isolated from pig intestine and showing about the same β-lactam resistance as the R40 mutant, contained large amounts of a PBP with low β-lactam affinity which had the same apparent size as PBP 3.

Identification of PBP 3R trypsin cleavage products was done as described above for those isolated from PBP 5. Two main soluble polypeptides originated from PBP 3R. Both were different from PBP 3 and PBP 5 derivatives (Fig. 4). tPBPs 3Ra and 3Rb (66.5 and 42 kDa, respectively) were still able to bind benzylpenicillin.

Up to now, no conditions were found under which tPBP 3Ra was released alone. Under the best conditions, tPBPs 3Ra and 3Rb were found in a ratio of 1:1 after 95% degradation of PBP 3R. The membranes were treated with 4% trypsin for 30 min in 1 mM MgCl$_2$–5% glycerol–40 mM phosphate buffer (pH 7). A slight change in the buffer allowed the isolation of tPBP 3Rb alone. About 80% of PBP 3R was degraded when membranes suspended in 50 mM phosphate buffer, pH 7.8, were subjected to 4% trypsin treatment for 30 min. No tPBP 3Ra and almost no PBP 3R were observed. The influence of glycerol on PBP 3R cleavage is under current investigation.

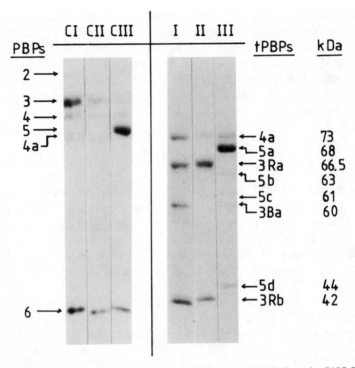

Figure 4. Identification of the trypsin degradation products of PBP 3R from the S185 β-lactam-resistant mutant. Membranes (1 mg/90 μl) were labeled with 5 × 10⁻⁴ M benzyl[¹⁴C]penicillin for 1 h (sample I) or preincubated with 10⁻⁶ M benzylpenicillin for 30 min and then labeled with 3 × 10⁻⁴ M benzyl[¹⁴C]penicillin for 1 h (sample II). Samples were collected after labeling as controls (CI and CII, respectively). The remaining membranes were treated with 4% trypsin for 30 min and then centrifuged (40,000 × *g*, 30 min). The supernatants were collected and submitted, with the control samples, to SDS-PAGE and fluorography. Membranes from the R40 mutant, fully labeled (CIII) as for sample I, were treated with 0.2% trypsin for 5 min. Soluble tPBPs collected after centrifugation (III) were treated like the other samples. The separating gel was 7.2% acrylamide.

Conclusions

To avoid the difficulties of purification of detergent-extracted PBPs from *E. hirae*, we decided to explore the potentialities of membrane proteolysis for the isolation of hydrosoluble, active PBP derivatives. The results presented here clearly show that trypsin proteolysis can be used to isolate such soluble PBP derivatives in good yields. This is certainly the case for PBPs 3B, 4, 5, and 6 from either strain ATCC 9790 or its resistant mutant R40, as well as for PBP 3R from the resistant strain S185. Under the conditions used, the majority of tPBPs were still able to bind β-lactams, some of them even after a 45 to

50% reduction of size. These results are in accordance with those published by others concerning PBPs from different species (6, 8, 9).

Our investigations show in addition that some tPBPs were obtained under such specific conditions that the trypsin lysates contained only the derivatives of one or two PBPs. These lysates are good candidates for a purification of PBP derivatives.

Finally, the PBPs present striking differences in their mode of association with the membranes of *E. hirae*. If PBPs 4 and 5 appear to be anchored via a small terminal peptide, PBPs 2, 3, and probably 6 are inserted into the membranes through a large hydrophobic domain. PBP 1 is the most hydrophobic of them. Indeed, tPBP 1c, the smallest PBP 1 derivative observed, was still associated with trypsin-treated membranes even though it had lost a 50-kDa terminal portion.

ACKNOWLEDGMENTS. This work was supported by the Fonds de Recherche de la Faculté de Médecine, the Fonds de la Recherche Scientifique Médicale, Brussels (contracts 3.4522.86 and 3.4507.83), the Gouvernement belge (action concertée 86/91-90), and the Région wallonne (C2/C16/conv. 246/20428).

LITERATURE CITED

1. **Canepari, P., M. Del Mar Lleo, G. Satta, R. Fontana, G. D. Shockman, and L. Daneo-Moore.** 1983. Division blocks in temperature-sensitive mutants of *Streptococcus faecium* (*S. faecalis* ATCC 9790). *J. Bacteriol.* **156:**1046–1051.
2. **Clément, J. M.** 1983. On the insertion of proteins into membranes. *Biochimie* **65:**325–338.
3. **Coyette, J., J. M. Ghuysen, and R. Fontana.** 1978. Solubilization and isolation of the membrane-bound DD-carboxypeptidase of *Streptococcus faecalis* ATCC9790. *Eur. J. Biochem.* **88:**297–305.
4. **Coyette, J., J. M. Ghuysen, and R. Fontana.** 1980. The penicillin-binding proteins in *Streptococcus faecalis* ATCC9790. *Eur. J. Biochem.* **110:**445–456.
5. **Coyette, J., A. Somzé, J. J. Briquet, and J. M. Ghuysen.** 1983. Function of the penicillin-binding protein 3 in *Streptococcus faecium*, p. 523–530. *In* R. Hakenbeck, J. V. Höltje, and H. Labischinski (ed.), *The Target of Penicillin.* Walter De Gruyter, Berlin.
6. **Ellerbrok, H., and R. Hakenbeck.** 1984. Penicillin-binding proteins of *Streptococcus pneumoniae*: characterization of tryptic-peptides containing the β-lactam-binding site. *Eur. J. Biochem.* **144:**637–641.
7. **Fontana, R., R. Cerini, P. Longoni, A. Grossato, and P. Canepari.** 1983. Identification of a streptococcal penicillin-binding protein that reacts very slowly with penicillin. *J. Bacteriol.* **155:**1343–1350.
8. **Waxman, D. J., and J. L. Strominger.** 1979. Cleavage of a COOH-terminal hydrophobic region from D-alanine carboxypeptidase, a penicillin-sensitive bacterial membrane enzyme. *J. Biol. Chem.* **254:**4863–4875.
9. **Waxman, D. J., and J. L. Strominger.** 1981. Limited proteolysis of the penicillin-sensitive

D-alanine carboxypeptidase purified from *Bacillus subtilis* membranes. Active, water-soluble fragments generated by cleavage of a COOH-terminal membrane anchor. *J. Biol. Chem.* **256:**2059–2066.

10. **Waxman, D. J., and J. L. Strominger.** 1983. Penicillin-binding proteins and the mechanism of action of β-lactam antibiotics. *Annu. Rev. Biochem.* **52:**825–869.

Chapter 37

Purification and Characterization of a β-Lactam-Resistant Penicillin-Binding Protein from *Enterococcus hirae (Streptococcus faecium)*

Philippe Jacques
Aboubaker El Kharroubi
Bernard Joris
Bernard Dussenne
Fabienne Thonon
Graziella Piras
Jacques Coyette
Jean-Marie Ghuysen

The natural β-lactam resistance of enterococci is a critical problem in clinical chemotherapy. It is thus of interest to better understand the biochemical basis of this resistance.

The resistant strain R40, provided by R. Fontana (2), was isolated in vitro from *Enterococcus hirae (Streptococcus faecium)* ATCC 9790 by successive selections in the presence of increasing concentrations of benzylpenicillin. This strain overproduces PBP 5, the PBP most resistant to benzylpenicillin. Thus, we planned to purify and characterize the PBP 5 to better understand the resistance phenomenon.

Philippe Jacques, Aboubaker El Kharroubi, Bernard Joris, Bernard Dussenne, Fabienne Thonon, Graziella Piras, Jacques Coyette, and Jean-Marie Ghuysen • Service de Microbiologie, Université de Liège, Institut de Chimie, B6, B-4000 Sart Tilman (Liège 1), Belgium.

Table 1. Characterization of the tPBP 5 derivatives

Derivative	Molecular mass (daltons)	Yield (%)[a]
tPBP 5a	68,000	79
tPBP 5b	63,000	7.4
tPBP 5c	61,000	1.1
tPBP 5d	44,000	3.5

[a]Related to the total amount of PBP 5 in the membrane.

The R40 mutant and the parent strain, like most of other *E. hirae* strains, contain seven membrane proteins able to form stable complexes with benzyl-penicillin (penicillin-binding proteins [PBPs]) (1). These PBPs are numbered in the order of decreasing molecular weights.

If the solubilization of the native protein was easy (90% was extracted from the cells with 0.05% Triton X-100), the purification of the soluble material was very difficult. Three chromatographic techniques were tested to separate the PBPs from the detergent extracts: hydrophobic interaction, gel filtration, and anion exchange. In all three experiments, the PBPs behaved as mixed lipid-protein micelles. Another approach was thus necessary to obtain great quantities of these membrane proteins with a high degree of purity.

Treatment of a membrane preparation from *Enterococcus hirae* R40 with 0.125% (wt/wt) trypsin for 15 min at 37°C released four soluble derivatives

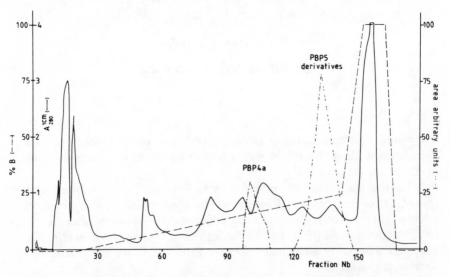

Figure 1. Anion-exchange chromatography. The supernatant obtained after trypsin treatment (371 mg of proteins) was fractionated on an anion-exchange column (Q Sepharose FF) with the FPLC system. The buffer used was 25 mM Tris-borate (pH 8), and the elution was done with increasing amounts of the Tris-borate buffer containing 1 M NaCl.

Table 2. Purification of soluble tryptic products

Step	Protein (mg)	Recovery (%)		Sp act (%, wt/wt)	
		tPBP 5a alone	Total tPBPs 5	tPBP 5a alone	Total tPBPs 5
Membranes	725	100	100	0.92	0.92
Proteolysis	371	79	91	1.42	1.64
Q Sepharose FF	8.29	39	59	31.37	47.47
Phenyl-Superose	2.13	9	32	28.18	100

(tPBPs) of PBP 5 (75,000 daltons) (Table 1). The four tPBP 5 derivatives were purified from the other proteins up to about 100% purity by using two chromatographic steps on a fast polypeptide liquid chromatography system (FPLC). The first step (Fig. 1) was an anion exchange, and the second step (Fig. 2) was a hydrophobic chromatography. The yields of this purification are given in Table 2.

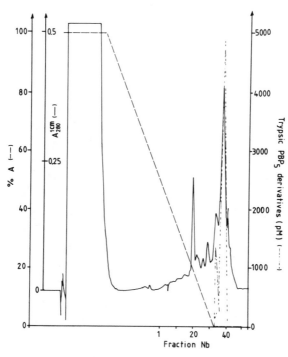

Figure 2. Hydrophobic chromatography. All the fractions collected during the anion-exchange chromatography, containing tPBP 5 derivatives, were pooled, concentrated, and deposited on a hydrophobic column (Phenyl-Superose HR 5/5, FPLC). The buffer used was 25 mM Tris-borate (pH 8), and the elution was done with decreasing percentages of the A buffer (Tris-borate) containing 1.7 M ammonium sulfate.

The improvement of the purity at the different steps of purification is presented in Fig. 3, which shows the Coomassie blue stain and the fluorogram of the sodium dodecyl sulfate-polyacrylamide gel electrophoresis made after the purification. The degradation of tPBP 5a continued during the purification, but after the last step, the relative amount of the tPBP 5 derivatives remained stable.

The pure fraction was used to define the isoelectric point (4.65) and the affinity for benzylpenicillin (3×10^{-4} M) of tPBP 5a. The latter value was identical to that of the native protein. In addition, the active-site tryptic peptide of tPBP 5a was isolated, purified, and sequenced: Gly-Leu-Glu-Ile-(Ser)-Asn-Leu-Lys. The serine was not seen in the results of the Sequenator but could be deduced on the basis of the amino acid composition.

In conclusion, we have established an original procedure to obtain, in good yield, almost pure, soluble tryptic products of PBP 5, which confers β-lactam resistance in *E. hirae*.

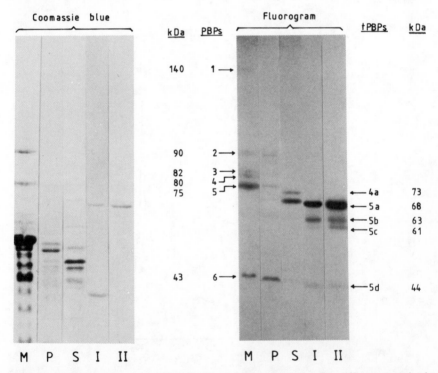

Figure 3. Coomassie blue stain and fluorogram of the gel showing the different steps in the purification of tPBP5 derivatives. M, Membrane preparation; P, pellet after ultracentrifugation of the membrane preparation treated with trypsin; S, supernatant after ultracentrifugation of the membrane preparation treated with trypsin; I, tPBP 5 derivatives after the anion-exchange chromatography; II, tPBP 5 derivatives after the hydrophobic chromatography. kDa, Kilodaltons.

The membranes of *E. hirae* were first treated with trypsin to give tPBPs 5a, 5b, 5c, and 5d, four soluble products of the native PBP which still had the capacity to bind benzylpenicillin. The tPBP 5 derivatives were purified to a level of 100% purity, and the purification yield determined on the basis of the membrane amount was 32%. The tPBP 5a isoelectric point was determined by isoelectrofocusing, and its affinity for benzylpenicillin was found to be identical to that of PBP 5.

ACKNOWLEDGMENTS. This work was supported by the Institut pour l'Encouragement de la Recherche Scientifique dans l'Industrie et l'Agriculture (grant to P.J.), the Gouvernement belge (action concertée 86/91-90), the Fonds de Recherche de la Faculté de Médecine, the Fonds de la Recherche Scientifique Médicale, Brussels (contracts 3.4522.86 and 3.4507.83), and the Région wallonne (C2/C16/conv. 246/20428).

LITERATURE CITED

1. **Coyette, J., J. M. Ghuysen, and R. Fontana.** 1980. The penicillin-binding proteins in *Streptococcus faecalis* ATCC9790. *Eur. J. Biochem.* **110:**445–456.
2. **Fontana, R., R. Cerini, P. Longoni, A. Grossato, and P. Canepari.** 1983. Identification of a streptococcal penicillin-binding protein that reacts very slowly with penicillin. *J. Bacteriol.* **155:**1343–1350.

Chapter 38

Shifting of the Target for the Inhibitory Action of β-Lactams in Gram-Positive Streptococci

Roberta Fontana
Pietro Canepari
Maria del Mar Lleò
Giuseppe Satta

Extensive studies on penicillin-binding proteins (PBPs) have been carried out for many years now, and much information on their role in cell physiology and the mechanism of action of β-lactams has been accumulated, particularly in *Escherichia coli*. It has been suggested that, in this microorganism, some PBPs perform a key role in cell growth and are essential, whereas others are not involved in specific functions and are nonessential (13–18). However, recent studies of streptococci have suggested that PBPs do not necessarily each perform a fixed, constant function in cell wall metabolism and that the functions of the essential PBPs may be taken over, under certain conditions, by one of the nonessential PBPs (2, 6–10).

In view of the substantial interest that this hypothesis may have for future studies on PBP functions, particularly if applicable to other bacteria, including gram-negative rods, in this chapter we review our studies on *Streptococcus faecium* PBPs which corroborate the validity of the above-mentioned suggestion and present new data to confirm it, obtained by analysis of mutants with conditional or constitutive PBP defects.

The strains used for these studies are described in Table 1. Bacteria were

Roberta Fontana, Pietro Canepari, and Maria del Mar Lleò • Istituto di Microbiologia dell'Università di Verona, 37134 Verona, Italy. **Giuseppe Satta** • Istituto di Microbiologia dell'Università di Siena, Siena, Italy.

grown either in a chemically defined medium (CDM) (12) or in SB (8) or BH (8). Different temperatures were used: in general, bacteria were grown in a given medium both at optimal temperature (that is, the temperature allowing the fastest growth rate) and at one or more suboptimal temperatures (that is, temperatures causing a 30% or more decrease in growth rate), as described by Fontana et al. (8). Isolation and characterization of penicillin-resistant mutants have been described (7, 8). Temperature-sensitive cell division mutants with altered PBPs have been isolated and described by Canepari et al. (3).

DNA, protein, and peptidoglycan synthesis was evaluated in CDM at 30 and 42°C as incorporation of labeled precursors into trichloroacetic acid-precipitable fractions (3). Before radioactivity was counted in a Beckman L57000 counter, samples for peptidoglycan synthesis were incubated with pronase, as described by Boothby et al. (1).

Membranes were prepared and treated with radioactive penicillin, as described elsewhere (4, 8). In some experiments, the in vivo PBP binding procedure was used (7). After reaction with radioactive penicillin, the PBPs were separated by sodium dodecyl sulfate-polyacrylamide gel electrophoresis and detected by fluorography (7, 9).

Susceptibility of S. faecium PS to Penicillin and Identification of the Lethal Target(s)

The MIC of penicillin for *S. faecium* is significantly influenced by medium and temperature of growth (7, 8). In CDM at 45°C (optimal temperature), the MIC was 0.04 μg/ml, i.e., 200 times lower than that determined in the same medium at 30°C (suboptimal temperature). In SB at 37°C (optimal temperature), the MIC was 0.25 μg/ml, i.e., 32 times lower and 62 times higher than those found at 30 and 45°C (suboptimal temperatures), respectively. Similar results were obtained when a third medium (BH) was used (8).

Quantitative alterations in PBPs were observed in the various media at the different temperatures: in particular, a relative decrease in the amounts of PBPs 1, 2, and 3 in CDM at suboptimal temperatures of 30 and 32°C. However, no qualitative alterations in affinity of individual PBPs for penicillin were found that depended on medium or temperature used for growth. As a consequence, under certain growth conditions (for instance, in CDM at 45°C; MIC, 0.04 μg/ml), the MIC bound or saturated only one PBP, which could therefore be considered as the primary target for the mechanism of action of penicillin, but in others (for instance, in CDM at 30°C; MIC, 8 μg/ml), saturation of the same protein did not cause growth inhibition (6). These observations suggest that the activities of some PBPs are essential under certain growth conditions, but not in others, and that the target for penicillin action might shift from one PBP to another.

Table 1. S. faecium strains used

Strain	Relevant properties	Source
PS	Wild type, in previous studies designated as *S. faecalis (faecium)* ATCC 9790	Coyette et al. (5); Fontana et al. (6–10)
R5, R20, R40	Penicillin-resistant mutants of PS hyperproducing PBP 5	Fontana et al. (9)
Rev14	Penicillin-hypersusceptible revertant of R40 lacking PBP 5	Fontana et al. (10)
NT1/27	Temperature-sensitive cell division mutant of PS with altered PBP 1	Canepari et al. (3)
NT1/20	Temperature-sensitive cell division mutant of PS with altered PBP 2	Canepari et al. (3)
NT1/119	Temperature-sensitive cell division mutant of PS with altered PBP 3 and hyperproducing PBP 5	Canepari et al. (3)
NT1/37	Temperature-sensitive cell division mutant of PS with altered PBP 3	Canepari et al. (3)
NT1/131	Temperature-sensitive cell division mutant of PS with altered PBP 5	Canepari et al. (3)
NT1/41	Temperature-sensitive cell division mutant of PS with altered PBP 6	Canepari et al. (3)

Isolation and Properties of Penicillin-Resistant and -Hypersusceptible Mutants of *S. faecium*

Penicillin-resistant mutants of *S. faecium* were isolated by serial culture in media containing increasing penicillin concentrations (9). Three strains were selected for analysis: R5 (MIC, 20 μg/ml), R20 (MIC, 40 μg/ml), and R40 (MIC, 80 μg/ml). All these strains showed an increased amount of PBP 5, which was proportional to the level of resistance reached. This protein was found to possess a very low affinity for penicillin, requiring 60 min to become saturated. This forced us to modify the standard assay for PBP analysis by increasing the incubation time of membranes or whole cells with penicillin from 10 to 60 min (9). When growing cells of both the wild type and resistant mutants were treated for 60 min with various penicillin concentrations, it was found that the MIC saturated PBPs 1, 2, and 3 in the wild type but not PBPs 4, 5, and 6. Resistant strains, on the other hand, grew normally in the presence of penicillin concentrations that saturated all PBPs except PBP 5 and were inhibited by concentrations saturating this protein, too, which thus became the target for penicillin action (9).

From R40 treated with novobiocin, a strain showing hypersusceptibility to penicillin (MIC, 0.015 μg/ml in SB) was isolated. This strain grew normally, did not show any alteration in cell permeability, and did not synthesize

PBP 5 (10). It was hypersusceptible to all β-lactams, and the MICs were not influenced by medium or temperature of growth. In this strain, PBP 3 was the target for this antibiotic action in all conditions.

Isolation and Properties of Temperature-Sensitive Cell Division Mutants of S. faecium PS Showing Alterations in PBPs

Analysis of PBPs of 209 temperature-sensitive cell division mutants resulted in the identification of nine strains that showed either constitutive or temperature-sensitive conditional damage to PBPs (3). The main properties of six of these mutants are reported in Table 2. The effects of nonpermissive temperature on macromolecular synthesis (protein, DNA, and peptidoglycan), residual divisions, and the PBP pattern of revertants selected for growth capability at nonpermissive temperature were determined for each strain. This last approach was used to establish in which mutants PBP damage was re-

Table 2. Some properties of temperature-sensitive cell division mutants with altered PBPs

| Mutant | PBP alteration | Time (min), after the shift to 42°C, required to inhibit: | | | | PBP pattern of revertants able to grow at 42°C |
		No. increase	DNA	Protein	Peptidoglycan	
NT1/27	PBP 1 with a lower mol wt at 30 and 42°C	30	60	50	120	Same as the wild type
NT1/20	PBP 2 lacking penicillin-binding ability at 30 and 42°C	0	60	60	60	Same as the wild type
NT1/119	PBP 3 lacking penicillin-binding ability at 42°C; PBP 5 overproduced	150	150	150	150	
NT1/37	PBP 3 lacking penicillin-binding ability at 42°C	30	60	60	120	Same as the wild type
NT1/131	PBP 5 lacking penicillin-binding ability at 30 and 42°C	30	0	0	150	Same as the parent
NT1/41	PBP 6 split into two bands	15	0	0	150	Same as the parent

sponsible for inability to grow at nonpermissive temperature and whether a single mutation governed the phenotype observed. In general, in all mutants, cell division and protein and DNA synthesis stopped within the first hour of incubation at nonpermissive temperatures. By contrast, peptidoglycan synthesis continued at 42°C in almost all strains for at least 2 h. Cell numbers continued to increase for at least 30 min or more in all strains, but not in PBP 2-altered mutants, in which cell division was blocked immediately after shifting to 42°C.

All revertants isolated from mutants with altered PBP 1, 2, or 3 showed a normal PBP pattern, suggesting that these mutants might carry a single mutation governing the phenotype observed and that alterations in these PBPs might be directly responsible for the inability of the cells to grow at 42°C. In addition, normal PBPs 1 and 2 seemed to be unnecessary for growth at suboptimal temperature. By contrast, all revertants isolated from mutants lacking PBP 5 or carrying an altered PBP 6 maintained the altered PBP pattern. In these mutants, more than one mutation might be responsible for the phenotype observed, and inability to grow at 42°C might be unrelated to alteration in PBPs 5 and 6.

Of particular interest was the finding that a mutant lacking PBP 3 and overproducing PBP 5 grew normally at 42°C for at least 2 h (3). As overproduction of PBP 5 was also found in penicillin-resistant strains of S. faecium, isolation of spontaneous penicillin-resistant derivatives from mutants NT1/20, NT1/37, and NT1/43 was attempted. With the exception of the first of these, several strains with relatively high-level resistance to penicillin were obtained. All overproduced PBP 5 and grew at 42°C, though still lacking PBPs 2 and 3 like the respective parent strains.

Conclusion

The effects of the β-lactam antibiotics on the growth and morphology of S. faecium, when analyzed in conjunction with the properties of temperature-sensitive cell division mutants with altered PBPs, allow us to identify some of the functions that each PBP may perform in the physiology of this microorganism (Table 3). The essential role of PBPs 1, 2, and 3 in cell growth at optimal temperature is suggested by the finding that the MIC of penicillin saturated one (in CDM at 45°C) or all of these PBPs (in SB at 37°C), but did not bind PBPs 4, 5, and 6 at all; by contrast, cells grew at suboptimal temperature in the presence of antibiotic concentrations saturating all these PBPs. The above suggestion is also supported by the properties of temperature-sensitive cell division mutants showing constitutive or conditional alterations in these PBPs. These mutants did not grow at 42°C, and revertants that divided normally at this temperature regained the normal PBP pattern. In addition, PBP 1- and PBP 2-altered mutants showed constitutive damage in these PBPs, thus confirming

Table 3. Some properties of PBPs of *S. faecium*

PBP	Mol wt	Relative affinity for β-lactams	Role in cell physiology
PBP 1	140,000	High	Essential for growth at optimal temperature
PBP 2	90,000	High	Essential for growth at optimal temperature; involved in a late stage of septum formation and, maybe, in peripheral surface growth
PBP 3	85,000	High	Essential for growth at optimal temperature; involved in an early stage of septum formation
PBP 4	80,000	Low	Nonessential
PBP 5	75,000	Very low	Nonessential, multifunctional; responsible for resistance to β-lactams
PBP 6	43,000	Low	Nonessential; carboxypeptidase activity

that these proteins are not essential for growth at 30°C, i.e., at suboptimal temperature.

With regard to the possible specific role played by each of the three PBPs in cell surface growth during the cell division cycle, studies with β-lactams suggest that PBP 3 is involved in septum formation, while PBP 2 is necessary for cell size control or for peripheral surface growth (6, 7, 11). In addition, PBP 3 seems to operate at an early stage in the cell division cycle, whereas PBP 2 is presumably involved in a later stage. The results obtained with cell division mutants have confirmed the importance of PBPs 2 and 3 in cell division, but do not indicate whether PBP 2 plays a role in peripheral wall growth. No hypothesis can be suggested regarding the role of PBP 1, as no β-lactam selectively binding this protein is as yet available and PBP 1-altered cell division mutants showed no suggestive properties.

PBP 4 appears to be a nonessential PBP, mainly on the basis of studies with β-lactams indicating that the affinity of this protein for these antibiotics showed no correlation with MIC in any growth conditions (4, 8). Genetic studies have not been possible, as mutants lacking PBP 4 or showing alteration in this protein have not been found, probably because mutations affecting the PBP 4 gene occurred with a low frequency.

PBP 5 is the PBP with the lowest affinity for β-lactams. It is nonessential for cell viability, as mutants lacking this protein grew normally in all conditions. However, it plays an important role in the mechanism of resistance to β-lactams, as it appears to be capable of taking over the function of the other PBPs when these are saturated by β-lactams. This conclusion is supported by (i) the finding that, under certain conditions, *S. faecium* grew normally in the presence of penicillin concentrations saturating all PBPs except PBP 5; (ii) the isolation of a mutant which lacked PBP 3, hyperproduced PBP 5, and grew normally; (iii) the isolation of penicillin-resistant derivatives of PBP 1- and PBP 2-altered cell division mutants which hyperproduced PBP 5 and grew normally even at nonpermissive temperature in spite of the fact that they maintained the altered PBP phenotype; and (iv) the fact that susceptibility to β-lactams of mutants lacking PBP 5 was not influenced by cultural conditions and PBP 3 was always identified as the target for growth inhibition in these strains.

PBP 6, like the lowest-molecular-weight PBPs of other species, has carboxipeptidase activity (5). It is nonessential for cell growth because some β-lactams, such as cefoxitin, saturate this protein and totally inhibit its enzymatic activity at concentrations far below the MIC. In conclusion, the PBP pattern of *S. faecium* is similar in its physical, chemical, and physiological properties to that of all gram-positive and gram-negative bacteria studied to date. However, a new aspect of PBP function appears from our studies: PBPs do not necessarily each perform a fixed, constant function in cell wall metabolism; their functions may vary depending on the physiological status of the cells. Consequently, the target(s) for growth inhibition by β-lactams may change from a high-affinity to a low-affinity PBP, thus profoundly influencing the susceptibility of *S. faecium* to these antibiotics.

LITERATURE CITED

1. **Boothby, D., L. Daneo-Moore, and G. D. Shockman.** 1971. A rapid, quantitative and selective estimation of radioactively labeled peptidoglycan in Gram-positive bacteria. *Anal. Biochem.* **44:**645–653.

2. **Canepari, P., M. M. Lleò, R. Fontana, G. Cornaglia, and G. Satta.** 1986. In *Streptococcus faecium* penicillin binding protein 5 alone is sufficient for cell growth at sub-maximal but not at maximal rate. *J. Gen. Microbiol.* **132:**625–631.

3. **Canepari, P., M. M. Lleò, R. Fontana, and G. Satta.** 1987. *Streptococcus faecium* mutants that are temperature sensitive for cell growth and show alterations in penicillin-binding proteins. *J. Bacteriol.* **169:**2432–2439.

4. **Coyette, J., J.-M. Ghuysen, and R. Fontana.** 1980. The penicillin binding proteins in *Streptococcus faecalis* ATCC 9790. *Eur. J. Biochem.* **110:**445–456.

5. **Coyette, J., A. Somzè, J. J. Briquet, J.-M. Ghuysen, and R. Fontana.** 1983. Function of penicillin binding protein 3 in *Streptococcus faecium*, p. 523–530. *In* R. Hakenbeck, J. V. Höltje, and H. Labischinski (ed.), *The Target of Penicillin*. Walter de Gruyter, Berlin.

6. **Fontana, R., G. Bertoloni, G. Amalfitano, and P. Canepari.** 1984. Characterization of penicillin-resistant *Streptococcus faecium* mutants. *FEMS Microbiol. Lett.* **9:**263–266.

7. **Fontana, R., P. Canepari, G. Satta, and J. Coyette.** 1980. Identification of the lethal target of benzylpenicillin in *Streptococcus faecalis* by in vivo penicillin binding studies. *Nature* (London) **287:**70–72.

8. **Fontana, R., P. Canepari, G. Satta, and J. Coyette.** 1983. *Streptococcus faecium* ATCC 9790 penicillin-binding proteins and penicillin sensitivity are heavily influenced by growth conditions: proposal for an indirect mechanism of growth inhibition by β-lactams. *J. Bacteriol.* **154:**916–923.

9. **Fontana, R., R. Cerini, P. Longoni, A. Grossato, and P. Canepari.** 1983. Identification of a streptococcal penicillin-binding protein that reacts very slowly with penicillin. *J. Bacteriol.* **155:**1343–1350.

10. **Fontana, R., A. Grossato, L. Rossi, Y. R. Cheng, and G. Satta.** 1985. Transition from

resistance to hypersusceptibility to beta-lactam antibiotics associated with loss of a low-affinity penicillin-binding protein in a *Streptococcus faecium* mutant highly resistant to penicillin. *Antimicrob. Agents Chemother.* **28**:678–683.

11. **Pucci, M. J., E. T. Hinks, D. T. Dicker, M. L. Higgins, and L. Daneo-Moore.** 1986. Inhibition by beta-lactam antibiotics at two different times in the cell cycle of *Streptococcus faecium* ATCC 9790. *J. Bacteriol.* **165**:682–688.

12. **Shockman, G. D.** 1963. Aminoacids, p. 567–573. *In* F. Kavanagh (ed.), *Analytical Microbiology.* Academic Press, Inc., New York.

13. **Spratt, B. G.** 1975. Distinct penicillin binding proteins involved in the division, elongation and cell shape of *Escherichia coli* K-12. *Proc. Natl. Acad. Sci. USA* **72**:2999–3003.

14. **Spratt, B. G.** 1977. Temperature-sensitive cell division mutants of *Escherichia coli* with thermolabile penicillin-binding proteins. *J. Bacteriol.* **131**:293–305.

15. **Spratt, B. G.** 1980. Deletion of the penicillin-binding protein 5 gene of *Escherichia coli.* *J. Bacteriol.* **144**:1190–1192.

16. **Suzuki, H., H. Nishimura, and Y. Hirota.** 1978. On the process of cellular division in *Escherichia coli* altered in the penicillin-binding proteins. *Proc. Natl. Acad. Sci. USA* **75**:664–668.

17. **Tamaki, S., S. Nakajima, and S. Matsuhashi.** 1977. Thermosensitive mutation in *Escherichia coli* simultaneously causing defect in penicillin binding protein 1b and in enzyme activity for peptidoglycan synthesis in vitro. *Proc. Natl. Acad. Sci. USA* **74**:5472–5476.

18. **Waxman, D. J., and J. L. Strominger.** 1983. Penicillin binding proteins and mechanism of action of beta-lactam antibiotics. *Annu. Rev. Biochem.* **52**:825–869.

Chapter 39

Targets of β-Lactams in *Streptococcus pneumoniae*

Regine Hakenbeck
Thomas Briese
Heinz Ellerbrok
Götz Laible
Christiane Martin
Claudia Metelmann
Hans-Martin Schier
Spassena Tornette

Investigations of penicillin-binding proteins (PBPs) are addressed to two major aspects. One concerns the structural relationship between PBPs, of the same as well as of different species. After all, despite their diverse physiological roles, which have been clearly demonstrated for *Escherichia coli* PBPs, they share the property of interacting, i.e., recognizing β-lactam antibiotics, indicating a certain common three-dimensional arrangement of the enzymatic active center (for review, see references 3, 16). The other question concerns the functional importance of a PBP within the specific set that is contained in one organism. Since satisfactory enzymatic test systems are not available, the problem has often been approached by investigating which PBP becomes acylated at the β-lactam concentration at which the antibiotic exhibits its lethal effect.

This chapter deals with PBPs in *Streptococcus pneumoniae*. This organism

Regine Hakenbeck, Thomas Briese, Heinz Ellerbrok, Götz Laible, Christiane Martin, Hans-Martin Schier, and Spassena Tornette • Max-Planck Institut für Molekulare Genetik, D 1000 Berlin 33, Federal Republic of Germany. **Claudia Metelmann** • Institut für Medizinische Mikrobiologie der Freien Universität Berlin, D 1000 Berlin 45, Federal Republic of Germany.

contains six PBPs (PBP 1a, 1b, 2x, 2a, 2b, and 3) with M_rs ranging from 94,000 to 43,000. Since no antibiotic is known so far that binds exclusively to only one PBP in this bacterium, which would enable study of the defects of cell growth and murein biochemistry caused by the inhibition of this and only this particular enzyme, the search for penicillin targets in pneumococci has to rely upon more indirect experiments. Not much is known about the function of pneumococcal PBPs. The low-molecular-weight PBP 3, like many other low-molecular-weight PBPs from different organisms, acts in vitro as DD-carboxypeptidase (6). It might not be essential for the cell since several β-lactams bind to PBP 3 far below their respective MICs (17). For PBP 2b, on the other hand, several results indicate that it could be the primary penicillin target: the MICs of a series of β-lactams correlated with 50% saturation of PBP 2b (17), and the amount of acylated PBP 2b paralleled the residual murein synthesis during penicillin treatment (18). However, in these studies PBP 2x has not been taken into account as a result of the different gel systems used. Recently it was shown that β-lactams which do not bind to PBP 2b are nonlytic for pneumococci, suggesting that the lytic response observed with most other penicillins is based on inhibition of this particular PBP (8).

Another approach is based on the idea that bacteria become resistant to a β-lactam by altering the penicillin affinity of its primary target. For many years it has been known that β-lactam resistance in pneumococci, when induced in the laboratory, is a multistep process (14). Only recently has it become clear that this type of β-lactamase-independent "intrinsic resistance," which occurs also in a number of other bacterial species, has its cause in successive alterations of different PBPs, in that a decrease in penicillin affinity correlates with an increase of resistance (7, 12, 19). One can expect that in strains with low-level resistance the primary essential PBP will be changed, and with increasing resistance the next important PBP will be affected. Thus, by establishing the order of which PBP has become altered first, second, etc., during the pathway of penicillin resistance, one might obtain the functional hierarchy of these enzymes. Penicillin-resistant pathogenic pneumococci have been isolated with increasing frequency (11). Apart from affinity changes, the M_r of various high-molecular-weight PBPs appeared altered, and the relation to the R6 PBPs was not clear. In a series of transformation experiments using DNA from a highly penicillin-resistant strain as donor, Tomasz and co-workers showed that transformation of resistance indeed correlated with appearance of altered PBPs and that the changes were distinct for each resistance level (15). Alterations at lower resistance levels resembled each other when clinical isolates, pathogenic penicillin-resistant mutants isolated in the laboratory, and transformants with low-level resistance were compared, in that affinity decreases for penicillin were observed in the same groups of high-molecular-weight PBPs and not in PBP 3 (9). The relationship of totally altered PBPs from three highly penicillin-resistant strains to PBPs 1a and 2b of the R6 strain was established recently

(5), suggesting that one should focus on these two proteins when studying the development of resistance. However, it should be pointed out that the strains represent independent isolates from all over the world, and it is therefore so far impossible to determine whether the highly penicillin-resistant strains originate from strains with lower-level resistance containing the very same PBP alterations as have been described in isolates with low and intermediate resistance.

We have investigated the individual PBPs of the laboratory strain R6 biochemically and immunologically to obtain an experimental basis for studying PBP alterations in various penicillin-resistant strains. In this paper we first focus on different experimental approaches for identification of PBPs and comment on the problems that are involved with each of the methods. In a second part, alterations of PBPs associated with β-lactam resistance in laboratory mutants derived from the R6 strain are described. The results are compared with PBP changes found in various penicillin-sensitive and low-level resistant clinical isolates.

PBPs in the R6 Strain

S. pneumoniae R6, a penicillin-sensitive, noncapsulated laboratory strain contains six PBPs with M_rs ranging from 94,000 to 43,000. Early investigations revealed only three PBPs, called PBPs 1, 2 and 3. After improving the conditions for sodium dodecyl sulfate (SDS)-polyacrylamide gel electrophoresis, however, the PBP 1 band could be resolved into two and the PBP 2 band into three PBPs, leading to the somewhat misleading nomenclature depicted in Fig. 1. Quantitation of PBPs is often done by densitometry tracing of the X-ray film or by determination of radioactivity in PBP-containing gel slices. Because the molecular weight range of the PBPs in group 1 and especially group 2 pneumococci is very narrow, resolution of the peaks is not very satisfying and they can only be roughly estimated. Complications arise when cells are labeled and centrifuged prior to analysis or when membranes instead of cell lysates are used for analysis. Centrifugation after labeling results in different quantitative distribution as a result of loss of PBPs (and other membrane components), especially PBPs 1a, 2a, and 3. This is even more pronounced when long incubation times are used. When membranes are used instead of cells, PBP 2x often appears with a lower M_r as in cell lysates for unknown reasons, so that PBPs 2a and 2x cannot be distinguished on a one-dimensional gel system. Furthermore, PBP 1a becomes degraded upon prolonged storage of membranes, even at $-80°C$, and a PBP 1a derivative of similar M_r as PBP 2b appears. Therefore, when comparing PBP profiles in different strains, concentrated cell suspensions are used that are labeled and lysed at the same time.

We routinely use [^{3}H]propionylampicillin (^{3}H-PA), synthesized as de-

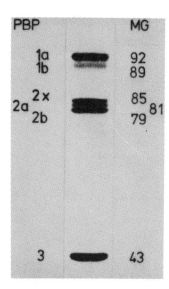

Figure 1. PBPs of *S. pneumoniae* R6. Cells from 1 ml of an exponentially growing culture (3×10^7 cells per ml) were collected by centrifugation and suspended in 10 μl of 20 mM phosphate buffer (pH 7.4). ^{3}H-PA (1 μCi or approximately 5 pmol) and Triton X-100 (final concentration, 0.2% [wt/vol]) were added, and the preparation was incubated at 37°C for 30 min. Proteins were separated on 7.5% SDS-polyacrylamide gels, and PBPs were revealed after fluorography. The exposure time was 3 days. The molecular sizes in kilodaltons are indicated on the right.

scribed by Schwarz et al. (13), for PBP labeling because of its superior specific activity. This is an advantage when PBPs with very high affinity corresponding to the high sensitivity of the bacteria are investigated, as in pneumococci. Affinity to PBPs and turnover rates are naturally different from those of benzylpenicillin, and this has to be kept in mind when experimental results are compared with benzylpenicilloylated PBPs. In Fig. 2a, the different affinities of the R6-PBPs for ^{3}H-PA are documented, with PBP 3 having the highest affinity and PBP 2a having the lowest affinity. For identification of PBPs even in crude fractions like cell lysates, a two-dimensional gel system involving partial V8-protease digestion can be applied (2). Each PBP gives rise to a specific pattern of penicilloyl-peptides that can be recognized after fluorography. Figure 2b shows the result obtained with cell lysates of R6. Thus, it can be demonstrated not only by their different affinities, but also by their penicilloyl-peptide patterns, that the six PBPs of pneumococci are structurally unrelated proteins.

In addition to radioactive labeling, we have produced various antibody probes for visualizing the PBPs. With purified PBPs 1a, 2b, and 3 as antigens, specific antisera were obtained (5; T. Briese, Ph.D. dissertation, Freie Universität, Berlin, Federal Republic of Germany, 1986; H. Ellerbrok, Ph.D. dissertation, Freie Universität, Berlin, Federal Republic of Germany, 1986); also, monoclonal anti-PBP 1a antibodies were produced (Briese, Ph.D. dissertation). Another assay system is based on the reactivity of anti-β-lactam antibodies on Western blots with corresponding β-lactam-PBP complexes. With this method one does not depend on radioactive penicillin, so that high concentrations of a β-lactam for PBP labeling can be reached easily (4; Briese et al., this vol-

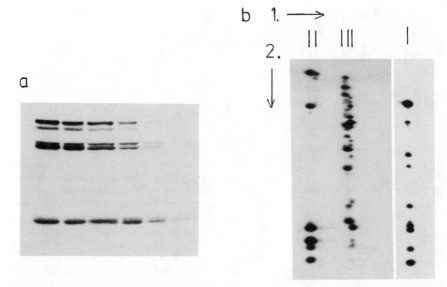

Figure 2. (a) Affinities of pneumococcal PBPs for ³H-PA. Cells were incubated in 15-μl volumes with threefold dilutions of ³H-PA, starting with 0.6 μCi. (b) Penicilloyl-peptides of R6 after partial proteolysis with V8-protease. After a first-dimension SDS-polyacrylamide gel as described in the legend to Fig. 1, a lane was cut and layered on top of a second 15% SDS-polyacrylamide gel, with V8-protease included in the stacking gel. The arrows indicate the direction of first- and second-dimension runs. The PBP pattern after the first dimension is schematically indicated at the top of panel b.

ume). This is important when using PBPs with low affinities or when in vivo experiments using large volumes are performed. However, not all benzylpenicilloyl (BPO)-PBP complexes are recognized by anti-BPO antibodies like, for example, the pneumococcal PBP 3. Figure 3 shows pneumococcal PBPs stained on Western blots with the help of different antisera.

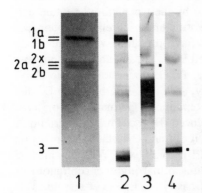

Figure 3. Reactivity of PBPs with different antisera. (Lane 1) PBPs in a cell lysate were labeled with benzylpenicillin (10 μg/ml). After Western blotting, the immunostain was developed with rabbit anti-BPO antiserum. (Lanes 2–4) Western blots of membrane preparations were stained with the help of (2) rabbit anti-PBP 1a antiserum, (3) mouse anti-PBP 2b antiserum, and (4) rabbit anti-PBP 3 antiserum. The PBP position is indicated on the left.

PBPs in β-Lactam-Resistant Laboratory Mutants

Two β-lactams differing in their physiological effect were chosen for selection of resistant mutants. Piperacillin induces rapid lysis in pneumococci and resembles benzylpenicillin in that it binds to all PBPs within a narrow concentration range. Cefotaxime belongs to a group of antibiotics with a similar R1 side chain that do not induce lysis in pneumococci, probably as a result of an inability to react with PBP 2b (8). The mutants were named according to (i) the selected β-lactam (P = piperacillin; C = cefotaxime); (ii) the mutant obtained after the first selection (C001, C002, etc.); and (iii) the step isolated (C101, C201, C501, etc., thus represent members of a mutant family, with C501 as the highest cefotaxime-resistant mutant obtained after the fifth round of selection from C101). All mutants were isolated and grown at 30°C. For piperacillin, resistance levels between 8 (P504) and 16 (P506 and P408) times the MIC for R6 were determined. The C mutants showed increases in MIC of a factor of 16 (C505), 32 (C501 and C604), and 64 (C506 and C503). Figure 4 shows the PBP profiles from these mutants of the highest resistance level obtained, revealed as BPO complexes on Western blots with anti-BPO antiserum. Raising the penicillin concentration to 100 μg/ml gave the same result, indicating that all PBPs are saturated under the conditions. It can clearly be seen that either PBP 2x or 2a cannot be labeled in several C mutants; i.e., they have apparently lost the capacity to bind penicillin completely. The P mutants on the other hand revealed different alterations: PBP 1a or PBP 2b (or both) is stained less intensely, indicating a reduction in the amount of bound β-lactam or in the amount of protein. Surprisingly, P mutants not only differed from C mutants, but each group was heterogeneous in itself. An estimation of affinity changes in the mutants is shown in Table 1.

With the methods described in the preceding section, we have further analyzed the PBPs in the different mutant families (G. Laible and R. Hakenbeck,

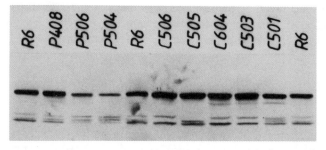

Figure 4. PBP profile in piperacillin- and cefotaxime-resistant laboratory mutants. Samples of each mutant, as indicated at the top, containing equal amounts of cells were labeled with benzylpenicillin (10 μg/ml). After Western blotting, PBPs were visualized by using anti-BPO antiserum.

Table 1. Factor of affinity decrease for the selective β-lactam in C and P mutants[a]

PBP	Affinity decrease in mutants:							
	C501	C503	C604	C505	C506	P504	P506	P408
1a	0	0	0	0	0	?[b]	?	0
1b	10	0	0	0	0	0	0	0
2x	100	100	100	/[c]	/	0	8	3
2a	0	/	/	0	/	0	0	0
2b	0	0	0	0	0	?	?	3
3	0	0	10	0	0	0	0	0

[a]The numbers represent estimates obtained from fluorograms after preincubation of cells with different concentrations of cefotaxime or piperacillin and postlabeling with ^{3}H-PA. For P mutants, immunoblots obtained from piperacillin-PBPs and antipiperacillin antiserum were evaluated additionally.
[b]Decrease in total amount, and therefore not estimated.
[c]No binding of ^{3}H-PA under the conditions used.

Mol. Microbiol., in press). All mutants contained multiple PBP alterations. It became evident that major affinity shifts occurred at different resistance levels depending on the mutant family. In P mutants, PBP 2b was always primarily affected in terms of decreased penicillin affinity; a reduced amount of PBP 1a was determined in P504 and P506 either with anti-PBP 1a antiserum or by measuring the amount of bound β-lactam. In C mutants, PBP 2x and in some cases PBP 2a were affected.

From this, one might conclude that these two PBPs are the respective primary targets for the two β-lactams used for selection. However, the situation turned out to be more complex after analyzing the cross-resistance pattern. Mutant C501 was rather specifically resistant to cefotaxime, and P408 was more resistant to penicillins than to cephalosporins. On the other hand, C506 as well as P506 showed a wide range of cross-resistance to the same degree, irrespective of their quite different PBP alterations. This means that resistance against a β-lactam can be the result of various genotypes. Therefore, it seems difficult (if possible at all) to deduce the selective β-lactam even after determination of the resistance pattern and the PBP alterations.

PBPs in Penicillin-Sensitive and Low-Level Resistant Clinical Strains

In the Institute for Medical Microbiology in Berlin, an increasing number of strains with low penicillin resistance have been isolated on the basis of their decreased oxacillin susceptibility. The 10 strains investigated in our laboratory represent an arbitrary selection of sensitive and low-level resistant isolates, with MICs of penicillin up to 16 times the MIC for R6, which is somewhat below the range of the laboratory mutants described above. Again, PBP profiles and affinities for β-lactams were tested by different methods. Figure 5 shows the PBP pattern of seven strains, after labeling with aztreonam, Western blotting, and staining with anti-aztreonam antiserum. The antiserum was a generous gift from N. F. Adkinson (1). PBP 2b cannot be revealed this way, since aztreonam does not bind to PBP 2b. From this and further experiments, it became clear that all but one of the strains (641) contained one or more PBPs that differed in M_r compared with the R6 PBPs. The following alterations were observed in various combinations: a smaller PBP 1a (e.g., in strain 208 or 343), a somewhat larger PBP 2b so that PBPs 2a and 2b could not be resolved on a one-dimensional gel (e.g., Sp1), and in two cases not included in Fig. 5 a larger PBP 3. The PBP which appears in some cases above PBP 2x of R6 has a somewhat different penicilloylpeptide pattern as revealed after partial V8-proteolysis and thus presumably represents a derivative of PBP 2x. It is probably identical to PBP 2′ noted in earlier publications (7, 12, 19). None of these changes was observed with the laboratory mutants, and we suggest that they are due to the nonisogenicity of the strains.

Apart from the different M_rs, the strains also varied in their PBP affinities. Figure 6 shows two examples, one case with altered PBP 2b and 2′ affinity and another strain in which PBP 2′ is not altered but instead PBP 1a and PBP 2b are altered. It should be mentioned that both strains contain PBPs 2a and 2b which can hardly be distinguished on this gel system, so that densitometry tracing cannot be applied. The MIC of penicillin for the two strains was eightfold higher compared with R6.

In summary, the clinical strains with low-level penicillin resistance showed alterations in affinity for benzylpenicillin or propionylampicillin, but in different PBPs, depending on the strain investigated. Only strains of low resistance were used, and in several cases the electrophoretic mobility of some of the PBPs, especially of 2a and 2b, was very similar or even the same; therefore, a quantitative evaluation was not made. The diversity of affinity shifts, superimposed onto changes in M_r, suggests that resistance in pneumococci can be achieved in multiple ways. A common pathway for resistance seems unlikely,

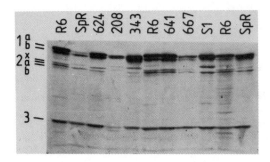

Figure 5. Aztreonam-PBP complexes of various clinical isolates. Cell samples were labeled with aztreonam (50 μg/ml), followed by SDS-polyacrylamide gel electrophoresis and immunoblotting. The immunostain was stained by using antiaztreonam antiserum. The strains are indicated at the top, and the PBP profile of R6 is shown on the left.

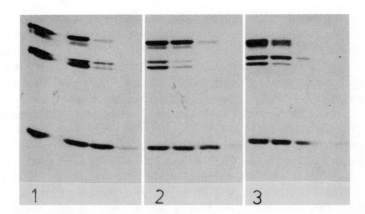

Figure 6. Affinity of PBPs for ³H-PA in two clinical isolates with low-level penicillin resistance. Cells were labeled with eightfold dilutions of ³H-PA, starting with 1 μCi in a 20-μl volume. Lanes: (1) R6; (2) SpR; (3) 343. PBPs were visualized after fluorography.

at least in terms of a determined sequence. However, it should be pointed out that the alterations observed here were restricted to PBPs 1a, 2x (or 2′), and 2b, in agreement with the finding of Handwerger and Tomasz, who reported on alterations within PBP groups 1 and 2 (9).

In a broad sense, the clinical strains also resembled the piperacillin-resistant laboratory mutants in that PBPs 1a and 2b were often altered, and PBPs 2a and 2x (or 2′) could be perfectly labeled, unlike in the cefotaxime-resistant laboratory mutants. Since pneumococcal infections are treated with penicillins rather than with cephalosporins, therapy might well play a role during the process of resistance development. Molecular genetics will certainly be the method of choice in determining which changes occurred under laboratory conditions and how far they resemble those of pathogenic strains. By knowing the structure of the wild-type PBPs and determining the kind of PBP alterations that lead to affinity decreases, one can hope to gain more insight into the structure of the active center of PBPs and to learn which amino acids are important for interaction with the β-lactam moiety. For *E. coli* PBP 3, this type of analysis has shown that only few amino acids were substituted in mutants with reduced affinity (10).

LITERATURE CITED

1. **Adkinson, N. F., E. A. Swabb, and A. A. Sugerman.** 1984. Immunology of the monobactam aztreonam. *Antimicrob. Agents Chemother.* **25:**93–97.
2. **Cleveland, D. W., S. G. Fischer, M. W. Kirschner, and U. K. Laemmli.** 1977. Peptide mapping by limited proteolysis in sodium dodecyl sulfate and analysis by gel electrophoresis. *J. Biol. Chem.* **252:**1102–1106.

3. **Frère, J.-M., and B. Joris.** 1985. Penicillin-sensitive enzymes in peptidoglycan biosynthesis. *Crit. Rev. Microbiol.* **11**:299–396.

4. **Hakenbeck, R., T. Briese, and H. Ellerbrok.** 1986. Antibodies against the benzylpenicilloyl moiety as a probe for penicillin-binding proteins. *Eur. J. Biochem.* **157**:101–106.

5. **Hakenbeck, R., H. Ellerbrok, T. Briese, S. Handwerger, and A. Tomasz.** 1986. Penicillin-binding proteins of penicillin-susceptible and -resistant pneumococci: immunological relatedness of altered proteins and changes in peptides carrying the β-lactam binding site. *Antimicrob. Agents Chemother.* **30**:553–558.

6. **Hakenbeck, R., and M. Kohiyama.** 1982. Purification of penicillin-binding protein 3 from *Streptococcus pneumoniae. Eur. J. Biochem.* **127**:231–236.

7. **Hakenbeck, R., M. Tarpay, and A. Tomasz.** 1980. Multiple changes of penicillin-binding proteins in penicillin-resistant clinical isolates of *Streptococcus pneumoniae. Antimicrob. Agents Chemother.* **17**:363–371.

8. **Hakenbeck, R., S. Tornette, and N. F. Adkinson.** 1987. Interaction of non-lytic β-lactams with penicillin-binding proteins in *Streptococcus pneumoniae. J. Gen. Microbiol.* **133**:755–760.

9. **Handwerger, S., and A. Tomasz.** 1986. Alterations in penicillin-binding proteins of clinical and laboratory isolates of pathogenic *Streptococcus pneumoniae* with low levels of penicillin resistance. *J. Infect. Dis.* **153**:83–89.

10. **Hedge, P. J., and B. G. Spratt.** 1985. Amino acid substitutions that reduce the affinity of penicillin-binding protein 3 of *Escherichia coli* for cephalexin. *Eur. J. Biochem.* **151**:111–121.

11. **Malouin, F., and L. E. Bryan.** 1986. Modification of penicillin-binding proteins as mechanisms of β-lactam resistance. *Antimicrob. Agents Chemother.* **30**:1–5.

12. **Percheson, P. B., and L. E. Bryan.** 1980. Penicillin-binding components of penicillin-susceptible and -resistant strains of *Streptococcus pneumoniae. Antimicrob. Agents Chemother.* **12**:390–396.

13. **Schwarz, U., K. Seeger, F. Wegenmayer, and H. Strecker.** 1981. Penicillin-binding proteins of *Escherichia coli* identified with a ^{125}I-derivative of ampicillin. *FEMS Microbiol. Lett.* **10**:107–109.

14. **Shockley, T. E., and R. D. Hotchkiss.** 1970. Stepwise introduction of transformable penicillin resistance in pneumococcus. *Genetics* **64**:397–408.

15. **Tomasz, A., S. Zighelboim-Daum, S. Handwerger, H. Liu, and H. Qian.** 1984. Physiology and genetics of intrinsic beta-lactam resistance in pneumococci, p. 393–397. *In* L. Leive and D. Schlessinger (ed.), *Microbiology—1984.* American Society for Microbiology, Washington, D.C.

16. **Waxman, D. J., and J. L. Strominger.** 1983. Penicillin-binding proteins and the mechanism of action of beta lactam antibiotics. *Annu. Rev. Biochem.* **52**:825–869.

17. **Williamson, R., R. Hakenbeck, and A. Tomasz.** 1980. *In vivo* interaction of β-lactam antibiotics with the penicillin-binding proteins of *Streptococcus pneumoniae. Antimicrob. Agents Chemother.* **18**:629–637.

18. **Williamson, R., and A. Tomasz.** 1985. Inhibition of cell wall synthesis and acylation of the penicillin binding proteins during prolonged exposure of growing *Streptococcus pneumoniae* to benzylpenicillin. *Eur. J. Biochem.* **151**:475–483.

19. **Zighelboim, S., and A. Tomasz.** 1980. Penicillin-binding proteins of multiply antibiotic-resistant South African strains of *Streptococcus pneumoniae. Antimicrob. Agents Chemother.* **17**:434–442.

Chapter 40

Binding and Release of Penicillin by Minimal Fragments of Pneumococcal Penicillin-Binding Proteins

Heinz Ellerbrok
Regine Hakenbeck

Penicillin-Binding Proteins as Membrane-Bound Enzymes

Penicillin-binding proteins (PBPs) are membrane-bound enzymes that are involved in final steps of murein biosynthesis. Their inhibition by β-lactam antibiotics evidently follows a general scheme in which a serine residue of the enzymatic active center becomes penicilloylated (4, 6, 11). Not much is known about the organization of PBPs within the cytoplasmic membrane. Carboxypeptidases of *Bacillus stearothermophilus* and *B. subtilis* appear to be ectoproteins, inserted into the membrane by a short C-terminal peptide (10), but the *Escherichia coli* carboxypeptidase PBP 5 may be inserted in a different way (9). Conformational data are restricted to two soluble proteins, the *Streptomyces* sp. strain R61 DD-carboxypeptidase and the β-lactamase of *B. licheniformis*. Despite different activity and little sequence homology, the arrangement of α-helixes and β-sheets is amazingly similar (7, 8). Because of their hydrophobicity, crystallization of entire PBPs has not been possible so far. Hydrophilic peptides which are still enzymatically active should provide a tool for obtaining structural information covering at least the active center of these membrane proteins.

Streptococcus pneumoniae contains six PBPs with M_rs ranging from 92,000

Heinz Ellerbrok and Regine Hakenbeck • Max-Planck Institut für Molekulare Genetik, D-1000 Berlin 33, Federal Republic of Germany.

to 43,000. In this species, not only the low-molecular-weight PBP 3, but also the high-molecular-weight PBP 2b seem to be anchored to the membrane by a short terminal peptide, since trypsin treatment of membranes results in the formation of hydrophilic derivatives with M_rs that are 2,000 to 3,000 lower (3). PBP 1a, on the other hand, is largely hydrophobic, and the mode of membrane insertion appears different. After extensive trypsin treatment of native solubilized PBPs, hydrophilic, so-called minimal fragments of similar size (M_r, around 30,000) are generated which are still able to bind penicillin (3). We have studied the interaction of fragments Ia, IIb, and III with penicillin in detail and have compared these fragments with the corresponding PBPs 1a, 2b, and 3, to see to what extent the enzymatic activity (and therefore the conformational arrangement) has remained intact.

β-Lactam Binding of PBP 1a Fragments

Binding to minimal fragments and PBPs of benzylpenicillin and ^{3}H-propionylampicillin was studied with purified PBP preparations. After incubation with different concentrations of the respective β-lactam, fluorograms of different exposure times were prepared and evaluated by microdensitometry. PBP 2b and fragment IIb contained very similar β-lactam binding activity, and this was also found for PBP 3 and the III peptide. Fragment Ia lost binding activity rather rapidly (unpublished results). Therefore, the experiment was repeated under conditions of partial proteolysis (Fig. 1), resulting in about 50% PBP Ia fragment and equal amounts of a somewhat larger fragment Ia* (see lane K).

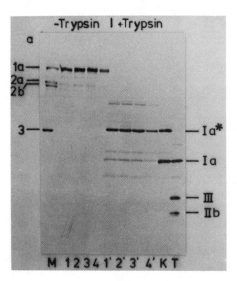

Figure 1. Labeling of PBP 1a and minimal fragment Ia with different concentrations of ^{3}H-propionylampicillin. Purified intact PBP 1a (−Trypsin) and trypsin-digested PBP 1a (+Trypsin) were incubated with different concentrations of ^{3}H-propionylampicillin for 15 min at 37°C. The reaction was terminated by boiling in sodium dodecyl sulfate-sample buffer, proteins were separated on sodium dodecyl sulfate-polyacrylamide gel electrophoresis, and β-lactam-protein complexes were visualized by fluorography. A control sample (K) represents fragments obtained from prelabeled PBP 1a. The position of intact PBPs is indicated on the left, and minimal fragments are indicated on the right side. ^{3}H-propionylampicillin concentrations (picomoles per milliliter): 1, 420; 2, 105; 3, 25; 4, 6.5. Ia* represents the smallest hydrophobic tryptic derivative of PBP 1a.

It can be seen, that fragment Ia hardly binds ^{3}H-propionylampicillin, although the peptide itself is quite stable. However, binding activity is perfectly retained in Ia*, which is the smallest hydrophobic tryptic peptide that can be obtained. This suggests that in PBP 1a hydrophobic parts are essential for maintaining the functional (and thereby structural) integrity of the active center.

Turnover Rates of Minimal Fragments

For determination of the rate of penicillin release, a partially purified PBP preparation was used which contained predominantly PBPs 1a, 2b, and 3. On fluorograms, the proteins were sufficiently separated to allow quantitation with densitometry. Several experiments were performed, some including hydroxylamine for stimulation of the reaction (Table 1). Since PBP 1a released less than 20% of bound β-lactams in 24 h, this protein was not investigated further. PBP 3 and fragment III showed considerable release only with benzylpenicillin, and in both cases similar values were obtained. However, the IIb peptide showed turnover rates twice as fast as those of PBP 2b, independent of the β-lactam used and also in the presence of hydroxylamine. The reason for this phenomenon is unknown.

In summary, the minimal fragments IIb and III both were able to enzymatically bind and release penicillins as the intact PBPs 2b and 3, despite the loss of about 50 kilodaltons in the case of IIb and 15 kilodaltons in the case of III. The enhanced turnover rate observed with the IIb fragment remains obscure, but could be related to differences between PBPs and fragments that became evident when the reaction products were analyzed by thin-layer chromatography (H. Ellerbrok and R. Hakenbeck, *Eur. J. Biochem.*, in press). The intact PBPs seemed to use two reaction pathways, since penicilloic acid (the

Table 1. Deacylation rates of PBPs and minimal fragments[a]

PBP or fragment	50% release of bound β-lactam (min)		
	PA	BP	PA + NH$_2$OH (0.2 M)
2b	250, 240	280, 260, 315	4, 4
IIb	140, 120	185, 170, 165	2, 2
3	—[b]	360, 410	—[c]
III	—[b]	390, 410	—[c]

[a]Partially purified PBPs were labeled with ^{3}H-propionylampicillin (PA) or radioactive benzylpenicillin (BP), and one-half of the sample was digested with trypsin to generate minimal fragments. Nonradioactive benzylpenicillin (100 μg/ml) and in some cases hydroxylamine (NH$_2$OH) were added, and incubation at 37°C was continued. Samples were removed at various times, and β-lactam-PBP complexes on fluorograms were quantified by microdensitometry. The values shown were obtained from different experiments.
[b]No significant release during 24 h.
[c]No significant release during 1 h.

hydrolysis product) as well as phenylacetylglycine (which is formed during the fragmentation pathway [5]) could be identified, as has also been described for *E. coli* PBPs 5 and 6 (1), whereas the fragments produced penicilloic acid only. It would be interesting to know whether this apparent loss of complexity in the minimal fragments can also be detected with the actual in vivo substrates. Enzymatically active tryptic peptides (which might not necessarily correspond to minimal fragments) have been generated from various other PBPs (2, 10). A careful analysis of enzymatic activity is certainly necessary for determining whether hydrophilic native peptides are suitable objects for crystallization to resolve the three-dimensional arrangement of the β-lactam binding site.

LITERATURE CITED

1. **Amanuma, H., and J. L. Strominger.** 1984. Simultaneous release of penicilloic acid and phenylacetyl glycine by penicillin-binding proteins 5 and 6 of *Escherichia coli*. *J. Bacteriol.* **160:**822–823.
2. **Coyette, J., J.-M. Ghuysen, and R. Fontana.** 1980. Solubilization and isolation of the membrane-bound DD-carboxypeptidase of *Streptococcus faecalis* ATCC 9790. *Eur. J. Biochem.* **88:**297–305.
3. **Ellerbrok, H., and R. Hakenbeck.** 1984. Penicillin-binding proteins of *Streptococcus pneumoniae*: characterization of tryptic peptides containing the β-lactam binding site. *Eur. J. Biochem.* **144:**637–641.
4. **Frère, J.-M., C. Duez, J.-M. Ghuysen, and J. Vanderkerckhove.** 1976. Occurrence of a serine residue in the penicillin-binding site of the exocellular DD-carboxypeptidase-transpeptidase from *Streptomyces* R61. *FEBS Lett.* **70:**257–260.
5. **Frère, J.-M. J.-M. Ghuysen, J. Degelaen, A. Loffet, and H. R. Perkins.** 1975. Fragmentation of benzylpenicillin after interaction with the exocellular DD-carboxypeptidase-transpeptidase of *Streptomyces* R61 and R39. *Nature* (London) **258:**168–170.
6. **Glauner, B., W. Keck, and U. Schwarz.** 1984. Analysis of penicillin-binding sites with a micro-method—the binding site of PBP 5 from *Escherichia coli*. *FEMS Microbiol. Lett.* **23:**151–155.
7. **Kelly, J. A., O. Dideberg, P. Charlier, J. P. Wery, M. Libert, P. C. Moews, J. R. Knox, C. Duez, C. Fraipont, B. Joris, J. Dusart, J.-M. Frère, and J.-M. Ghuysen.** 1986. On the origin of bacterial resistance to penicillin: comparison of a β-lactamase and a penicillin target. *Science* **231:**1429–1431.
8. **Samraoui, B., B. J. Sutton, R. J. Todd, P. J. Artymiuk, S. G. Waley, and D. C. Phillips.** 1986. Tertiary structural similarity between a class A β-lactamase and a penicillin-sensitive D-alanyl carboxypeptidase-transpeptidase. *Nature* (London) **320:**378–380.
9. **Spratt, B. G.** 1983. Penicillin-binding proteins and the future of β-lactam antibiotics. *J. Gen. Microbiol.* **129:**1247–1260.
10. **Waxman, D. J., and J. L. Strominger.** 1979. Cleavage of a COOH-terminal hydrophobic region from D-alanine carboxypeptidase, a penicillin-sensitive bacterial membrane enzyme. *J. Biol. Chem.* **254:**4863–4875.
11. **Yocum, R. R., D. J. Waxman, J. R. Rasmussen, and J. L. Strominger.** 1979. Mechanism of penicillin action: penicillin and substrate bind covalently to the same active site serine in two bacterial D-alanine carboxypeptidases. *Proc. Natl. Acad. Sci. USA* **76:**2730–2734.

Chapter 41

Reactivity of Anti-β-Lactam Antibodies with β-Lactam-Penicillin-Binding Protein Complexes

Thomas Briese
Heinz Ellerbrok
Hans-Martin Schier
Regine Hakenbeck

Penicillin-binding proteins (PBPs) are the target enzymes of β-lactam antibiotics. They share the property of binding β-lactams enzymatically, forming an inactive PBP-antibiotic complex via a serine residue located in the active center (4, 6, 15). Therefore, PBPs can easily be detected after incubation with a radioactive β-lactam, separation on sodium dodecyl sulfate-polyacrylamide gels, and subsequent fluorography (13). The main disadvantage of this technique is the restriction to benzylpenicillin and radioactive derivatives of ampicillin (12); binding of other β-lactams has to rely upon indirect labeling techniques (5). Further problems arise with PBPs of low affinity for β-lactams or during in vivo labeling experiments.

Recently we have established a different method for detection of PBPs which is based on the reactivity of anti-β-lactam antibodies with β-lactam-PBP complexes on immunoblots (7). Antibodies against the benzylpenicilloyl (BPO) determinant have long been known to occur in penicillin-allergic patients. Anti-BPO antibodies can easily be induced in animals (10, 11). For immunization,

Thomas Briese, Heinz Ellerbrok, Hans-Martin Schier, and Regine Hakenbeck • Max-Planck Institut für Molekulare Genetik, D-1000 Berlin 33, Federal Republic of Germany.

the hapten benzylpenicillin is commonly coupled via amino groups (primarily of lysine) to a protein carrier, e.g., bovine immunoglobulin G (bovine gamma globulin; BGG), and roughly 30 BPO residues can be linked to one BGG molecule. The antisera have been shown in several cases to contain antibodies specific for the BPO determinant. The BPO group also occurs in BPO-PBP complexes. However, it is bound via a serine-ester linkage, and a different epitope density is presented as a result of only one BPO-binding site per PBP. It was therefore questionable whether the anti-BPO antibodies would react with BPO-PBP complexes.

Specificity of Anti-β-Lactam Antibodies

Antisera against BGG conjugates of benzylpenicillin (BPO; 30 residues per BGG), mecillinam (6 residues per BGG), piperacillin (20 residues per BGG), and 6-amino penicillanic acid (6-APA; 3 residues per BGG) were raised in rabbits. Specificity of the sera was determined first with the BGG conjugates in an enzyme-linked immunosorbent assay. Both anti-BPO and antimecillinam sera gave a strong cross-reaction with the 6-APA conjugate, indicating the presence of antibodies against the β-lactam thiazolidine ring system as the antigenic determinant, whereas the antipiperacillin serum reacted only with the corresponding piperacillin conjugate, showing a major specificity for the complex side chain of this antibiotic. With the anti-6-APA serum a fairly good reaction against all conjugates was observed.

To elucidate the specificity of anti-BPO antibodies in more detail, mouse monoclonal antibodies were produced (anti-BPO monoclonal antibodies) (T. Briese, Ph.D. dissertation, Freie Universität, Berlin, 1986). Two determinants could be characterized by monoclonal antibody epitopes: the ring system (covered by anti-BPO monoclonal antibody a311) and the side chain (covered by a504). Both epitopes seem to overlap in the area of the amide bond between β-lactam ring and side chain, but exhibit a surprisingly great independence for a 350-dalton hapten. Similar epitopes on the BPO moiety have been reported by De Haan et al., who used a different set of anti-BPO monoclonal antibodies (2). These antibodies gave only negative results with BPO-PBP complexes on immunoblots (unpublished data), probably because their reaction is restricted to higher substituted conjugates (P. de Haan, personal communication).

Reactivity with β-Lactam-PBP Complexes

The antiserum specificity for various β-lactams was further investigated with β-lactam-PBP complexes after Western blotting. In principle, a strong reaction was observed with all β-lactams that were positive in the enzyme-linked immunosorbent assay when presented as BGG conjugates. Surprisingly, none of the antisera reacted with ampicillin-PBPs, although they should contain the thiazolidine ring as in the recognized 6-APA-PBP complexes. As expected, the piperacillin antiserum reacted also with cefoperazone-PBP complexes as a result of the same side chain as in piperacillin, resembling the specificity for the antibiotic side chain of antiaztreonam antiserum (1), which showed cross-reactivity to ceftazidime-PBP complexes (8). The activity of antimecillinam, antipiperacillin, and antiaztreonam antisera is documented in Fig. 1, using *Escherichia coli* PBPs labeled with the respective antibiotics. It has been shown by competition experiments that mecillinam binds exclusively to PBP 2 (13), and piperacillin (9) as well as aztreonam (14) at lower concentrations bind to PBP 3, which is in perfect agreement with the results obtained in the experiment shown in Fig. 1.

An unexpected result concerns the discrimination of certain β-lactam-PBP complexes: all antisera against penicillins failed to give strong reactions with, for example, pneumococcal PBP 3, although this is the most abundant PBP in this species. In contrast, this was not true for aztreonam (8). Figure 2 shows this phenomenon for anti-BPO-monoclonal antibody a311 and a variety of bacterial species. Although most PBPs can be easily identified, some BPO-PBP

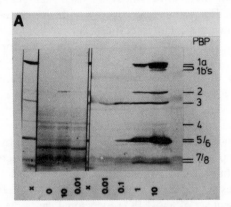 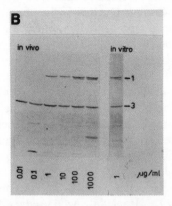

Figure 1. Immunoblot of *E. coli* PBPs after labeling with mecillinam, piperacillin, or aztreonam. (A) Membranes (150 μg) were incubated with mecillinam (0, 10, and 0.01 μg/ml as indicated). Samples were prepared for sodium dodecyl sulfate-polyacrylamide gel electrophoresis as described (13). Lanes indicated with *x* had been labeled with benzylpenicillin (100 μg/ml). After Western blotting, strips were cut and treated with the corresponding anti-β-lactam antisera which had been preadsorbed to *E. coli* cells to reduce background staining. The antimecillinam antibodies had been further purified by affinity chromatography on mecillinam-Sepharose. (B) In vivo labeling of *E. coli* PBPs with aztreonam. Exponentially growing *E. coli* C600 was incubated with aztreonam at the concentrations indicated at the bottom. After 10 min, cells were centrifuged, and membranes were prepared after sonication of the cells and differential centrifugation. Membranes from untreated control cells were labeled in vitro. An immunoblot was prepared and developed with antiaztreonam antiserum.

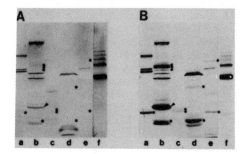

Figure 2. BPO-PBP complexes of various bacterial species visualized with anti-BPO monoclonal antibody a311. Proteins of [³⁵S]benzylpenicillin-labeled membrane preparations of *Streptococcus pneumoniae* (a), *Bacillus subtilis* (b), *Neisseria meningitidis* (c), *Thermos thermophilus* (d), *Staphylococcus aureus* (e), and *Streptococcus pyogenes* (f) were separated on sodium dodecyl sulfate-polyacrylamide gels. After Western blotting, PBP complexes were revealed with monoclonal antibody a311, purified by fast protein-liquid chromatography. *, PBPs which are detected on the autoradiogram but not on the immunoblot. (A) Immunoblot; (B) autoradiogram of the same blot.

complexes, especially the low-molecular-weight PBPs, failed to react completely. Irrespective of whether monoclonal antibodies or polyclonal sera were used, the picture remained essentially the same. The reason for this is totally unclear. It is possible that a partial renaturation of antigens occurs during the blocking procedure of immunoblots so that the antibodies cannot reach their determinant in the PBP complexes as a result of steric hindrance. This idea is supported by the observation that antisera with specificities for a pronounced side chain of the respective β-lactam, like aztreonam, reacted perfectly well with pneumococcal aztreonam-PBP 3, whereas all antisera with main specificities for the thiazolidine ring failed to do so.

This assumption is also compatible with results shown in Fig. 3. The recognition of BPO-PBP complexes was tested in solution, using native intact proteins, native minimal fragments containing the β-lactam binding site, or denatured BPO-PBPs. The minimal fragments are products obtained after trypsin treatment of native PBPs, which are still able to bind β-lactams, suggesting that the three-dimensional arrangement of the active center is greatly preserved (3; Ellerbrok and Hakenbeck, this volume). In a radioimmunoprecipitation assay, the BPO antibodies did not recognize native PBPs, with the exception of some PBP 2b and the corresponding minimal fragment IIb. After denaturation, however, all BPO-PBPs could be sedimented.

In summary, the immunological technique for determining PBPs can be readily applied by using denatured β-lactam-PBP complexes. It does not depend on radioactive antibiotics and is superior in sensitivity to the commonly used fluorographic technique, especially when high concentrations of the antibiotic are needed for PBP labeling. Since the antibodies readily react with β-lactam-containing peptides, they should be also useful in isolating the β-lactam binding site of various PBPs.

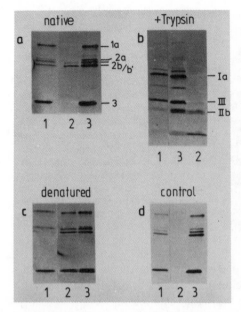

Figure 3. Radioimmunoprecipitation assay of BPO-PBPs. Solubilized, partially purified PBPs from *S. pneumoniae* were used. In this preparation, PBPs 2x and 2a appear with the same electrophoretic mobility, and PBP 2b′ is a degradation product of PBP 2b. After labeling with [^{35}S]benzylpenicillin, unbound antibiotic was removed by microstep exclusion chromatography. Anti-BPO antiserum was added to (a) native, (b) trypsin-treated, and (c) denatured protein samples. (d) Native proteins (control); no antibodies added. After incubation for 18 h at 4°C, antigen-antibody complexes were sedimented with protein A-Sepharose. PBPs were analyzed by fluorography after the microstep (lane 1), in the sediment (lane 2), and in the supernatant (lane 3). Intact PBPs 1a, 2b, and 3 and the corresponding minimal fragments Ia, IIb, and III are indicated.

LITERATURE CITED

1. **Adkinson, N. F., E. A. Swabb, and A. A. Sugerman.** 1984. Immunology of the monobactam aztreonam. *Antimicrob. Agents Chemother.* **25**:93–97.

2. **De Haan, P., A. J. R. de Jonge, T. Verbrugge, and D. H. Boorsma.** 1985. Three epitope-specific monoclonal antibodies against the hapten penicillin. *Int. Arch. Allergy Appl. Immunol.* **76**:42–46.

3. **Ellerbrok, H., and R. Hakenbeck.** 1984. Penicillin-binding proteins of *Streptococcus pneumoniae*: characterization of tryptic peptides containing the β-lactam binding site. *Eur. J. Biochem.* **144**:637–641.

4. **Frère, J.-M., C. Duez, J.-M. Ghuysen, and J. Vanderkerckhove.** 1976. Occurrence of a serine residue in the penicillin-binding site of the exocellular DD-carboxypeptidase-transpeptidase from *Streptomyces* R61. *FEBS Lett.* **70**:257–260.

5. **Frère, J.-M., and B. Joris.** 1985. Penicillin-sensitive enzymes in peptidoglycan biosynthesis. *Crit. Rev. Microbiol.* **11**:299–396.

6. **Glauner, B., W. Keck, and U. Schwarz.** 1984. Analysis of penicillin-binding sites with a micro-method—the binding site of PBP 5 from *Escherichia coli*. *FEMS Microbiol. Lett.* **23**:151–155.

7. **Hakenbeck, R., T. Briese, and H. Ellerbrok.** 1986. Antibodies against the the benzylpenicilloyl moiety as a probe for penicillin-binding proteins. *Eur. J. Biochem.* **157**:101–106.

8. **Hakenbeck, R., S. Tornette, and N. F. Adkinson.** 1987. Interaction of non-lytic β-lactams with penicillin-binding proteins in *Streptococcus pneumoniae*. *J. Gen Microbiol.* **133**:755–760.

9. **Iida, K., S. Hirata, S. Nakamuta, and M. Koike.** 1978. Inhibition of cell division of *Escherichia coli* by a new synthetic penicillin, piperacillin. *Antimicrob. Agents Chemother.* **14**:257–266.

10. **Levine, B. B.** 1960. Studies on the mechanism of the formation of the penicillin antigen. I. Delayed allergic cross-reactions among penicillin G and its degradation products. *J. Exp. Med.* **112:**1131–1154.
11. **Locher, G. W., C. Schneider, and A. L. de Weck.** 1969. Hemmung allergischer Reaktionen auf Penicillin durch Penicilloyl-Amide und chemisch verwandte Substanzen. *Z. Immunitaetsforsch. Allerg. Klin. Immunol.* **138:**299–323.
12. **Schwarz, J., K. Seeger, F. Wegenmayer, and H. Strecker.** 1981. Penicillin-binding proteins of *E. coli* identified with a ^{125}J-derivative of ampicillin. *FEMS Microbiol. Lett.* **10:**107–109.
13. **Spratt, B. G.** 1975. Distinct penicillin-binding proteins involved in the division, elongation, and shape of *Escherichia coli*. *Proc. Natl. Acad. Sci. USA* **72:**2999–3003.
14. **Sykes, R. B., D. P. Bonner, K. Bush, and N. H. Georgopapadakou.** 1982. Azthreonam (SQ 26,776), a synthetic monobactam specifically active against aerobic gram-negative bacteria. *Antimicrob. Agents Chemother.* **21:**85–92.
15. **Yocum, R. R., D. J. Waxman, J. R. Rasmussen, and J. L. Strominger.** 1979. Mechanism of penicillin action: penicillin and substrate bind covalently to the same active site serine in two bacterial D-alanine carboxypeptidases. *Proc. Natl. Acad. Sci. USA* **76:**2730–2734.

Chapter 42

Development of Resistance to Penicillin G in *Streptococcus mutans* GS5

O. Massidda
L. Daneo-Moore

Intrinsic resistance to penicillin G in streptococci usually involves alterations in the binding of penicillin by one or more penicillin-binding proteins (PBPs) (for recent reviews, see references 2, 7). These alterations may be a decrease in the affinity of one or more PBPs for penicillin (3, 8) or the apparent or actual loss of PBPs (5, 9). When penicillin resistance develops in a laboratory environment, it usually occurs in several steps (1, 3, 6), each exhibiting an increasing level of resistance. These steps could involve alterations in amount or affinity of a single PBP for each step, or one step may be accompanied by alterations of more than one PBP. In a study of progressively more resistant transformants of *Streptococcus pneumoniae,* Zighelboim and Tomasz observed that more than one PBP could be altered in the process of acquisition of several levels of resistance (9). The progressively more resistant transformants were not characterized genetically.

In the present study we examined the kinetics of transformation of *Streptococcus mutans* GS5 with DNA obtained from a penicillin-resistant laboratory derivative. This method permits inferences on the number of DNA fragments required to obtain a given level of resistance (4). The results of our study are consistent with the view that a single transformation event can produce alterations of the affinity or amounts, or both, of more than one PBP.

O. Massidda and L. Daneo-Moore • Department of Microbiology and Immunology, Temple University School of Medicine, Philadelphia, Pennsylvania 19140.

410

A penicillin-resistant derivative of *S. mutans*, GS5-M30, was obtained by sequential plating of 10^9 cells on increasing levels of penicillin G. The MICs of penicillin G in Todd-Hewitt broth (2% glucose) were 0.04 μg/ml for GS5 and 0.32 μg/ml for the GS5-M30 derivative. The penicillin-binding profiles of these two organisms are shown in Fig. 1 (lanes A and B). In the resistant strain M30 the most striking change was a decreased affinity for penicillin of a PBP with an M_r of 76,000, which we call PBP 3 (Table 1).

When 2.5 μg of DNA from M30 per ml was used to transform *S. mutans* GS5, transformants growing on 0.08 μg of penicillin G per ml occurred at a frequency of about 5.1×10^{-4}. At different DNA concentrations, the kinetics of transformation had a slope of 1.00, indicating a single transformation event (Fig. 2). DNase I (1,000 μg/ml) plus 0.01 M magnesium sulfate (final concentration) were added 30 min after the DNA. The cells were incubated for an additional 90 min to allow phenotypic expression and were plated on penicillin G.

The transformants resistant to 0.08 μg/ml were replica plated on increasing concentrations of penicillin G. The second curve shown in Fig. 2 gives the

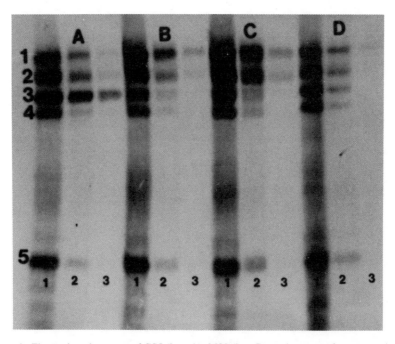

Figure 1. Electrophoretic pattern of GS5 (lane A), M30 (lane B), and two transformants resistant to 0.32 (lane C) and 0.08 (lane D) μg of penicillin G per ml. Each strain was incubated with 2.5 (lanes 1), 0.5 (lanes 2), and 0.1 (lanes 3) μg of ³H-penicillin G per ml.

Table 1. PBPs of S. mutans GS5

Protein	Apparent mol wt	% found in isolate membranes[a]
PBP 1	84,000	21.5 ± 5.7
PBP 2	80,000	24.8 ± 0.1
PBP 3	76,000	22.9 ± 0.0
PBP 4	73,000	15.2 ± 0.0
PBP 5	46,000	15.3 ± 4.2

[a]Determined at 12.5 μg of penicillin G per ml.

results obtained with 0.20 μg of penicillin G per ml and has a slope of 1.4. The identical slope was obtained when plating on 0.24 μg of penicillin G per ml. In a separate experiment, direct selection on 0.08 and 0.16 μg/ml, re-

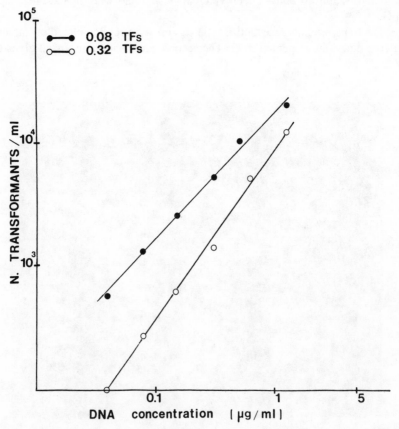

Figure 2. Transformation in S. mutans GS5 using different concentrations of donor DNA and selecting on different concentrations of penicillin G. The slope is 1.0 for the first step (0.08 μg/ml) and 1.4 for the second step (0.32 μg/ml).

spectively, gave slopes of 1.0 and 1.3. Transformation with sheared DNA gave a slope of about 2 for transformants selected on 0.24 μg/ml. The transformants resistant to 0.16 μg/ml were replica plated on various concentrations of penicillin and found to grow at up to 0.32 μg/ml, the resistance level of the M30 donor. The GS5 recipient was transformed to resistance to 0.32 μg/ml with a frequency of 6.2×10^{-5}. The MICs in Todd-Hewitt broth for representatives of the first- and second-level transformants were 0.08 and 0.32 μg/ml.

The results shown in Fig. 2 suggest that two transformation events are required to obtain resistance to 0.32 μg of penicillin per ml, and one event is required to obtain resistance to 0.08 μg/ml. The two genes tentatively named *pen-1* and *pen-2* appear to be linked.

Preliminary analyses of the PBPs of the recipient, the donor, and the two transformants are shown in Fig. 1 for three penicillin concentrations and in Fig. 3 for a variety of concentrations. The apparent molecular weights (M_r) are given in Table 1. To compare data from various binding experiments, we used a standard consisting of recipient membranes incubated with 2.5 μg of penicillin G per ml in each gel. Quantitation was done with an image analyzer (Technal Tc-1; Technal Corp, Englewood, N.J.) which was programmed to convert densities in the autoradiograph to density per 100 μg of membrane protein and to relate these data to the standard. We usually analyzed two autoradiographs developed for 7 and 15 days, to correct for overdevelopment; a computer program was also used.

PBP 3 of *S. mutans* GS5 was saturated at the lowest penicillin concentration (Fig. 3A) whereas PBP 4 was not saturated even at the highest penicillin concentration tested. In the first-step transformants (Fig. 3B), PBP 2 saturated at a lower level, and PBP 3 had a greatly decreased affinity for penicillin. At the second step of penicillin resistance (Fig. 3C), PBP 3 decreased further in affinity for penicillin, and PBPs 4 and 5 were decreased in saturation levels. Comparison of the data in Fig. 3C and D indicates that the second-step transformant is essentially identical to the donor (see also Fig. 2).

These data are preliminary and require further analysis at additional penicillin G concentrations, to define saturation levels and binding affinities more precisely. However, the data suggest that more than one PBP is altered by any one single transformation event. The observed alterations are consistent with a model in which transformation of an altered structural gene is followed by secondary compensatory changes in other PBPs. Alternatively, transformation to penicillin resistance may involve alterations of a gene or genes responsible for the processing of more than one PBP. While the molecular mechanism remains unclear, the present data are consistent with the observations of Zighelboim and Tomasz in *S. pneumoniae* (9) and with observations in *Streptococcus sanguis* (E. Zito and L. Daneo-Moore, *J. Gen. Microbiol.*, in press). The present study complements the study of *S. pneumoniae* in that transformation kinetics are used to define the number of transformation events.

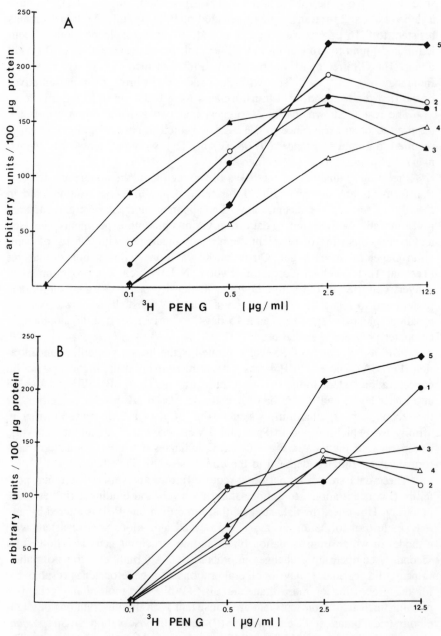

Figure 3. Binding of penicillin G to PBPs of the GS5 recipient (A), the transformants resistant to 0.08 (B) and 0.32 μg (C) of penicillin G per ml, and the M30 donor (D).

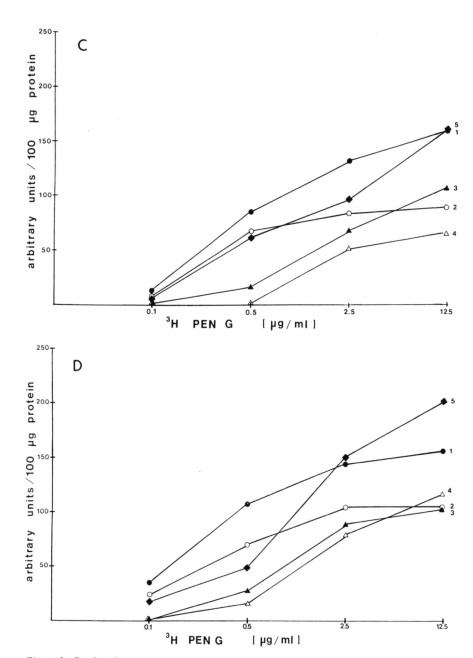

Figure 3. Continued

ACKNOWLEDGMENTS. This research was supported by Public Health Service grant DE 051-80-08 from the National Institutes of Health.

LITERATURE CITED

1. **Daneo-Moore, L., M. Pucci, E. Zito, and M. Ferrero.** 1983. Studies of β-lactam resistance in *Streptococcus faecium,* p. 493–497. *In* R. Hakenbeck, J. V. Höltje, and L. Labischinski (ed.), *The Target of Penicillin.* Walter De Gruyter, Berlin.
2. **Fontana, R.** 1985. Penicillin-binding proteins and the intrinsic resistance to β-lactams in gram-positive cocci. *J. Antimicrob. Chemother.* **16:**412–416.
3. **Fontana, R., R. Cerini, P. Longoni, A. Grossato, and P. Canepari.** 1983. Identification of a streptococcal penicillin-binding protein that reacts very slowly with penicillin. *J. Bacteriol.* **155:**1343–1345.
4. **Goodal, G. H., and R. M. Herriott.** 1961. Studies on transformation of *Hemophilus influenzae. J. Gen. Physiol.* **44:**1229.
5. **Hakenbeck, R., H. Ellerbrok, T. Briese, S. Handwerger, and A. Tomasz.** 1986. Penicillin-binding proteins of penicillin-susceptible and -resistant pneumococci: immunological relatedness of altered proteins and changes in peptides carrying the β-lactam binding sites. *Antimicrob. Agents Chemother.* **30:**553–558.
6. **Handwerger, S., and A. Tomasz.** 1986. Alteration in penicillin binding proteins of clinical and laboratory isolates of pathogenic *Streptococcus pneumoniae* with low levels of penicillin resistance. *J. Infect. Dis.* **153:**83–89.
7. **Malouin, F., and L. E. Bryan.** 1986. Modification of penicillin-binding proteins: a mechanism of β-lactam resistance. *Antimicrob. Agents Chemother.* **30:**1–5.
8. **Reynolds, P. E., and D. F. J. Brown.** 1985. Penicillin-binding proteins of beta-lactam-resistant strains of *Staphylococcus aureus. FEBS Lett.* **192:**28–32.
9. **Zighelboim, S., and A. Tomasz.** 1980. Penicillin-binding proteins of multiply antibiotic resistant South African strains of *Streptococcus pneumoniae. Antimicrob. Agents Chemother.* **17:**437–442.

VI.   OUTER MEMBRANE PERMEABILITY

Chapter 43

Significance of the Outer Membrane Barrier in β-Lactam Resistance

Hiroshi Nikaido
Kalle Gehring

It is well known that penicillin G shows poor activity against gram-negative bacteria. Indeed, Fleming (5) justified his work on penicillin on the basis of its possible utility as an ingredient of selective media for gram-negative bacilli. It has now become clear that this intrinsic resistance of gram-negative organisms is at least partially due to the presence of the outer membrane barrier, a permeation barrier that is unique to these bacteria (9, 11). Another factor that is often equally important in the intrinsic resistance of gram-negative bacteria toward β-lactams is the nearly ubiquitous presence of β-lactamase in the periplasmic space of these organisms (15). Thus, in the gram-negative cell envelope, the β-lactams must go through two successive barriers, first the outer membrane barrier and then the β-lactamase "barrier" in the periplasm, to reach their targets, which are the penicillin-binding proteins (PBPs) located on the outer surface of the cytoplasmic membrane. Although each of these barriers has been studied rather extensively, the synergistic interaction of these two barriers has not been examined seriously up to this time. In this chapter we present a theoretical model for the interplay of these two barriers and discuss how each of them contributes to produce a very effective composite barrier.

Hiroshi Nikaido and Kalle Gehring • Department of Microbiology and Immunology, University of California, Berkeley, California 94720.

The Outer Membrane Barrier: Diffusion of β-Lactams through Porin Channels

The lipopolysaccharide-phospholipid bilayer of the outer membrane is relatively impermeable to molecules of moderate hydrophobicity such as β-lactams, and thus the major pathway of diffusion of these compounds appears to be through the water-filled channels of porins, a special class of proteins that exist in very large numbers, for example, 100,000 copies per *Escherichia coli* cell (9, 11). The major role of porins has been confirmed for the transport of many cephalosporins by measuring the rate of diffusion across the outer membrane directly by coupling it to their hydrolysis in the periplasmic space, using the method introduced by Zimmermann and Rosselet (18). Such studies (17) showed that porins were responsible for at least 90% of the penetration through the outer membrane by cephaloridine, cefamandole, cephalothin, ampicillin, and 6-aminopenicillanic acid, although very recently β-lactams designed to penetrate the outer membrane by utilizing specific iron-chelator receptor pathways have appeared (12, 16b).

Although the porin channels are totally nonspecific, the rate of penetration of solutes is greatly influenced by the gross physicochemical properties of the solute, and larger, more hydrophobic, or negatively charged solutes tend to diffuse more slowly through the channel (11). Furthermore, the extent of such discrimination is affected by the diameter of the channel, and enteric bacilli characteristically produce one or more porin species in response to the particular environmental conditions (11).

The permeability of a number of β-lactam compounds through the OmpF porin of *E. coli* has been determined in the reconstituted liposome system (17). By normalizing these results with the permeability coefficients of several cephalosporins determined in intact cells (10), we can get the interpolated permeability coefficients for a wide range of compounds (Table 1). It can be seen that the permeability per se has little obvious relationship to the efficacy of the compound.

The β-Lactamase Barrier

Even when the β-lactam molecule successfully penetrates through the outer membrane barrier, it may get degraded by the periplasmic β-lactamase before it reaches the target, PBPs. Almost all gram-negative bacteria seem to produce endogenous, presumably chromosomally coded β-lactamases (class C enzyme of Ambler [1]) (15), and many resistant strains contain plasmids that code for class A enzymes, represented by the TEM enzyme. The properties of various β-lactamases have been studied extensively, but a few pertinent points must be emphasized here.

Table 1. Relevant parameters in several *E. coli* strains

| β-Lactam | P^a | c_{inh} (μM) | Chromosomal (AmpC) enzyme | | | | | TEM enzyme from strain JF701(R$_{471a}$) | | |
| | | | From strain LA5 (wild type) | | | From strain TE18 (overproducer) | | | | |
			K_m (μM)	V_{max} (nmol mg⁻¹ s⁻¹)	TAI[b]	V_{max} (nmol mg⁻¹ s⁻¹)	TAI[b]	K_m (μM)	V_{max} (nmol mg⁻¹ s⁻¹)	TAI[b]
Penicillin G	0.2	3	1.9	0.07	0.02	7.1	0.0002	18	115	0.00006
Ampicillin	9.8	2	0.9	0.007	5.5	0.7	0.053	37	127	0.001
Cephalothin	1.1	2	22	0.263	0.13	27.3	0.001	170	11	0.028
Cephaloridine	35.7	6	230	0.16	70	16.6	0.67	350	81	0.26
Cefoxitin	3.7	10	0.22	0.00016	280	0.016	2.71	3,600	0.004	5,400[c]
Cefotaxime	1.8	0.1	0.16	0.00006	9.7	0.0065	0.09	9,500	1.96	14[c]
Ceftazidime	0.96	0.1	0.16	0.00036	56	0.037	0.54	30,000	0.02	2,500[c]
Aztreonam	0.33	0.1	0.01	<0.00001	75	0.00007	0.72	3,100	0.16	10

[a]Permeability coefficient in 10⁻⁵ cm s⁻¹.

[b]Target access index, reflecting the probability of a β-lactam molecule reaching the target without getting hydrolyzed. See text for definition.

[c]These are "theoretical" values based solely on the properties of the TEM enzyme. Since plasmid-containing cells also produce chromosomal enzyme, the values of TAI in these cells are actually much lower (see reference 9a and Table 2).

421

(i) Because the affinity of various β-lactam substrates toward the enzymes can be very different, and K_m values can range from less than 0.05 to more than 30,000 μM, reports of activities determined at only a single substrate concentration often do not give us any indication of the real specificity of the enzyme, especially in view of the importance of the activity at low concentrations (see below).

(ii) Because it is more convenient, practically all assays are carried out at substrate concentrations of 0.1 mM or higher. However, PBPs are irreversibly inactivated by very low concentrations of β-lactams (6). Although it is difficult to predict precisely the concentration needed for inactivation (6), we can take, as a first approximation, the I_{50} concentration, or the concentration that inhibits, by 50%, the binding of the penicillin G to one of the "essential" PBPs, i.e., PBP 1b, PBP 2, and PBP 3. The lowest I_{50} value among these three usually corresponds rather well with the MIC of that particular β-lactam for mutants defective either in the outer membrane barrier (4) or in the β-lactamase barrier (9a) and thus is likely to reflect the periplasmic β-lactam concentration that is just necessary to inhibit the target (c_{inh} of Table 1). The values of c_{inh} are usually in the range of 0.1 to 0.6 μM (4) (see also Table 1), and what is important for the understanding of the bacterial resistance is the behavior of the β-lactamase at these low concentrations (7, 16), because the bacteria will have been killed a long time ago if the periplasmic concentration rises to levels much higher than c_{inh}.

We wanted to analyze the interaction between the outer membrane barrier and the β-lactamase barrier in *E. coli*, because the properties of its outer membrane barrier have been studied in detail as described above. However, the properties of its β-lactamases, especially those of its chromosomally coded enzyme, have not been studied in recent years. We therefore determined the kinetic properties of the *E. coli* chromosomal enzyme as well as of a TEM-type enzyme (9a). Some of these data are shown in Table 1.

The Synergism between the Two Barriers

Because the β-lactam molecules must pass through the outer membrane barrier and the β-lactamase barrier in succession, these two barriers must work synergistically. If we consider the probability of any β-lactam molecule reaching the target at the far end of the periplasm, this should then be proportional to the product of the probability of passage across the outer membrane and the probability of diffusing unhydrolyzed through the periplasm. We recently defined an index, the target access index (TAI), which reflects the probability mentioned above. One property required for an index denoting probability is that it should be dimensionless. Our definition of TAI was aided greatly by the theoretical treatment proposed by S. G. Waley (16a) to analyze the interaction between the outer membrane barrier and β-lactamase-catalyzed degradation. Waley defined a dimensionless index, called permeability number, as follows:

$$PN = (P \cdot A)/(V_{max}/K_m) \tag{1}$$

where *PN*, *P*, and *A* represent permeability number, permeability coefficient of the outer membrane, and the area of the outer membrane per unit mass of cells, and V_{max} and K_m correspond to the usual kinetic constants of the periplasmic β-lactamase. We follow a very similar approach, but use, instead of the physiological efficiency (V_{max}/K_m) (13) of the enzyme, a parameter that is related to the rate of hydrolysis at the β-lactam concentration just sufficient to inhibit the target PBPs, i.e., c_{inh}. Thus:

$$TAI = (P \cdot A) \cdot [V_{max}/(K_m + c_{inh})]^{-1} \tag{2}$$

An advantage of using the TAI is that it allows one to predict MICs for any organism, as long as one knows how permeable the outer membrane is to the drug and how sensitive the drug is to the β-lactamase. We assume that the cells will be killed when the steady-state concentration of the drug in the periplasm reaches c_{inh}, and that at steady state the net influx across the outer membrane is balanced by the hydrolytic degradation in the periplasm (18). Then the external concentration of the drug that will bring about this situation, i.e., the MIC, can be predicted by a simple equation (9a):

$$MIC = (TAI^{-1} + 1)c_{inh} \tag{3}$$

We carried out calculations of this type for five strains of *E. coli* producing widely different levels of two types of β-lactamases, with 13 different β-lactams. In 44 of these 65 predictive attempts, the predicted MIC values were within ±1 tube of the values obtained in serial twofold dilution assays. Furthermore, two compounds, cefsulodin and carbenicillin, consistently gave observed MICs that are widely different from the predicted ones, and we suspect that some of the parameters used for these compounds must be in error. If we exclude these two compounds, the rate of success of this quantitative MIC prediction approach was 43 of 55, i.e., 78%. These results suggest that the theoretical model is a correct one and further indicate that the data used in the calculation, such as the kinetic parameters of the enzymes and c_{inh} values, were not far off.

How Important Is the Outer Membrane Barrier in Comparison with the β-Lactamase Barrier?

The TAI reflects the probability with which β-lactam molecules reach the target, given the presence of the two successive barriers. However, sometimes we would like to know the relative importance of each of the barriers. Although the first term $(P \cdot A)$ in equation 2 is related to the penetration rate across the outer membrane, and the second term $[V_{max}/(K_m + c_{inh})]^{-1}$ is roughly related to the probability of the β-lactamase molecules escaping enzymatic degradation, one cannot compare the numerical values of these two terms directly, because their dimensions are different, corresponding to $cm^3\ s^{-1}\ mg^{-1}$ and $cm^{-3}\ s\ mg$, respectively. What we need here is a set of conversion factors to convert these terms into dimensionless numbers. Let us call the two parameters generated by multiplying the two terms of equation 2 by the appropriate conversion factors, k and k', the access index (outer membrane), AI_{om}, and the access index (periplasm), AI_{peri}:

$$AI_{om} = k \cdot (P \cdot A) \tag{4}$$
$$AI_{peri} = k' \cdot [V_{max}/(K_m + c_{inh})]^{-1} \tag{5}$$

We want to define k and k' so that:

$$TAI = AI_{om} \cdot AI_{peri} \quad \text{or} \quad k \cdot k' = 1 \tag{6}$$

We also wish to define the conversion factors in such a way that the parameters of the same magnitude will produce the effects of the same magnitude on the cell. As an example of such an "effect," we can take the decrease in the concentration of β-lactam at the target site, produced purely by the action of one of the barriers.

If we have only the outer membrane barrier, we can assume that the decrease in the periplasmic concentration will occur mainly by the dilution due to the increase in cellular volume during growth. If C_o, C_p, V_p, μ, and t represent the drug concentration in the outside medium, that in the periplasm, the periplasmic volume, the growth rate (i.e., the rate of increase of periplasmic volume, also), and a very short time, respectively, then under steady-state conditions:

$$P \cdot A \cdot (C_o - C_p) \cdot t = (\mu \cdot V_p \cdot t) \cdot C_p \tag{7}$$

because the influx represented by the left-hand side, according to Fick's first law of diffusion, must be balanced by the amount needed to fill the increased periplasmic space during this time period, shown on the right-hand side of the equation. From this equation and equation 4, we get equation 8 if we denote the ratio, C_o/C_p, by R.

$$\text{AI}_{om}/k = (\mu \cdot V_p)/(R - 1) \tag{8}$$

A similar analysis can be carried out for the hypothetical situation in which the β-lactamase molecules would exist at exactly the same location as in the real cell but the cell would be totally devoid of the outer membrane barrier. In such a situation, theoretical analysis shows that the concentration at the surface of the cytoplasmic membrane, or at the target, C_t, is:

$$C_t = C_o - (a^2 \cdot v)/(2 \cdot D \cdot V_p) \tag{9}$$

where a, v, and D denote the depth of the periplasmic space, the rate of hydrolysis of the β-lactam by the enzyme present in cells of unit mass, and the diffusion coefficient of the β-lactam (see Appendix). If we consider the situation where $C_t = c_{inh}$, then $v = (V_{max} \cdot c_{inh})/(K_m + c_{inh})$. Using equation 5:

$$\text{AI}_{peri}/k' = (a^2/2 \cdot D \cdot V_p)/(R - 1) \tag{10}$$

Since we assume that equal values of AI_{om} and AI_{peri} will generate an identical value of R, dividing both sides of equation 10 with both sides of equation 8 gives:

$$\begin{aligned} k/k' &= a^2/(2 \cdot \mu \cdot D \cdot V_p^2) \\ &= 1/(2 \cdot \mu \cdot D \cdot A^2) \end{aligned} \tag{11}$$

because $V_p = a \times A$, A being the surface area of unit mass (1 mg) cells. This equation conveniently enables us to avoid making arbitrary assumptions on parameters with uncertain values, such as a and V_p.

For computation, we use the following values: assuming the doubling time of 30 min, $\mu = (\ln 2)/(30 \times 60) = 3.85 \times 10^{-4}$ s^{-1}; assuming that most β-lactams will have diffusion coefficients lower than that of sucrose (342 daltons) $(5.21 \times 10^{-6}$ cm^2 s$^{-1})$ (8) and higher than that of raffinose (504 daltons) $(4.34 \times 10^{-6}$ cm^2 s$^{-1})$ (8), $D = 5 \times 10^{-6}$ cm^2 s^{-1}; surface area of the cells, 132 cm^2 mg^{-1} (14). Substituting these values into equation 11, and also from equation 6, we obtain: $k = 120.9$ cm^{-3} s mg and $k' = 0.00827$ cm^3 s^{-1} mg^{-1}.

Some of the values of AI are shown in Table 2. As can be seen, in wild-type *E. coli* in which the periplasmic hydrolysis is slow, the outer membrane contributes more (i.e., AI_{om} is lower than AI_{peri}) toward the resistance, and the converse situation is often seen in strains overproducing either the TEM enzyme or the chromosomal enzyme. This comparison also allows us to estimate, quantitatively, the factors involved in the improvement of the efficacy of β-

Table 2. Relative contribution of the outer membrane barrier and the β-lactamase barrier toward the resistance

β-Lactam	AI_{OM}	AI_{peri} for strain:		
		LA5 (wild type)	TE18 (AmpC overproducer)	JF701(R_{471a}) (TEM producer)
Penicillin G	0.032	0.59	0.006	0.002
Ampicillin	1.56	3.51	0.033	0.003
Cephalothin	0.18	0.75	0.007	0.13
Cephaloridine	5.70	12.2	0.12	0.036
Cefoxitin	0.59	483	4.66	451[a]
Cefotaxime	0.29	34.1	0.33	18[a]
Ceftazidime	0.15	372	3.59	361[a]
Aztreonam	0.053	1,440	13.9	156

[a]Much of the hydrolysis occurs by the chromosomally coded enzyme.

lactams. For example, by changing penicillin G into ampicillin, the access through the outer membrane was indeed improved by a factor of 50, but at the same time the resistance to the endogenous, chromosomally coded enzyme was also improved significantly (nearly sixfold). Also, in comparing cephaloridine with penicillin G, we see a strong improvement in permeability (a nearly 180-fold increase in AI_{om}), but at the same time also a 20-fold better chance of survival of cephaloridine in the periplasmic space. These examples show us that resistance to hydrolysis by endogenous β-lactamase is sometimes very important in producing an effective β-lactam antibiotic. It should also be pointed out that the chromosomal β-lactamases of enteric bacilli are much more efficient toward penicillins than toward the classical cephalosporins as a result of their much higher affinity for the former compounds; in this sense, calling these enzymes cephalosporinases is quite misleading.

Is There a "Limiting Step" in the β-Lactam Penetration through Cell Envelope?

When one of the parameters, either AI_{om} or AI_{peri}, is much smaller than the other, can we consider the corresponding penetration step to be the "limiting step"? Before we attempt to answer this question, we have to define what the limiting step is. One definition is the step that alone essentially determines the final outcome. When we consider the fluxes of β-lactams, either the outer membrane or the periplasmic degradation can easily become the rate-limiting step. Thus, the rate of hydrolysis by intact cells of compounds of high permeability such as cephaloridine may be nearly equal to the rate of hydrolysis by cell extracts under certain conditions, producing "crypticity" of 1 (9): clearly the β-lactamase barrier is the limiting step here. Similarly, with poorly penetrating compounds, we can easily find a situation in which increases in the level of periplasmic enzyme do not substantially affect the rate of hydrolysis by intact cells, because the whole process is limited by the diffusion rate across the outer membrane (9).

We see, however, a very different situation when we consider the steady-state periplasmic concentration of the drug, rather than the fluxes of the drug, as the "outcome" of the process. When we calculate the β-lactam concentration that will be achieved under the hypothetical conditions of the total absence of the outer membrane barrier or the total absence of the β-lactamase barrier, according to equations 9 and 7, respectively, the results (Table 3) indicate clearly that even when some of the least permeable and most β-lactamase-sensitive compounds interact with cells overproducing huge amounts of enzyme, one single barrier is nearly totally useless, and the synergism of the two barriers is absolutely essential. (We note, however, that the values in Table 3 were calculated by using diffusion coefficients determined in water. If small molecules also diffuse in the periplasm as slowly as macromolecules [2], the β-lactamase barrier may become much more effective even in the absence of the outer membrane barrier.) Thus, it appears that it is a misleading and futile exercise to try to determine the "rate-limiting" step if we are interested in the final periplasmic concentration of the agent and the bactericidal action of the drug. It is only the product of the two access indices, i.e., the TAI, that matters, and changes in the outer membrane permeability and those in the efficiency of the β-lactamase should always have an equivalent effect on the MIC.

In some cases, however, the alterations in the AI_{om} or AI_{peri} do not produce much visible change in the resistance level, and in others they do. This is due to the form of equation 2. When the TAI is much larger than 1, then even severalfold changes in the values of AI and therefore of TAI are unlikely to produce more than minimal changes in the MIC, but very large changes accompany the alterations of AI when TAI is much smaller than 1. Thus, in wild-type *E. coli* the decrease in the porin content and therefore AI_{om} produces large

Table 3. Calculated steady-state β-lactam concentration (*C*) at the target under various conditions

Condition	*C* at target (μg/ml)
JF701(R$_{471a}$) (containing TEM enzyme)	
in 4,096 μg of penicillin G per ml	
With both barriers (real cell)	0.8
With no outer membrane (hypothetical)	4,013
With no β-lactamase barrier (hypothetical)	4,092
TE18 (chromosomal enzyme overproducer)	
in 1,024 μg of cephalothin per ml	
With both barriers (real cell)	1
With no outer membrane (hypothetical)	1,020
With no β-lactamase barrier (hypothetical)	1,024

changes in the MIC of compounds such as cephalothin (TAI = 0.13) but only very small changes in the MIC of cefotaxime, ceftazidime, or aztreonam (TAI > 9). Although such an observation was considered evidence that the latter compounds do not pass through porin pathways (3), these results are explainable on the basis of diffusion of the compounds through the porin channel followed by hydrolysis by the periplasmic β-lactamase.

Appendix. Calculation of the Effect of the β-Lactamase Barrier

We consider a layer of β-lactamase-containing space, with a thickness of a. If we consider an infinitesimally thin slice that is x centimeters away from the bottom, where the targets, PBPs, are located, the flux of the β-lactam J_x through the slice of unit area due to diffusion is

$$J_x = -D \cdot (dC(x)/dx)$$

where D is the diffusion coefficient and $C(x)$ is the concentration of the β-lactam x centimeters away from the bottom. Since the system is at steady state, this flux must be balanced by the hydrolysis that occurs in the volume with a cross-section of unit area and the thickness of x (see Fig. 1). If we let v represent the rate of hydrolysis of the β-lactam by all the enzymes present in the cells of unit mass, then only a fraction $[(x/a) \times (1/A) = x/V_p$, A being the area of the surface of cells of unit mass and V_p being the total periplasmic volume of these cells] of these enzymes participate in the hydrolysis. Thus, if we assume that the rate of hydrolysis stays more or less constant throughout the periplasm, the rate of hydrolysis occuring in this volume $H(x)$ is:

$$H(x) = J_x = -v \cdot x/V_p$$

When we eliminate J_x from these equations and integrate, we get:

$$(x^2/2)(v/V_p) = D \cdot C(x) + \text{const}$$

where const is the integration constant. At the surface of this layer, where $x = a$, $C(x) = C_o$. Substituting these values, we get:

$$C(x) = C_o - (a^2 - x^2) \cdot (v/V_p)/(2 \cdot D)$$

Thus, at the target, where $x = 0$

$$C(0) = C_t = C_o - (a^2 \cdot v)/(2 \cdot V_p \cdot D)$$

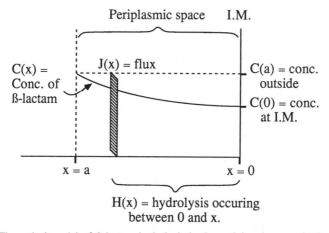

Figure 1. Theoretical model of β-lactam hydrolysis in the periplasmic space in the absence of outer membrane barrier. Through a plane x centimeters away from the surface of the inner membrane (I.M.) the flux $J(x)$ is determined by the diffusion of the drug down the gradient of its concentration $C(x)$. At the same time, steady-state conditions dictate that the influx of the drug should be precisely balanced by the hydrolysis $H(x)$ that takes place by enzymes located in the layer with a thickness of x centimeters.

Calculations (Table 3) were performed by assuming the following: $a = 5 \times 10^{-6}$ cm and $V_p = 132 \times 5 \times 10^{-6} = 6.6 \times 10^{-4}$ cm^3 mg^{-1}. The values obtained showed that we can expect less than 2% difference in the β-lactam concentration between the outside medium and at target, and thus the simplifying assumption of constant hydrolysis rate, v, was justified. If $C(x)$ is not much larger than the K_m of the enzyme, and if v changes depending on $C(x)$, the difference between C_t and C_o will be even less.

LITERATURE CITED

1. **Ambler, R. P.** 1980. The structure of β-lactamases. *Philos. Trans. R. Soc. London* **289:**321–331.
2. **Brass, J. M., C. F. Higgins, M. Foley, P. A. Rugman, J. Birmingham, and P. B. Garland.** 1986. Lateral diffusion of proteins in the periplasm of *Escherichia coli. J. Bacteriol.* **165:**787–794.
3. **Curtis, N. A. C., R. L. Eisenstadt, K. A. Turner, and A. J. White.** 1985. Porin-mediated cephalosporin resistance in *Escherichia coli* K-12. *J. Antimicrob. Chemother.* **15:**642–644.
4. **Curtis, N. A. C., D. Orr, G. W. Ross, and M. G. Boulton.** 1979. Affinities of penicillins and cephalosporins for the penicillin-binding proteins of *Escherichia coli* K-12 and their antibacterial activity. *Antimicrob. Agents Chemother.* **16:**533–539.
5. **Fleming, A.** 1929. On the antibacterial action of cultures of a *Penicillium*, with special reference to their use in the isolation of *B. influenzae. Br. J. Exp. Pathol.* **10:**226–236.
6. **Frère, J.-M., and B. Joris.** 1985. Penicillin-sensitive enzymes in peptidoglycan biosynthesis. *Crit. Rev. Microbiol.* **4:**299–396.

7. **Livermore, D. M.** 1984. Kinetics and significance of the activity of the Sabbath and Abraham's β-lactamase of *Pseudomonas aeruginosa* against cefotaxime and cefsulodin. *J. Antimicrob. Chemother.* **11**:169–179.

8. **Longworth, L. G.** 1953. Diffusion measurements, at 25°, of aqueous solutions of amino acids, peptides and sugars. *J. Am. Chem. Soc.* **20**:5705–5709.

9. **Nikaido, H.** 1985. Role of permeability barriers in resistance to β-lactam antibiotics. *Pharmacol. Ther.* **27**:197–231.

9a. **Nikaido, H., and S. Normark.** 1987. Sensitivity of *Escherichia coli* to various β-lactams is determined by the interplay of outer membrane permeability and degradation by periplasmic β-lactamases: a quantitative predictive treatment. *Mol. Microbiol.* **1**:29–36.

10. **Nikaido, H., E. Y. Rosenberg, and J. Foulds.** 1983. Porin channels in *Escherichia coli*: studies with β-lactams in intact cells. *J. Bacteriol.* **153**:232–240.

11. **Nikaido, H., and M. Vaara.** 1985. Molecular basis of bacterial outer membrane permeability. *Microbiol. Rev.* **45**:1–32.

12. **Ohi, N., B. Aoki, T. Shinozaki, K. Moro, T. Noto, T. Nehashi, H. Okazaki, and I. Matsunaga.** 1986. Semisynthetic β-lactam antibiotics. I. Synthesis and antibacterial activity of new ureidopenicillin derivatives having catechol moieties. *J. Antibiot.* **39**:230–241.

13. **Pollock, M. R.** 1965. Purification and properties of penicillinases from two strains of *Bacillus licheniformis. Biochem. J.* **94**:666–675.

14. **Smit, J., Y. Kamio, and H. Nikaido.** 1975. Outer membrane of *Salmonella typhimurium*: chemical analysis and freeze-fracture studies with lipopolysaccharide mutants. *J. Bacteriol.* **124**:942–958.

15. **Sykes, R. B., and M. Matthew.** 1976. The β-lactamases of Gram-negative bacteria and their role in resistance to β-lactam antibiotics. *J. Antimicrob. Chemother.* **2**:115–157.

16. **Vu, H., and H. Nikaido.** 1985. Role of β-lactam hydrolysis in the mechanism of resistance of a β-lactamase-constitutive *Enterobacter cloacae* strain to expanded-spectrum β-lactams. *Antimicrob. Agents Chemother.* **27**:393–398.

16a. **Waley, S. G.** 1987. An explicit model for bacterial resistance: application to β-lactam antibiotics. *Microbiol. Sci.* **4**:143–146.

16b. **Watanabe, N., T. Nagasu, K. Katsu, and K. Kitoh.** 1987. E-0702, a new cephalosporin, is incorporated into *Escherichia coli* cells via the *tonB*-dependent iron transport system. *Antimicrob. Agents Chemother.* **31**:497–504.

17. **Yoshimura, F., and H. Nikaido.** 1985. Diffusion of β-lactam antibiotics through the porin channels of *Escherichia coli* K-12. *Antimicrob. Agents Chemother.* **27**:84–92.

18. **Zimmerman, W., and A. Rosselet.** 1977. Function of the outer membrane of *Escherichia coli* as a permeability barrier to beta-lactam antibiotics. *Antimicrob. Agents Chemother.* **12**:368–372.

Chapter 44

Enhanced Penetration of Externally Added Macromolecules through the Outer Membrane of Gram-Negative Bacteria

Hans J. P. Marvin
Bernard Witholt

Many studies require a temporary removal of the barrier function of the outer membrane of gram-negative bacteria to introduce large molecules such as plasmids and proteins into the bacterium. Chelating agents such as EDTA are often used to accomplish this permeability increase (4, 6). An EDTA treatment is accompanied by a loss of lipopolysaccharide (LPS) (ca. 50%) and minor amounts of proteins and phospholipids (3, 6). When lysozyme, a glycosidic-bond-hydrolyzing enzyme, is introduced into the periplasmic space by means of an EDTA treatment with or without an additional osmotic shock (2, 9), it degrades the murein and turns the rod-shaped cells to spheroplasts. The spheroplasts formed by the procedures described to date are damaged to various extents and are unable to recover growth. However, recovering spheroplasts would be useful to improve our understanding of cell envelope synthesis, assembly, and repair.

In this chapter we present a highly efficient procedure to translocate lysozyme and other macromolecules across the outer membrane of gram-negative bacteria. The resulting spheroplasts sustain very little damage as judged from their quick recovery in liquid media. The LPS loss which also occurs in this procedure is visualized by immuno-gold electron microscopy.

Hans J. P. Marvin and Bernard Witholt • Department of Biochemistry, University of Groningen, Nijenborgh 16, 9747 AG Groningen, the Netherlands.

431

Spheroplast Preparation

The results described below were obtained in a number of experiments with several *Pseudomonas* and *Escherichia coli* strains, including W3110 and C600.

The spheroplast formation techniques used in this paper have been described elsewhere (6a) and depend on the following parameters: pH of the medium, duration of the heat shock, and lysozyme and EDTA concentrations (data not shown). Under optimum conditions the conversion of *E. coli* cells to spheroplasts was achieved with 10- to 100-fold less lysozyme than in the most efficient procedures used previously (2, 9). The effectiveness of the spheroplasting procedure also depended on the cell density. This dependency is related to

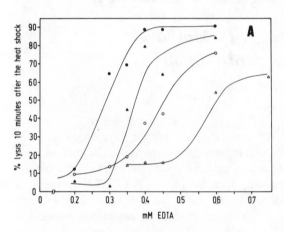

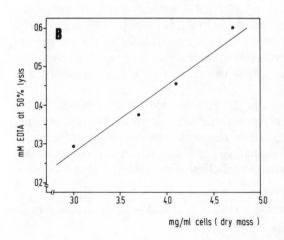

Figure 1. Dependency of the lysozyme activity on the cell/EDTA ratio. The spheroplasting procedure was performed with *E. coli* C600 and has been described in detail before (6a). Briefly, cells were grown to the early stationary phase, cooled on ice, harvested, and washed with ice-cold 100 mM $CaCl_2$. These cells were kept on ice for 2 h, washed once with 100 mM Tris hydrochloride (pH 8.0) containing 10 mM glycerol, and suspended in the same buffer at a cell concentration of 3.7 mg/ml. Immediately thereafter, lysozyme was added (0.38 μg/mg of cells [dry weight]). After 10 min of incubation on ice, EDTA (pH 8.0) was added. The cell suspension was kept on ice for 1 min, followed by 1 min of heat shock (42°C), and stabilized by the addition of 50 mM $MgCl_2$. (A) Percentage of lysis at 21-fold dilution 10 min after heat shock. Values were determined for several EDTA concentrations. The following cell concentrations were used: 3.0 mg/ml (●), 3.7 mg/ml (▲), 4.1 mg/ml (○), and 4.7 mg/ml (△). (From reference 6a, with permission of the publisher.) (B) Plot of the EDTA concentration which causes 50% cell lysis (from panel A) against the cell concentration.

the EDTA concentration. The relationship between the cell/EDTA ratio used in the spheroplasting procedure and the extent of lysis at a 21-fold dilution of the spheroplasting cells is illustrated in Fig. 1. A nearly linear relationship between the EDTA concentration and the cell density was found. This means that about 100 to 110 nmol of EDTA is required per mg of cells to allow 50% lysis of the spheroplasting cells after an osmotic shock. If a small percentage of this EDTA chelates envelope-bound Ca^{2+}, about 10^5 to 10^6 Ca^{2+} ions will be removed per cell, perhaps stoichiometrically with LPS and other outer membrane components. Leive (5) has reported that EDTA causes the release of LPS, ions, and other outer membrane components. This process may be stimulated by prior Ca^{2+} treatment (1).

Divalent cations are known to influence bilayer stability and transition temperatures (7). We therefore assume that the complexing of divalent cations by EDTA alters the susceptibility of this membrane to a heat shock. The resulting sudden changes of the fluidity of the outer membrane could cause gaps through which lysozyme can diffuse.

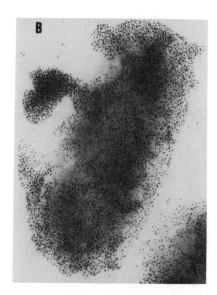

Figure 2. Immuno-gold labeling of LPS molecules of (A) untreated and (B) Ca^{2+}-EDTA-treated *E. coli* W3110 cells. Cells were grown in minimal medium to early stationary phase. A part of the batch (3.7 mg of cells) was fixed overnight at 4°C with 0.1% OsO_4. The remaining cells were treated with the Tris buffer and EDTA (0.8 mM) as described in the legend to Fig. 1. Immediately after the EDTA incubation, the cells were added to the fixative (0.1% OsO_4, 4°C) and kept for 1 night at 4°C. Before the monoclonal antibody (directed against LPS) was added, the cells were washed several times with PBS. The cells were further prepared for electron microscopy as described by Vos-Scheperkeuter et al. (8). Visualization of the surface-bound antibodies was performed with protein A-gold particles with an average diameter of 8 nm in a Philips EM 201 microscope at 60 kV.

Visualization of LPS by Immunoelectron Microscopy of Intact and EDTA-Treated E. coli Cells

By labeling LPS molecules with monoclonal antibodies and 8-nm protein A-gold particles before and after an EDTA treatment, we could demonstrate the well-known loss of LPS (see Fig. 2). We propose that the EDTA-induced, LPS-poor patches seen on the EDTA-treated cells are the weak spots through which macromolecules can penetrate.

Viability of Lysozyme-Induced Spheroplasts

The EDTA treatment had little effect on the viability of the cells (Fig. 3A). When lysozyme was introduced into the periplasmic space and the cells were converted to spheroplasts, the cells could fully recover both on brain heart infusion agar plates and in fluid media (Fig. 3), indicating that it is possible

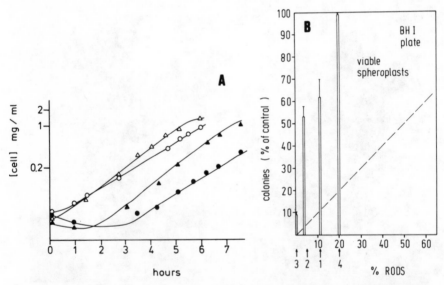

Figure 3. Recovery of spheroplasts in fluid media and on agar plates. (A) Spheroplasts were prepared as described in the legend of Fig. 1. The control cells were treated in the same way except that no lysozyme was added. A sample of 100% *E. coli* C600 spheroplasts (●, ▲) or control cells (○, △) was added to a growth medium at time zero, after which the recovery was recorded. Symbols: ○, ●, minimal medium; △, ▲, minimal medium plus amino acids. (B) Spheroplasts were prepared as described in the legend of Fig. 1. To determine the percentage of rods remaining, samples were photographed under a phase-contrast microscope and rods and spheroplasts were counted. To determine viable cells (spheroplasts), samples and controls were plated on 1.5% agar plates (2% brain heart infusion) and grown overnight at 37°C. The arrows indicate four different experiments. The broken line indicates the expected number of colonies if only rods were to recover. (From reference 6a, with permission of the publisher.)

to alter the peptidoglycan network with external agents while retaining full viability of the altered cells. The number of viable spheroplasts on the agar plates and the duration of the recovery period in the liquid media depended on the rate at which these spheroplasts were formed. The existence of a manipulable recovery period of the spheroplasts before they start to divide provides an opportunity to study protein transport and murein synthesis during envelope repair.

ACKNOWLEDGMENTS. We thank Nico H. J. Panman for drawing the graphs and Klaas Gilissen for providing the photographs.

This investigation was supported by the Foundation for Fundamental Biological Research (BION), which is subsidized by the Netherlands Organization of Pure Research (ZWO).

LITERATURE CITED

1. **Bayer, M. E., and L. Leive.** 1977. Effect of ethylenediaminetetraacetate upon the surface of *Escherichia coli. J. Bacteriol.* **130**:1364–1381.

2. **Birdsell, D. C., and E. H. Cota-Robles.** 1967. Production and ultrastructure of lysozyme and ethylenediaminetetraacetate-lysozyme spheroplasts of *Escherichia coli. J. Bacteriol.* **93**:427–437.

3. **Leive, L.** 1965. Release of lipopolysaccharide by EDTA treatment of *E. coli. Biochem. Biophys. Res. Commun.* **21**:290–296.

4. **Leive, L.** 1968. Studies on the permeability change produced in coliform bacteria by EDTA. *J. Biol. Chem.* **243**:2373–2380.

5. **Leive, L.** 1974. The barrier function of gram-negative envelope. *Ann. N.Y. Acad. Sci.* **238**:109–129.

6. **Leive, L., U. K. Shovlin, and S. E. Mergenhagen.** 1968. Physical, chemical, and immunological properties of lipopolysaccharide release from *E. coli* by ethylenediaminetetraacetate. *J. Biol. Chem.* **243**:6384–6391.

6a. **Marvin, H. J. P., and B. Witholt.** 1987. A highly efficient procedure for the quantitative formation of intact and viable lysozyme spheroplasts from *Escherichia coli. Anal. Biochem.* **164**:320–330.

7. **Vasilenko, I., B. De Kruijff, and A. J. Verkleij.** 1982. Polymorphic phase behaviour of cardiolipin from bovine heart and from *Bacillus subtilis* as detected by ^{31}P-NMR and freeze-fracture techniques: effects of Ca^{2+}, Mg^{2+}, Ba^{2+} and temperature. *Biochim. Biophys. Acta* **684**:282–286.

8. **Vos-Scheperkeuter, G. H., E. Pas, G. J. Brakenhoff, N. Nanninga, and B. Witholt.** 1984. Topography of the insertion of LamB protein into the outer membrane of *E. coli* wild-type and *lac-lamB* cells. *J. Bacteriol.* **159**:440–447.

9. **Witholt, B., M. Boekhout, M. Brock, J. Kingma, H. van Heerikhuizen, and L. De Leij.** 1976. An efficient and reproducible procedure for the formation of spheroplasts from variously grown *Escherichia coli. Anal. Biochem.* **74**:160–170.

Chapter 45

Identification of Mutants with Altered Outer Membrane Permeability

Rajeev Misra
Barbara A. Sampson
James L. Occi
Spencer A. Benson

The gram-negative bacterium *Escherichia coli* has a trilaminar envelope structure composed of an inner membrane, a peptidoglycan layer, and an outer membrane which serves as the first permeability barrier to the environment.

While the composition of the outer membrane is known, the molecular bases for various properties associated with the outer membrane are only now coming to light (for review, see references 6 and 7). One especially important property of the outer membrane is its selective permeability. This property is a function of both the composition and the organization of the outer membrane. To a large degree the permeability of the outer membrane to hydrophilic substances is controlled by a specialized class of proteins termed the porins (3). These proteins form proteinaceous, water-filled channels which allow small (<600 daltons [Da]) molecules to cross the outer membrane. *E. coli* K-12 has two major porin species, OmpF and OmpC, which are coordinately regulated by the *ompR* and *envZ* genes (5). Large substrates, such as maltodextrins, and charged molecules, such as phosphates and nucleotides, have specialized porins which allow these molecules to cross the outer membrane barrier (7).

A number of genetic screens have been devised to obtain mutations which alter the permeability of the outer membrane (1, 4, 8). We have developed a

Rajeev Misra, Barbara A. Sampson, James L. Occi, and Spencer A. Benson • Department of Molecular Biology, Princeton University, Princeton, New Jersey 08544.

novel selection for mutants which allow maltodextrin to cross the outer membrane in the absence of the LamB maltoporin. Many of the mutants have increased permeability to a variety of compounds.

The Maltodextrin Selection System

The LamB maltoporin facilitates the uptake of maltose and maltodextrins up to a chain length of 7 glucose units (9). LamB$^-$ strains can grow on maltose and maltotriose, but not on larger maltodextrin substrates which are unable to cross the outer membrane in the absence of the LamB protein. By starting with a strain carrying a large internal *lamB* deletion and selecting for growth on a mixture of maltodextrins, it is possible to isolate a variety of mutants which have altered permeability (Table 1).

Effect of Porin Composition on Maltodextrin Growth

One fact that may influence the types of mutants obtained in this selection is that the OmpF porin forms a slightly larger pore than the OmpC porin (3). To determine whether this difference affects the ability to grow on specific maltodextrins, growth tests were done using commercially available defined maltodextrin substrates from Boehringer Mannheim Biochemicals, Indianapolis, Ind. (Fig. 1). LamB$^-$ OmpC$^+$ OmpF$^+$ strains have a reduced growth rate on maltotriose compared with a LamB$^+$ strain. In contrast, LamB$^-$ OmpC$^+$ OmpF$^-$ strains do not grow on maltotriose. For these growth substrates there is a clear difference between the exclusion limit of the OmpF and OmpC pores.

Table 1. Selection for spontaneous Dex$^+$ mutants

Strain	Porin composition	No. of Dex$^+$ mutants isolated	Characteristics[a]	Genetic location[b]
MCR106	B$^-$ C$^+$ F$^+$	189	Dex$^+$	*ompF* (43)
		22	Dex$^+$ Hyp	*ompF* (16)
		1	Dex$^+$ Imp	*imp* (1)
RAM105	B$^-$ C$^+$ F$^-$	26	Dex$^+$ Det^s	*ompC* (6)
		1	Dex$^+$ Hyp	*ompC* (1)
		1	Dex$^+$ Imp	*imp* (1)
		17	Dex$^+$	Unknown

[a]Dex$^+$, Ability to grow on dextrins as a sole carbon source; Hyp, increased sensitivity to various detergents, dyes, and antibiotics, due to an alteration in the porin protein; Imp, increased sensitivity to various detergents, dyes, and antibiotics due to an alteration in the *imp* gene; Det^s, slight sensitivity to sodium dodecyl sulfate.
[b]The genetic location was determined by P1 mapping as described previously. The number in parentheses indicates the number of isolates tested.

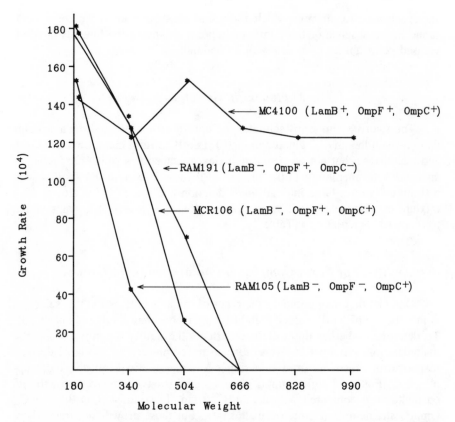

Figure 1. Effect of the porin composition on the growth rate when defined maltodextrins were used. Strains were grown in minimal media containing the indicated sugars as carbon source: glucose, 180 Da; maltose, 340 Da; maltotriose, 504 Da; maltotetrose, 666 Da; maltopentose, 828 Da; and maltohexose, 990 Da. The growth rate is shown as the inverse of the doubling time.

Characteristics of the Dex⁺ Mutants

The type of mutants obtained is affected by the genetic composition of the starting strain. MCR106 (LamB⁻ OmpC⁺ OmpF⁺) almost exclusively yields mutants which have alterations in the *ompF* gene (Table 1). The frequency of spontaneous Dex⁺ mutants for this strain is between 5×10^{-8} and 1×10^{-9}. Approximately 10% of the Dex⁺ mutants obtained exhibit greatly increased sensitivity to a variety of compounds, including detergents, dyes, and antibiotics. We refer to this phenotype as hyperpermeability (Hyp). In contrast, RAM105 (LamB⁻ OmpC⁺ OmpF⁻) yields a variety of Dex⁺ mutants. The frequency of spontaneous Dex⁺ mutants is approximately 10-fold lower than that observed with strain MCR106. The majority of the mutants tested contain *ompC* mutations which confer a Dex⁺ phenotype. These mutants are of two types,

Dex⁺ and Dex⁺ Hyp. The Dex⁺ nonhyperpermeable mutants, unlike the analogous *ompF* mutants, exhibit a slight increase in sensitivity to sodium dodecyl sulfate and β-lactam antibiotics. The Dex⁺ Hyp isolate, like the *ompF* Dex⁺ Hyp mutants, has greatly increased sensitivity to a variety of compounds, including detergents, dyes, and antibiotics.

With both MCR106 and RAM105 we have been able to obtain *ompC* and *ompF* mutations that result in a hyperpermeable phenotype. We have also obtained two isolates with this phenotype (Dex⁺ and increased sensitivity to a variety of compounds) which carry mutations that map to neither *ompF* nor *ompC*. These mutations map to the 1′ region of the *E. coli* chromosome (B. Sampson, unpublished data). We have tentatively designated these mutations as *imp* mutations (for increased membrane permeability).

Lastly, with strain RAM105 we obtained a class of Dex⁺ mutations that do not map to the *ompF*, *ompC*, or *imp* regions of the chromosome (Table 1; R. Misra, unpublished data). Like the nonhyperpermeable *ompF* mutations, these strains do not show any detectable increase in detergent or antibiotic sensitivity.

Nature of the Mutations in the ompC Gene

To provide insights into the nature of the porin alterations that confer the ability to grow on dextrins in the absence of the *lamB* gene product, we have characterized six of the *ompC* mutations that confer a Dex⁺ phenotype. All six mutants have nearly identical growth profiles on defined maltodextrin substrates (Fig. 2). They grow significantly faster on maltotriose and maltotetrose than do the parent strain RAM105 (Fig. 2) and the OmpF⁺ strain MCR106. None of the mutants is able to grow on maltopentose or larger oligomers. These mutants have a slightly increased sensitivity to β-lactam antibiotics and the detergent sodium dodecyl sulfate as compared with the parent strain RAM105 (Table 2).

DNA sequence analysis of these six mutants shows that a single region of the gene has been altered. In five of the six mutants the genetic alteration results in the arginine residue at position 74 (R74) of the mature protein being changed to a serine residue (RAM119, RAM120, RAM125, and RAM126) or a glycine residue (RAM124). The remaining mutant (RAM121) contains a 6-base-pair insertion that inserts a valine and an alanine residue immediately following the R74 residue.

Concluding Remarks

At least six types of Dex⁺ mutants have been isolated. Both the *ompF* and *ompC* genes can be altered in at least two ways to yield a Dex⁺ phenotype. The majority of the mutations do not dramatically increase the permeability of

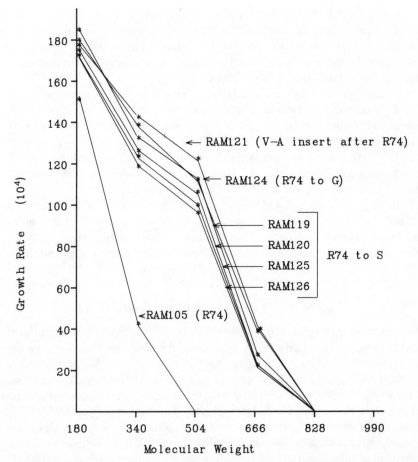

Figure 2. Growth rate of Dex$^+$ *ompC* mutants on defined maltodextrins. Growth rates were determined as described in the legend of Fig. 1. RAM105 is the parent strain for the mutants RAM119, RAM120, RAM124, RAM125, and RAM126. The amino acid alterations present in the OmpC porins of the mutants are shown in parentheses. R, Arginine; G, glycine; S, serine; V, valine; and A, alanine.

the outer membrane to other substances. Five to ten percent of the spontaneous *ompF* Dex$^+$ and *ompC* Dex$^+$ mutations result in a hyperpermeable phenotype. We have previously identified this type of mutant in an analogous selection (2). In addition to the mutations which map to the porin genes, at least two other types of Dex$^+$ alterations have been isolated. The mutations that we designate as *imp* map to the 1' region of the *E. coli* chromosome and, like the *ompF* and *ompC* hyperpermeable mutations, result in a greatly increased sensitivity to a variety of agents, including detergents, dyes, and antibiotics. The molecular basis for this increased permeability phenotype is not known. The

Table 2. Sensitivity of *ompC* mutants to various compounds[a]

Strain[b]	Sensitivity to:		
	Ampicillin (5 µg)	Penicillin G (10 U)	Sodium dodecyl sulfate (1 mg)
MCR106	8	0	0
RAM105	8	0	0
RAM119	14	9	13
RAM120	14	9	13
RAM125	13	9	13
RAM126	13	9	13
RAM124	15	9	15
RAM121	17	15	19

[a]Sensitivity was determined by disk test as described previously (2) and is given as the size of the zone of growth inhibition in millimeters.
[b]All strains are derived from strain MCR106 (2). Strain RAM105 is a Δ*ompF* derivative of MCR106. RAM119 through RAM126 are Dex[+] nonhyperpermeable mutants derived from RAM105.

final class of Dex[+] mutants, like the majority of the *ompF* and *ompC* Dex[+] mutants, do not exhibit a Hyp phenotype. Their Dex[+] phenotype is independent of both the OmpF and OmpC porins (Misra, unpublished data).

DNA sequence analysis of the six Dex[+] *ompC* mutations shows that a single residue (R74) of the mature protein is altered. In five cases, the amino acid is changed from a charged residue to an uncharged residue. However, we do not believe that simply the loss of the charge is responsible for the acquired Dex[+] property of the porin, since one of the mutants contains a small insert which does not remove the charged residue. The fact that six independent mutants alter this region of the mature protein strongly suggests that the R74 residue is involved in pore function. One possibility is that these alterations result in a mutant protein which has an enlarged pore that allows saccharides with a molecular weight of greater than 600 but less than 828 to cross the outer membrane barrier.

LITERATURE CITED

1. **Aline, R. F., Jr., and W. S. Reznikoff.** 1975. Bacteriophage Mu-1-induced permeability mutants in *Escherichia coli*. *J. Bacteriol.* **124:**578–581.
2. **Benson, S. A., and A. DeCloux.** 1985. Isolation and characterization of outer membrane permeability mutants in *Escherichia coli* K-12. *J. Bacteriol.* **161:**361–367.
3. **Benz, R.** 1985. Porin from bacterial and mitochondrial outer membranes. *Crit. Rev. Biochem.* **19:**145–190.
4. **Coleman, W. G., Jr., and K. Deshpande.** 1985. New *cysE-pyrE*-linked *rfa* mutation in *Escherichia coli* K-12 that results in a heptoseless lipopolysaccharide. *J. Bacteriol.* **161:**1209–1214.
5. **Hall, M. N., and T. J. Silhavy.** 1981. Genetic analysis of the *ompB* locus in *Escherichia coli* K-12. *J. Mol. Biol.* **151:**1–15.

6. **Nakae, T.** 1986. Outer-membrane permeability of bacteria. *Crit. Rev. Microbiol.* **13**:1–62.
7. **Nikaido, H., and M. Vaara.** 1985. Molecular basis of bacterial outer membrane permeability. *Microbiol. Rev.* **49**:1–32.
8. **Richmond, M. H., D. C. Clark, and S. Wotton.** 1976. Indirect method for assessing the penetration of beta-lactamase-nonsusceptible penicillins and cephalosporins in *Escherichia coli* strains. *Antimicrob. Agents Chemother.* **10**:215–218.
9. **Wandersman, C., M. Schwartz, and T. Ferenci.** 1979. *Escherichia coli* mutants impaired in maltodextrin transport. *J. Bacteriol.* **140**:1–3.

Chapter 46

Contribution of Outer Membrane Permeability and of β-Lactamases to β-Lactam Resistance in *Pseudomonas maltophilia*

Helmut Mett
Sabine Rosta
Barbara Schacher

β-Lactam Resistance of Pseudomonads

Pseudomonas maltophilia is a pathogen isolated from seriously ill patients with increasing frequency. It is generally highly resistant to most antibiotics, including the latest β-lactams. The resistance in pseudomonads has been attributed to (i) low outer membrane permeability (exopolysaccharides, capsules, smooth lipopolysaccharide, porins in a "closed" status; see references 2 and 3), (ii) one or two chromosomally coded inducible β-lactamases (1, 9, 10), (iii) various different plasmid-borne β-lactamases (4), and (iv) modified penicillin-binding proteins (6).

In this chapter we refer to our investigation of the contribution of β-lactamases and outer membrane permeability to the β-lactam resistance of clinical isolates as well as laboratory mutants of *P. maltophilia*. This organism possesses two inducible β-lactamases, a Zn-containing penicillinase (10) and an active-site-serine cephalosporinase (9); taken together, their hydrolytic activities span the entire range of β-lactams available to date (Table 1).

The *P. maltophilia* strains used in our study are listed in Table 2. Strain

Helmut Mett, Sabine Rosta, and Barbara Schacher • Research Laboratories, Pharmaceuticals Division, CIBA-GEIGY Ltd., CH-4002 Basel, Switzerland.

Table 1. Cleavage of β-lactams by β-lactamases L1 and L2
from *P. maltophilia* and by type Id β-lactamase
from *P. aeruginosa*

Substrate	V_{max} (relative to cephaloridine = 100)		
	L1[a]	L2[a]	Id
Cephalothin	83	7	250
Cefsulodin	0	38	0[b]
Cefotaxime	117	2	0.4[c]
Ceftizoxime	17	29	0
Aztreonam	0	12	0[b]
Benzylpenicillin	1,670	32	13[c]
Carbenicillin	770	3	0
Imipenem	400	0	0[b]

[a]Data from Saino et al. (9, 10).
[b]Traces of hydrolytic activity detected.
[c]Because of low enzyme activity, initial rates of hydrolysis were compared instead of uncalculated maximal rates.

IID1275/873 (10) was obtained from S. Waley; strains G-219 and G-234, isolated in our own laboratories, are mutants of IID1275/873; all other strains were fresh clinical isolates from hospitals in the Basel region (R. Frei, W. Vischer). *P. aeruginosa* 18S (inducible type Id β-lactamase [1]; nomenclature

Table 2. Quantitative determination of induction of β-lactamases by cefotaxime and permeability of the outer membrane for cephaloridine

Bacterial strain	Sp act[a] of β-lactamase (nmol/mg of protein per min)	Inducibility[b] (fold)	Permeability coefficient (ml/mg of protein per min)	
			Uninduced[c]	Induced
P. maltophilia				
IID1275/873	3.5	340	0.076	0.13
G-219	6,100	1.4	0.41	0.32
G-234	5.3	71	ND	0.78
G-244	1	110	0.057	0.11
G-269	13	440	0.038	0.093
G-272	10	180	0.11	0.066
G-290	3.3	680	ND	0.035
G-291	4.1	470	ND	0.068
G-292	14	75	0.19	0.043
G-295	2.1	190	0.037	0.15
G-296	2.1	390	0.027	0.021
P. aeruginosa 18S	4.8	140	ND	0.56
E. coli 255	760	0.64	0.70	1.3

[a]Suspension of uninduced homogenized bacteria.
[b]Increase of specific β-lactamase activity of homogenized bacteria (v_h) after growth in the presence of inducer (50 to 100 μg of cefotaxime per ml).
[c]V_{max} of the uninduced bacterial homogenate was in most cases calculated from V_{max} of the induced culture as described in the text. ND, Not detectable.

of Richmond and Sykes [8]) and *Escherichia coli* 255 (constitutive type Ib β-lactamase [13]) were used for comparison in some experiments. *P. maltophilia* was grown with vigorous shaking at 30°C, and the other strains were grown without shaking at 37°C in brain heart infusion broth (BHI).

β-Lactamase Regulation

The activity of β-lactamases in bacterial cultures was qualitatively detected with filter papers impregnated with PADAC (Calbiochem; 1 mg/ml in distilled water). Bacteria expressing inducible β-lactamases caused a color change of the filter only after growth in BHI containing subinhibitory concentrations of β-lactam but not after growth in drug-free medium; β-lactamase-constitutive strains produced a color change also after growth in drug-free medium, and apparently β-lactamase-deficient strains did not cause a color change after growth in either medium. β-Lactamase activity of intact bacteria or bacterial homogenates (bacterial suspension adjusted to an optical density of 20 at 650 nm in phosphate-buffered saline, pH 7.2, and subjected to two consecutive treatments at 14,000 lb/in^2 in a prechilled French pressure cell [Aminco]) was assayed spectrophotometrically (Uvikon 810, Kontron) at ambient temperature at 260 nm as described by Waley (14) with 0.1 mM cephaloridine as substrate.

All clinical isolates had inducible β-lactamases; some of the strains had an elevated basal level of enzyme activity (strains G-269, G-272, and G-292; Table 2). Cefotaxime showed the highest inducing potency, followed by imipenem, whereas cefoxitin showed significant inducing potency only in two strains with elevated basal β-lactamase activity (G-272 and G-291; Table 3).

Table 3. Growth-inhibitory activity (MIC) and β-lactamase-inducing potency (MINC) of β-lactams for *P. maltophilia*[a]

P. maltophilia strain and properties	Cefoxitin		Cefotaxime		Imipenem	
	MIC	MINC	MIC	MINC	MIC	MINC
IID1275/873, clin., Bla^i	1,024	>512	512	16	512	8
G-219, mutant of G-217, Bla^c	1,024	0	512	0	512	0
G-234, mutant of G-217, Bla^i	1,024	>512	512	64	512	>256
G-244, clin., Bla^{ci}	256	ND	512	ND	512	ND
G-269, clin., Bla^i	512	>256	256	1	512	>256
G-272, clin., Bla^{ci}	256	4	256	≤2	512	≤2
G-290, clin., Bla^i	512	>256	256	≤2	512	≤2
G-291, clin., Bla^{ci}	128	8	1,024	≤2	64	≤2
G-292, clin., Bla^{ci}	512	>256	1,024	2	256	≤2
G-295, clin., Bla^i	512	64	256	≤2	256	≤2
G-296, clin., Bla^i	512	32	1,024	≤2	512	≤2

Antibiotic concn (μg/ml)

[a]Abbreviations: clin., clinical isolate; Bla^i, inducible β-lactamase; Bla^c, constitutive β-lactamase; Bla^{ci}, inducible β-lactamase with elevated basal activity; ND, not detectable.

Outer Membrane Permeability

The method of Zimmermann and Rosselet (15) was adopted to determine outer membrane permeability. The specific β-lactamase activity of intact bacteria (v_i), of supernatant of intact bacteria (v_u), and of bacterial homogenate (v_h) was determined: K_m and V_{max} were measured with bacterial homogenates. For calculation of the outer membrane permeability, we ignored the bacterial cell surface area and applied the formula:

$$\text{permeability parameter } C = \frac{v_{ic}}{S_o - S_{peri}}$$

$$= \frac{v_{ic} \cdot (V_{max} - v_{ic})}{S_o \cdot (V_{max} - v_{ic}) - K_m \cdot v_{ic}}$$

where v_{ic} equals $v_i - v_u$ (activity of intact bacteria corrected for contribution by spontaneously released extracellular β-lactamase), S_o is the extracellular substrate concentration (0.1 mM cephaloridine), S_{peri} is the periplasmic substrate concentration (unknown), and V_{max} and K_m were determined as mentioned above. If uninduced bacteria with low β-lactamase activity were to be measured, K_m was taken from induced cultures and V_{max} was calculated as:

$$V_{max}\text{ (uninduced)} = \frac{v_h\text{ (uninduced)}}{v_h\text{ (induced)}} \cdot V_{max}\text{ (induced)}$$

We calculated the permeability coefficients for uninduced and cefotaxime-induced bacteria (Table 2). The data are averages of two to four experiments; in some strains the β-lactamase activity of intact cells was, after subtraction of the activity of extracellular enzyme, zero or negative. Permeability of these strains was labeled as nondetectable (Table 2). Generally, permeabilities of uninduced bacteria were about half those of induced cells.

MIC and MINC

MICs were determined in microtiter plates. Antibiotic was serially diluted (twofold) in BHI, and the wells were inoculated with 10^6 CFU of bacteria from a fresh overnight culture per ml. After 20 h of incubation at 30°C, the MIC was recorded as the lowest drug concentration preventing visible bacterial growth. PADAC (40 μl of 0.25 mM solution in 75% ethanol-saline per 0.1 ml of culture) was then added to each well, and after 10 to 60 min of incubation at ambient temperature, color changes were recorded. The minimum β-lactamase-inducing concentration (MINC) was defined as the lowest drug concentration revealing within 60 min a significant color change as compared with drug-free control cultures.

All clinical isolates were similarly resistant to cefoxitin and cefotaxime (Table 3); G-291 was the only strain with some sensitivity to imipenem (MIC, 64 µg/ml).

Isolation of β-Lactamase Mutants

Exponentially growing bacteria (10^8 CFU/ml) were mixed with various concentrations of ethyl methanesulfonate and incubated for 30 min at 30°C. Viable counts were then determined by plating dilutions on BHI agar. The bacteria were collected by centrifugation, suspended in 10 volumes of fresh medium, and grown overnight at 30°C. Cultures derived from suspensions in which ethyl methanesulfonate treatment had reduced the viability of the bacteria by at least 99% were diluted and plated on BHI agar plates to produce ca. 100 colonies per plate. These were replicated onto PADAC-containing filter papers, and β-lactamase-constitutive (Bla^c) and β-lactamase-deficient (Bla$^-$) colonies were isolated (see above). Colonies of the Bla^c phenotype were found with a frequency of about 10^{-3}, whereas the phenotype for all colonies initially identified as Bla$^-$ could not be confirmed upon further testing (see Tables 2 and 3; frequency less than 3×10^{-5}).

As already mentioned, all mutants initially identified as Bla$^-$ were shown still to express inducible β-lactamase. G-234 is shown as an example (Tables 2 and 3); its induction system had a lower affinity to the inducers cefotaxime and imipenem, and under standard conditions the enzyme activity of induced bacteria was lower than that of the parent strain. Constitutive mutants, e.g., G-219, were fully derepressed, and the addition of an inducer did not significantly enhance enzyme activity (Table 2). Both of these mutants had the same antibiotic sensitivity as the parent strain.

Concluding Remarks

We have not been able to isolate true β-lactamase-deficient mutants. The indicator substrate PADAC is hydrolyzed by only L2 β-lactamase (H. Mett, unpublished data); therefore, only mutants deficient for this enzyme would have been detected. Since both enzymes became constitutive in Bla^c mutants, it appears that L1 and L2 are, at least to some extent, regulated by a common system. Much more work is needed to understand the regulation of β-lactamase expression in *P. maltophilia,* which is only beginning to be understood in other bacteria (5, 11). The fact that we could not find β-lactamase-deficient mutants may indicate that these enzymes have other important functions in the absence of β-lactams.

Since β-lactam resistance of inducible strains was already very high, it is not surprising that resistance in Bla^c mutants did not increase measurably. The increased sensitivity of G-291 to imipenem was not studied further. The outer membrane permeability of G-291 for cephaloridine was not significantly different from that of imipenem-resistant strains; however, imipenem may penetrate outer membranes through channels different from the normal route of β-lactam uptake (7). If so, our permeability measurements for imipenem would be misleading.

All clinical isolates of *P. maltophilia* showed inducible β-lactamase activity with varying basal enzyme levels. Cefoxitin, known to be an efficient inducer of enterobacterial β-lactamases (12), appeared not to induce the *P. maltophilia* enzymes. However, this compound is a potent inhibitor (K_i = 3 and 9 μM, respectively, for L1 and L2 [9, 10]). Hence, it is possible that β-lactamases were induced by cefoxitin but could not be detected because they were inactivated by cefoxitin.

Permeability measurements have to be done very carefully, especially with strains of *Pseudomonas* spp. which, because of low outer membrane permeability, show only low enzyme activities in intact cells. Because the reliability of V_{max} determinations with low-specific-activity enzymes from uninduced bacteria is low, we have calculated these permeability coefficients using an approximation based on V_{max} of induced cells and the observed induction factor. The general trend that uninduced bacteria were less permeable than induced ones needs more experimental work for substantiation. In this context, we are starting to investigate the composition of outer membranes from uninduced and induced bacteria.

In conclusion, it appears that the high β-lactam resistance of *P. maltophilia* is due to inducible broad-spectrum β-lactamases in combination with a very low outer membrane permeability.

LITERATURE CITED

1. **Flett, F., N. A. C. Curtis, and M. H. Richmond.** 1976. Mutant of *Pseudomonas aeruginosa* 18S that synthesizes type Id β-lactamase constitutively. *J. Bacteriol.* **127**:1585–1586.
2. **Godfrey, A. J., M. S. Shahrabadi, and L. E. Bryan.** 1986. Distribution of porin and lipopolysaccharide antigens on a *Pseudomonas aeruginosa* permeability mutant. *Antimicrob. Agents Chemother.* **30**:802–805.
3. **Hancock, R. E. W.** 1986. Intrinsic antibiotic resistance of *Pseudomonas aeruginosa*. *J. Antimicrob. Chemother.* **18**:653–656.
4. **Jacoby, G. A., and L. Sutton.** 1979. Activity of β-lactam antibiotics against *Pseudomonas aeruginosa* carrying R plasmids determining different β-lactamases. *Antimicrob. Agents Chemother.* **16**:243–245.
5. **Lindberg, F., S. Lindquist, and S. Normark.** 1986. Induction of chromosomal β-lactamase expression in enterobacteria. *J. Antimicrob. Chemother.* **18**(Suppl. C):43–50.
6. **Malouin, F., and L. E. Bryan.** 1986. Modification of penicillin-binding proteins as mech-

anisms of β-lactam resistance. *Antimicrob. Agents Chemother.* **30:**1–5.

7. **Quinn, J. P., E. J. Dudek, K. A. DiVincenzo, D. A. Lucks, and S. A. Lerner.** 1986. Emergence of resistance to imipenem during therapy for *Pseudomonas aeruginosa* infections. *J. Infect. Dis.* **154:**289–294.

8. **Richmond, M. H., and R. B. Sykes.** 1973. The β-lactamases of Gram-negative bacteria and their possible physiological role. *Adv. Microb. Physiol.* **9:**31–88.

9. **Saino, Y., M. Inoue, and S. Mitsuhashi.** 1984. Purification and properties of an inducible cephalosporinase from *Pseudomonas maltophilia* GN12873. *Antimicrob. Agents Chemother.* **25:**362–365.

10. **Saino, Y., F. Kobayashi, M. Inoue, and S. Mitsuhashi.** 1982. Purification and properties of inducible penicillin β-lactamase isolated from *Pseudomonas maltophilia. Antimicrob. Agents Chemother.* **22:**564–570.

11. **Salerno, A. J., and J. O. Lampen.** 1986. Transcriptional analysis of β-lactamase regulation in *Bacillus licheniformis. J. Bacteriol.* **166:**769–778.

12. **Sanders, C. C., W. E. Sanders, Jr., and R. V. Goering.** 1982. In vitro antagonism of β-lactam antibiotics by cefoxitin. *Antimicrob. Agents Chemother.* **21:**968–975.

13. **Sawai, T., T. Yoshida, T. Tsukamoto, and S. Yamagishi.** 1981. A set of bacterial strains for evaluation of β-lactamase stability of β-lactam antibiotics. *J. Antibiot.* **34:**1318–1326.

14. **Waley, S. G.** 1974. A spectrophotometric assay for β-lactamase action on penicillins. *Biochem. J.* **139:**789–790.

15. **Zimmermann, W., and A. Rosselet.** 1977. Function of the outer membrane of *Escherichia coli* as a permeability barrier to β-lactam antibiotics. *Antimicrob. Agents Chemother.* **12:**368–372.

VII. β-LACTAMASES

Chapter 47

β-Lactamases: Mechanism, Inhibition, and Role

Stephen G. Waley

Since our current work on mechanisms and inhibition of β-lactamases was summarized only a few months ago (S. J. Cartwright, I. E. Crompton, C. Little, P. J. Madgwick, J. Monks, S. G. Waley, B. W. Cuthbert, G. Lowe, R. Bicknell, A. Schäffer, and J. S. Auld, *in Frontiers of Antibiotic Research*, 4th Takeda Science Foundation Symposium, 1987, in press), the approach in this chapter is rather different. The first part consists of an exceedingly brief general review, and the second part is an account of new work on quantitative aspects of the role of β-lactamases.

Mechanism and Inhibition of β-Lactamases

Two of the more sinister aspects of β-lactamases are their diversity and efficiency. Both aspects can be discussed in terms of mechanism. The main generalization about the mechanisms of β-lactamases is that (as far as we know) they are either serine enzymes or zinc enzymes.

Serine β-Lactamases

Knowledge of enzyme active sites often comes in stages: inhibitors, then sluggish substrates, then good substrates. Inhibitors reacted with a serine residue (about one-sixth of the way in from the N terminus) (7, 16), now called serine 70 (1), and then the acyl-enzyme mechanism was put on a firm footing

---

Stephen G. Waley • Sir William Dunn School of Pathology, University of Oxford, Oxford OX1 3RE, United Kingdom.

by the use of sluggish substrates (10, 15), but we still have few results with good substrates (15). Site-directed mutagenesis has confirmed the importance of serine 70: only cysteine can replace it (8, 26). There must be other groups that collaborate with the serine. When the highly conserved lysine 73 is replaced by arginine, the activity of β-lactamase I from *Bacillus cereus* drops markedly (P. J. Madgwick and S. G. Waley, *Biochem. J.*, in press). Perhaps this lysine functions as a proton donor. Carboxyl groups are good candidates for active-site groups (20, 27), but there are no conserved histidine residues in these enzymes. Serine β-lactamases are entirely different from, and completely unrelated to, serine proteases.

Zinc β-Lactamases

There are several zinc β-lactamases, of at least two kinds. β-Lactamase II from *B. cereus* (9, 23) functions by a branched pathway involving noncovalent intermediates (4, 5). No fewer than five candidate active-site groups were identified from experiments in solution, and these have now been confirmed by crystallography (B. J. Sutton, P. J. Artymiuk, A. E. Cordero-Borboa, C. Little, D. C. Phillips, and S. G. Waley, *Biochem. J.*, in press). β-Lactamase II is distinguished by its breadth of action; few β-lactams can withstand it. The zinc β-lactamase from *Pseudomonas maltophilia* (24) differs markedly from β-lactamase II (3).

Inhibitors of β-Lactamases

Serine β-lactamases are irreversibly inhibited by various reactive β-lactams (6, 17). Broadly speaking, these inhibitors do to β-lactamases what β-lactams do to transpeptidases: acylation without deacylation. The functional relationship between β-lactamases and transpeptidases is important. There are no effective inhibitors (other than chelating agents) of zinc β-lactamases; it is as well that the occurrence of zinc β-lactamases in pathogens is relatively restricted.

Structural Evidence on the Origin of β-Lactamases

The first structural information on β-lactamases came from amino acid sequences (1). These showed that the plasmid-encoded RTEM β-lactamase from *Escherichia coli* could be assigned to a class (class A) that comprised β-lactamases from gram-positive bacteria. This class now numbers some 10 members, of diverse properties. On the other hand, the chromosomally encoded (*ampC*) β-lactamase from *E. coli* had so different a sequence that it was regarded as evolutionarily unrelated, and a new class (class C) was created for it (12).

Three-dimensional structures tell us more about relationships than sequences do. Two structures have been reported, one of β-lactamase I (25) and the other of the β-lactamase from *Bacillus licheniformis* (13). Both structures showed striking similarity to that of the D-alanyl carboxypeptidase-transpeptidase from *Streptomyces* sp. R61, which had been determined earlier (14). The extensive region of common tertiary structure shows that serine β-lactamases and transpeptidases have probably evolved by divergence from a common ancestor. Both the amino acid sequence (J.-M. Frere, personal communication) and the mechanism of the R61 transpeptidase resemble class C rather than class A β-lactamases, and thus the separate evolutionary origin of class C and class A β-lactamases now seems less likely.

For the zinc β-lactamases, on the other hand, there is no comparable unifying theme. At the moment, the evolutionary origin of zinc β-lactamases is a complete mystery.

Implications

When β-lactam antibiotics interact with transpeptidases they acylate them, and the acyl-enzyme is relatively inert. When β-lactams interact with serine β-lactamases they acylate them, but the acyl-enzyme is short-lived. Hence the difficulty of producing active β-lactam antibiotics that are resistant to β-lactamases.

Role of β-Lactamases and Transport into Gram-Negative Bacterial Cells

There turn out to be close connections between the part that β-lactamases play in resistance to β-lactam antibiotics on the one hand and the growth of bacteria on low concentrations of nutrients on the other. At first sight these may seem somewhat disparate topics, but in fact they are linked in two ways. The first way is that diffusion of small polar molecules through the outer membranes of gram-negative bacteria takes place through porins (11, 21). The second way is that the same equation is used to express the balance between uptake and either breakdown (by β-lactamase) or active transport. This equation was derived independently (2, 31) for the two situations mentioned above. The rest of the present paper is mainly concerned with the implications and applications of this equation. Initially, I want to stress the usefulness of the permeability number, P_n.

Permeability Number

The permeability number, P_n, is a dimensionless parameter that expresses the importance of permeability in a given situation (28, 29). It arises from the Zimmermann-Rosselet equation in the context of β-lactamases as follows:

$$P \cdot A \cdot (C_o - C_p) = C_p \cdot V_{max}/(C_p + K_m) \tag{1}$$

where C_o and C_p denote the outer and periplasmic concentrations of antibiotic, P is the permeability coefficient, A is the area of the outer membrane, and V_{max} and K_m are the kinetic parameters for the β-lactamase. We may rearrange this equation as follows:

$$\frac{C_o - C_p}{C_p} = \left(\frac{V_{max}}{P \cdot A \cdot K_m}\right)\left(\frac{1}{1 + C_p/K_m}\right) \tag{2}$$

Now the term on the left-hand side of this equation is dimensionless, as is the second term on the right-hand side; thus, the first term on the right-hand side must also be dimensionless. It is well known that it is a great help in characterizing phenomena quantitatively to use parameters which do not vary with the units employed, i.e. dimensionless parameters. As a measure of the importance of permeability, it seems more logical to have P_n increasing as P increases, so we invert the permeability term and define the permeability number as follows:

$$P_n = \frac{P \cdot A}{V_{max}/K_m} \equiv \frac{\text{uptake}}{\text{breakdown}} = \frac{P \cdot A \cdot K_m}{V_{max}} \tag{3}$$

This is the provenance of the permeability number. But what use is it?

MIC and the Permeability Number

A simplified model for the effect of a β-lactam antibiotic relates cell killing to the value of a second-order rate constant (k_T) for reaction with a transpeptidase (28, 29). This then enables the MIC to be related to the permeability number:

$$\text{MIC} = \frac{L}{k_T}\left(1 + \frac{Q}{P_n}\right)$$

where L is a number (units time^{-1}) whose value depends on the assumptions about cell killing, and Q is a dimensionless number between 0 and 1 given by

$$Q = \frac{1}{1 + L/k_T K_m}$$

The upshot is that when P_n is much less than unity, then permeability is important. Indeed, the MIC may be raised by a factor of $1/P_n$. The interplay of k_T and P_n decides the effectiveness of an antibiotic. Similarly, it is the joint effect of low permeability and high β-lactamase activity that increases the MIC; neither permeability nor β-lactamase activity can be sensibly considered in isolation. Graphs illustrating these points have already been given (28).

A related parameter, the barrier index (BI), has been used to great effect to predict MIC values (H. Nikaido *in Frontiers of Antibiotic Research*, 4th Takeda Science Foundation Symposium, 1987, in press). The barrier index is just the left-hand side of equation 2; i.e.,

$$\text{BI} = (C_o - C_p)/C_p \quad \text{or} \quad C_o = C_p(1 + \text{BI})$$

If the periplasmic concentration of antibiotic required to kill the cell is written C_{inh} and the corresponding outer concentration is the MIC, then

$$\text{MIC} = C_{inh}(1 + \text{BI}) \tag{4}$$

The barrier index is related to the permeability number as follows:

$$\text{BI} = \frac{1}{P_n(1 + C_{inh}/K_m)} \tag{5}$$

When C_{inh} is $<<K_m$, then BI equals $1/P_n$. Otherwise, the barrier index will vary with C_{inh}; the permeability number does not depend on the "killing" reaction, but the barrier index does.

Determination of the Permeability Number

The permeability number may be obtained from measurements of the rate (v) of hydrolysis of β-lactams by intact gram-negative bacterial cells. There are, in fact, several different ways of extracting values of P_n from such measurements. The next three sections describe such methods, after which the use of progress curves is discussed.

Straight-line method. When applicable, the straight-line method is the simplest method. We use equation 1 in two ways. First:

$$P \cdot A \cdot (C_o - C_p) = v$$

which can be written as

$$\frac{C_o}{K_m} - \frac{C_p}{K_m} = \frac{V_{\max}}{K_m} \cdot \frac{v}{V_{\max}} \cdot \frac{1}{P \cdot A}$$

or

$$c - c^* = \frac{w}{P_n} \tag{6}$$

where c equals C_o/K_m, c^* equals C_p/K_m, and w equals $v/V_{\max}$.
    Second:

$$v = \frac{V_{\max} \cdot C_p}{K_m + C_p}$$

and so

$$w = \frac{c^*}{1 + c^*} \qquad \text{or} \qquad c^* = \frac{w}{1 - w}$$

Substitute for c^* in equation 6 to obtain

$$c = \frac{w}{1 - w} + \frac{w}{P_n} \tag{7}$$

Now divide through by w:

$$\frac{c}{w} = \frac{1}{1 - w} + \frac{1}{P_n}$$

and then multiply by $K_m/V_{\max}$:

$$\frac{C_o}{v} = \frac{K_m}{V_{\max} - v} + \frac{K_m}{V_{\max}} \cdot \frac{1}{P_n} \tag{8}$$

Thus, C_o/v is plotted against $1/(V_{\max} - v)$; $V_{\max}$ has to be obtained from measurements with broken cells. A set of 1,000 simulated experiments was carried out; the Gaussian errors had mean zero and standard deviation of 0.01 in $v/V_{\max}$. The results (Fig. 1) show that the linear plot is useful only over a limited

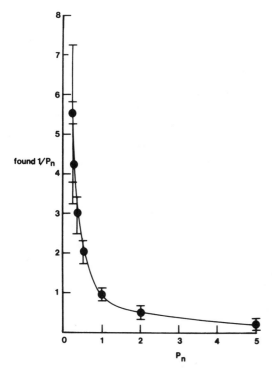

Figure 1. Errors in determination of the permeability number from the linear plot. Equation 8 was used to estimate P_n from the intercept, which is $1/P_n$. The random errors had mean zero and standard deviation of 0.01 in v/V_{max}, corresponding to about 1 to 10%. The resulting errors in $1/P_n$ are shown and represent the standard deviation from a large set of simulated experiments.

range of values of P_n; this range is approximately 0.25 to 5. Thus, at the lower end, when P_n was 0.25, the found value was 0.23 ± 0.06. At the upper end, when P_n was larger than 5, then $1/P_n$ became indistinguishable from zero. Thus, for practical purposes there was no detectable effect of permeability when P_n was greater than 5.

An example of the use of equation 8 is given in Fig. 2 for measurements on the rate of hydrolysis of a galactoside by *E. coli* (30). The drawback of this method is that the line is more or less defined by the points at which v is more than 50% of V_{max}. When v is less than 15% of V_{max}, however, then v is approximately proportional to C_o, and the plot of v against C_o (Fig. 3) can be used to get P_n from the slope, which is $(V_{max}/K_m)/(1 + 1/P_n)$.

Nonlinear regression. From equation 7 we obtain a quadratic for w:

$$w^2 - bw + P_n c = 0 \tag{9}$$

where b is $1 + P_n(1 + c)$ and hence w is $[b - (b^2 - 4P_n c)^{0.5}]/2$. Although the most obvious procedure is to use nonlinear regression, the simplest method is to utilize the fact that equation 9 is linear in P_n; thus, define

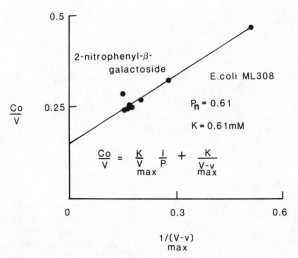

Figure 2. Linear plot applied to flux of a galactoside into *E. coli*. Equation 8 in the text applied to 2-nitrophenyl-β-galactoside and *E. coli* ML308 (30). The intercept is $K_m/V_{max} \cdot P_n$, and the slope is K_m; V_{max} was taken as 6.7 nmol mg^{-1} (dry weight) s^{-1}, K_m is then 0.61 mM, and P_n is 0.61. (Data from reference 30.)

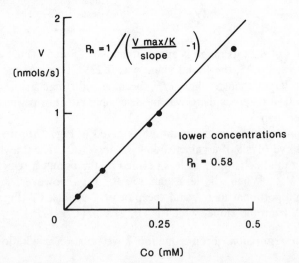

Figure 3. Lower rates in the flux of the galactoside into *E. coli*. The same data were used as for Fig. 2 (30). The plot of v against C_o approximates a straight line through the origin when v is less than 15% of V_{max}. The values of V_{max} and K_m given in the legend to Fig. 2 were used to find P_n from the slope; P_n was 0.58. (Data from reference 30).

$$g_i = w_i(1 + c_i) - c_i$$

for each of i data points, and then find P_n from

$$P_n = \frac{\Sigma g_i w_i^2 - \Sigma g_i w_i}{\Sigma g_i^2}$$

Relative rates in intact and broken cells. The relative rate (r) of hydrolysis of a β-lactam, when intact and broken cells are compared, is

$$r = 0.5\left(\frac{1 + c}{c}\right)[b - (b^2 - 4P_n c)^{0.5}] \tag{10}$$

When c is less than 0.5 and the relative rate, r, is less than 0.7, then equation 10 can be approximated by

$$r = \frac{P_n(1 + c)}{1 + P_n(1 + c)} \quad \text{and} \quad P_n = \frac{r}{(1 + c)(1 - r)} \tag{11}$$

The error introduced in this approximation may be up to 30% in P_n, but for many purposes the simplicity of equation 11 will recommend itself. In fact, a good way of finding P_n from these relative rates is to measure the relative rates at several concentrations of substrate and then calculate approximate values of $1/P_n$ from

$$\frac{1}{P_n} = \left(\frac{1 - r}{r}\right)\left(1 + \frac{C_o}{K_m}\right)$$

When these approximate values of $1/P_n$ are plotted against C_o, the ordinate intercept is then (accurately) $1/P_n$.

Measurements of the relative rates with intact and broken cells have been found useful and have been expressed in terms of the "crypticity" (22). The crypticity is $1/r$. Thus, equation 10 provides the formula for the reciprocal of the crypticity, calculated on the basis of the Zimmermann-Rosselet equation. Clearly, the crypticity depends on C_o/K_m as well as on P_n; however, when C_o is $<<K_m$, then the crypticity is approximately $1 + 1/P_n$, and a high crypticity implies a low P_n. As already mentioned, the MIC, like the crypticity, may be raised by a factor of $1/P_n$.

Progress curves for the hydrolysis of β-lactams by intact cells. It turns out that equation 9 can be integrated, so that progress curves can be calculated. The rate in equation 9 is the relative rate, which may be written as

$$w = \frac{v}{V_{max}} = -\frac{1}{V_{max}} \cdot \frac{dC_o}{dt} = -\frac{K_m}{V_{max}} \cdot \frac{dc}{dt}$$

Now define T as $(V_{max}/K_m)t$, and so

$$T = -2 \int_{c_{initial}}^{c} \frac{dc}{b - \sqrt{b^2 - 4P_n c}} = -\frac{2}{P_n} \int_{b_{initial}}^{b} \frac{db}{b - \sqrt{b^2 - 4P_n c}}$$

The way to integrate this is to multiply numerator and denominator by $b + \sqrt{b^2 - 4P_n c}$; then add $u = P_n c$ and $q = 1 + P_n$, so that

$$T = \frac{1}{2P_n} \int_{u_o}^{u} \left(\frac{1}{u}\right) [q + u + \sqrt{(u + q)^2 - 4u}] du$$

and hence

$$T = \frac{1}{2P_n} \left[u_o - u + 2q \ln \left(\frac{u_o}{u}\right) + \sqrt{X_o} - \sqrt{X} \right.$$
$$\left. + Q \ln \frac{\sqrt{X_o} + u_o + Q}{\sqrt{X} + u + Q} - q \ln \left(\frac{q\sqrt{X_o} + Qu_o + q^2}{q\sqrt{X} + Qu + q^2}\right) \right] \tag{12}$$

where X equals $u^2 + 2Qu + q^2$; X_o equals $u_o^2 + 2Qu_o + q^2$; and Q equals $q - 2$. The shape of such progress curves is shown in Fig. 4; at low values of c the curves resemble those of a first-order reaction.

Although it is satisfactory to have a theoretical equation describing the form of a progress curve, equation 12 is a trifle cumbersome as it stands. It can, however, be used to test methods for obtaining initial rates. Previous workers have used rates but given little guidance on how to measure them. Fortunately, one of the simplest methods turns out to be accurate. Plot $\Delta C_o/t$ against ΔC_o, where ΔC_o is the change in the concentration of the (outer) substrate over time t; the ordinate intercept then gives the initial rate.

Another use of equation 12 is to see whether P_n can be obtained from the half-life $(t_{1/2})$ of the reaction. Now the relation between P_n and $t_{1/2}$ becomes simple at a low enough value of C_o and is given by

$$P_n = 1 \left/ \left[\left(\frac{V_{max}}{K_m}\right) \frac{t_{1/2}}{\ln 2} - 1 \right] \right. \tag{13}$$

It turns out that a satisfactory value of P_n can be found by plotting the apparent values found from equation 13 against C_o and taking the ordinate intercept. There is one proviso: that is, there must be enough effect of permeability. This can be ascertained by measuring $t_{1/2}$ with broken cells. The value of $t_{1/2}$ with

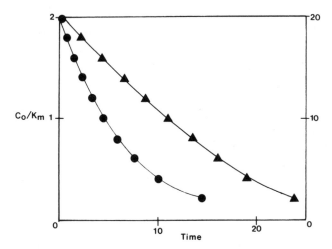

Figure 4. Progress curves showing the drop in concentration of substrate. Equation 12 was used, with P_n equal to 0.2. The ordinate shows C_o/K_m (●, left-hand scale; ▲, right-hand scale), and the abscissa shows normalized time, $(V_{max}/K_m) \times t$.

intact cells must be at least 20 to 30% greater than the value with broken cells for the above method to be accurate.

Permeability number and resistance. Several different ways of determining the permeability number have been described above. I am concerned that the essential simplicity of the concept of the permeability number may have got drowned in a welter of equations. The permeability number expresses the balance between uptake and breakdown, and as mentioned earlier (28), a good antibiotic should have a high value of k_T and P_n. The permeability number derives directly from the Zimmermann-Rosselet equation and may be expressed in terms of the periplasmic concentration of antibiotic (C_p) as

$$P_n = \frac{C_o/C_p - 1}{C_o/K_m + 1}$$

although the definition in terms of measurable parameters (equation 3) will generally be preferred.

Kinetics of Growth of Gram-Negative Bacteria in Low Concentrations of Nutrient

The same equation (equation 1) turns up when considering the part played by diffusion in the growth of bacteria on low concentrations of nutrients (18). In fact, the growth of *E. coli* on low concentrations of glucose is better represented by the equations based on equation 1 than by those based on the more customary Monod equation (19). The equations for batch growth are

$$\frac{dW}{dt} = rW \quad \text{and} \quad \frac{dS}{dt} = -rW/Y$$

where W is the weight of bacteria, S is the concentration of limiting nutrient, and Y is the yield. The same symbols are used as in reference 18, except that K/J is replaced by P_n, to illustrate the use of the permeability number in this context. If the initial concentration of nutrient is C and the initial weight is W_o, then S is equal to $C - (W - W_o)/Y$. Thus, the specific growth rate, r, may be expressed in terms of W and the growth equation can be written as

$$\frac{dW}{dt} = r_{\max}W \left[A + BW - \sqrt{(A + BW)^2 - 4(a_1 + BW)} \right] \tag{14}$$

The three constants, A, B, and a_1, depend on the parameters of the system (K is the "transport equivalent" of K_m) and are given by $A = a_1 + P_n + 1$, $a_1 = P_n(C + W_o/Y)/K$, and $B = -P_n/YK$. The solution of equation 14 follows by the use of the same method that was used to obtain equation 12. This leads to the following integral:

$$t = \frac{1}{2r_{\max}} \int \frac{A + BW + \sqrt{R}}{W(a_1 + BW)} \, dW + \text{constant}$$

where R equals $a + bW + cW^2$, a equals $A^2 - 4a_1$, b is $2B(A - 2)$, and c equals B^2. This in turn leads to equation 15 (I am indebted to E. Süli for help here):

$$t = \left[A \ln \left| \frac{BW}{a_1 + BW} \right| + a_1 \ln |a_1 + BW| - \sqrt{a} \ln \left| \frac{\sqrt{a} + \sqrt{R}}{W} + \frac{b}{2\sqrt{a}} \right| \right.$$

$$+ p\sqrt{c} \sin h^{-1} \left(\frac{2cW + b}{\sqrt{4ac - b^2}} \right) \tag{15}$$

$$\left. + \sqrt{d} \ln \left| \frac{\sqrt{R} + \sqrt{d}}{W + p} + \frac{b - 2pc}{2\sqrt{d}} \right| \right] \Big/ 2a_1 r_{\max} + \text{constant}$$

where p equals a_1/B and d equals $a - bp + cp^2$. This somewhat fearsome equation is easy enough to program, and an example, similar to the numerical solution in Fig. 3 of reference 18, is shown in Fig. 5. In practice, there are

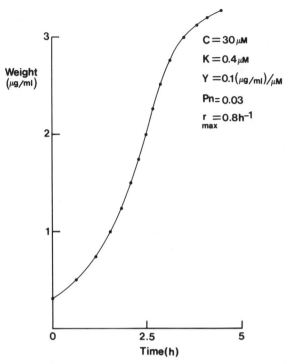

Figure 5. Plot of equation 15. The initial weight was 0.3 μg/ml, and the initial concentration of limiting nutrient was 30 μM. Values: r_{max}, 0.8 h^{-1}; K, 0.4 μM; Y, 0.1 (μg/ml)/μM; P_n, 0.03. (See reference 19.)

four parameters (r_{max}, K, Y, and P_n) to evaluate from the growth curve, and further work is needed to find out how precisely this can be done.

ACKNOWLEDGMENTS. The support of the Medical Research Council is gratefully acknowledged; I am a member of the Oxford Enzyme Group.

LITERATURE CITED

1. **Ambler, R. P.** 1980. The structure of β-lactamases. *Philos. Trans. R. Soc. London* Ser. B **289:**321–331.
2. **Best, J. B.** 1955. The inference of intracellular enzymatic properties from kinetic data obtained on living cells. *J. Cell. Comp. Physiol.* **46:**1–27.
3. **Bicknell, R., E. L. Emanuel, J. Gagnon, and S. G. Waley.** 1985. The production and molecular properties of the zinc β-lactamase of *Pseudomonas maltophilia* IID 1275. *Biochem. J.* **229:**791–797.
4. **Bicknell, R., A. Schaffer, S. G. Waley, and D. S. Auld.** 1986. Changes in the coordination

466 VII • β-Lactamases

geometry of the active-site metal during catalysis of benzylpenicillin hydrolysis by *Bacillus cereus* β-lactamase II. Biochemistry **25**:7208–7215.

5. **Bicknell, R., and S. G. Waley.** 1985. Cryoenzymology of *Bacillus cereus* β-lactamase II. *Biochemistry* **24**:6876–6887.

6. **Cartwright, S. J., and S. G. Waley.** 1984. β-Lactamase inhibitors. *Med. Res. Rev.* **3**:341–382.

7. **Cohen, S. A., and R. F. Pratt.** 1980. Inactivation of *Bacillus cereus* β-lactamase I by 6β-bromopenicillanic acid: mechanism. *Biochemistry* **19**:3996–4003.

8. **Dalbadie-McFarland, G., J. J. Neitzel, and J. H. Richards.** 1986. Active-site mutants of β-lactamase: use of an inactive double mutant to study requirements for catalysis. *Biochemistry* **25**:332–338.

9. **Davies, R. B., and E. P. Abraham.** 1974. Metal cofactor requirements of β-lactamase II. *Biochem. J.* **143**:129–135.

10. **Fisher, J., J. G. Belasco, S. Khosla, and J. R. Knowles.** 1980. β-Lactamase proceeds via an acyl-enzyme intermediate. Interaction of the *Escherichia coli* RTEM enzyme with cefoxitin. *Biochemistry* **19**:2895–2901.

11. **Hancock, R. E. W.** 1987. Role of porins in outer membrane permeability. *J. Bacteriol.* **169**:929–933.

12. **Jaurin, B., and T. Grundstrom.** 1981. *ampC* cephalosporinase of *Escherichia coli* K-12 has a different evolutionary origin from that of β-lactamases of the penicillinase type. *Proc. Natl. Acad. Sci. USA* **8**:4897–4901.

13. **Kelly, J. A., O. Dideberg, P. Charlier, J. P. Wery, M. Libert, P. C. Moews, J. R. Knox, C. Duez, C. Fraipont, B. Joris, J. Dusart, J.-M. Frere, and J.-M. Ghuysen.** 1986. On the origin of bacterial resistance to penicillin: comparison of a β-lactamase and a penicillin target. *Science* **231**:1429–1431.

14. **Kelly, J. A., J. R. Knox, P. C. Moews, G. J. Hite, J. B. Bartolone, H. Zhao, B. Joris, J.-M. Frere, and J.-M. Ghuysen.** 1985. 2.8-Å structure of penicillin-sensitive D-alanyl carboxypeptidase-transpeptidase from *Streptomyces* R61 and complexes with β-lactams. *J. Biol. Chem.* **260**:6449–6458.

15. **Knott-Hunziker, V., S. Petursson, S. G. Waley, B. Jaurin, and T. Grundstrom.** 1982. The acyl-enzyme mechanism of β-lactamase action: the evidence for class C β-lactamases. *Biochem. J.* **207**:315–322.

16. **Knott-Hunziker, V., S. G. Waley, B. S. Orlek, and P. G. Sammes.** 1979. Penicillinase active sites: labelling of serine-44 in β-lactamase I by 6β-bromopenicillanic acid. *FEBS Lett.* **99**:59–61.

17. **Knowles, J. R.** 1985. Penicillin resistance: the chemistry of β-lactamase inhibition. *Acc. Chem. Res.* **18**:97–104.

18. **Koch, A. L.** 1985. The macroeconomics of bacterial growth. *In* M. Fletcher and G. D. Floodgate (ed.), *Bacteria in Their Natural Environments*, p. 1–42. Society of General Microbiology Symposium 16. Academic Press, Inc., London.

19. **Koch, A. L., and C. H. Wang.** 1982. How close to the theoretical diffusion limit do bacterial uptake systems function? *Arch. Microbiol.* **131**:36–42.

20. **Little, C., E. L. Emanuel, J. Gagnon, and S. G. Waley.** 1986. Carboxy groups as essential residues in β-lactamases. *Biochem. J.* **240**:215–219.

21. **Nikaido, H., and M. Vaara.** 1985. Molecular basis of bacterial outer membrane permeability. *Microbiol. Rev.* **49**:1–32.

22. **Richmond, M. H., and R. B. Sykes.** 1973. The β-lactamases of Gram-negative bacteria and their possible physiological role. *Adv. Microb. Physiol.* **9**:31–88.

23. **Sabath, L. D., and E. P. Abraham.** 1966. Zinc as a cofactor for cephalosporinase from *Bacillus cereus* 569. *Biochem. J.* **98**:11–13.

24. **Saino, Y., F. Kobayashi, M. Inoue, and S. Mitsuhashi.** 1982. Purification and properties

of inducible penicillin β-lactamase isolated from *Pseudomonas maltophilia. Antimicrob. Agents Chemother.* **22:**564–570.

25. **Samraoui, B., B. J. Sutton, R. J. Todd, P. J. Artymiuk, S. G. Waley, and D. C. Phillips.** 1986. Tertiary structural similarity between a class A β-lactamase and a penicillin-sensitive D-alanyl carboxypeptidase-transpeptidase. *Nature* (London) **320:**378–380.

26. **Sigal, I. S., W. F. DeGrado, B. J. Thomas, and S. R. Petteway, Jr.** 1984. Purification and properties of thiol β-lactamase. *J. Biol. Chem.* **259:**5327–5332.

27. **Waley, S. G.** 1975. The pH-dependence and group modification of β-lactamase I. *Biochem. J.* **149:**547–551.

28. **Waley, S. G.** 1987. An explicit model for bacterial resistance: application to β-lactam antibiotics. *Microbiol. Sci.* **4:**143–146.

29. **Waley, S. G.** 1987. The kinetics of uptake and breakdown: antibiotic resistance. *Biochem. Soc. Trans.* **15:**235–236.

30. **West, I. C., and M. G. P. Page.** 1984. When is the outer membrane of *Escherichia coli* rate-limiting for uptake of galactosides? *J. Theor. Biol.* **110:**11–19.

31. **Zimmerman, W., and A. Rosselet.** 1977. Function of the outer membrane of *Escherichia coli* as a permeability barrier to β-lactam antibiotics. *Antimicrob. Agents Chemother.* **12:**368–372.

Chapter 48

β-Lactamase-Induced Resistance

Jean-Marie Frère
Bernard Joris

Increased bacterial resistance to β-lactam antibiotics has usually been attributed to three causes: (i) a decreased sensitivity of the target enzyme(s), the membrane-bound DD-peptidases, or the synthesis of new, more resistant DD-peptidases; (ii) the synthesis and excretion through the cytoplasmic membrane of enzymes capable of hydrolyzing the amide bond of the β-lactam nucleus, i.e, β-lactamases; and (iii) the hindering of free diffusion of the antibiotic by the outer membrane of gram-negative bacteria (permeability barrier). The outer membrane also prevents the leaking of the β-lactamase into the culture supernatant, as happens with gram-positive strains.

The elegant work of Nikaido and his colleagues (7, 8, 14, 15) allowed estimations of the diffusion rates of various antibiotics through the outer membrane of *Escherichia coli* and other gram-negative bacteria. A first important conclusion of that work was that the permeability barrier was significant only if a β-lactamase was also present in the periplasm. Indeed, with *E. coli*, first-order rate constants for the penetration of the antibiotics in the periplasm ranged from 0.14 to 7 s^{-1}, values indicating a very rapid equilibration. Assuming a penetration rate of only 0.03% of the highest value (2×10^{-3} s^{-1}), a periplasmic concentration representing 50% of the external one would still be reached after only 6 min.

In this chapter we analyze the interplay between outer membrane permeability and β-lactamase activity in determining the periplasmic concentration of the antibiotic. The model (Fig. 1) is based on a realistic pathway (4, 5) for the β-lactamase-catalyzed hydrolysis of penicillins. It also assumes that the

Jean-Marie Frère and Bernard Joris • Laboratoire d'Enzymologie and Service de Microbiologie, Université de Liège, Institut de Chimie, B6, B-4000 Sart Tilman (Liège 1), Belgium.

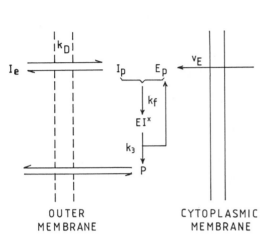

Figure 1. Complete model for the study of the interplay between outer membrane permeability and β-lactamase activity. I_e and I_p represent, respectively, the antibiotic concentrations in the culture medium and in the periplasm, E_p is the free enzyme in the periplasm, EI^* is the acyl enzyme intermediate, P is the hydrolyzed β-lactam, k_D is the first-order rate constant for the diffusion through the outer membrane, v_E is the rate of synthesis and excretion of the enzyme through the cytoplasmic membrane, k_f is the apparent second-order rate constant for acyl enzyme formation, and k_3 is the first-order rate constant for its hydrolysis. A more complete pathway for the β-lactamase-catalyzed reaction is:

$$E + I \underset{}{\overset{K}{\rightleftharpoons}} EI \overset{k_2}{\rightarrow} EI^* \overset{k_3}{\rightarrow} E + P$$

In this analysis, we have assumed that I_p is $<<K$, in which case the concentration of EI remains negligible and the rate of formation of EI^* is given by $k_f(I_p)(E_p)$, where $k_f = k_2/K$. The units used in the simulations are as follows: concentrations, micromolar; k_f, micromolar^{-1} second^{-1}; k_3, second^{-1}; and v_E, micromolar second^{-1}.

experimental conditions are those which prevail during the determination of an MIC, i.e., that the cell density is very low and that throughout the experiment the external concentration of β-lactam, I_e, remains virtually constant.

Preliminary simulations (see below) indicated that two very different situations could arise. For high values of k_3, the value of I_p rapidly reached a steady state, and the synthesis of new enzyme was not relevant ($v_e \simeq 0$). The simple equations derived by Zimmermann and Rosselet (16) and Vu and Nikaido (14) supplied an adequate description of the phenomena. For low values of k_3 ($<10^{-2}$ s^{-1}), the appearance of new enzyme could no longer be neglected, a true steady state was never really attained, and the value of I_p could present considerable variations over periods corresponding to the generation time of the bacterium.

Steady-State Situation

The model was simplified by neglecting v_E. For the interaction with the enzyme, the usual Henri-Michaelis parameters were used. Some interesting features arise from a close study of equations giving $I_p = f(I_e)$ and $I_e = f(I_p)$.

Equation A: $I_e = f(I_p)$

As emphasized by Vu and Nikaido (14), the equation allows the computation of the external concentration of antibiotic, I_e, which is necessary to yield a predetermined concentration in the periplasm.

$$I_e = I_p + \frac{k_{cat}E_0 \cdot I_p}{k_D(K_m + I_p)} \tag{1}$$

where E_0 is the total concentration of β-lactamase. For a fixed value of I_p, I_e increases linearly with $k_{cat}E_0$ (i.e., V_{max} of the enzyme). Note that if I_p/K_m is low and k_D is high, the slope of the line can be rather shallow. I_e also increases linearly with $(k_D)^{-1}$. The variation with K_m is hyperbolic with upward concavity, and the equation of the horizontal asymptote is $I_p = I_e$. The most interesting influence is that of I_p itself. The variation is hyperbolic with downward concavity, and the equation of the oblique asymptote is:

$$I_e = I_p + \frac{k_{cat}E_0}{k_D} \tag{2}$$

The curve approaches the asymptote when I_p becomes $>>K_m$; I_e then varies linearly with I_p. In contrast, when I_p is not much larger than K_m, it must be emphasized that the relationship between I_e and I_p is not linear. This can be better illustrated by plotting I_e/I_p versus I_p/K_m (Fig. 2). In fact, the lower the desired I_p/K_m, the larger the I_e/I_p ratio necessary to reach that desired value of I_p. Thus, under such conditions, an increase in the sensitivity of penicillin-binding proteins (PBPs) does not proportionally decrease the MIC, and this effect is more pronounced if $k_{cat}E_0/k_D$ is large.

Equation B: $I_p = f(I_e)$

Equation 3 has been used by Zimmerman and Rosselet (16) to study the outer membrane permeability to β-lactams:

$$I_p = \frac{-N \pm \sqrt{N^2 + 4k_D^2 I_e K_m}}{2k_D} \tag{3}$$

where N equals $k_{cat}E_0 + k_D K_m - k_D I_e$.

Some extreme cases result in very simple equations. If $k_{cat}E_0$ is much larger than $k_D K_m$ and $k_D I_e$, I_p is directly proportional to I_e (equation 4).

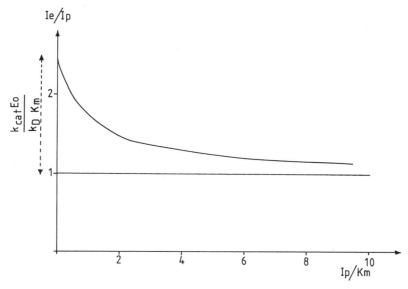

Figure 2. Influence of I_p/K_m on the ratio I_e/I_p. K_m, 1 μM; $k_{cat}E_0$, 10 μM s^{-1}; k_D, 7 s^{-1}.

$$I_p = \frac{k_D K_m}{k_{cat} E_0} I_e \tag{4}$$

This situation prevails in the presence of large concentrations of a very active enzyme ($k_{cat}E_0$ might be as high as 10^6 μM s^{-1}).

If I_e is much larger than K_m, I_p decreases linearly with E_0 (equation 5).

$$I_p = I_e - \frac{k_{cat}}{k_D} E_o \tag{5}$$

If, in addition, $k_D I_e$ is much larger than $k_{cat}E_0$, I_p equals I_e.

With four variable parameters ($k_{cat}E_0$, k_D, K_m, and I_e), a detailed discussion would be too long. We will only point out the most striking feature.

Figure 3 shows the influence of $k_{cat}E_0$ on I_p for various values of k_D. Depending upon the value of $k_{cat}E_0$, it can be seen that the effect of k_D can be nearly negligible (compare curves I and II for $k_{cat}E_0$ of <5 μM s^{-1}) or quite spectacular (compare curves II and III for values of $k_{cat}E_0$ between 5 and 10 μM s^{-1}).

Since large concentrations of periplasmic β-lactamase (100 to 900 μM) have been observed in some superproducing strains (7; H. Martin, personal communication), rather large values of $k_{cat}E_0$ can be obtained even if k_{cat} is relatively low. Moreover, when the activity of the β-lactamase on a "resistant"

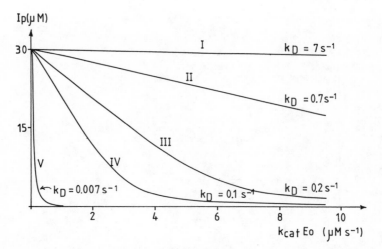

Figure 3. Variation of I_p with $k_{cat}E_0$ for various values of k_D. Other conditions: K_m, 1 μM; I_e, 30 μM.

substrate is less than 1% of that observed on a good substrate, it has often been considered as negligible. However, since k_{cat} for good substrates can be in the range of 1,000 to 3,000 s^{-1}, a rather high value of $k_{cat}E_0$ can be obtained even if k_{cat} is only 0.1% of that of the good substrate. It is thus important to measure accurately the kinetic parameters of the β-lactamase-substrate interaction even when the enzyme seems to transform the substrate very slowly.

Non-Steady-State Situation

Equations and Simulations

From the model presented in Fig. 1, the following differential equations can be deduced:

$$\frac{d(I_p)}{dt} = k_D(I_e - I_p) - k_f(I_p)(E_p) \tag{6}$$

$$\frac{d(E_p)}{dt} = v_E + k_3(EI^*) - k_f(I_p)(E_p) \tag{7}$$

$$\frac{d(EI^*)}{dt} = k_f(I_p)(E_p) - k_3(EI^*) \tag{8}$$

On that basis, simulations were performed with the help of a numerical integration program.

The value of v_E was estimated from the initial enzyme concentration in the periplasm (E_0) and the generation time of the bacterium (t_g). Indeed:

$$v_E = \frac{E_0}{2t_g} \qquad (9)$$

However, the equations as written do not take account of the increase of the periplasmic volume in growing cells. This problem will be discussed below (see *The Problem of v_E*). In most cases, high values of E_0 (100 to 950 μM) were chosen to exacerbate the effects of the presence of the β-lactamase.

Reaching the Steady State: Importance of k_3

As stated above, k_3 appears to be the most important factor in determining the time before a situation approaching the steady state (computed with the assumption that v_E equals 0) is reached. Very low values of k_D and k_f do not significantly delay the reaching of a value of I_p close to $(I_p)_{ss}$ if k_3 is relatively high (0.2 s^{-1}) (Table 1). The values of $t_{0.5}$ and $t_{0.75}$ increase when k_f decreases but remain very low compared with the generation time even with a value of k_f as low as 10^{-4} μM^{-1} s^{-1}. Observed values of k_f for serine β-lactamases are usually in the range of 10^{-2} to 10 μM^{-1} s^{-1} (see, for example, references 1, 3, and 6).

The Problem of v_E

If v_E is not zero in equation 7, a true steady state is never observed. Figure 4 depicts the effects of increasing values of v_E (i.e., decreasing generation times) for three different values of k_D. A significant effect is observed only for the

Table 1. Influence of low values of k_D and k_f on the time required to reach I_p values representing 50% ($t_{0.5}$) and 75% ($t_{0.75}$) of the steady-state value[a]

k_D (s^{-1})	k_f (μM^{-1} s^{-1})	I_e (μM)	$(I_p)_{ss}$[b] (μM)	$t_{0.5}$ (s)	$t_{0.75}$ (s)
1×10^{-3}	10^{-2}	10	1.07×10^{-3}		<1
5×10^{-3}	10^{-2}	100	0.053		<1
5×10^{-3}	10^{-3}	100	0.525	<1	<2
$5 \times 10^{-3\,c}$	10^{-4}	100	5	7	14

[a]In all cases, E_0 is 950 μM, v_E is 0.2 μM s^{-1} (t_g = 42 min), and k_3 is 0.2 s^{-1}.
[b]$(I_p)_{ss}$, Value of I_p at the steady state, assuming v_E is 0.
[c]Represents 3% of the lowest value observed in *E. coli* by Nikaido and co-workers (7, 8).

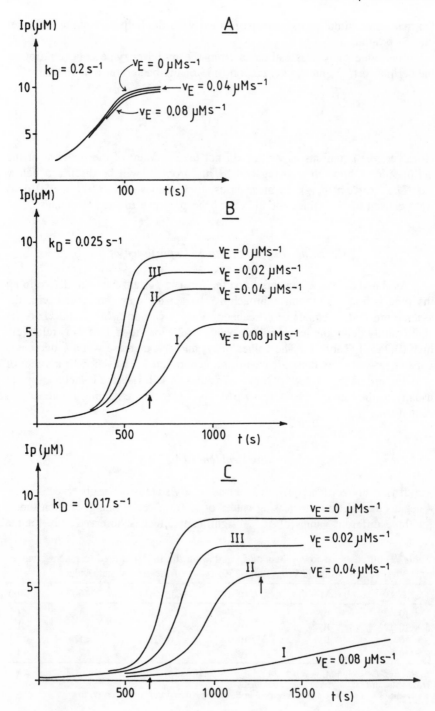

smallest k_D, and this effect becomes drastic only when a rather unrealistic generation time of about 10 min is assumed (Fig. 4C). However, when k_D is 0.017 s^{-1} and t_g is 21 min, the periplasmic concentration of β-lactam is important only for a short time before cell division, and that might not be sufficient to inactivate the essential PBPs. Finally, with a generation time of 42 min, an I_p value of 7.4 μM (i.e., 0.74 I_e) prevails during 1,500 s before cell division.

As stated above, one important factor is neglected in the present approach: as new enzyme is synthesized and excreted, the cell is also growing and the periplasmic volume is increasing. The total enzyme concentration ($E_{tot} = E_p + EI^*$) in the periplasm thus remains constant and equal to E_0, whereas our simple model assumes an increase of E_{tot} ($= E_0 + v_E \cdot t$).

The building of a complete model which will include those more complicated factors and extend into several generations is one of our future goals, but the present simple model can supply very useful information. Indeed, as can be seen by comparing the three panels of Fig. 4, the influence of v_E is important only when k_D is small, and it is sufficient to remember that the real situation is intermediate between those based on the following assumptions: $v_E = 0$; $v_E = E_0/2t_g$.

Influence of k_D and k_3

For large values of k_D (>0.5 s^{-1}), I_p values representing 75% of $(I_p)_{ss}$ (computed with $v_E = 0$) are reached within a few seconds or a few minutes. Figure 5 depicts situations where k_D is low and the enzyme concentration is extremely high. A rather stable value of I_p is reached within at most 1,000 s ($t_g = 42$ min) if k_3 is $\leq 10^{-3}$ s^{-1}. When k_3 is smaller than 0.5×10^{-4} s^{-1}, the curves can nearly be superimposed on curve I, indicating that within 600 s all the enzyme is immobilized as EI^* and the periplasmic concentration of β-lactam is virtually in equilibrium with the external one. The value of I_p after 1,000 s, computed with $v_E = 0.2$ μM s^{-1}, is significantly different from $(I_p)_{ss}$ (computed with $v_E = 0$) only when k_3 is 1×10^{-3} to 2×10^{-3} s^{-1}. But when k_3 is larger (5×10^{-3} s^{-1}), $(I_p)_{ss}$ is smaller, and after only 140 s, I_p becomes $0.75(I_p)_{ss}$.

For Fig. 6, a low value of k_3 has been chosen, and k_D is varied from 0.07 to 7 s^{-1}. As seen above under steady-state conditions, k_D can have a disproportionate influence on I_p: after 500 s, for instance, I_p is 0.9 when $k_D = 0.2$

Figure 4. Influence of v_E on the value of I_p. In all cases, E_0 is 100 μM, k_f is 0.01 μM^{-1} s^{-1}; k_3 is 2×10^{-4} s^{-1}; and I_e is 10 μM. (A) $k_D = 0.2$ s^{-1}; (B) $k_D = 0.025$ s^{-1}; (C) $k_D = 0.017$ s^{-1}. The arrows indicate the time of cell division, assuming the cycle started at $t = 0$. The values of v_E correspond to the following generation times: (A) $v_E = 0.02$, $t_g = 42$ min; (B) $v_E = 0.04$, $t_g = 21$ min; (C) $v_E = 0.08$, $t_g = 10.5$ min.

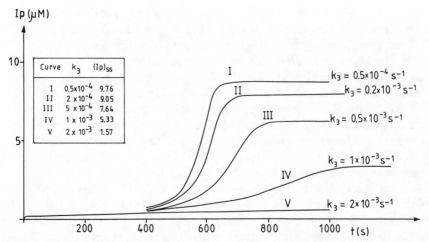

Figure 5. Influence of k_3 on I_p. In all cases, E_0 is 950 μM, k_f is 0.01 μM^{-1} s^{-1}, k_D is 0.2 s^{-1}, I_e is 10 μM, and $v_E = 0.2$ μM s^{-1} ($t_g = 42$ min). The insert shows the $(I_p)_{ss}$ values computed with v_E equal to 0.

s^{-1} and 7.5 when $k_D = 0.3$ s^{-1}. Thus, a 1.5-fold increase of k_D results in an eight-fold increase of I_p! Again, it can be noticed that v_E exhibits a strong influence only when k_D is small [compare curves VI and VII with the corresponding values of $(I_p)_{ss}$ in the insert].

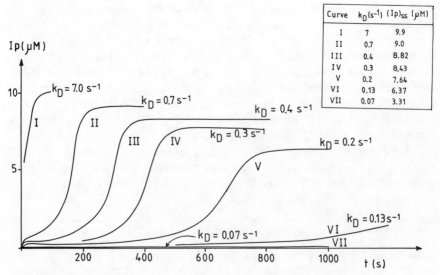

Figure 6. Influence of k_D. Conditions are as in Fig. 5, with k_3 equal to 5×10^{-4} s^{-1}.

Influence of k_f and E_0

Figure 6 represents simulations which were all performed using a k_f value of 0.01 μM^{-1} s^{-1}. In the cases of curves I, II, and III, a burst of antibiotic concentration in the periplasm is observed during the first few seconds. The size of the burst can be increased by increasing k_D (Fig. 6), decreasing k_f (Fig. 7), or decreasing E_0. Under the conditions described in the legend of Fig. 7, a k_f value of 10^{-4} μM^{-1} s^{-1} results in an I_p/I_e ratio of 0.7 after a few seconds, even in the presence of an enormous (0.95 mM) concentration of β-lactamase. This example shows that an intact β-lactam can coexist with a very high periplasmic concentration of β-lactamase if k_f is low.

In examining the influence of E_0 (Fig. 8), the value of v_E was proportionally varied so that the generation time remained constant (ca. 42 min). Under some conditions, E_0 also has a disproportionate influence on the value of I_p: after 300 s, I_p is 0.4 μM when $E_0 = 950$ μM and 8.9 μM when E_0 is 400 μM. On the contrary, after 800 s, the influence of E_0 is negligible.

Trapping or Not Trapping?

Various authors (9, 13) have suggested that the periplasmic β-lactamase might prevent poor substrates from reaching their targets by trapping them as acyl enzyme complexes. This hypothesis has generated a lot of controversy (11). Nikaido (7) has also noted that the outer membrane barrier could reduce the rate of penetration so that it would become similar to the rate of new β-

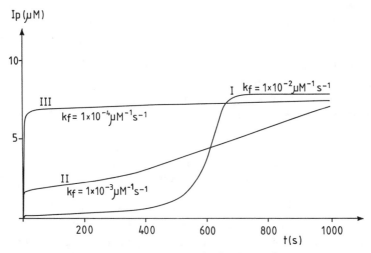

Figure 7. Influence of k_f. E_0, 950 μM; k_3, 2 × 10^{-4} s^{-1}; k_D, 0.2 s^{-1}; I_e, 10 μM; v_E, 0.2 μM s^{-1}.

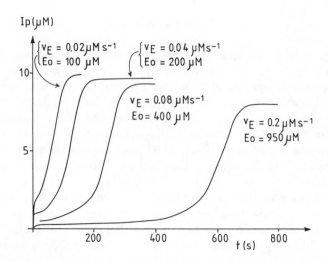

Figure 8. Influence of E_0 and v_E. k_f, 10^{-2} μM^{-1} s^{-1}; k_3, 2×10^{-4} s^{-1}; I_e, 10 μM; k_D, 0.2 s^{-1}.

lactamase synthesis. In the light of the results presented above, the following points might be stressed.

(i) High concentrations of β-lactamase result in rather high values of $k_{cat}E_0$ even if k_{cat} is as low as 10^{-2} s^{-1}, and hydrolysis, not trapping, might then be the relevant phenomenon.

(ii) When k_D is high, it is the β-lactamase which is rapidly trapped, and within a few seconds or a few minutes, I_p becomes similar to I_e (Fig. 6, curves I and II). The MIC then represents the intrinsic sensitivity of the essential target DD-peptidase.

(iii) When k_D is low, trapping can become a significant phenomenon (Fig. 9). The effect of the synthesis of new β-lactamase is also illustrated in Fig. 9 (compare curve I with curve II and curve III with curve IV). Again, small variations of k_D can induce rather large variations of I_p (compare curves II and IV). However, the restrictions outlined above about the increase of total periplasmic volume should be kept in mind. A complete answer must await the utilization of a more realistic model, and the simulation should be extended over several generations.

Another interesting question concerns the apparently nonessential, low-molecular-weight PBPs: could they perform trapping of the antibiotics and be responsible for resistance to low levels of antibiotics by protecting the essential, high-molecular-weight PBPs? In *E. coli* (2), only PBP 4 exhibits a value of k_f (0.7×10^{-2} μM^{-1} s^{-1}) considerably higher than that of PBPs 1, 2, and 3 (1 $\times 10^{-4}$ to 3×10^{-4} μM^{-1} s^{-1}). But the low number of PBP 4 molecules per cell (140 copies [12]) would not make the mechanism efficient. PBPs 5 and 6 are much more abundant (1,800 and 570 copies [12]), but in their cases, k_f is

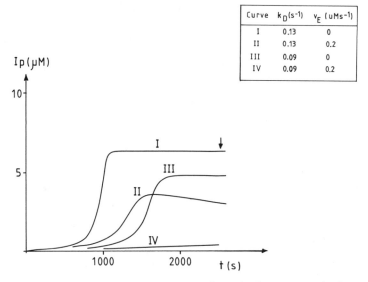

Figure 9. Influence of k_D and v_E. E_0, 950 μM; k_f, 10^{-2} μM^{-1} s^{-1}; k_3, 5×10^{-4} s^{-1}; I_e, 10 μM. The arrow indicates the time of cell division.

quite small (1.75×10^{-4} μM^{-1} s^{-1}), and it has been shown above that, with such a low value, even a very large number of "trapping" enzyme molecules per cell does not prevent a rapid accumulation of antibiotic in the periplasm (Fig. 7, curve III).

Unfortunately, quantitative data concerning the kinetic constants which characterize the interactions between PBPs and β-lactams are rather scarce. A possible candidate for trapping might be PBP 4 of *Proteus mirabilis* (k_f, 0.2 to 0.3 μM^{-1} s^{-1}; k_3, 4×10^{-5} s^{-1}), and a candidate for hydrolysis might be PBP 5 of the L form of the same species (k_f, 0.02 to 0.08 μM^{-1} s^{-1}; k_3, 1.6×10^{-3} s^{-1}) (2, 10). However, it is quite probable that the small number of those proteins present per cell might only marginally account for resistance (the values given above for the number of copies of PBPs per cell should be compared with values of the order of 10^5 and more copies per cell for β-lactamases) (7).

ACKNOWLEDGMENTS. This work was supported by the Fonds de Recherche de la Faculté de Médecine, the Fonds de la Recherche Scientifique Médicale, Brussels (contracts 3.4522.86 and 3.4507.83), the Gouvernement belge (action concertée 86/91-90), and the Région wallonne (C2/C16/conv. 246/20428). B.J. is chargé de recherches of the Fonds National de la Recherche Scientifique, Brussels.

We thank Michel Crine and Jean-Marie Ghuysen for their interest in this work.

LITERATURE CITED

1. **Emanuel, E. L., J. Gagnon, and S. J. Waley.** 1986. Structural and kinetic studies on β-lactamase K1 from *Klebsiella aerogenes*. *Biochem. J.* **234**:343–347.
2. **Frère, J. M., and B. Joris.** 1985. Penicillin-sensitive enzymes in peptidoglycan biosynthesis. *Crit. Rev. Microbiol.* **11**:299–396.
3. **Joris, B., F. De Meester, M. Galleni, S. Masson, J. Dusart, J. M. Frère, J. Van Beeumen, K. Bush, and R. B. Sykes.** 1986. Properties of a class C β-lactamase from *Serratia marcescens*. *Biochem. J.* **239**:581–586.
4. **Kelly, J. A., O. Dideberg, P. Charlier, J. P. Wéry, M. Libert, P. C. Moews, J. R. Knox, C. Duez, C. Fraipont, B. Joris, J. Dusart, J. M. Frère, and J. M. Ghuysen.** 1986. On the origin of bacterial resistance to penicillin: comparison of a β-lactamase and a penicillin target. *Science* **231**:1429–1431.
5. **Knott-Hunziker, V., S. Petursson, S. G. Waley, B. Jaurin, and T. Grundström.** 1982. The acyl-enzyme mechanism of β-lactamase action. The evidence for class C β-lactamases. *Biochem. J.* **207**:315–322.
6. **Labia, R., M. Barthélémy, C. Fabre, M. Guionie, and J. Peduzzi.** 1979. Kinetic studies of three R-factor mediated β-lactamases, p. 429–442. *In* J. M. F. Hamilton-Miller and J. T. Smith (ed.), *β-Lactamases*. Academic Press, Inc., New York.
7. **Nikaido, H.** 1985. Role of permeability barriers in resistance to β-lactam antibiotics. *Pharmacol. Ther.* **27**:197–231.
8. **Nikaido, H., and M. Vaara.** 1985. Molecular basis of bacterial outer membrane permeability. *Microbiol. Rev.* **49**:1–32.
9. **Sanders, C. C.** 1983. Novel resistance selected by the new expanded-spectrum cephalosporins: a concern. *J. Infect. Dis.* **47**:585–589.
10. **Schilf, W., and H. H. Martin.** 1980. Purification of two DD-carboxypeptidases/transpeptidases with different penicillin sensitivities from *Proteus vulgaris*. *Eur. J. Biochem.* **105**:361–370.
11. **Seeberg, A. H., R. M. Tolxdorff-Neutzling, and B. Wiedemann.** 1983. Chromosomal β-lactamases of *Enterobacter cloacae* are responsible for resistance to third generation cephalosporins. *Antimicrob. Agents Chemother.* **23**:918–925.
12. **Spratt, B. G.** 1977. Properties of the penicillin-binding proteins of *Escherichia coli* K12. *Eur. J. Biochem.* **72**:341–352.
13. **Takahashi, I., T. Sawai, T. Ando, and S. Yamagishi.** 1980. Cefoxitin resistance by a chromosomal cephalosporinase in *Escherichia coli*. *J. Antibiot.* **33**:1037–1042.
14. **Vu, M., and H. Nikaido.** 1985. Role of β-lactam hydrolysis in the mechanism of resistance of a β-lactamase constitutive *Enterobacter cloacae* strain to expanded-spectrum β-lactams. *Antimicrob. Agents Chemother.* **27**:393–398.
15. **Yoshimura, F., and H. Nikaido.** 1985. Diffusion of β-lactam antibiotics through the porin channels of *Escherichia coli*. *Antimicrob. Agents Chemother.* **27**:84–92.
16. **Zimmerman, W., and A. Rosselet.** 1977. The function of the outer membrane of *Escherichia coli* as a permeability barrier to β-lactam antibiotics. *Antimicrob. Agents Chemother.* **12**:368–372.

Chapter 49

Induction of β-Lactamase in *Bacillus* spp., from Penicillin Binding to Penicillinase

J. Oliver Lampen
Nancy J. Nicholls
Anthony J. Salerno
Matthew J. Grossman
Ying Fang Zhu
Tetsuo Kobayashi

The induction of β-lactamase synthesis in *Bacillus* species was one of the first microbial regulatory systems to be investigated (14) and has continued to attract interest because of its unusual properties. Induction is delayed and protracted even though the half-life of the specific mRNA for the enzyme is short (4, 16). The continued presence of the inducer (a β-lactam) is not essential, and the level of inducer added determines the amplitude of the response, but not the timing (3). Furthermore, it has not been possible to induce production of the enzyme in protoplasts. Because of these features it was of considerable interest to identify and characterize the various components of the regulatory system.

    Bacillus cereus and *Bacillus licheniformis* have received the greatest attention. The time course of induction is similar in the two species, although *B. licheniformis* forms only a single enzyme whereas *B. cereus* 569 makes three different β-lactamases that are coordinately regulated. We have focused

J. Oliver Lampen, Nancy J. Nicholls, Anthony J. Salerno, Matthew J. Grossman, Ying Fang Zhu, and Tetsuo Kobayashi • Waksman Institute of Microbiology, Rutgers University, The State University of New Jersey, Piscataway, New Jersey 08855-0759.

on the single enzyme in *B. licheniformis* with the expectation that the process in *B. cereus* will be parallel but with some elaborations.

In *B. licheniformis* the structural gene *blaP* (*penP* [16]) has been cloned and sequenced (12). A negative regulatory element, *blaI,* is 90% linked with *blaP,* as determined by transformation (18). Both genes are carried on a 4.2-kilobase-pair (kbp) *Eco*RI fragment. A second regulatory element (R1) was 50% linked, and a third (R2) was unlinked.

There was indirect evidence that *blaI* was located 3' to *blaP* on a polycistronic message (8, 18), but the major *blaP* transcript, in vivo and in vitro, was only 1.2 kilobases (kb) and terminated about 60 bases beyond the COOH terminus of the enzyme protein (11, 17). Although a few percent of the in vivo *blaP* transcripts continue to about 3 kb, there is no evidence that the region 3' to *blaP* has any role in β-lactamase synthesis or regulation.

Location of *blaI* in *B. licheniformis*

Our effort to locate *blaI,* the presumed repressor gene, began with the observation that the repressor was functional when carried on a plasmid in *Escherichia coli.* When the 4.2-kbp *Eco*RI fragment from strain 749, known to carry *blaI* and *blaP* (7, 19), was present, β-lactamase production was low (ca. 50 U/ml). The constitutive mutant 749/C does not form an active repressor, and the 4.2-kbp fragment from this mutant gave high expression of β-lactamase (10,000 U/ml). A two-plasmid system was devised; one plasmid contained the 4.2-kbp fragment from 749/C as the target for the repressor, and the second carried segments of the *Eco*RI fragment from the inducible strain 749 (Fig. 1). These tests showed *blaI* to lie in the 939-bp *Pst*I-*Sau*3A1 segment just 5' to *blaP* (12a).

The target area for the *blaI* product is within the region between *blaI* and *blaP* which contains both promoters. If the target fragment contained only the *blaP* promoter, partial repression occurred, but full repression was obtained when both promoters were present (1.6-kbp *Bcl*I-*Bcl*I segment). On this basis we propose that the repressor can interact with both promoters and probably must do so to produce full repression. Himeno et al. (6), who sequenced the *blaI* region of *B. licheniformis* 9945A, have made the same suggestion on the basis of the similarity of the two promoter sequences.

The *blaI* region of strain 749 has also been sequenced (Fig. 2) (12a) and is transcribed divergently from *blaP* and on the opposite strand. Another open reading frame follows *blaI* after a gap of only one nucleotide. There is a good Shine-Dalgarno sequence within the 3'-terminal segment of *blaI,* but no obvious promoter region. Presumably this open frame, which we term *blaR1,* is translated along with *blaI* on a polycistronic message.

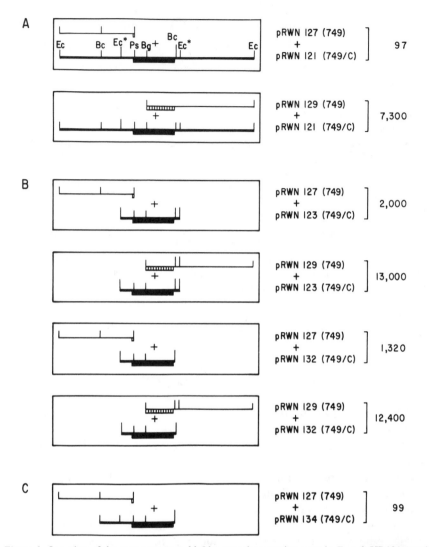

Figure 1. Location of the repressor gene *blaI* by complementation tests in *E. coli* HB101 (modified from Fig. 1 of Nicholls and Lampen [12a]). (A) Cells were cotransformed with pRWN121 containing the mutant 749/C *blaP* region on the 4.2-kbp *Eco*RI fragment and with either the 5' 1.6-kbp *Eco*RI-*Pst*I fragment on pRWN127 or the 3' 2.3-kbp *Bgl*II-*Eco*RI fragment from strain 749 on pRWN129. (B) The 1.4-kbp *Eco*RI*-*Eco*RI* fragment on pRWN123 and the 1.2-kbp *Eco*RI*-*Bcl*I fragment on pRWN132 (both containing mutant 749/C *blaP*) along with pRWN127 or pRWN129. (C) The 1.6-kbp *Bcl*I-*Bcl*I fragment containing mutant 749/C *blaP* (on pRWN134) along with pRWN127. Solid bars and thick lines, DNA from mutant 749/C; hatched bars and thin lines, DNA from strain 749 or 9945A. Values at right: Units of β-lactamase per milliliter at stationary phase. Abbreviations for restriction nuclease sites: Bc, *Bcl*I; Bg, *Bgl*II; Ec, *Eco*RI; Ec*, *Eco*RI*, Ps, *Pst*I.

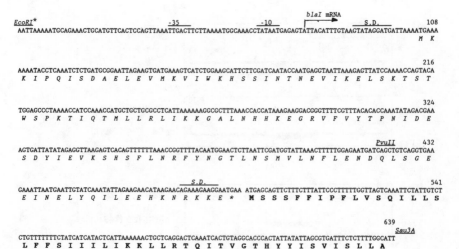

Figure 2. The *bla* region and the sequence of *blaI*. The nucleotide sequence of *blaI* and the deduced amino acid residues are shown. The *blaI* product (residues 1 through 128) is in italics. The potential second protein, which begins at nucleotide 491, is in boldface for contrast. The Shine-Dalgarno sequence and the promoter regions (−35 and −10) are indicated, as are the *EcoRI**, *PvuII*, and *Sau3A* sites. *, Terminator. One-letter codes for amino acids: A, Ala; C, Cys; D, Asp; E, Glu; F, Phe; G, Gly; H, His; I, Ile; K, Lys; L, Leu; M, Met; N, Asn; P, Pro; Q, Gln; R, Arg; S, Ser; T, Thr; V, Val; W, Trp; Y, Tyr.

Nature of the Repressor (BlaI)

The sequence of the repressor from strain 749 is identical to that from strain 9945A (6). The hydrophilic protein of 128 amino acid residues has an M_r of 15,036. Its secondary structure as predicted by the principles of Chou and Fasman (2) is shown in Fig. 3. There are two α-helix-β-turn-α-helix structures of the type present in most repressors, and a highly charged nine-residue COOH-terminal segment. Thus, the repressor for the β-lactamase of *B. licheniformis* has the general characteristics of repressors from gram-negative microorganisms (13, 15), but the helix-turn-helix regions do not exhibit convincing homology with the consensus sequences for other repressors (10, 13).

The identification of the *blaI* product, BlaI, as the repressor for the β-lactamase gene (*blaP*) was confirmed by sequencing the *blaI* region from the constitutive mutant 749/C (5a). An amber codon is present at codon 32, and this may lead to a nonfunctional truncated protein. For production of the repressor in quantity, the promoter region of *blaI* was removed, and the gene was expressed in the two-plasmid T7 RNA polymerase-promoter system of Tabor and Richardson (20). Heat induction of this system in *E. coli* K38 gave BlaI as 5 to 10% of the total soluble protein. The repressor was purified and sequenced by repetitive Edman degradation. The NH₂-terminal 28 residues were

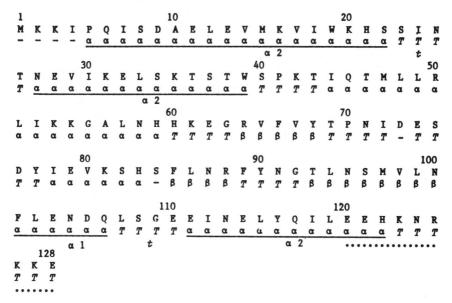

Figure 3. Probable secondary structure of the repressor for β-lactamase from strain 749 (originally published as Fig. 3 by Nicholls and Lampen [12a]). Predicted according to the principles of Chou and Fasman (2). α, Alpha-helical; *T*, β-turn; β, beta-sheet; . . . , charged COOH-terminal sequence. Two helix-turn-helix structures are indicated.

identical with those predicted from the DNA sequence. To demonstrate that BlaI is a specific DNA binding protein, the competitive DNA binding gel electrophoresis procedure (5) was employed. BlaI selectively bound to DNA fragments carrying the promoters for *blaI* and *blaP*.

In Vivo Transcription of the bla Region

After addition of inducer (a β-lactam) to a culture of *B. licheniformis* 749, the *blaP* promoter produces rightward transcripts (mainly 1.2 kb, as mentioned earlier) and the *blaI* promoter produces large leftward mRNAs (2.3 kb and some 1.9 kb) as well as small mRNAs (0.75 to 0.95 kb) (A. J. Salerno and J. O. Lampen, submitted for publication). The small mRNAs encode only *blaI*, while the long mRNA is probably polycistronic, carrying both *blaI* and the additional gene indicated to lie beyond *blaI*. The large transcripts are present only during the initial period of induction. Consequently, the product, BlaR1, is likely to function at an early stage in induction.

Production of the small (0.75 to 0.95 kb) *blaI* mRNAs continues for about another hour, rising and declining approximately in parallel with the mRNAs for the structural gene *blaP* (17). This suggests that BlaI can also serve, in some way, as a positive effector.

Cloning of blaR1

The 5.2-kbp *Sph*I fragment spanning *blaP, blaI,* and the entire 2.3-kb transcript (and thus, presumably, containing the entire *blaR1* gene as well) was cloned into *E. coli* and into *B. subtilis*. The presence of this fragment did not make *E. coli* inducible, but *B. subtilis* showed an essentially normal inductive response (9). Sequencing of the *blaR1* region revealed that the unfinished open frame mentioned earlier ultimately codes for a protein of 601 amino acid residues (M_r, 68,425). BlaR1 appears to be a membrane protein with five *trans*-membrane regions followed by a COOH-terminal segment of about 260 residues which is predicted to be on the outer surface of the membrane. Within this external segment, the sequence around Ser-402 is typical of penicillin-binding proteins and shows extensive homology to penicillin-binding protein 2 of *E. coli* (1). BlaR1 might be a membrane receptor whose binding of a β-lactam initiates the process of induction. It might also function by interacting directly with the repressor. To test these possibilities, the gene has been cloned in an expression vector for preparation of BlaR1 and direct testing of its role in induction.

ACKNOWLEDGMENTS. This work was supported in part by Public Health Service grant AI-23096 from the National Institute for Allergy and Infectious Diseases, by the Charles and Johanna Busch fund, and by Miles Laboratories, Inc.

LITERATURE CITED

1. **Asoh, S., H. Matsuzawa, F. Ishino, J. L. Strominger, M. Matsuhaśhi, and T. Ohta.** 1986. Nucleotide sequence of the *pbpA* gene and characteristics of the deduced amino acid sequence of penicillin-binding protein 2 of *Escherichia coli* K12. *Eur. J. Biochem.* **160**:459–472.

2. **Chou, P. Y., and G. O. Fasman.** 1978. Empirical predictions of protein conformation. *Annu. Rev. Biochem.* **17**:251–276.

3. **Collins, J. F.** 1979. The *Bacillus licheniformis* β-lactamase system, p. 351–368. *In* M. T. Hamilton-Miller and J. T. Smith (ed.), β-*Lactamases*. Academic Press, Inc., London.

4. **Davies, J. W.** 1969. The stability and cell content of penicillinase messenger RNA in *Bacillus licheniformis*. *J. Gen. Microbiol.* **59**:171–184.

5. **Fried, M., and D. M. Crothers.** 1984. Equilibrium studies of the cyclic AMP receptor protein-DNA interaction. *J. Biol. Chem.* **172**:241–262.

5a. **Grossman, M. J., and J. O. Lampen.** 1987. Purification and DNA binding properties of the *blaI* gene product, repressor for the β-lactamase gene, *blaP*, of *Bacillus licheniformis*. *Nucleic Acids Res.* **15**:6049–6062.

6. **Himeno, T., T. Imanaka, and S. Aiba.** 1986. Nucleotide sequence of the penicillinase repressor gene *penI* of *Bacillus licheniformis* and regulation of *penP* and *penI* by the repressor. *J. Bacteriol.* **168**:1128–1132.

7. **Imanaka, T., T. Tanaka, H. Tsunekawa, and S. Aiba.** 1981. Cloning of the genes for penicillinase, *penP* and *penI*, of *Bacillus licheniformis* in some vector plasmids and their expression in *Escherichia coli, Bacillus subtilis,* and *Bacillus licheniformis. J. Bacteriol.* **147:**776–786.

8. **Kelly, L. E., and W. J. Brammar.** 1973. The polycistronic nature of the penicillinase structural and regulatory genes in *Bacillus licheniformis. J. Mol. Biol.* **80:**149–154.

9. **Kobayashi, T., Y. F. Zhu, N. J. Nicholls, and J. O. Lampen.** 1987. A second regulatory gene, *blaR1,* encoding a potential penicillin-binding protein required for induction of β-lactamase in *Bacillus licheniformis. J. Bacteriol.* **169:**3873–3878.

10. **Laughon, A., and M. P. Scott.** 1984. Sequence of *Drosphila* segmentation gene: protein structure homology with DNA binding proteins. *Nature* (London) **310:**25–31.

11. **McLaughlin, J. R., S.-Y. Chang, and S. Chang.** 1982. Transcriptional analyses of the *Bacillus licheniformis penP* gene. *Nucleic Acids. Res.* **10:**3905–3919.

12. **Neugebauer, K., R. Sprengel, and H. Schaller.** 1981. Penicillinase from *Bacillus licheniformis*: nucleotide sequence of the gene and implications for the biosynthesis of a secretory protein in a gram-positive bacterium. *Nucleic Acids Res.* **9:**2577–2588.

12a. **Nicholls, N. J., and J. O. Lampen.** 1987. Repressor gene, *blaI,* for *Bacillus licheniformis* 749 β-lactamase. *FEBS Lett.* **221:**179–183.

13. **Pabo, C. O., and R. T. Sauer.** 1984. Protein-DNA recognition. *Annu. Rev. Biochem.* **53:**293–321.

14. **Pollock, M. R.** 1950. Penicillinase adaption in *B. cereus*: adaptive enzyme formation in the absence of free substrate. *Br. J. Exp. Pathol.* **31:**739–753.

15. **Rosenberg, M., and D. Court.** 1979. Regulatory sequence involved in the promotion and termination of RNA transcription. *Annu. Rev. Genet.* **13:**319–353.

16. **Salerno, A. J., and J. O. Lampen.** 1986. Transcriptional analysis of β-lactamase regulation in *Bacillus licheniformis. J. Bacteriol.* **166:**769–778.

17. **Salerno, A. J., N. J. Nicholls, and J. O. Lampen.** 1985. Penicillinase synthesis in *Bacillus licheniformis*: nature of induction, p. 110–116. *In* J. H. Hoch and P. Setlow (ed.), *Molecular Biology of Microbial Differentiation.* American Society for Microbiology, Washington, D.C.

18. **Sherratt, D. J., and J. F. Collins.** 1973. Analysis by transformation of the pencillinase system in *Bacillus licheniformis. J. Gen. Microbiol.* **76:**217–230.

19. **Sprengel, R., B. Reiss, and H. Schaller.** 1983. Translationally coupled initiation of protein synthesis in *Bacillus subtilis. Nucleic Acids Res.* **13:**893–909.

20. **Tabor, S., and C. C. Richardson.** 1985. A bacteriophage T7 RNA polymerase/promoter system for controlled exclusive expression of specific genes. *Proc. Natl. Acad. Sci. USA* **32:**1074–1078.

Chapter 50

Molecular Mechanisms of β-Lactamase Induction

Frederik Lindberg
Susanne Lindqvist
Staffan Normark

In principle, the efficiency with which a β-lactam antibiotic exerts its lethal effect on a given gram-negative bacterium is determined by the rate of diffusion through the outer membrane, the affinity of the β-lactam for the target, and the rate of hydrolysis of the compound by a periplasmic β-lactamase (10, 16, 17). Certain gram-negative species such as *Citrobacter freundii*, *Enterobacter cloacae*, *Serratia marscescens*, *Pseudomonas* species, and indole-positive *Proteus* species express β-lactamases that are inducible by β-lactams (14). The ability of a certain β-lactam to induce β-lactamase expression also affects the level of resistance of the bacterium to that particular drug. The inducible β-lactamases have been found to be chromosomally encoded and structurally as well as functionally related to the noninducible AmpC β-lactamase of *E. coli* K-12 (5, 6, 9). The frequency at which gram-negative species with inducible β-lactamase mutate to β-lactam resistance is unusually high (10^{-5} to 10^{-7}). Since most of these resistant variants result from mutations affecting the regulation of the chromosomal β-lactamase rather than the structural gene itself, there has been considerable interest in understanding the molecular details of β-lactamase induction (12).

It is generally believed that β-lactam antibiotics exert their biological effect by covalently binding to penicillin-binding proteins at the outer surface of

Frederik Lindberg, Susanne Lindqvist, and Staffan Normark • Department of Microbiology, University of Umeå, S-901 87 Umeå, Sweden.

the inner membrane (15). There are no indications that β-lactams can penetrate the cytoplasmic membrane and enter the cytoplasm of bacteria. Thus, an inducer on the outside of the inner membrane is able to cause an increased expression from a β-lactamase gene in the cytoplasm. We have shown that induction in *C. freundii* and *E. cloacae* requires the expression of at least two proteins, AmpR and AmpD (8, 11, 12). In both *C. freundii* and *E. cloacae* the *ampR* gene is a key factor in induction, positioned next to the *ampC* β-lactamase gene. Probably as a result of a deletion, the *ampR* gene is physically absent in species producing noninducible AmpC β-lactamase such as *Escherichia coli* and *Shigella* spp. (1, 4, 12). The *ampD* gene has been found in both noninducible and inducible species (8, 12). Its inactivation leads to semiconstitutive or constitutive expression of AmpC β-lactamase.

The ampR Gene Codes for a Transcriptional Regulator

Expression of noninducible AmpC β-lactamase in *E. coli* is directed from a promoter positioned in the end of the fumarate reductase (*frd*) operon, i.e., within the last part of the *frdD* coding sequence. The low expression is due to poor promoter activity and to the fact that the transcriptional terminator for the *frd* operon is positioned between the *ampC* promoter and the structural gene, where it functions as an attenuator of *ampC* transcription. Increased expression of *ampC* β-lactamase occurs at low frequency through mutations increasing the strength of the *ampC* promoter or decreasing that of the *ampC* attenuator (for a review, see reference 10).

E. coli clones harboring *ampR* and *ampC* from either *C. freundii* or *E. cloacae* express the respective enzymes inducibly. In both cases, genetic inactivation of *ampR* leads to low constitutive expression of β-lactamase. Wild-type *ampR* supplied in *trans* on a separate plasmid complements the effect of the *ampR* mutation back to inducible enzyme expression. Therefore, *ampR* must code for a *trans*-acting regulatory function. The *ampR* genes of both *C. freundii* and *E. cloacae* have been sequenced (4; unpublished data) and are transcribed in a direction opposite that of *ampC*. The intercistronic region between *ampR* and *ampC* in *E. cloacae* consists of only 131 base pairs (4). This stretch of DNA is totally missing in *E. coli,* providing a rational explanation for the lack of effect of *ampR* on the expression of the *E. coli ampC* β-lactamase (10). By primer extension studies and S1 nuclease mapping, the 5' ends of both the *ampR* and *ampC* mRNA were identified (4). The data showed that transcription of the two genes was initiated from two partially overlapping promoters located in the *ampR-ampC* intercistronic region. Induction with a β-lactam affected neither *ampR* transcription (4) nor the amount of AmpR protein produced when measured in minicells (8, 12). In contrast, induction resulted in a marked increase in the production of *ampC* mRNA as measured in *E. cloacae* by primer extension (4) and in *C. freundii* by *ampC-lacZ* transcriptional fusions (unpublished data). Thus, it is clear that β-lactamase induction occurs at the level of transcriptional initiation.

The AmpR proteins from *E. cloacae* and *C. freundii* show a high (more than 90%) direct amino acid homology (4; unpublished data), which might be expected since the *C. freundii ampR* gene can complement a mutation in *E. cloacae ampR* and vice versa (11). However, some quantitative differences in the induction levels were found when homologous and heterologous *ampR* genes were used in the *trans* complementation experiments. Hence, synthesis of *C. freundii* β-lactamase is considerably lower, both uninduced and induced, when expressed in the presence of the homologous *ampR* gene than with the *ampR* gene from *E. cloacae*. The reversed situation was seen with the *E. cloacae ampC* gene (11). These differences could be due to the structural differences between the respective AmpR proteins and intercistronic regions in the two species. No actual binding studies between AmpR and AmpC promoter DNA have been reported. At this stage we can only speculate that after induction AmpR facilitates *ampC* transcription by binding to sequences close to the *ampC* promoter. In the absence of inducer, β-lactamase expression from both *C. freundii* and *E. cloacae ampC* is twice as high in an *ampR* mutant as in the wild type. Thus, also in the absence of inducer, AmpR affects *ampC* expression, this time as a repressor (11, 12).

Mutations in ampD Cause Semiconstitutive Overproduction of Inducible C. freundii and E. cloacae β-Lactamase

In species expressing inducible AmpC β-lactamase, mutants that are considerably more resistant to most penicillins and cephalosporins than the wild type arise at a high frequency. One class of these mutants expresses β-lactamase constitutively or semiconstitutively (2, 12).

When the *ampR ampC* region was cloned from semiconstitutive or constitutive mutants of either *C. freundii* or *E. cloacae* into *E. coli*, expression of β-lactamase was low and inducible (8, 12). This result implies that the lesion leading to a loss of inducibility has affected a gene elsewhere on the chromosome. Since induction was possible in *E. coli*, we argued that this species might contain a similar gene. Therefore, we selected cefotaxime-resistant mutants of *E. coli* harboring the *ampR* and *ampC* genes of *C. freundii* on a recombinant plasmid (8). Spontaneous cefotaxime-resistant mutants (50 μg/ml) were obtained at an incidence of 1×10^{-7} to 3×10^{-7} per viable cell. Of 45 mutants analyzed, all but 1 carried lesions on the *E. coli* chromosome and not on the recombinant $ampR^+$ $ampC^+$ plasmid. Two classes of resistant mutants, semiconstitutive and inducible (8; unpublished data), appeared at roughly equal frequency. *E. coli* SN0302 is a representative of the semiconstitutive class. The *ampD2* mutation in this strain was mapped by conjugation and phage P1 transductions to 2.4 min on the *E. coli* chromosome, close to the *aroP* and *nadC* genes; *aroP* and *nadC* had been cloned previously on phage λG78N by Guest and his collaborators (3). A 10-kilobase *Eco*RI fragment from λG78N was subcloned on a pACYC184-based vector. The resulting plasmid carried *aroP* and *nadC* as well as the ability to *trans* complement *E. coli* SN0302 to full inducibility. It therefore contains the wild-type allele of *ampD*. By subcloning, *ampD* could be mapped between *aroP* and *nadC* (8). The *ampD2* mutation from SN0302 was cloned by reciprocal recombination between *ampD2* on the chromosome and wild-type *ampD* on a plasmid as well as by direct cloning with selection for NadC$^+$ clones. By DNA sequencing, *ampD2* was shown to be caused by an insertion of IS*1* into the *ampD* gene.

Interspecies trans Complementation of ampD

A plasmid carrying the *E. coli ampD* wild-type allele was transformed into both the inducible *C. freundii* wild type (OS60) and a semiconstitutive mutant (OS61) (8). It was noted that *E. coli ampD* reduced the expression of β-lactamase under full induction when present in *C. freundii* OS60. In the semiconstitutive mutant OS61, *ampD* in *trans* complemented OS61 to low, fully inducible β-lactamase expression. This suggested that *C. freundii* OS61 was mutated in a *C. freundii ampD* gene. *C. freundii ampD* from OS60 and OS61 has been cloned by direct selection for NadC$^+$ transformants in an *E. coli nadC* mutant. *C. freundii ampD* behaves as the corresponding gene in *E. coli* (unpublished data). It complements *E. coli* SN0302 harboring either *C. freundii* or *E. cloacae ampR* and *ampC* genes to full inducibility. More recently, *ampD* has also been cloned from *E. cloacae* by selection for NadC$^+$ (unpublished data). Also, the *E. cloacae ampD* gene could fully complement the *E. coli ampD* mutation. The conclusion is that *ampD* from *E. coli*, *C. freundii*, and *E. cloacae* behave similarly in all combinations of *trans* complementation. This is in contrast to *ampR*, for which quantitative differences were found between different complementation pairs. The *ampD* genes from *E. coli* and *C. freundii* complemented all semiconstitutive cefotaxime-resistant *E. coli* mutants to sensitivity and low, fully inducible β-lactamase synthesis. Hence, all these mutants are likely to have *ampD* mutations (unpublished data).

As yet, it is not known whether all semiconstitutive and constitutive clinical isolates of *C. freundii* and *E. cloacae* have *ampD* mutations. However, most isolates investigated in this laboratory could be *trans* complemented to the wild-type level of β-lactamase expression by both the *E. coli* and *C. freundii ampD* genes (unpublished data).

The Connection between ampD and ampR

There is an intimate functional connection between *ampD* and *ampR*. Thus expression from both genes is required for normal β-lactamase induction. Strains with *ampD* mutations expressing β-lactamase semiconstitutively require an intact *ampR* gene (8). Thus, it appears that AmpD is a protein that directly or indirectly affects the transcriptional regulator AmpR, rather than directly binding to the *ampC* control region. Since *ampD* is present also in enterobacteria lacking *ampR* and inducible β-lactamase, we believe that its gene product has a normal function not related to β-lactamase regulation. At present, we have no clue as to what that function could be. None of the known genes for penicillin-binding proteins have been mapped to the same chromosomal position as *ampD*. Still, it is possible that β-lactams may bind to AmpD covalently or noncovalently. Our working hypothesis is that AmpD is an enzyme that catalyzes a reaction in cell wall synthesis. Accumulation of its substrate would occur when AmpD is inhibited by β-lactams or destroyed by mutations. This substrate could then be the AmpR coactivator leading to *ampC* induction. Alternatively, the disappearance of the AmpD product could be the induction signal. To further investigate this hypothesis, we have to identify the function for AmpD and define the effector that activates AmpR. Our hypothesis also suggests that the ability of a β-lactam to induce AmpC β-lactamase is related to its ability to directly or indirectly interfere with AmpD. There is, on the other hand, no relation between the ability of a β-lactam to induce AmpC β-lactamase and its sensitivity to hydrolysis by the enzyme. Both cephalothin and imipenem are good inducers, although one is rapidly hydrolyzed by the β-lactamase whereas the other is almost totally resistant to the action of the enzyme.

β-Lactamase Induction in P. aeruginosa

It is not yet known whether β-lactamase induction in *P. aeruginosa* is regulated in the same way as in *C. freundii* and *E. cloacae*. The inducible enzyme of *P. aeruginosa* 1822S/H shows structural similarities to other AmpC β-lactamases in the active-site region (7). Several loci that are involved in β-lactamase induction have been identified in *P. aeruginosa* PAO (13), and *cis*-dominant mutations leading to semiconstitutive β-lactamase production were localized to a locus designated *blaJ*, which maps close to the structural gene *blaP*. It was thought that *blaJ* defined an operator region for *blaP*. Mutations in two other loci, *blaI* and *blaK*, caused semiconstitutive β-lactamase expression. A complete constitutive phenotype was obtained in *blaI blaK* double mutants. It will be interesting to see which of these two loci is similar to *ampD*.

ACKNOWLEDGMENTS. The work reported here from our laboratory was supported by grants from the Swedish Medical Research Council (Dnr 4769) and the Board of Technological Development (Dnr 81-3384B).

LITERATURE CITED

1. **Bergström, S., F. Lindberg, O. Olsson, and S. Normark.** 1983. Comparison of the overlapping *frd* and *ampC* operons of *Escherichia coli* with the corresponding DNA sequences in other gram-negative bacteria. *J. Bacteriol.* **155:**1297–1305.

2. **Gootz, T. D., D. B. Jackson, and J. C. Sherris.** 1984. Development of resistance to cephalosporins in clinical strains of *Citrobacter* spp. *Antimicrob. Agents Chemother.* **25:**591–595.

3. **Guest, J. R., and P. E. Stephens.** 1980. Molecular cloning of the pyruvate dehydrogenase complex genes of *Escherichia coli*. *J. Gen. Microbiol.* **121:**277–292.

4. **Honoré, N., M. H. Nicolas, and S. T. Cole.** 1986. Inducible cephalosporinase production in clinical isolates of *Enterobacter cloacae* is controlled by a regulatory gene that has been deleted from *Escherichia coli*. *EMBO J.* **13:**3709–3714.

5. **Jaurin, B., and T. Grundström.** 1981. *ampC* cephalosporinase of *Escherichia coli* K-12 has a different evolutionary origin from that of β-lactamases of the penicillinase type. *Proc. Natl. Acad. Sci. USA* **78:**4897–4901.

6. **Joris, B., F. de Meester, M. Galleni, G. Recklinger, J. Coyette, J.-M. Frère, and J. van Beeumen.** 1985. The β-lactamase of *Enterobacter cloacae* P99. Chemical properties, N-terminal sequence, and interaction with 6β-halogenopenicillanates. *Biochem. J.* **228:**241–248.

7. **Knott-Hunziker, V., S. Petursson, G. S. Jayatilake, S. G. Waley, B. Jaurin, and T. Grundström.** 1982. Active sites of β-lactamases: the chromosomal β-lactamases of *Pseudomonas aeruginosa* and *Escherichia coli*. *Biochem. J.* **201:**621–627.

8. **Lindberg, F., S. Lindqvist, and S. Normark.** 1987. Inactivation of the *ampD* gene causes semiconstitutive overproduction of the inducible *Citrobacter freundii* β-lactamase. *J. Bacteriol.* **169:**1923–1928.

9. **Lindberg, F., and S. Normark.** 1986. Sequence of the *Citrobacter freundii* OS60 *ampC* β-lactamase gene. *Eur. J. Biochem.* **156:**441–445.

10. **Lindberg, F., and S. Normark.** 1986. Contribution of chromosomal β-lactamases to β-lactam resistance in enterobacteria. *Rev. Infect. Dis.* **8**(Suppl. 3):292–304.

11. **Lindberg, F., and S. Normark.** 1987. Common mechanism of AmpC β-lactamase induction in enterobacteria: regulation of the cloned *Enterobacter cloacae* P99 β-lactamase gene. *J. Bacteriol.* **169:**758–763.

12. **Lindberg, F., L. Westman, and S. Normark.** 1985. Regulatory components in *Citrobacter freundii ampC* β-lactamase induction. *Proc. Natl. Acad. Sci. USA* **82:**4620–4624.

13. **Matsumoto, H., and Y. Terawaki.** 1982. Chromosomal location of the genes participating in the formation of β-lactamase in *Pseudomonas aeruginosa*, p. 207–211. *In* S. Mitsuhashi (ed.), *Drug Resistance in Bacteria*. Japan Scientific Press, Tokyo.

14. **Sykes, R. B., and M. Matthew.** 1976. The β-lactamases of gram-negative bacteria and their role in resistance to β-lactam antibiotics. *J. Antimicrob. Chemother.* **2:**115–157.

15. **Tomasz, A.** 1986. Penicillin-binding proteins and the antibacterial effectiveness of β-lactam antibiotics. *Rev. Infect. Dis.* **8**(Suppl. 3):260–278.

16. **Vu, H., and H. Nikaido.** 1985. Role of β-lactam hydrolysis in the mechanism of resistance of a β-lactamase constitutive *Enterobacter cloacae* strain to expanded-spectrum β-lactams. *Antimicrob. Agents Chemother.* **27:**393–398.

17. **Yoshimura, F., and H. Nikaido.** 1985. Diffusion of β-lactam antibiotics through the porin channels of *Escherichia coli* K-12. *Antimicrob. Agents Chemother.* **27:**84–92.

Chapter 51

Initiation of Induction of Chromosomal β-Lactamase in Gram-Negative Bacteria by Binding of Inducing β-Lactam Antibiotics to Low-Molecular-Weight Penicillin-Binding Proteins

Hans H. Martin
Bärbel Schmidt
Sandra Bräutigam
Hiroshi Noguchi
Michio Matsuhashi

In different gram-positive and gram-negative bacteria, synthesis of chromosomally coded β-lactamase is inducible by β-lactam antibiotics (2, 17). The primary mechanism by which β-lactams function as inducers is unknown. There is little information on the transport of the antibiotics into the bacterial cytoplasm and on their direct action as regulators of gene expression. An exogenous interaction of the inducer with a membrane receptor has been postulated on the basis of early reports on the cytoplasmic membrane as exclusive site of penicillin binding in the bacterial cell (1, 3).

Hans H. Martin, Bärbel Schmidt, and Sandra Bräutigam • Institut für Mikrobiologie, Technische Hochschule, D-6100 Darmstadt, Federal Republic of Germany. Hiroshi Noguchi • Takarazuka Research Center, Biotechnology Laboratory, Sumitomo Chemical Co., Ltd., Takarazuka, Hyogo-ken, Japan. Michio Matsuhashi • Institute of Applied Microbiology, University of Tokyo, Bunkyo-ku, Tokyo, Japan.

Bacterial membrane binding sites for β-lactam antibiotics are now well characterized as multiple penicillin-binding proteins (PBPs) with various enzymatic functions in the final stage of cell wall peptidoglycan synthesis (10, 20). It is attractive to speculate that, in a given bacterium, one or several of the PBPs may also function as receptors for β-lactamase induction.

A lead toward the identity of the hypothetical induction receptors is provided by observations on the very different β-lactamase-inducing potency of structurally different β-lactam antibiotics in several gram-negative bacteria. The 7-α-methoxycephalosporin cefoxitin was identified early as a particularly potent inducer, e.g., in *Enterobacter cloacae, Pseudomonas aeruginosa,* and indole-positive *Proteeae,* and other 7-α-methoxycephalosporins were similar in this respect (5, 6, 11, 18, 21). Initially, the high stability of cefoxitin to hydrolysis by β-lactamases was thought to be the essential property for β-lactamase induction. This notion was untenable, however, in view of the low inducing efficiencies of various structurally different β-lactam compounds with comparably high β-lactamase stability. Thus, good inducers must be distinguished by another specific feature from β-lactams which induce poorly or only at high concentrations. Such a property has been recognized for cefoxitin and related compounds in their high binding affinity for low-molecular-weight PBPs, PBP 5 in particular, in for example, *Escherichia coli, Proteus mirabilis,* and *P. aeruginosa* (4, 9, 15, 16, 19).

Therefore, the possible correlation of β-lactamase induction with in vivo binding of inducers to PBP 5 was studied in *P. aeruginosa* and *E. cloacae.* Furthermore, the dependence of β-lactamase induction on the intact function of PBP 5 was investigated in a temperature-sensitive, PBP 5-defective mutant of *P. aeruginosa* (12).

Concentration Dependence of β-Lactamase Induction

Induction of β-lactamase in *P. aeruginosa* and *E. cloacae* was studied at different concentrations of a variety of β-lactam antibiotics with the aim of determining the lowest concentration of each antibiotic at which induction took place at a significant level above constitutive enzyme synthesis in uninduced bacteria. It was expected that from these threshold concentrations upwards a correlation of β-lactamase induction with the binding of antibiotics to specific PBPs could be recognized.

In confirmation and extension of previous reports (5, 11, 21), structurally different antibiotics were found to vary greatly in their inducing potencies in both bacteria (Fig. 1). The carbapenem compound imipenem surpassed all other antibiotics tested in producing the highest levels of induction in a range of low concentrations from 0.05 to 3 μg/ml. Among other antibiotics, groups of good inducers (cefoxitin and other 7-α-methoxycephalosporins, cefotetan, cefmeta-

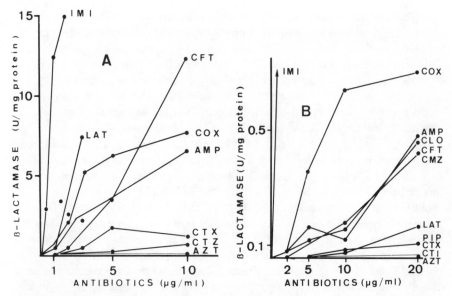

Figure 1. Induction of β-lactamase by structurally different β-lactam antibiotics in *E. cloacae* NCTC 10005 (A) and *P. aeruginosa* ATCC 27853 (B). Antibiotics tested were imipenem (IMI), cefoxitin (COX), cefotetan (CFT), cefmetazole (CMZ), latamoxef (LAT), ampicillin (AMP), cephaloridine (CLO), piperacillin (PIP), cefotaxime (CTX), ceftizoxime (CTI), ceftazidime (CTZ), and aztreonam (AZT). Growth media were Penassay broth (Difco Laboratories) for *E. cloacae* and Isosensitest broth (Oxoid Ltd.) for *P. aeruginosa*. Induction at different antibiotic concentrations and preparation of crude extracts from induced bacteria were performed as described by Minami et al. (11). Photometric assays of β-lactamase activity (14) were carried out with the substrates cephalothin (0.3 mmol/liter) and nitrocefin (0.1 mmol/liter) at wavelengths of 262 and 484 nm, respectively. One unit of β-lactamase designates the amount of enzyme which hydrolyzes 1 μmol of substrate per min at 25°C.

zole, and latamoxef, particularly in *E. cloacae*) and poor inducers, even at high concentration (aminothiazole cephalosporins, cefotaxime, ceftizoxime and ceftazidime, the monobactam aztreonam, and the acylureidopenicillin piperacillin), were distinguished. Of particular interest was an intermediate group (ampicillin and cephaloridine in both bacteria and cefotetan, cefmetazole, and latamoxef, particularly in *P. aeruginosa*), for which a concentration-dependent increase of induction from 1 to 20 μg of inducer per ml was typical.

In Vivo Binding of β-Lactams to PBPs

In the range of the lowest β-lactamase-inducing concentrations determined above, selective binding of antibiotics to PBPs in living bacteria was measured quantitatively. A survey of all PBPs was made in *P. aeruginosa* (Table 1).

Table 1. In vivo binding to PBPs in *P. aeruginosa* ATCC 27853 of β-lactam antibiotics with high, intermediate, and low potency for β-lactamase induction[a]

Potency group	Antibiotic	% of indicated PBPs not bound to antibiotics[b]					
		1A	1B	2	3	4	5
High potency	Imipenem,	29	21	39	50	28	9
	0.25 μg/ml	25	21	34	47	30	10
	Cefoxitin						
	5 μg/ml	118	49	81	53	53	16
		118	64	117	65	73	26
	20 μg/ml	73	39	61	44	54	11
Intermediate potency	Cephaloridine						
	5 μg/ml	110	32	101	64	22	77
		103	29	97	70	23	112
	10 μg/ml	67	28	53	48	22	19
	20 μg/ml	74	31	61	52	29	22
	Cefotetan						
	5 μg/ml	126	70	107	43	55	56
		126	59	99	48	54	76
	20 μg/ml	101	41	87	51	53	25
		77	34	56	35	49	18
Low potency	Cefotaxime,	57	30	95	44	26	94
	50 μg/ml	59	24	112	39	36	89
	Piperacillin,	41	68	99	44	25	106
	50 μg/ml	49	56	92	39	22	98
	Aztreonam,						
	50 μg/ml	55	23	94	41	28	89

[a]Shake cultures of bacteria grown to an optical density at 547 nm of 0.8 in 300 ml of Iso-Sensitest broth (Oxoid Ltd.) at 30°C were supplemented with different antibiotics at the indicated final concentrations and incubated further for 2 h. The cells were then cooled rapidly to 4°C, sedimented, suspended in phosphate buffer (0.05 mol/liter, pH 7.0), and fragmented by sonication. In membranes recovered from sonic extracts (samples with 650 μg of protein), those fractions of PBPs which had not bound β-lactam antibiotic during the previous in vivo incubation were labeled with [^{3}H]benzylpenicillin (specific activity, 148 GBq/mmol; Amersham Corp.). After separation of solubilized PBPs (220 μg of protein) by sodium dodecyl sulfate-poly-acrylamide gel electrophoresis, individual labeled PBPs were eluted with Protosol (E. I. du Pont de Nemours & Co., Inc.) and counted by scintillation spectrometry.
[b]In comparison with amounts of PBPs in control bacteria which were not preincubated with the respective antibiotics. Single or double measurements were made.

Considerable in vivo binding to one or several PBPs, especially PBPs 1B, 3, and 4, was observed with all antibiotics, including the poorly inducing compounds at high concentrations and the concentration-dependent inducers at low, noninducing concentrations. However, a general correlation was evident between the ability of antibiotics to induce β-lactamase and to bind to PBP 5. The coincident concentration dependence of induction and binding to PBP 5 of *P. aeruginosa* is shown in more detail for latamoxef (Table 2) and for im-

Table 2. Correlation of concentration-dependent in vivo binding to PBPs and β-lactamase induction by latamoxef in *P. aeruginosa* ATCC 27853[a]

Latamoxef concn (μg/ml)	% of indicated PBPs not bound to latamoxef						β-Lactamase activity (mU/mg of protein)
	1A	1B	2	3	4	5	
5	61	67	108	35	39	95	0
10	57	50	96	33	36	72	85
20	43	36	99	32	32	59	143
50	35	27	80	36	40	35	286

[a]Conditions for β-lactamase induction and PBP binding as described in the legend of Fig. 1 and Table 1, footnote *a*.

ipenem in a narrow range of low concentrations (Fig. 2).

A comparable connection between β-lactamase induction and PBP binding was made likely by preliminary experiments with *E. cloacae*. At antibiotic concentrations of 2 μg/ml (0.5 μg/ml for imipenem), the percentage amounts of PBP 5 bound to antibiotic in vivo were 91, 94, 88, and 73, respectively, with the good inducers imipenem, latamoxef, cefoxitin, and cefotetan; 23 with the intermediate inducer ampicillin; and 2 and 0, respectively, with the poor inducers ceftizoxime and aztreonam.

Temperature Dependence of β-Lactamase Induction

Strain hs-257 was isolated by Noguchi et al. (12) as a temperature-sensitive mutant from the parent strain *P. aeruginosa* PAO 2142. In the mutant, PBP 5 was shown to be also temperature sensitive and unable to bind penicillin at the nonpermissive temperature of 40°C (12, 13).

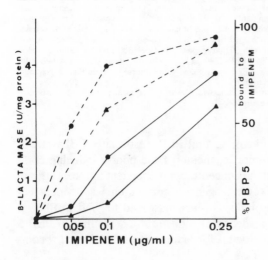

Figure 2. Correlation of concentration-dependent in vivo binding to PBP 5 (----) and β-lactamase induction (——) by imipenem in *P. aeruginosa* ATCC 27853 (▲) and DSM 50071 (●). Conditions for induction and PBP binding were as described in the legend to Fig. 1 and Table 1, footnote *a*.

Induction of β-lactamase by the good inducers imipenem and cefoxitin was studied in mutant hs-257 and its PBP 5-nondefective parent PAO 2142 at the permissive and nonpermissive temperatures of 30 and 40°C (Table 3). In the mutant induction took place at the permissive temperature, although at a lower level than in the parent strain. This agreed with the previous finding that penicilllin binding to PBP 5 is less effective in the mutant than in the parent, even at 30°C (13). However, induction was inhibited or at least fivefold decreased when the β-lactam-binding function of hs-257 PBP 5 was inactivated by incubation at 40°C during the induction period. In the parent strain β-lactamase induction took place at 30 and 40°C, although the enzyme level reached only 60 to 80% of the low-temperature value at elevated temperature.

Viable counts showed that 80% of the mutant cells and 100% of the parent bacteria survived the 2-h induction period at 40°C as colony formers. This excluded the possibility that the lack of induction at 40°C was due to extensive killing of mutant bacteria by the combined action of inducing antibiotic and elevated temperature.

Conclusions

The reported observations on the correlation of specific PBP binding and β-lactamase induction and on the temperature dependence of β-lactamase induction in the temperature-sensitive, PBP 5-defective mutant of *P. aeruginosa*

Table 3. Temperature dependence of β-lactamase induction by cefoxitin and imipenem in the *P. aeruginosa* PAO 2142 parent strain and mutant hs-257[a]

| | | | β-Lactamase (mU/mg of protein) | | | |
| | | | 30°C | | 40°C | |
Inducer	Strain	Inducer (μg/ml)	C	N	C	N
Cefoxitin	Parent	0	0	14	0	5
		20	250	337	158	263
		50	357	491	238	425
	Mutant	0	0	0	0	0
		20	59	102	9	14
		50	64	94	0	4
		100	178	302	0	0
Imipenem	Parent	0.1	269	371	122	189
	Mutant	0.1	95	161	20	31

[a]The substrates were cephalothin (C) and nitrocefin (N), and the induction temperatures were 30 and 40°C. Induction and assay of β-lactamase were as described in the legend to Fig. 1.

support the assumption that binding of the inducing β-lactam antibiotic to PBP 5 is the initial step in the induction of chromosomal β-lactamase in *P. aeruginosa, E. cloacae,* and presumably also other gram-negative bacteria.

Regulation of the inducible synthesis of AmpC β-lactamase in enterobac teria has been shown to involve the participation of a regulatory effector, AmpR, and a sensor, AmpD. In the proposed model of induction, binding of the inducing β-lactam antibiotic to AmpD inactivates the control function of this sensor on the effector AmpR, which is then free to stimulate the expression of AmpC enzyme (7, 8). So far, direct interaction of AmpD with β-lactam antibiotics has not been demonstrated. With a molecular weight of 25,000, the AmpD protein is considerably smaller than any known enterobacterial PBP (Lindberg, Lindqvist, and Normark, this volume). As a working hypothesis we would like to propose an extension of the existing model in which inactivation of AmpD is brought about not by direct binding of inducer but by direct or indirect interaction of AmpD with the initially formed complex of PBP 5 and inducing β-lactam antibiotic.

LITERATURE CITED

1. **Abraham, E. P., and S. G. Waley.** 1979. β-Lactamases from *Bacillus cereus,* p. 315. *In* J. M. T. Hamilton-Miller and J. T. Smith (ed.), β-*Lactamases.* Academic Press, Inc., London.
2. **Citri, N., and M. R. Pollock.** 1966. The biochemistry and function of β-lactamase (penicillinase). *Adv. Enzymol.* **28:**237–323.
3. **Collins, J. F.** 1969. The *Bacillus licheniformis* β-lactamase system, p. 357–360. *In* J. M. T. Hamilton-Miller and J. T. Smith (ed.), β-*Lactamases.* Academic Press, Inc., London.
4. **Curtis, N. A. C., D. Orr, G. W. Ross, and M. Boulton.** 1979. Competition of β-lactam antibiotics for the penicillin-binding proteins of *Pseudomonas aeruginosa, Enterobacter cloacae, Klebsiella aerogenes, Proteus rettgeri,* and *Escherichia coli:* comparison with antibacterial activity and effects upon bacterial morphology. *Antimicrob. Agents Chemother.* **16:**325–328.
5. **Gootz, T. D., and C. Sanders.** 1983. Characterization of β-lactamase induction in *Enterobacter cloacae. Antimicrob. Agents Chemother.* **23:**91–97.
6. **Hoffman, T. A., T. J. Cleary, and D. H. Bercuson.** 1980. Effects of inducible beta-lactamase and antimicrobial resistance upon the activity of newer beta-lactam antibiotics against *Pseudomonas aeruginosa. J. Antibiot.* **34:**1334–1340.
7. **Lindberg, F., S. Lindquist, and S. Normark.** 1987. Inactivation of the *ampD* gene causes semiconstitutive overproduction of the inducible *Citrobacter freundii* β-lactamase. *J. Bacteriol.* **169:**1923–1928.
8. **Lindberg, F., and S. Normark.** 1987. Common mechanism of *ampC* β-lactamase induction in *enterobacteria*: regulation of the cloned *Enterobacter cloacae* P99 β-lactamase gene. *J. Bacteriol.* **169:**758–763.
9. **Martin, H. H., M. Tonn-Ehlers, and W. Schilf.** 1980. Cooperation of benzylpenicillin and cefoxitin in bacterial growth inhibition. *Philos. Trans. R. Soc. London* Ser. B **289:**265–367.
10. **Matsuhashi, M., F. Ishino, S. Tamaki, S. Nakajima-Iijima, S. Tomioka, J. Nakagawa, A. Hirata, B. G. Spratt, T. Tsuruoka, S. Inouye, and Y. Yamada.** 1982. Mechanism

of action of β-lactam antibiotics: inhibition of peptidoglycan transpeptidases and novel mechanisms of action, p. 99–114. *In* H. Umezawa, A. L. Demain, T. Hata, and C. R. Hutchinson (ed.), *Trends in Antibiotic Research*. Japan Antibiotics Research Association, Tokyo.

11. **Minami, S., A. Yotsuji, M. Inoue, and S. Mitsuhashi.** 1980. Induction of β-lactamase by various β-lactam antibiotics in *Enterobacter cloacae*. *Antimicrob. Agents Chemother.* **18:**382–385.

12. **Noguchi, H., M. Fukasawa, T. Komatsu, S. Iyobe, and S. Mitsuhashi.** 1980. Isolation of two types of *Pseudomonas aeruginosa* mutants highly sensitive to a specific group of β-lactam antibiotics and with defect in penicillin-binding proteins. *J. Antibiot.* **33:**1521–1526.

13. **Noguchi, H., M. Fukasawa, T. Komatsu, S. Mitsuhashi, and M. Matsuhashi.** 1985. Mutation in *Pseudomonas aeruginosa* causing simultaneous defects in penicillin-binding protein 5 and in enzyme activities of penicillin release and D-alanine carboxypeptidase. *J. Bacteriol.* **162:**849–851.

14. **O'Callaghan, C. H., P. W. Muggleton, and G. W. Ross.** 1969. Effects of β-lactamase from gram-negative organisms on cephalosporins and penicillins, p. 57–63. *Antimicrob. Agents Chemother.* 1968.

15. **Ohya, S., M. Yamazaki, S. Sugawara, and M. Matsuhashi.** 1979. Penicillin-binding proteins in *Proteus* species. *J. Bacteriol.* **137:**473–479.

16. **Ohya, S., M. Yamazaki, S. Sugawara, S. Tamaki, and M. Matsuhashi.** 1978. New cephamycin antibiotic, CS-1170: binding affinity to penicillin-binding proteins and inhibition of peptidoglycan cross-linking reactions in *Escherichia coli*. *Antimicrob. Agents Chemother.* **14:**780–785.

17. **Richmond, M. H., and R. B. Sykes.** 1972. The β-lactamase of gram-negative bacteria and their possible physiological role. *Adv. Microb. Physiol.* **9:**31–88.

18. **Sanders, C., and W. E. Sanders, Jr.** 1979. Emergence of resistance to cefamandole: possible role of cefoxitin-inducible beta-lactamase. *Antimicrob. Agents Chemother.* **15:**792–797.

19. **Spratt, B.** 1977. Properties of the penicillin-binding proteins of *Escherichia coli* K 12. *Eur. J. Biochem.* **72:**341–352.

20. **Spratt, B. G.** 1983. Penicillin-binding proteins and the future of β-lactam antibiotics. *J. Gen. Microbiol.* **129:**1247–1260.

21. **Yotsuji, A., S. Minami, Y. Araki, M. Inoue, and S. Mitsuhashi.** 1982. Inducer activity of β-lactam antibiotics for the β-lactamase of *Proteus rettgeri* and *Proteus vulgaris*. *J. Antibiot.* **35:**1590–1593.

VIII.   NON-β-LACTAMS

Chapter 52

New Cell Wall Antibiotic Targets

Nafsika H. Georgopapadakou

Antibiotic targets are by definition processes that are essential, and their inhibition lethal, to the microorganism. The targets of many clinically useful antibacterials are in addition unique and universal to bacteria, the inhibitors having low or no mammalian toxicity and broad-spectrum antibacterial activity. Biosynthesis of peptidoglycan, the major and most common component of bacterial cell wall, is such a classical target.

Peptidoglycan consists of linear sugar chains substituted with short peptide strands which are cross-linked at a D-alanine residue either directly or via a peptide bridge (7). The sugar chains are composed invariably of alternating N-acetylglucosamine and N-acetylmuramic acid residues. The amino acid composition of the peptide strands alternates between D and L isomers and is generally conserved except for the third residue, which varies from species to species. The peptidoglycan structures of *Escherichia coli,* a typical gram-negative rod, and *Staphylococcus aureus,* one of the most prominent gram-positive pathogens, are shown in Fig. 1. In *E. coli,* peptidoglycan is anchored to the outer membrane through a covalent linkage between *meso*-diaminopimelic acid of the peptide strand and the carboxy-terminal lysine of the lipoprotein. In *S. aureus,* peptidoglycan is covalently linked to teichoic acids through the muramic acid of the sugar chain (4, 5).

Peptidoglycan biosynthesis proceeds in three stages (7): (i) synthesis of the N-acetylmuramic acid-pentapeptide monomer in the cytoplasm; (ii) transfer of the monosaccharide-pentapeptide unit to the cytoplasmic membrane by attachment to a C_{55} lipid carrier, addition of N-acetylglucosamine and species-specific modifications; and (iii) transfer of the disaccharide-peptide unit outside the cytoplasmic membrane and release from the lipid carrier followed by cross-linking, through both sugar and peptide moieties, to sugar and peptide groups in the growing peptidoglycan.

Nafsika H. Georgopapadakou • Roche Research Center, Nutley, New Jersey 07110.

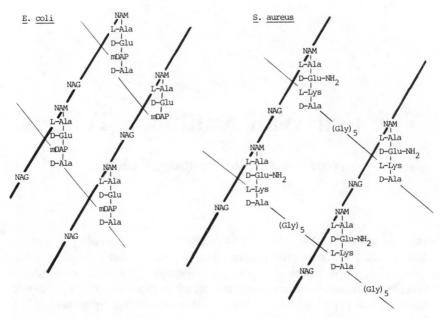

Figure 1. Peptidoglycan structure in *E. coli* and *S. aureus.*

Chapters 54 and 55 focus on the early, cytoplasmic steps of peptidoglycan biosynthesis. These steps, and the sites of inhibition by some known antibiotics (8), are shown in Fig. 2. Briefly, *N*-acetylglucosamine from the metabolic pools forms an ether linkage with pyruvate, which is subsequently reduced to the lactate forming *N*-acetylmuramic acid, the first intermediate found exclusively in bacteria. Phosphomycin inhibits the first step, catalyzed by a transferase, by acting as a phosphoenolpyruvate analog (2, 9). The enzyme activates C-2 of phosphomycin, which then attaches covalently to a cysteine residue. The lack of inhibition of mammalian phosphoenolpyruvate-requiring reactions by phosphomycin is due to the fact that in the mammalian reactions phosphoenolpyruvate is a substrate exclusively of kinases.

A series of ligases add L- and D-amino acids to *N*-acetylmuramic acid, several of which (D-glutamic acid, *meso*-diaminopimelic acid, D-alanine) are found exclusively in bacteria. Predictably, the biosynthesis of these amino acids can also be an antibiotic target. β-Haloalanines inhibit both D-alanine and D-glutamic acid biosynthesis, in reactions that involve alanine and pyridoxal phosphate (3). β-Haloalanines are mechanism-based inhibitors, binding covalently to the coenzyme pyridoxal phosphate. The active moiety in the dipeptide alafosfalin, a phosphonate analog of L-alanine, inhibits alanine racemase (1). Finally, D-cycloserine, an analog of D-alanine, inhibits competitively both alanine racemase and D-alanyl-D-alanine-ligase (3).

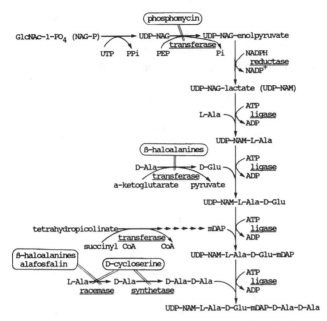

Figure 2. Biosynthesis of the monosaccharide-pentapeptide unit in *E. coli*. The sites of inhibition of some cell wall inhibitors are also shown.

The above inhibitors illustrate several important points. First, they are all relatively small molecules (molecular weight, <300). This is related to the fact that the early, cytoplasmic steps involve small molecules. By contrast, inhibitors of the later steps (Chapters 53, 56, 57, and 58), other than β-lactams (γ-lactams; Chapter 59), are relatively large (molecular weight, >1,500). Consequently, excellent inhibitors such as vancomycin are inactive against gram-negative bacteria because they do not penetrate the outer membrane. Second, they are actively transported into the cytoplasm, and resistance to them involves decreased penetration. Other forms of resistance include decreased affinity of the enzyme target (phosphomycin) and increased amount of the enzyme target (cycloserine). Third, they act by binding to the enzyme, reversibly (cycloserine) or irreversibly (phosphomycin), or to some other cofactor (β-haloalanines). Fourth, although they may be substrate analogs common to both procaryotes and eucaryotes (phosphoenolpyruvate), mechanistic differences still make possible selective toxicity (phosphomycin).

The molecular biology of D-alanine metabolism is reviewed in Chapter 54. Phosphinic acid inhibitors of D-alanyl-D-alanine ligase are described in Chapter 55.

Chapter 62 deals with teichoic acid biosynthesis as an antibiotic target. Tunicamycin, the only active compound so far, also inhibits peptidoglycan biosynthesis and, unfortunately, protein glycosylation in eucaryotes. However, other inhibitors of teichoic acid biosynthesis with more desirable properties may emerge as this pathway becomes elucidated.

The spectacular success of β-lactam antibiotics in the past 40 years has highlighted the biosynthetic steps, and eventually the enzymes, that these compounds inhibit (6). In contrast, progress in other areas of cell wall biosynthesis has been rather sluggish. The recent emergence of organisms intrinsically resistant to β-lactams, such as the methicillin-resistant *S. aureus,* has made alternative targets in peptidoglycan biosynthesis increasingly attractive. However, much work is needed to achieve the level of sophistication attained in the β-lactam area.

LITERATURE CITED

1. **Atherton, F. R., M. J. Hall, C. H. Hassall, R. W. Lambert, W. J. Lloyd, and P. S. Ringrose.** 1979. Phosphonopeptides as antibacterial agents: mechanism of action of alaphosphin. *Antimicrob. Agents Chemother.* **15:**696–705.
2. **Kahan, F. M., J. S. Kahan, P. J. Cassidy, and H. Kropp.** 1974. The mechanism of action of fosfomycin (Phosphonomycin). *Ann. N.Y. Acad. Sci.* **235:**364–386.
3. **Neuhaus, F. C., and W. P. Hammes.** 1981. Inhibition of cell-wall biosynthesis by analogues of alanine. *Pharmacol. Ther.* **14:**265–319.
4. **Rogers, H. J., J. B. Ward, and I. D. J. Burdett.** 1977. Structure and growth of the walls of gram-positive bacteria, p. 139–175. *In* R. Y. Stanier, H. J. Rogers, and J. B. Ward (ed.), *Relations between Structure and Function in the Prokaryotic Cell.* Cambridge University Press, Cambridge.
5. **Shockman, G. D., and J. F. Barrett.** 1983. Structure, function, and assembly of cell walls of gram-positive bacteria. *Annu. Rev. Microbiol.* **37:**521–527.
6. **Tipper, D. J.** 1985. Mode of action of β-lactam antibiotics. *Pharmacol. Ther.* **27:**1–35.
7. **Tipper, D. J., and A. Wright.** 1979. The structure and biosynthesis of bacterial cell walls, p. 291–426. *In* J. R. Sokatch and L. N. Ornston (ed.), *The Bacteria: a Treatise on Structure and Function,* vol. 7. Academic Press, Inc., Orlando, Fla.
8. **Ward, J. B.** 1984. Biosynthesis of peptidoglycan: points of attack by wall inhibitors. *Pharmacol. Ther.* **25:**327–369.
9. **Woodruff, H. B., J. M. Mata, S. Hernandez, S. Mochales, A. Rodriguez, E. O. Stapley, H. Wallich, A. K. Miller, and D. Hendlin.** 1977. Fosfomycin: laboratory studies. *Chemotherapy* (Tokyo) **23**(Suppl. 1):1–22.

Chapter 53

Glycopeptide Antibiotics: a Comprehensive Approach to Discovery, Isolation, and Structure Determination

Peter W. Jeffs
L. J. Nisbet

The search for new glycopeptide antibiotics has attracted increased interest over the past 5 years. The reason for this interest, at least as far as the pharmaceutical industry is concerned, has been the dramatic increase in the use of vancomycin, which is the sole representative of the glycopeptide antibiotics of this class approved for clinical use. There are a number of reasons for this increased use of vancomycin; however, the overriding factor has been its effectiveness in treating methicillin-resistant staphylococcal infections, particularly those that have been found to be resistant to the newer β-lactamase-resistant cephalosporins (22). In addition, glycopeptide antibiotics have found application as feed additives to promote growth in livestock and poultry (6).

Discovery of Novel Glycopeptide Antibiotics

In 1983 we embarked upon a program that was directed toward the discovery and characterization of novel glycopeptide antibiotics, with the ultimate goal of seeking compounds that could be developed for their therapeutic potential in human diseases, also recognizing that other compounds in this series

Peter W. Jeffs • Smith Kline & French Research Laboratories, Swedeland, Pennsylvania 19479.   **L. J. Nisbet** • Xenova Ltd., Slough SL1 4EQ, England.

might be useful in animal health. The initial goal was to devise a screening procedure for this class of antibiotics that would avoid the usual pitfalls associated with random screening that inevitably lead to the rediscovery of previously observed compounds.

The approach that we adopted was based on the premise that an ideal microbiological screen should have the following characteristics: it should be specific for the desired pharmacophore being sought and therefore preferably mechanism based, it should have built-in dereplication with a low incidence of false positive results originating from antibiotics with a different mechanism of action, and it should possess high sensitivity, with the ability to detect antibiotic concentrations of less than 25 μg/ml.

The foundation for understanding the activity of vancomycin and related glycopeptide antibiotics was established by the early studies of Reynolds (34) and Jordan (19), who clearly showed that vancomycin exerts antibacterial action by inhibiting peptidoglycan synthesis. This work was followed by the demonstration that the antibiotics bind to putative cell wall intermediates containing the nucleotide sugar-linked precursor UDP-*N*-acetylmuramylpentapeptide or its corresponding lipid-carrier complex, or both (31). The origin of the specificity exhibited by the antibiotics in binding to these cell wall intermediates was clearly demonstrated by Perkins and Nieto (28–30). Their quantitative studies showed that the binding of vancomycin and ristocetin with both natural and synthetic peptides involves recognition of the L-Lys-D-Ala-D-Ala terminus of these peptides. This mode of action makes these antibiotics particularly attractive candidates for clinical use since the target D-alanyl-D-alanine terminus exists in the cell walls of all important bacterial pathogens. Furthermore, the essentiality of this unique terminal D-Ala-D-Ala dipeptide and its evolutionary conservation ensure that resistance development at the target site is unlikely. Although glycopeptide antibiotics have the potential to act against all bacteria, they are too large (>1,100 daltons) to penetrate the outer membrane of gram-negative bacteria and are effective only against gram-positive bacteria.

A considerable body of evidence has been accumulated from extensive studies employing nuclear magnetic resonance (NMR) methods, principally by Williams and co-workers (20, 43) but with important contributions by others (2, 6), that have defined the details at the molecular level of the interaction of these antibiotics with model peptides terminating in L-Lys-D-Ala-D-Ala. NMR studies show that in the ristocetin (20) and aridicin (18) series, a ready-made binding site exists for the D-Ala-D-Ala terminus, whereas in vancomycin the binding site is less obvious. However, it is known that vancomycin, on binding to D-Ala-D-Ala peptides, undergoes a large conformational change in which the resultant bound form (43) closely resembles the "host-guest" complex found in the ristocetin-aridicin series. The hydrogen-bonding interactions that contribute to the stability of the complex are illustrated for an aridicinlike aglycone-Ac₂-L-Lys-D-Ala-D-Ala complex in Fig. 1. Examination of this model shows

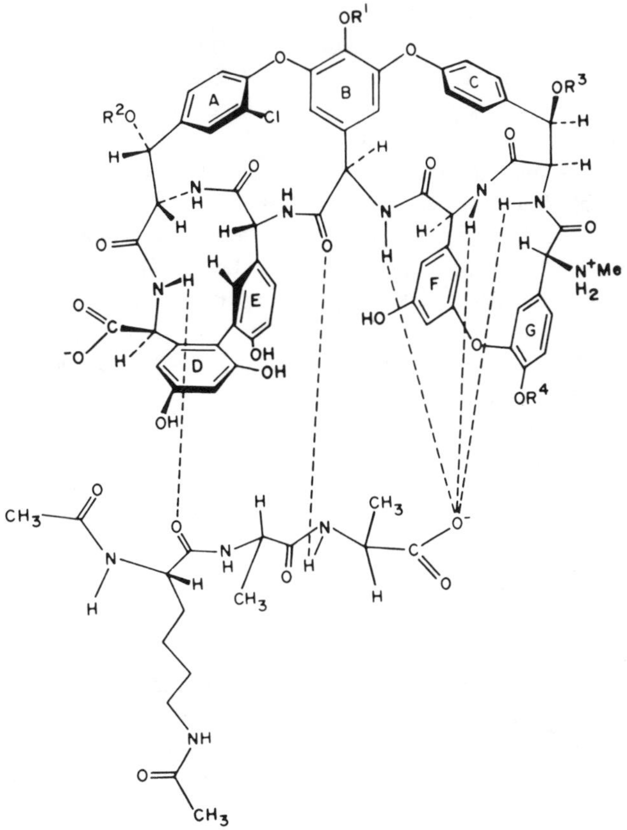

Figure 1. Representation of an aricidinlike Ac$_2$-L-Lys-D-Ala-D-Ala complex showing interpeptide hydrogen bonds.

(i) three NH groups of the antibiotic in close proximity and suitably located for binding a carboxylate ion and (ii) an amide carbonyl plus an amide NH at locations which permit hydrogen bonding with complementary amide functions on the D-Ala-D-Ala residue. These conclusions are supported by several observations, including the ability to detect all of the specific hydrogen bonds in both the antibiotic and the D-Ala-D-Ala peptide components of the complex (see Fig. 1). Some recent developments in deriving structures of these complexes will be discussed subsequently.

Screening Procedure

The mechanism described for the inhibition of peptidoglycan synthesis is unique to this class of antibiotics and as such offered an opportunity to develop what in retrospect has become a highly successful screening procedure. The full details of this screen, developed at Smith Kline & French Laboratories, have been described recently (33), and only a brief description will be provided here. The basis of the screen was to use the tripeptide diacetyl-L-Lys-D-Ala-D-Ala to specifically antagonize the activity of glycopeptide antibiotics against sensitive bacteria. A schematic of how the screen was operated is shown in Fig. 2. The left-hand side contains a plain disk loaded with 30 μl of clarified fermentation broth and placed on a *Bacillus subtilis* agar plate. The right-hand side shows a disk loaded with 100 μg of the specific antagonist, diacetyl-L-Lys-D-Ala-D-Ala, plus the same quantity of broth. Through appropriate tests with available standards and subsequent experience in screening over 4,200 cultures, criteria were established for selecting glycopeptide-producing cultures on the basis of the inhibition zone difference between the comparative disks. The screen was completely specific, and in our experience there have been no false-positive results.

Culture Selection and Media Conditions

During the course of the discovery phase of the program, a significant enhancement in the percentage of cultures producing glycopeptide antibiotics was achieved. This was primarily the result of a number of factors including optimization of media and culture conditions to maximize the biosynthetic potential for metabolite overproduction from a wide range of genera represented by *Streptomyces* spp., *Kibdelosporangium* spp. (36), *Actinoplanes* spp., *Nocardia* spp., and two unidentified genera of actinomycetes; and emphasis on selection of cultures in the genera *Actinoplanes* and *Nocardia,* which were previously found to produce glycopeptide antibiotics.

Isolation: Affinity-Based Separation Procedure

Three further important developments proved valuable in providing us with a rapid and reliable scheme for isolation and dereplication of glycopeptide antibiotics. The first of these involved an affinity chromatography procedure.

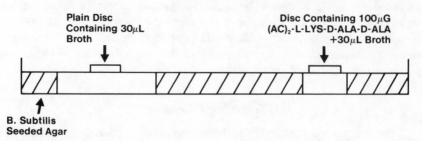

Figure 2. Schematic of the mechanism-based tripeptide antagonism screening procedure using *B. subtilis* plates.

Affinity chromatography (25), although well established as a tool for purification of proteins and other macromolecules, has few examples of use when the ligand and the ligate are of low molecular weight (32). This limitation has been governed by the need for the surface architecture of one of the components, usually the ligate, to provide a cleft or cavity containing the necessary distribution of functional groups which are complementary in both shape and charge to those of the ligand to permit binding. The nature of the molecular recognition phenomena exhibited by vancomycin and ristocetin for peptides terminating in D-alanyl-D-alanine suggested that it might be possible to devise an affinity-based separation method for the isolation of the glycopeptide antibiotics.

The reduction of this concept to practice was accomplished at Smith Kline & French Laboratories by Wasserman and co-workers (42) and, interestingly, independently by Corti and Cassani (3) at Dow-Lepetit Laboratories. In our procedure, Affi-gel 10, an activated support containing a 10-atom spacer terminating in an *N*-hydroxysuccinimide ester, was coupled with a variety of ligands terminating in D-alanyl-D-alanine. This is exemplified in Fig. 3 for the column of choice, Affi-gel 10–D-Ala-D-Ala. This dipeptide column provides an affinity support that has fulfilled the requirements for a general-purpose, long-life, reusable column showing high binding selectivity for all glycopeptides that we have encountered. This column exhibits a minimum of nonspecific binding and properties that allow an effective desorption procedure leading to high, usually >80%, recoveries in both analytical and preparative separations.

The effectiveness in providing a rapid one-step purification of glycopep-

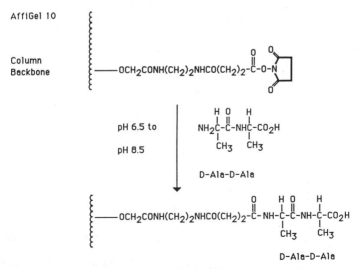

Figure 3. Reactions involved in preparation of the Affi-gel 10 affinity column.

tide antibiotics from fermentation broths in which they may be present at low concentration was demonstrated by loading 10 ml of broth containing 0.02 mg of vancomycin per ml onto 2 ml of gel. Figure 4A shows the analytical reversed-phase high-performance liquid chromatography (HPLC) of the broth containing vancomycin. The concentration of vancomycin is near the detection limit of the HPLC method and is hardly visible in the broad envelope of component peaks present in the fermentation milieu. Elution of the column to remove nonspecifically bound components with 0.02 M sodium phosphate buffer was followed by 0.1 M ammonium hydroxide containing 50% acetonitrile, which on lyophilization and recovery gave pure vancomycin, as shown by HPLC (Fig. 4B). Vancomycin obtained by this procedure was identical chromatographically, spectroscopically (UV, infrared, and fast atom bombardment [FAB] mass spectrometry), and biologically to an authentic sample.

Dereplication Methods Employing HPLC and FAB Mass Spectrometry

The application of reversed-phase HPLC procedures at two different pHs was found to resolve 12 glycopeptide antibiotics that were available to us at the outset of these studies. The system, developed by Sitrin and co-workers, has been described (38) and has continued to prove useful as our own collection of novel compounds has more than tripled this number. An update of the re-

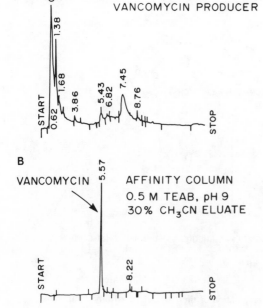

Figure 4. Illustration of the effectiveness and selectivity provided by the Affi-gel 10 affinity column in purification of vancomycin. (A) HPLC of broth of vancomycin producer. (B) HPLC of broth of vancomycin producer after Affi-gel 10 affinity column purification.

tention behavior of some of the new glycopeptide antibiotics available to us is shown in Table 1.

Although the HPLC analysis of affinity-purified antibiotics is in most cases sufficient to indicate whether the compounds are novel, we have routinely incorporated FAB mass spectrometry into the dereplication procedure. The application of FAB mass spectrometry has been shown to be a very powerful tool in providing not only molecular weights, but also elemental and carbohydrate composition on 5 to 10 μg of these antibiotics (35). An example of the information provided in the FAB mass spectrum is illustrated for synmonicin B (Fig. 5).

The combination of the HPLC retention data and the mass spectral information proved particularly useful in indicating the novelty of the synmonicins (A. K. Verma, R. Prakash, S. A. Carr, G. D. Roberts, and R. D. Sitrin, *Program Abstr. 26th Intersci. Conf. Antimicrob. Agents Chemother.*, abstr. no. 940, 1986). The HPLC profiles of the affinity column isolate of the synmon-

Table 1. HPLC retention times of glycopeptide antibiotics subjected to reversed-phase HPLC at pH 3.2 (system A) and pH 6.0 (system B)[a]

Antibiotic	Retention time (min)		Reference
	System A	System B	
A 477	11.5, *11.7*, 12.1	8.4, *8.9*, 10.3	—[b]
A 35512B	*5.2*, 6.2	*6.1*, 7.2	27
Actaplanin	6.1, *6.3*, 6.6, 10.7	7.5, 7.9, 17.2	15
Actinoidin A/B	*5.5*, 5.9, 6.1	6.5, 7.1, *8.1*	24
Actinoidin A₂	3.55, *4.42*	6.88	1a
Aridicin A	15.3	14.5	38
Aridicin B	15.9	15.2	38
Aridicin C	17.2	16.1	38
Avoparcin	4.8, *5.8*, 6.0	7.6, 7.7	5
Kibdelin A	15.0	14.6	41
Kibdelin B	15.8	15.6	41
Kibdelin C	16.7	16.4	41
OA-7653	10.31		21
Parvodicin	*17.15*, 17.7, 18.1, 18.3		12a
Ristocetin	3.6	5.4	1, 9
Synmonicins	4.9, *5.4*, 7.5	7.64, *7.99–8.26*, 9.76	—[c]
Teicoplanin A₂	13.1, *13.6*, 13.8	*13.8*, 14.0	16
Vancomycin	6.0	7.4	11, 26

[a] For both systems, the column was ultrasphere ODS, 5 μm, and was 4.6 by 150 mm; the flow was 1.5 ml/min; and the detection was at 220 nm in UV light. For system A the solvent was 7 to 34% acetonitrile (7% for 1 min and then ramp to 34% over 13 min and hold at 34%) in 0.1 M phosphate (pH 3.2); for system B the solvent was 5 to 35% acetonitrile (5% for 1 min and then ramp to 35% over 13 min and hold at 35%) in 0.025 M phosphate (pH 6.0). Values shown in italics represent the retention time of the major component in multicomponent mixtures. Variations in retention times are experienced from run to run.
[b] R. L. Hamill, M. E. Haney, and W. M. Stark, U.S. Patent 3780174-A, 17 December 1973.
[c] Verma et al., 26th ICAAC.

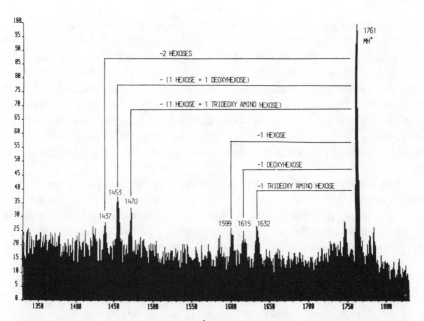

Figure 5. FAB mass spectrum of synmonicin B. The MH⁺ cluster at m/z 1,761 in the FAB mass spectrum of synmonicin B indicated a nominal mass of 1,760 daltons. The relative abundance of peaks in this cluster revealed the presence of one chlorine atom in the molecule. Fragment ions indicative of the number and types of sugars present also were observed. Mass differences between the parent and fragment ions provided a unique sugar composition of two hexoses (2 × 162), one deoxyhexose (146), and one trideoxy amino sugar (129).

icin-producing organism were virtually indistinguishable from those for the avoparcins (see Table 1). However, the protonated molecular ion patterns in the FAB spectra were clearly different, and in the case of the synmonicins, the values were unique for the members of this class and provided unambiguous evidence for their novelty.

A representation of the overall sequence of operations developed for screening, Affi-gel isolation, and dereplication procedures is shown in Fig. 6. This sequence has provided a rapid and reliable method for the discovery of many novel antibiotics of the glycopeptide family during the period since its introduction in 1983.

Structural Studies

Parent Antibiotics and Their Aglycones

The structures of the glycopeptide antibiotics are complex. It took more than 20 years of work involving several research groups (1, 9–11, 26, 37, 45) to solve the structures of vancomycin and ristocetin, and the final resolution

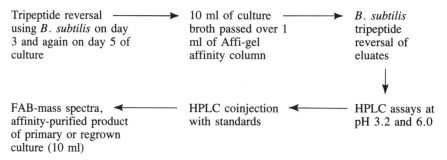

Figure 6. Summary of screening, isolation, and dereplication scheme.

became possible only after the introduction of modern spectroscopic methods. As the structures of new antibiotics in this series have been elucidated, the general features that have emerged indicate that they all contain a central heptapeptide core with a variable number of carbohydrates linked to this core, usually at one or more of the phenolic positions and also in some cases through the A-1′ hydroxyl. The five amino acids embodied in rings A through E of the core (see structure 1 in Fig. 7) are highly conserved, with the *F* and *G* residues being variable. In addition, other variations that occur in the aglycone are found at C-1′, where a hydroxyl may or may not be present and a variable number of chlorines appear at the 3,5 positions of rings A and C, at E-5 and F-2 (when *F* = aromatic). Our approach focused on deriving structural information on these antibiotics by making maximum use of sensitive spectroscopic and chromatographic methods both to minimize the amount of material consumed and to avoid the need to use manpower-intensive chemical degradation methods. Since these methods have been illustrated by reports of our work on the aridicins (18) and other antibiotics (1a, 12, 41), they will not be discussed in detail here. In general, the elucidation of the structure involves characterization of the neutral and amino sugars followed by separate studies to determine the more complex structures presented by the aglycones. The structure of the antibiotic can then be assembled by defining the glycosidic linkage sites and stereochemistry of each sugar residue.

Structure Determination of Heptapeptide Aglycone and Pseudoaglycone Core Structures

Structural studies of the heptapeptide aglycones today have the advantage of a large body of literature that has greatly simplified the task of elucidating the structures of these compounds. Two reactions, acid hydrolysis and oxidative degradation, are commonly employed. Hydrolysis of the heptapeptide core, under carefully controlled conditions, affords the ubiquitous actinoidinic

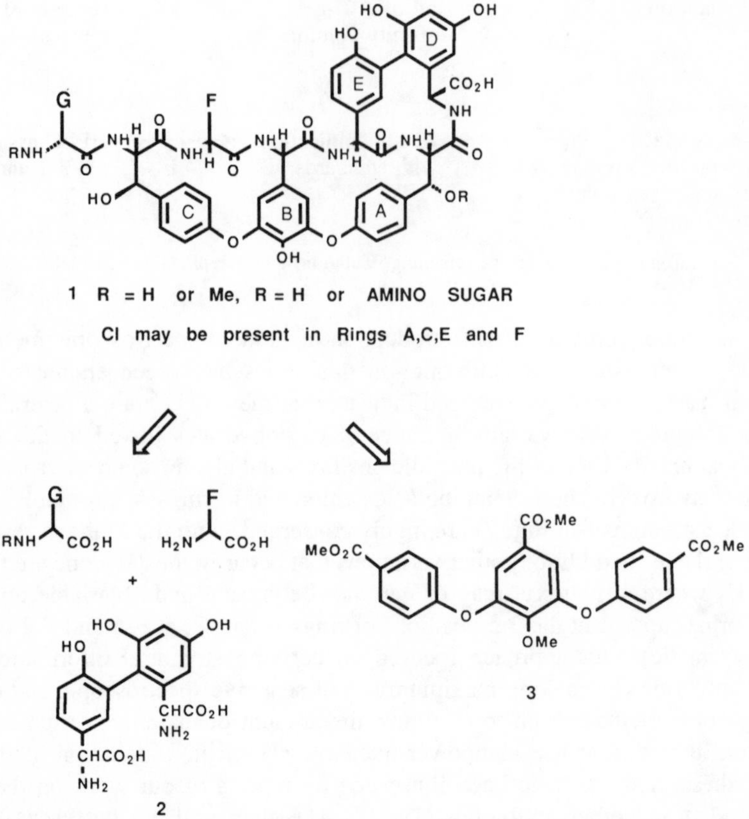

1 R = H or Me, R = H or AMINO SUGAR

Cl may be present in Rings A,C,E and F

Figure 7. Products from typical degradation scheme of the glycopeptide aglycones illustrating acid hydrolysis and oxidation.

acid and the amino acid residue or residues associated with the *F* and *G* segments (see structure 2 in Fig. 7). Unfortunately, the amino acid liberated from the ABC ring structure is unstable and cannot be isolated from this reaction. Conditions for characterization of the latter unit have been devised (39) through use of an oxidative degradation procedure that yields the A-B-C fragment as the tricarboxylic acid derivative (see structure 3 in Fig. 7). Although the hydrolysis procedure can be adapted to provide the amino acids in high optical purity for further study (17), it has proved most useful as a rapid analytical method for mapping the *F* and *G* amino acid residues through direct analysis of the acid hydrolysate of the antibiotic by using reverse-phase HPLC (R. D. Sitrin, unpublished data).

In some cases, it is now possible to deduce the structures of certain aglycones solely on the basis of spectral (NMR and FAB mass spectrometry) information. A recent example is provided by the aglycone of the antibiotic parvodicin (1a). A great deal of structural information is obtained from analysis of nuclear Overhauser effects (NOEs). The NOEs, which arise mainly from dipolar relaxation taking place between protons that are close (<0.5 nm) in space, have been employed previously in one-dimensional NOE difference experiments by Williamson (45) and Hunt (13, 14) and their respective collaborators to provide valuable structural information on this series of compounds. However, the information provided by the two-dimensional NOESY (nuclear Overhauser enhancement spectroscopy) spectrum has several advantages, one being that it is possible to observe many more NOEs than are provided by the one-dimensional experiment. The elucidation of the structure of the aglycone of the antibiotic aridicin (18) provided an example of the power of modern two-dimensional NMR methods for elucidation of the structures of this class of compounds. In the aglycone of aridicin, the pairwise NOE interactions observed in the NOESY were utilized in assembling the structural units (Fig. 8).

The development of the complete primary structure of aridicin aglycone, as shown in structure 1 of Fig. 7, required assignment of the stereochemistry at nine chiral centers and definition of the stereochemistry of the six amide bonds. A full discussion of how this was achieved and of its subsequent elaboration into a three-dimensional model (Fig. 9) by using distance information derived from quantitative NOESY data in conjunction with interactive modeling and energy refinement procedures has been presented earlier (18).

The topological features shown in Fig. 9 are in excellent agreement with those described in ristocetin (44) and the form of vancomycin as it occurs when it is bound to its tripeptide cell wall mimic. Given the structures of the aglycone and the associated carbohydrate moieties, elucidation of the structure of the antibiotic is completed by determining the sites and stereochemistries of attachment of the sugars to the heptapeptide core.

The particular strategy that we have employed in using NMR methods for solving this problem is dependent on the number and complexity of carbohydrates present in a particular antibiotic. A general requirement is that one proton signal from each carbohydrate unit in the antibiotic should be resolved from overlapping resonances. At 500 MHz this situation generally obtains for the anomeric proton signals in the one-dimensional spectrum. In a recent example, i.e., studies of the antibiotic actinoidin A (Fig. 10) (12a), in which the four monosaccharide units actinosamine, mannose, glucose, and rhamnose were known to be present from previous data (24, 40), the anomeric proton signals of each sugar are well resolved (Fig. 11). The assignment of each of these signals to a specific sugar was accomplished by using a recent modification of the homonuclear two-dimensional Hartman-Hahn experiment described by Davis and Bax (4). This experiment gives a separate family of cross peaks at the chemical shifts of the C-6 and ring protons of each monosaccharide unit and

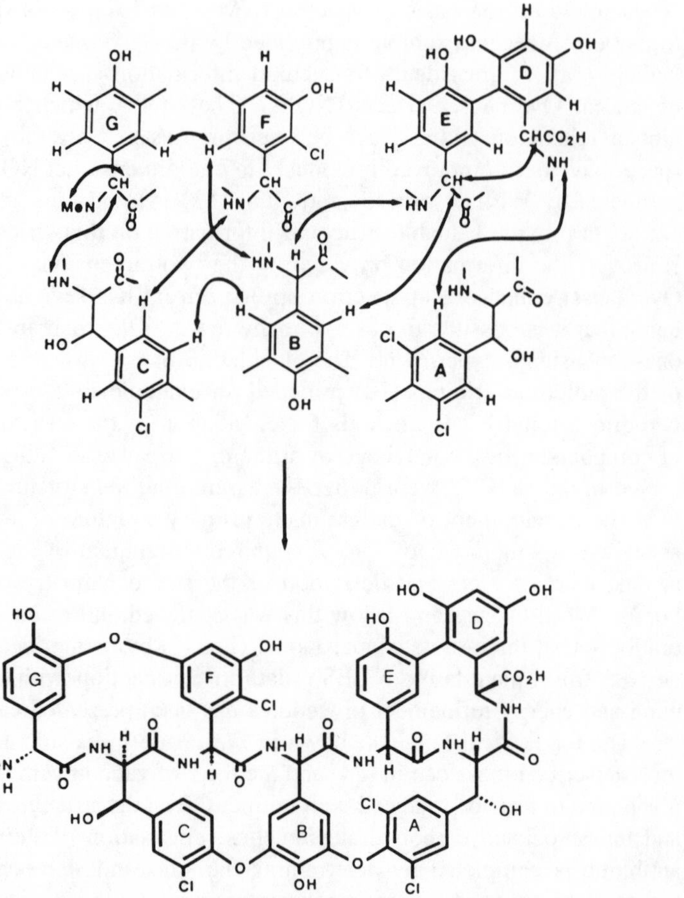

Figure 8. Schematic representation of the use of NOE interactions in assembling and sequencing the structural units of aridicin aglycone.

provides identification of the separate J networks associated with the individual carbohydrates. This procedure is a very powerful method for making assignments in severely overlapped regions and is likely to find wide application in NMR studies of other complex molecules.

Once assignments of the individual carbohydrate ring protons have been made, the stereochemistry of the glycosidic linkage in each sugar is determined either from J values or, if necessary, by resorting to information obtainable from a NOESY experiment. The latter experiment is often necessary to ascertain linkage information. Our investigations of the structure of actinoidin A_2 (12a) exemplify the utility of this approach. Cross sections taken through the anomeric proton signals originating from rhamnose and L-actinosamine in the

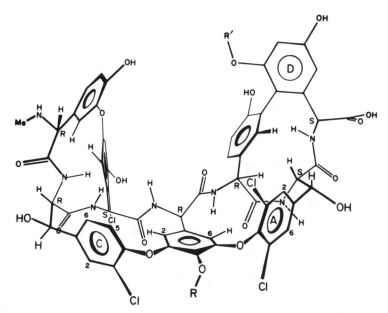

Figure 9. Three-dimensional model of the structure of aridicin aglycone.

NOESY spectrum are shown in Fig. 12. Interpretation of these data identifies the location of actinosamine at A-1′ (cross peak to A-1′) and establishes that rhamnose is linked to the 2 position of the glucose, as indicated in Fig. 10.

The application of the mass spectrometry and NMR methods described above has enabled us to define the structures of several new and one previously isolated antibiotic. The structures of these compounds are shown in Fig. 13. Included in this group is synmonicin B, which is the first member in which the variable *G* and *F* residues are represented by an aromatic and an aliphatic amino acid, respectively. Also, the structure of OA-7653, a compound which was isolated previously (21), has been elucidated.

Structures of Glycopeptide Complexes with Tri- and Pentapeptides

The structure of the vancomycin-tripeptide complex and the analogous structure for the ristocetin complex were deduced by Williams and co-workers (44) in a series of investigations in which distance information obtained by NOEs was interpreted with the aid of molecular models. Developments in computational methods and computer graphics have provided a means to undertake a more critical examination of many aspects of the structures of complex molecules. Using the computational approach (i) provides a quantitative measure

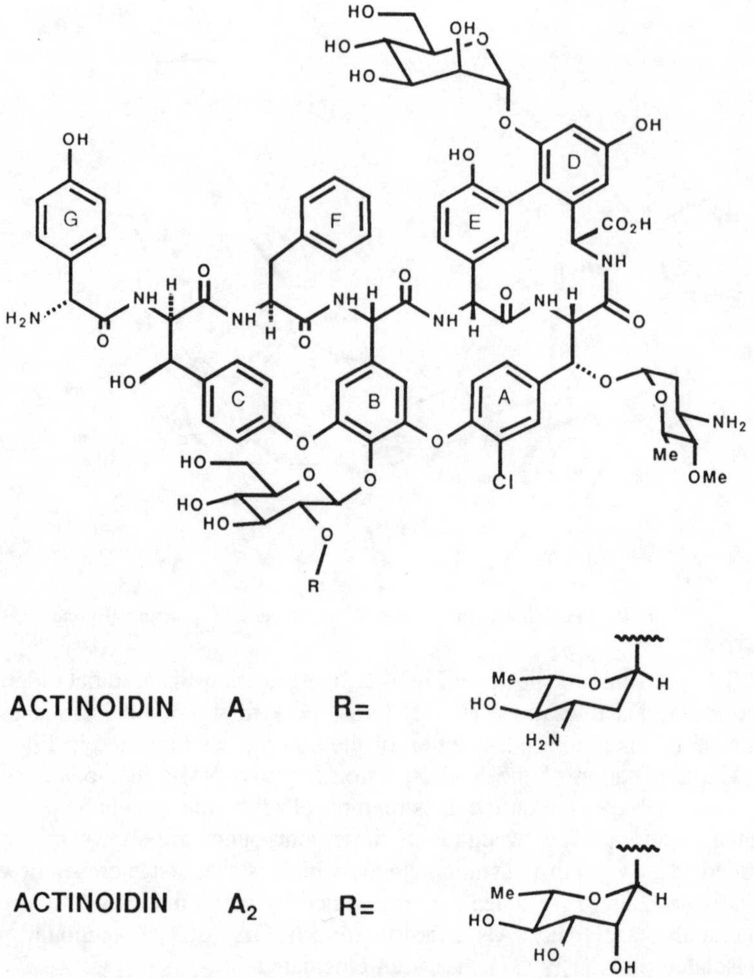

Figure 10. Structures of actinoidin A and actinoidin A₂.

of distances, surfaces, and volumes; (ii) enables comparisons of stability through quantitative measure of energies of conformations and through strength and direction of intramolecular forces; and (iii) allows modeling of complex flexible molecules with many degrees of internal freedom. These developments have made it possible to exploit the full potential of the information obtained from modern NMR methods and computer-assisted modeling to derive three-dimensional structures. The glycopeptide antibiotics and their complexes with tri- and pentapeptides are relatively rigid assemblies, which makes their structure analysis particularly amenable to the NMR-computer-assisted modeling approach.

Information provided by the appropriate two-dimensional NMR experi-

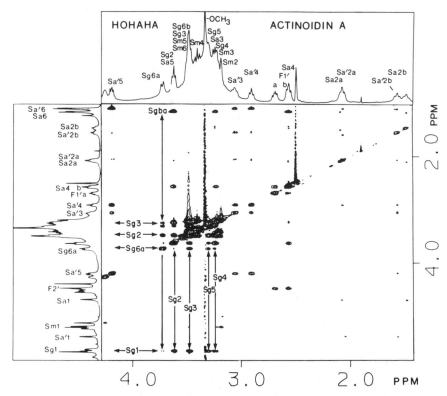

Figure 11. Hartman-Hahn (HOHAHA) ¹H spectrum of actinoidin A showing J networks associated with each of the four carbohydrate components, D-mannose (Sm), D-glucose (Sg), L-acosamine (Sa'), and actinosamine (Sa). Sets of cross peaks we identified for Sg6a and Sg1, showing the complete J networks of all of the protons, are readily discerned.

ments includes (i) short interproton distances (<0.4 nm) from analysis of quantitative NOEs, (ii) estimates of dihedral angle relationships from vicinal coupling constants, (iii) chemical exchange rates for exchangeable hydrogens bound to heteroatoms, and (iv) indications of local variations in mobility from carbon and proton T_1 values.

The information from the NMR experiments may be utilized in several ways to derive a three-dimensional structure(s). One approach that has been used involves (i) the construction of a model in the computer, (ii) interactive manipulation of the model to fit the experimental data obtained from NMR measurements to provide initial structures, and (iii) energy refinement of the initial structures in developing proposed models.

The energy refinement procedures that have been employed include molecular mechanics using semiempirical force fields and molecular dynamics employing valence force field calculations. Each of these methods is usually

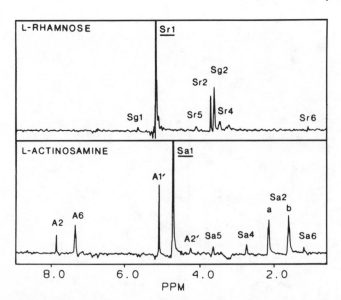

Figure 12. Cross sections through the anomeric ^{1}H resonances of rhamnose and L-actinosamine in the two-dimensional NOESY spectrum of actinoidin A$_2$. Cross peak between Sr1 and Sg2 indicates linkage of rhamnose through C-1 to the C-2 position of glucose, and the cross-peak between C-1 of L-actinosamine to A-1' indicates the site of attachment of this amino sugar to the A-1' position of the aglycone (see Fig. 10).

employed in conjunction with the incorporation of distance constraints from NMR data. Since the advantages and disadvantages of these approaches have been recently discussed (7), they will not be included here except as they apply to the specific example described below. Recent examples of these methods include the models of aridicin A (18) and the complex formed between risto-cetin pseudoaglycone and Ac$_2$-L-Lys-D-Ala-D-Ala (8).

As an alternative to the above procedures we have investigated the use of a distance geometry algorithm (23) to generate initial structures of aridicin agly-cone-tripeptide and aridicin aglycone-pentapeptide. A brief summary of the re-sults for the aridicin aglycone-acetyl-L-Ala-(γ)-D-Gln-L-Lys-(Ac)-D-Ala-D-Ala complex will be described here; full details will be provided elsewhere.

Distance geometry is an algorithm to generate conformations from distance constraints. The algorithm converts a distance matrix representation of a mol-ecule into a Cartesian coordinate representation. Distance geometry is well suited to modeling of molecular complexes and cyclic models (23). Also, it does not involve any operator bias, which is the major defect of generating structures by the interactive approach. The major disadvantage of distance geometry is that it can lead to energetically unfavorable structures. Provided the structure(s) generated is not too far away from a local minimum, force field energy min-imization is usually successful in providing models that are in agreement with

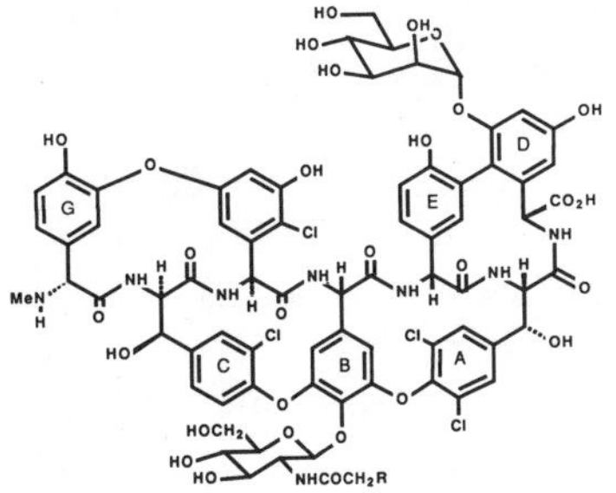

KIBDELINS

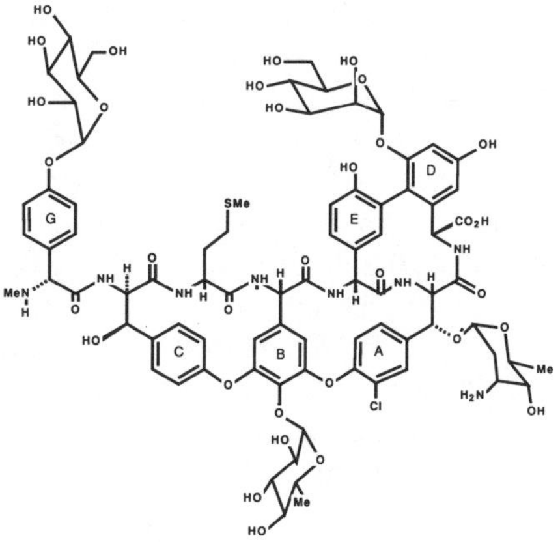

SYNMONICIN

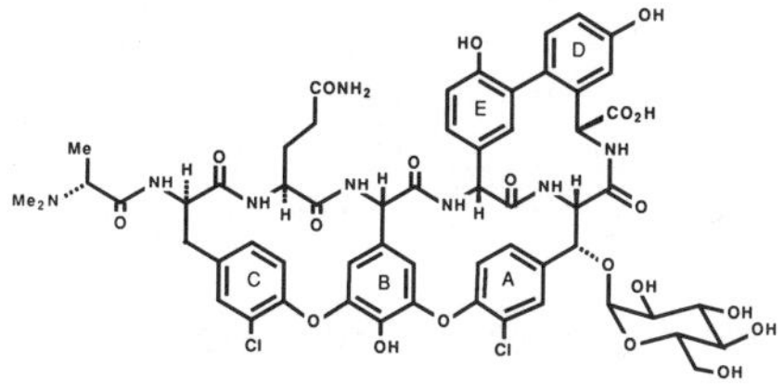

OA7653

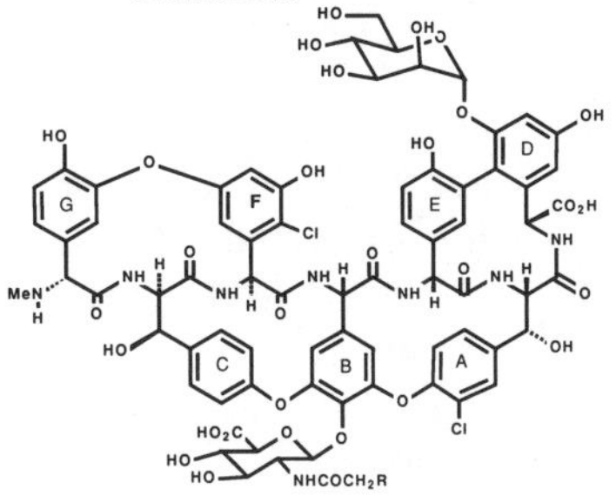

PARVODICINS

Figure 13. Structures of kibdelins, synmonicin, OA7653, and parvodicins.

the distance information, although molecular dynamics simulation is usually preferred if adequate computational power is available to run the current programs.

In undertaking experiments to derive model structures of the aridicin aglycone complexes, the NMR spectra were measured in both H_2O and D_2O. The NOESY experiments were performed in the phase-sensitive mode with a variety of mixing times between 200 and 400 ms. However, we did not derive NOE distance constraints from NOESY cross-peak buildup rates. Instead, the distances were calculated from individual NOESY spectra by using cross-peak intensities and employing the cross-peak intensities of the C5, C6 and G5, G6 *ortho*-hydrogens as internal distance markers. To minimize the effects of erroneous distance information arising from spin diffusion, only distances from the experiments with 200- and 300-ms mixing times were used.

Since NOEs obtained from distances of >0.35 nm tend to be biased to give shorter values as a result of the effect of spin diffusion, we compensated for this in the distance geometry program by providing asymmetric boundary limits. The upper and lower distance limits used were ±0.015 nm for distances of 0.2 to 0.3 nm, ±0.03 nm for distances of 0.3 to 0.35 nm, and +0.06 to −0.04 nm for distances of 0.35 to 0.4 nm. Only 12 distances exceed these constraint limits by more than 0.03 nm; 20 distances exceed these constraint limits by more than 0.005 nm, and all are within 0.055 nm.

Since the distances obtained by this procedure are less accurate than those obtained from NOE buildup rates, we made use of all NOESY connectivities that were identified to provide the maximum data set for the distance geometry calculation. The number of entries and the data employed included 150 distance constraints from the NOESY spectrum and additional structural constraints implicit in the primary covalent structures of both components of the complex:

$$\text{complex core} \equiv 174 \text{ atoms and } 183 \text{ bonds}$$
$$= \text{aglycone} + \text{Ac-L-Lys (Ac)-D-Ala-D-Ala}$$

with the Lys side chain clipped at the beta carbon. The distance constraints for the quantitative NOEs were as follows: aglycone, 75; Ac_2-L-Lys-D-Ala-D-Ala, 14; interpeptide, 14 (total, 103). Distance constraints for bond lengths (183) (five H bonds), planarity constraints (89), one to three atom distances (791), and chirality constraints (6 + 3) totaled 1,072. Aromatic side chains were treated as fused structural templates.

Seven conformers for the complex were generated in seven embedding cycles of the distance geometry algorithm. The aglycone region in three of the structures is the same, with a root mean square deviation of 0.02 nm, and all seven structures are within 0.06 nm. One of the structures, however, had an error in chirality at A-1', despite the fact that the input to the program had specified the correct chirality. Energy refinement of one of these structures was carried out using Gaustiger charges and MAXIMIN force field parameters, as implemented in version 3.3 of Sybyl (Tripos Associates, Clayton, Mo). The resulting structure, which is represented in Fig. 14, shows excellent agreement between the calculated and observed values for over 100 NOE determined distance constraints in the heptapeptide core region. The N-terminal region of the pentapeptide, which does not show NOE interactions back to the aglycone, was modeled by using systematic search routines, and the sterically allowed conformations were generated consistent with the internal NOEs. The NOEs in this region were all much weaker than those observed in the core region, reflecting increased motional freedom of the N-terminal amino acid segments.

Other conclusions derived from this study are that the differences reflected in the conformations generated by the separate embedding procedure may reflect the dynamic fluctuations that exist, since in one case a different orientation of the F and G rings is present, whereas the differences in the other structures are manifested in small changes of the relative orientations of the GFC segment

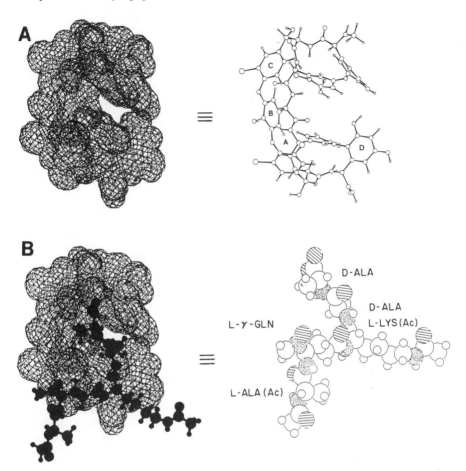

Figure 14. Computer-generated model showing space-filling representation of (A) aridicin agly-cone and (B) aridicin aglycone-diacetyl-L-Ala-(γ)-D-Gln-L-Lys (Ac)-D-Ala-D-Ala complex. Struc-tures were obtained using distance geometry with subsequent energy minimization.

relative to the AED segment. Both of these segments show a closer approach in all structures of the complex than is present in the uncomplexed aglycone, in accord with the notion that the aglycone "wraps around" the pentapeptide on complexing.

Finally, the overall features of the complex are in close agreement with those described by Fesik et al. (8) and Williams et al. (46) for the tripeptide complexes of ristocetin pseudoaglycone (in methanol) and ristocetin in water. In particular, the orientation of the lysine side chain over the D ring and the positioning of the isoglutamine chain adjacent to the A ring correspond well with the arrangement of the analogous regions of the tripeptide complexes in the ristocetin series.

Conclusions

The methodology and results described were evolved to provide an effective, comprehensive approach for the discovery, purification, and structural characterization of novel glycopeptide antibiotics that would be candidates with potential for therapeutic use in humans and also as feed additives to promote animal growth. On the basis of the many new compounds isolated and characterized (and many others discovered but not yet characterized), the research has been successful. The goal of obtaining an antibiotic for human use with advantages over vancomycin or an effective growth promotant for use in animals must await further developments.

ACKNOWLEDGMENTS. The work reported here is the result of a team effort involving both chemists and biologists. We express very special thanks to our colleagues J. Rake, G. Wasserman, R. Sitrin, M. Shearer, L. Mueller, S. Carr, S. Heald, S. Christensen, J. Hempel, A. Venkateswarlu, and A. Verma, whose names, with those of others, appear on the publications cited.

LITERATURE CITED

1. **Brazhnikova, M. G., N. N. Lomakina, M. F. Lavrova, I. V. Tolstykh, M. S. Yurina, L. M. Klyueva.** 1963. Isolation and properties of ristomycin. *Antibiotiki* (Moscow) **8:**392–396.
1a. **Christensen, S. B., H. S. Allaudeen, M. R. Burke, S. A. Carr, S. K. Chung, P. DePhillips, J. J. Dingerdissen, M. DiPaolo, A. J. Giovenella, S. L. Heald, L. B. Killmer, B. A. Mico, L. Mueller, C. H. Pan, B. Poehland, G. D. Roberts, M. C. Shearer, R. D. Sitrin, L. J. Nisbet, and P. W. Jeffs.** 1987. Parvodicin, a novel glycopeptide from a new species, *Actinomadura parvosata*: discovery, taxonomy, activity and structure elucidation. *J. Antibiot.* **40:**970–990.
2. **Convert, O., A. Bongini, and J. Feeney.** 1981. A ^{1}H nuclear magnetic resonance study of the interactions of vancomycin with N-acetyl-D-alanyl-D-alanine and related peptides. *J. Chem. Soc. Perkin Trans.* **2:**1262–1270.
3. **Corti, A., and G. Cassani.** 1985. Synthesis and characterization of D-alanyl-D-alanine-agarose. A new bioselective adsorbent for affinity chromatography of glycopeptide antibiotics. *Appl. Biochem. Biotechnol.* **11:**101–109.
4. **Davis, D. G., and A. Bax.** 1985. Assignment of complex ^{1}H NMR spectra via two-dimensional homonuclear Hartment-Hahn spectroscopy. *J. Am. Chem. Soc.* **107:**2820–2821.
5. **Ellstad, G. A., R. A. Leese, G. O. Morton, F. Barbatschi, W. E. Gore, W. R. McGahren, and I. M. Armitage.** 1981. Avoparcin and epiavoparcin. *J. Am. Chem. Soc.* **103:**622–624.
6. **Fesik, S. W., I. M. Armitage, G. A. Ellstad, and W. J. McGahren.** 1984. Nuclear magnetic resonance studies on the interaction of avoparcin with model receptors of bacterial cell walls. *Mol. Pharamcol.* **25:**281–286.
7. **Fesik, S. W., G. Bolis, H. L. Sham, and E. T. Olenjniczak.** 1987. Structure refinement of a cyclic peptide from 2D NMR data and molecular modeling. *Biochemistry* **26:**1851–1859.

8. **Fesik, S. W., T. J. O'Donnell, R. T. Gampe, Jr., and E. T. Olejniczak.** 1986. Determining the structure of a glycopeptide-Ac$_2$-Lys-D-Ala-D-Ala complex using NMR parameters and molecular modeling. *J. Am. Chem. Soc.* **108**:3165–3170.

9. **Grundy, W. E., A. C. Sinclair, R. J. Theriault, A. W. Goldstein, C. J. Rickher, H. B. Warren, T. J. Oliver, and J. C. Sylvester.** 1956–1957. Ristocetin, microbiologic properties. *Antibiot. Annu.* **1956–1957**:687–692.

10. **Harris, C. M., and T. M. Harris.** 1982. Structure of ristocetin A: configurational studies of the peptide. *J. Am. Chem. Soc.* **104**:363–365.

11. **Harris, C. M., H. Kopecka, and T. M. Harris.** 1983. Vancomycin: structure and transformation to CDP-I. *J. Am. Chem. Soc.* **105**:6915–6922.

12. **Heald, S. L., L. Mueller, and P. W. Jeffs.** 1987. Structural analysis of teicoplanin A$_2$ by 2D NMR. *J. Magn. Reson.* **72**:120–138.

12a. **Heald, S. L., L. Mueller, and P. W. Jeffs.** 1987. Actinoidins A and A$_2$: structure determination using 2D NMR methods. *J. Antibiot.* **40**:630–645.

13. **Hunt, A. H., M. Debono, K. E. Merkel, and M. Barnhart.** 1984. Structure of the pseudoaglycon of actaplanin. *J. Org. Chem.* **49**:635–640.

14. **Hunt, A. H., D. E. Dorman, M. Debono, and R. M. Malloy.** 1985. Structure of antibiotic A4130A. *J. Org. Chem.* **50**:2031–2035.

15. **Hunt, A. H., T. K. Elzey, K. E. Merkel, and M. Debono.** 1984. Structures of the actaplanins. *J. Org. Chem.* **49**:641–645.

16. **Hunt, A. H., R. M. Molloy, J. L. Occolowitz, G. G. Marconi, and M. Debono.** 1984. Structure of the major glycopeptide of the teicoplanin complex. *J. Am. Chem. Soc.* **106**:4891–4895.

17. **Jeffs, P. W., G. Chan, L. Mueller, C. DeBrosse, L. Webb, and R. Sitrin.** 1986. Aridicin aglycon: characterization of the amino acid components. *J. Org. Chem.* **51**:4272–4278.

18. **Jeffs, P. W., L. Mueller, C. DeBrosse, S. L. Heald, and R. Fisher.** 1986. Structure of aridicin A. An integrated approach employing 2D NMR, energy minimization and distance constraints. *J. Am. Chem. Soc.* **108**:3063–3075.

19. **Jordan, D. C.** 1961. Effect of vancomycin on the synthesis of the cell wall mucopeptide of *Staphylococuus aureus*. *Biochem. Biophys. Res. Commun.* **6**:167–170.

20. **Kalman, J. R., and D. H. Williams.** 1980. An NMR study of the structure of the antibiotic ristocetin A. The negative nuclear Overhauser effect in structure elucidation. *J. Am. Chem. Soc.* **102**:897–905.

21. **Kamogashira, T., T. Nishida, and M. Sugawara.** 1983. A new glycopeptide antibiotic, OA-7653, produced by *Streptomyces hygroscopicus* subsp. *Hiwasaensis*. *Agric. Biol. Chem.* **47**:499–506.

22. **Karchmer, A. W., G. L. Archer, and W. E. Dismukes.** 1983. *Staphylococcus epidermidis* causing prosthetic valve endocarditis: microbiologic and clinical observations as guides to therapy. *Ann. Intern. Med.* **98**:447–455.

23. **Kuntz, I. D., G. M. Crippen, and P. A. Kollman.** 1979. Application of distance geometry to protein tertiary structure calculations. *Biopolymers* **18**:939–957.

24. **Lomakina, N. N., I. A. Spirindonova, N. Yu, and T. F. Vlasova.** 1973. Structure of the amino sugars from the antibiotic actinoidin. *Khim. Prir. Soedin.* **9**:101–107.

25. **Lowe, C. R., and P. D. G. Dean.** 1974. Affinity chromatography. John Wiley & Sons, Inc., New York.

26. **McCormick, M. H., W. M. Stark, G. E. Pittinger, and R. C. McGuire.** 1955–1956. Vancomycin, a new antibiotic. I. Chemical and biological properties. *Antibiot. Annu.* **1955–1956**:606.

27. **Michel, K. H., R. M. Shah, and R. L. Hamill.** 1980. A35512, a complex of new antibacterial antibiotics produced by *Streptomyces candidus*. *J. Antibiot.* **33**:1397–1406.

28. **Nieto, M., and H. R. Perkins.** 1971. Physicochemical properties of vancomycin and io-dovancomycin and their complexes with diacetyl-L-lysyl-D-alanyl-D-alanine. *Biochem. J.* **123:**773–787.

29. **Nieto, M., and H. R. Perkins.** 1971. Modifications of the acyl-D-alanyl-D-alanine terminus affecting complex-formation with vancomycin. *Biochem. J.* **123:**789–803.

30. **Nieto, M., and H. R. Perkins.** 1971. The specificity of combination between ristocetins and peptides related to bacterial cell-wall mucopeptide precursors. *Biochem. J.* **124:**845–852.

31. **Perkins, H. R.** 1969. Specificity of combination between mucopeptide precursors and vancomycin or ristocetin. *Biochem. J.* **111:**195–205.

32. **Pirkle, W. H., M. H. Hyun, and B. Bank.** 1984. A rational approach to the design of highly-effective chiral stationary phases. *J. Chromatog.* **316:**585–604.

33. **Rake, J. B., R. Gerber, R. J., Mehta, D. J. Newman, Y. K. Oh, C. Phelen, M. C. Shearer, R. D. Sitrin, and L. J. Nisbet.** 1986. Glycopeptide antibiotics: a mechanism-based screen employing a bacterial cell wall receptor mimetic. *J. Antibiot.* **39:**58–67.

34. **Reynolds, P. E.** 1961. Studies on the mode of action of vancomycin. *Biochem. Biophys. Acta* **52:**403–405.

35. **Roberts, G. D., S. A. Carr, S. Rottschaefer, and P. W. Jeffs.** 1985. Structural characterization of glycopeptide antibiotics related to vancomycin by fast-atom bombardment mass spectrometry. *J. Antibiot.* **38:**713–720.

36. **Shearer, M. C., P. Actor, B. A. Bowie, S. F. Grappel, C. H. Nash, D. J. Newman, Y. K. Oh, C. H. Pan, and L. J. Nisbet.** 1985. Aridicins, novel glycopeptide antibiotics. I. Taxonomy, production and biological activities. *J. Antibiot.* **38:**555–560.

37. **Sheldrick, G. M., P. G. Jones, O. Kennard, D. H. Williams, and G. A. Smith.** 1978. Structure of vancomycin and its complex with acetyl-D-alanyl-D-alanine. *Nature* (London) **271:**223–225.

38. **Sitrin, R. D., G. W. Chan, J. J. Dingerdissen, W. Holl, J. R. E. Hoover, J. R. Valenta, L. Webb, and K. M. Snader.** 1985. Aridicins, novel glycopeptide antibiotics. II. Isolation and characterization. *J. Antibiot.* **38:**561–571.

39. **Smith, K. A., D. H. Williams, and G. A. Smith.** 1974. Structural studies on the antibiotic vancomycin; the nature of the aromatic rings. *J. Chem. Soc. Perkin Trans.* **1:**2369–2376.

40. **Starichkai, F., N. N. Lomakina, I. A. Spiridonova, M. S. Yurina, and M. Pushkash.** 1967. Determination of carbohydrates in actinoidins A and B. *Antibiotiki* (Moscow) **12:**126–129.

41. **Wasserman, G. F., B. L. Poehland, E. W-K. Yeung, D. Staiger, L. B. Killmer, K. Snader, and P. W. Jeffs.** 1986. Kibdelins (AAD-609), novel glycopeptide antibiotics. II. Isolation, purification and structure. *J. Antibiot.* **39:**1395–1406.

42. **Wasserman, G. F., R. D. Sitrin, F. Chapin, and K. M. Snader.** 1987. Affinity chromatography of glycopeptide antibiotics. *J. Chromatog.* **392:**225–238.

43. **Williams, D. H., and D. W. Butcher.** 1981. Binding site of the antibiotic vancomycin for a cell-wall peptide analogue. *J. Am. Chem. Soc.* **103:**5700–5704.

44. **Williams, D. H., M. P. Williamson, D. W. Butcher, and S. J. Hammond.** 1983. Detailed binding sites of the antibiotics vancomycin and ristocetin A: determination of intermolecular distances in antibiotic/substrate complexes by use of the time-dependent NOE. *J. Am. Chem. Soc.* **105:**1332–1339.

45. **Williamson, M. P., and D. H. Williams.** 1981. Structure revision of the antibiotic vancomycin. The use of nuclear Overhauser effect difference spectroscopy. *J. Am. Chem. Soc.* **103:**6580–6585.

46. **Williamson, M. P., and D. H. Williams.** 1985. ¹H NMR studies of the structure of ristocetin A and its complexes with bacterial cell wall analogues in aqueous solution. *J. Chem. Soc. Perkin Trans.* **1:**949–956.

Chapter 54

Enzymes of D-Alanine Metabolism: Cloning, Expression, Purification, and Characterization of *Salmonella typhimurium* D-Alanyl-D-Alanine Ligase

C. Walsh

E. Daub

L. Zawadzke

K. Duncan

The dipeptide D-alanyl-D-alanine is biosynthesized from the L-isomer of alanine by sequential action of two enzymes, alanine racemase and D-alanyl-D-alanine ligase. In turn D-alanyl-D-alanine is added to UDP-muramyl-L-alanine-D-glutamic acid-*meso*-diaminopimelic acid by the D-alanyl-D-alanine adding enzyme to yield the pentapeptide precursor of peptidoglycan (15). This pentapeptide, terminating in D-alanyl-D-alanine, thus contains two additional D-amino acids, D-glutamic acid and DL-diaminopimelic acid (*meso*). The enzymes involved in this D-amino acid metabolism are displayed in Fig. 1.

We previously determined that there are two genes in *Salmonella typhimurium* that encode alanine racemase isoenzymes, *dadB* (at min 37) (24) and *alr* (at min 91) (E. Daub, Ph.D. dissertation, Massachusetts Institute of Technology, Cambridge, 1986). The *ddl* gene, encoding D-alanyl-D-alanine ligase, and the *murF* gene, encoding the D-alanyl-D-alanine adding enzyme, map near

C. Walsh, E. Daub, L. Zawadzke, and K. Duncan • Departments of Biology and Chemistry, Massachusetts Institute of Technology, Cambridge, Massachusetts 02139.

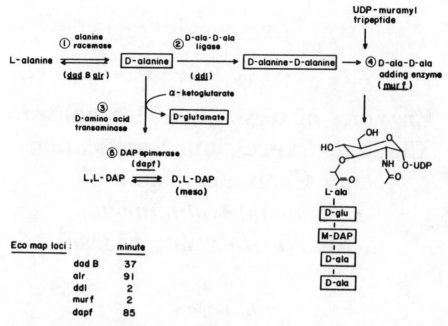

Figure 1. Bacterial enzymes involved in metabolism of D-amino acids in biosynthesis of the UDP-muramyl pentapeptide in peptidoglycan assembly.

min 2 on the *Escherichia coli* chromosome in a cluster of genes involved in cell envelope growth and division (9, 11). The *dapF* gene, encoding DAP epimerase, has recently been mapped to min 85 of the *E. coli* chromosome (17).

Enzyme Inhibitors

The alanine racemases are targets for natural antibacterial agents such as *O*-carbamyl D-serine (14) and for synthetic inhibitors such as β-D-fluoro- (7) and β-D-chloroalanine (10).

The ligase is a target for the antitubercular natural product D-cycloserine (14). There are no good inhibitors yet described for the *murF* or *dapB* gene products. D-Vinyl glycine (20) and D-β-haloalanines (19) are mechanism-based inactivators of D-amino acid transaminases from bacilli.

Alanine Racemase Status

We have demonstrated that these β-substituted alanines are suicide substrates for these pyridoxal 5′-phosphate-dependent racemases by an α-H-β-x elimination sequence to generate nascent aminoacrylate in the active site which can either yield product pyruvate or inactivate the enzyme by nucleophilic attack on the PLP-lysine aldimine to yield a covalent inactivator-pyridoxamine 5′-phosphate-enzyme species (3) (Fig. 2). This explains their molecular basis of action as antibacterial agents.

A distinct class of synthetic antibacterial agents acting on alanine racemases is represented by 1-aminoethylphosphonate and dipeptides such as L-alanine-L-aminoethylphosphonate (1) which show time-dependent inactivation of racemases from gram-positive bacteria (bacilli, staphylococci, and streptococci) but not from gram-negative bacteria (*Pseudomonas* spp., *Salmonella*, spp., and *E. coli*) (2, 4). The amino phosphonate in fact is a slow binding inhibitor (11a) specifically for the gram-positive alanine racemases with a half-time for dissociation of 2 to 3 weeks (2). *S. typhimurium dadB* and *alr* genes have both been cloned and sequenced, the encoded racemase enzymes have been overproduced, and each has been purified to homogeneity (5, 23). Similarly, the alanine racemase gene from the thermophilic gram-positive *Bacillus stearothermophilus* has been cloned, sequenced, and expressed, the encoded enzyme has been purified (6), and crystals of X-ray diffraction quality have been obtained (D. J. Neidhart, M. D. DiStefano, K. Tanizawa, K. Soda, C. Walsh, and G. A. Petsko, *J. Biol. Chem.*, in press).

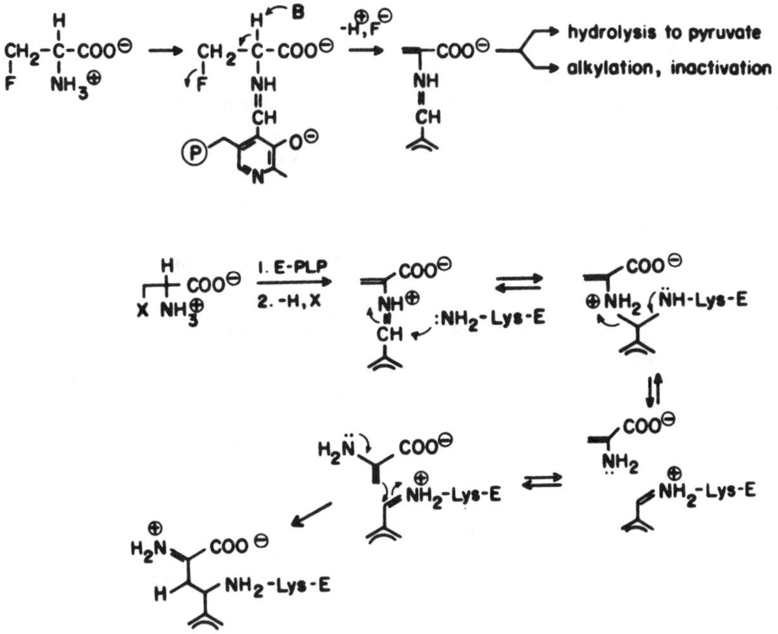

Figure 2. Partition of β-substituted alanine suicide substrates of alanine racemases between catalytic turnover to pyruvate and enzyme inactivation by covalent derivatization of the pyridoxal 5′-phosphate (PLP)-aldimine complex.

D-Alanyl-D-Alanine Ligase

Much less is known about the second enzyme of the pathway, the D-alanyl-D-alanine ligase, the in vivo target for D-cycloserine, catalyzing ATP-dependent synthesis of the dipeptide bond (Fig. 3). The enzyme best characterized in the literature is that from *Streptococcus faecalis* as a result of efforts by Neuhaus and colleagues on its purification and characterization some 20 years ago (12, 13, 16). Most of their kinetic work was done on partially purified enzyme, but homogeneous enzyme with a specific activity of ca. 8 μmol min^{-1} mg^{-1} was noted (16). Two discrete K_m values were obtained for D-alanine, 6 × 10^{-4} and 1 × 10^{-2} M, a 16-fold difference in affinity. The D-alanyl-D-alanine product also showed two binding sites (each with K_i of ca. 10^{-3} M) as a competitive inhibitor. That ligase showed high specificity for D-amino acids in the N-terminal site and lower specificity in the C-terminal site, such that D-alanyl-D-amino acid mixed dipeptides were produced. Essentially no information on enzyme size or structure was available from those studies.

Cloning of the Salmonella D-Alanyl-D-Alanine Ligase Gene

To gain insight into the structure and catalytic mechanism of the ligase, we have undertaken cloning of the ligase gene, from *S. typhimurium,* the organism used in our alanine racemase molecular biology studies. We used *E. coli* ST640, described as a temperature-sensitive mutant of ligase (8), as a host for a *Salmonella* DNA lambda library to select for complementing clones. The *Salmonella* gene was then subcloned into PBR322 to yield eventually a plasmid (pDS4) with a 2.5-kilobase insert. This DNA insert produces a functional D-alanyl-D-alanine ligase which was obtained as a homogeneous protein (molecular weight, 39,000) after ca. 100-fold purification, confirming reasonable overproduction. The sequence of the N-terminal 35 residues of the pure enzyme was obtained and has been used to locate the 5'-start of the encoding gene during DNA sequencing. The DNA sequence information is now complete and matches the observed protein N-amino-terminal sequence data, confirming that

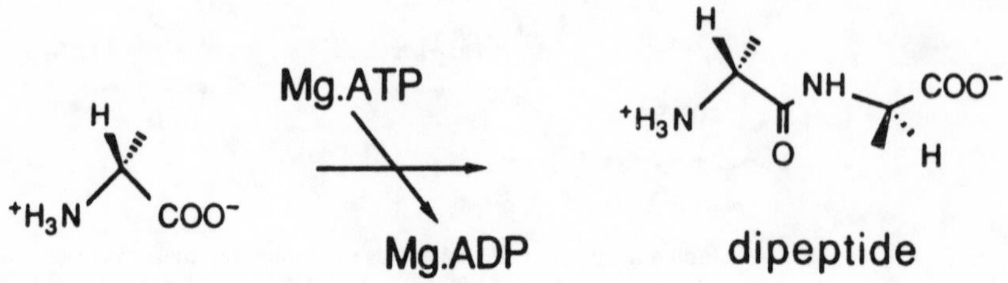

Figure 3. ATP-dependent synthesis of the dipeptide bond.

the structural gene (rather than a positive-regulatory gene; see below) for the overproduced ligase has been cloned. A predicted protein of 364 residues with a molecular weight of 39,271 is consistent with the observed molecular weight of purified enzyme.

While this work was in progress, the DNA sequence of the *ddl* gene of *E. coli* from min 2 of the chromosome was reported by Robinson et al. (18). The *E. coli ddl* gene is 918 nucleotides long and encodes a 306-residue protein of 33 kilodaltons, 10 kilodaltons less than the cloned *Salmonella* ligase. A comparison and alignment of the first 50 residues of the two ligases shows that the *E. coli* ligase has a stretch of 11/14 identity with *Salmonella* residues 11 through 22, but otherwise little detectable similarity (Fig. 4). Overall there are 118 matches of 306 residues, for a 38% homology. One expects a much greater homology for related genes between *E. coli* and *S. typhimurium,* and the sequence divergence raises the prospect of more than one ligase gene. This issue arose separately for us in *S. typhimurium* because when insertions in the ligase gene were transduced into the chromosome, no phenotypic effect was observed whereas we anticipated that a requirement for D-alanyl-D-alanine should have been generated (Daub, Ph.D. dissertation). A second route to D-alanyl-D-alanine could be the explanation. We are currently testing for a second ligase gene in *S. typhimurium.* For precedent we have our previous discovery (24; Daub, Ph.D. dissertation) of two alanine racemase genes, presumed to have anabolic and catabolic roles, in *S. typhimurium.* If there indeed are two ligase genes, evaluation of which is the more important in peptidoglycan biosynthesis may be important for future antibiotic design efforts targeted against ligase activity.

Substrate Specificity of the Salmonella Ligase

Given the availability of pure cloned *Salmonella* D-alanyl-D-alanine ligase, we have begun to characterize it catalytically. In particular, we have examined the antibacterial β-halo-D-alanines, utilizing a coupled spectrophotometric assay in which D-amino acid-dependent ATP cleavage activity is detected by coupling the ADP produced to reduced nicotinamide adenine dinucleotide consumption via pyruvate kinase and lactate dehydrogenase action. By this assay β-fluoro-D-alanine shows an eight-fold higher K_m but 55% of the V_{max} of D-alanine (Table 1), while β-chloro-D-alanine has 13-fold lower affinity and 17%

Figure 4. Alignment of the N-terminal protein sequences of the *Salmonella* D-alanyl-D-alanine ligase gene and the *E. coli ddl* gene. Homology is limited; two distinct genes?

of the V_{max}. One must yet identify D-haloalanine dipeptide products, but it is interesting that these two β-halo-D-alanines are suicide substrates for the racemase, the preceding enzyme in the pathway. A "self-reversal" of antibacterial action has been reported for high concentrations of D-β-fluoroalanine (F. Kahan and H. Kropp, *Program Abstr. 15th Intersci. Conf. Antimicrob. Agents Chemother.*, abstr. no. 100, 1975), with the suggestion that D-β-fluoroalanine was replacing D-alanine in peptidoglycan at those high levels, a notion validated by these data. Remarkably the β,β,β-trifluoro-D-alanine was also a substrate, with a 30-fold elevated K_m but a respectable 43% of V_{max}. Trifluoro-D-alanine is not antibacterial, probably as a result of lack of entry into cells, for it is a potent racemase suicide substrate (22). The amino group pK_a of 5.8 for trifluoroalanine is ca. 5 U depressed from that of alanine (pK_a 10.5) as a result of electronic effects of the three fluorines, yet that 10^5 effect is not expressed on ligase V_{max} (at least in this coupled assay). Synthesis of D-[^{14}C]trifluoroalanine is in progress to test for [^{14}C]hexafluorodipeptide formation.

As with the *Streptococcus faecalis* ligase (13), D-aminobutyrate is quite a good substrate, glycine is a very poor substrate (in terms of a 130-fold higher K_m), and no activity was detected with L-amino acids.

Inhibitor Studies with Salmonella Ligase

Initial studies with D-cycloserine, the naturally occurring antibiotic, confirm the anticipated competitive inhibition of D-alanine utilization (ATP consumption coupled assay) (K_i of ca. 15 μM). There is the possibility that D-cycloserine can serve as an alternate acceptor at the C-terminal site of the *Salmonella* ligase to make a D-alanine-D-cycloserine product, and this is under investigation. It has some precedent in the work of Neuhaus and colleagues with the *Streptococcus* enzyme, who found that aminooxy-D-alanine is converted by enzyme, in the presence of D-alanine, to the D-alanine-aminooxy-D-

Table 1. Activity of some D-amino acids on substrates for *Salmonella* D-alanyl-D-alanine ligase

Substrate	K_m (mM)	Relative V_{max}[a]
D-Alanine	0.5	1.0
D-Fluoroalanine	4.0	0.55
D-Trifluoroalanine	15.6	0.43
D-Chloroalanine	6.4	0.17
D-Aminobutyrate	8.7	0.86
Glycine	66	0.13

[a]The value of 1.0 for D-alanine represents 9.8 μmol min^{-1} mg^{-1}.

alanine dipeptide (Fig. 5). Thus, D-cycloserine may be an alternate substrate in the C-terminal site.

Recently, investigators at Merck Sharp & Dohme Research Laboratories reported that an alkyl aminophosphinate compound is a potent reversible inhibitor of the partially purified *S. faecalis* D-alanyl-D-alanine ligase (W. Parsons, A. Patchett, J. Davidson, W. Schoen, P. Combs, D. Taub, J. Springer, T. Mellin, R. Busch, W. Thornberry, W. Gadbusch, B. Weissberger, and M. Valiant, *Am. Chem. Soc. 193rd Natl. Meet.*, abstr. no. 63, 1987). With a gift sample of the indicated aminoethyl phosphinodecanoate and pure *Salmonella* ligase we have noted a time-dependent effect in which the addition of inhibitor to enzyme processing D-alanine induces a change to a lower steady level. This is characteristic of slow binding inhibitors where an initial collisional complex of enzyme and inhibitor is followed by an isomerization to an $E \cdot I^*$ complex, $E + I \rightleftarrows E \cdot I \rightleftarrows E \cdot I^*$, which accumulates and only slowly returns to free enzyme (11a). The initial K_i for the Merck phosphinate is 1.2 μM. We have not yet detected regain of catalytic activity to evaluate K_i^*. The structural basis of the stabilizing isomerization is unclear but may be a lead to pursue for development of more potent ligase inhibitors. In this connection it is likely that ligase catalysis involves activation of the first D-alanine as the acyl-phosphate, D-alanyl-PO_3, which suffers attack by the second D-alanine to yield dipeptide formation by way of the tetrahedral adduct indicated in Fig. 6. Perhaps the alkylphosphinate inhibitor is a mimic of this type of structure. We note that the 1-aminoethylphosphonate but not the 1-aminoethylphosphinate is a potent slow binding inhibitor of the gram-positive alanine racemases (2). We have

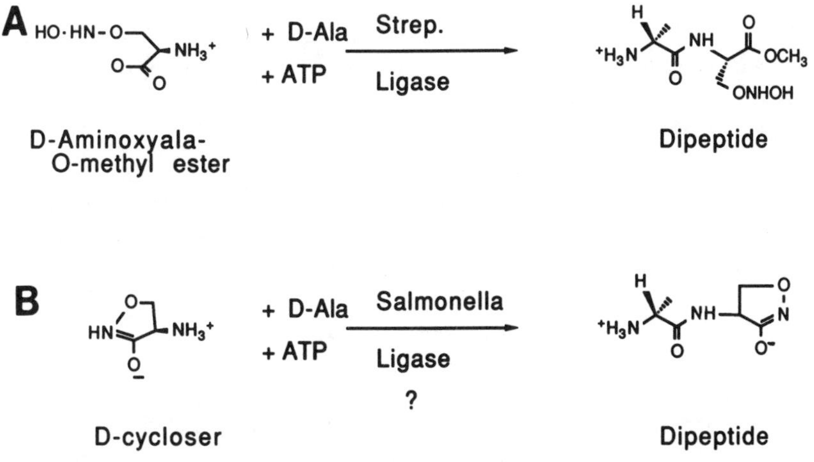

Figure 5. Inhibitors as alternative substrates (acceptor subsite). (A) Enzymatic utilization of aminooxy-D-alanine at the C-terminal site by the *S. faecalis* ligase. (B) Possible utilization of the inhibitor D-cycloserine as an alternative substrate at the C-terminal site of *Salmonella* ligase.

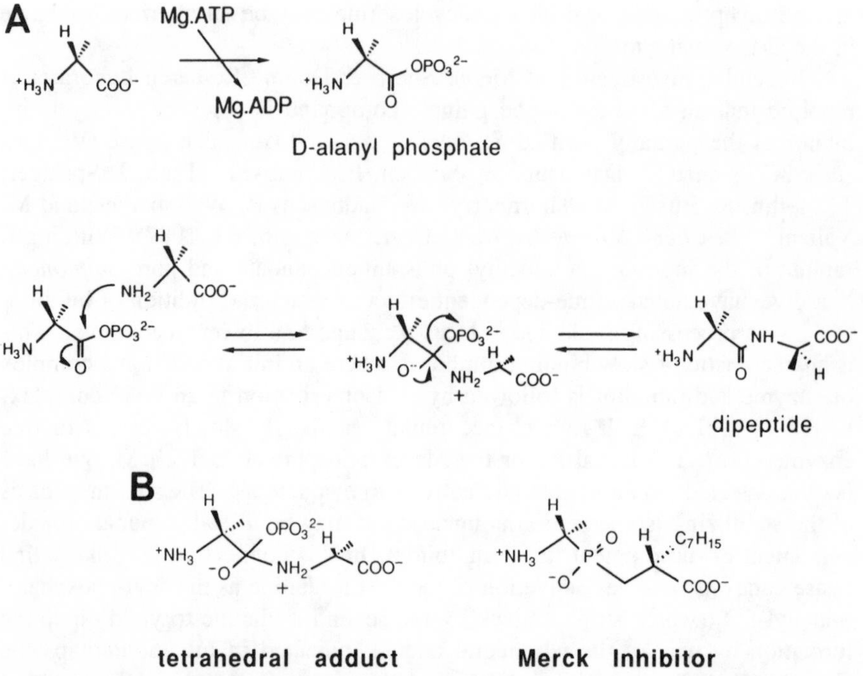

A

D-alanyl phosphate

dipeptide

B

tetrahedral adduct   Merck Inhibitor

Figure 6. (A) Mechanistic postulate for D-alanyl-D-alanine ligase catalysis. Activation of the carboxylate of the first D-alanine is likely to proceed via the mixed acylphosphoric anhydride, D-alanyl phosphate. Attack of the amino group of the second D-alanine would yield the indicated tetrahedral adduct on the way to DD-dipeptide product. (B) Structural comparison of the tetrahedral adduct with the aminoalkyl phosphinate slow binding inhibitor.

also determined that the phosphonate but not the phosphinate causes time-dependent inhibition of *Salmonella* ligase when preincubated with ATP, and the basis of this effect is under study.

ACKNOWLEDGMENTS. This work was supported in part by National Science Foundation grant PCM 8308969 and a European Molecular Biology Organisation Long-Term Fellowship (K.D.).

LITERATURE CITED

1. **Atherton, F., M. Hall, C. Hassall, R. Lambert, W. Lloyd, and P. Ringrose.** 1979. Phosphonopeptides as antibacterial agents: mechanism of action of alaphospholin. *Antimicrob. Agents Chemother.* **15:**696–705.
2. **Badet, B., K. Nagaki, K. Soda, and C. Walsh.** 1986. Time dependent inhibition of *Bacillus stearothermophilus* alanine racemase by 1-aminoethylphosphonate isomers by isomerization to noncovalent slowly dissociating complex. *Biochemistry* **25:**3275–3282.

3. **Badet, B., D. Roise, and C. Walsh.** 1984. Inactivation of the dadB *Salmonella typhimurium* alanine racemase by D- and L-isomers of β-substituted alanines: kinetics, stoichiometry, active site peptide sequencing, and reaction mechanism. *Biochemistry* **23**:5188–5194.

4. **Badet, B., and C. Walsh.** 1985. Purification of an alanine racemase from *Streptococcus faecalis* and analysis of its inactivation by 1-amino ethyl phosphonic acid enantiomers. *Biochemistry* **24**:1333–1341.

5. **Galakatos, N., E. Daub, D. Botstein, and C. Walsh.** 1986. Biosynthetic *alr* alanine racemase from *Salmonella typhimurium*. DNA sequence and protein sequence determination. *Biochemistry* **23**:3255–3260.

6. **Inagaki, K., K. Tanizawa, B. Badet, C. Walsh, H. Tanaka, and K. Soda.** 1986. Thermostable alanine racemase from *Bacillus stearothermophilus*: molecular cloning of the gene, enzyme purification, and characterization. *Biochemistry* **25**:3268–3274.

7. **Kollonitsch, J., L. Barash, F. Kahan, and H. Kropp.** 1973. New antibacterial agent via photofluorination of a bacterial cell wall constituent. *Nature* (London) **243**:346–347.

8. **Lugtenberg, E. J. J., and A. Van Schijndel-van Dam.** 1972. Temperature-sensitive mutants of *Escherichia coli* K12 with low activities of the L-alanine adding enzymes and the D-alanyl-D-alanine adding enzyme. *J. Bacteriol.* **110**:35–40.

9. **Luktenhaus, J. F., and H. C. Wu.** 1980. Determination of transcriptional units and gene products from the *ftsA* region of *Escherichia coli*. *J. Bacteriol.* **143**:1281–1288.

10. **Manning, J. M., N. Merrifeld, W. Jones, and E. Gotschalk.** 1974. Inhibition of bacterial growth by β-chloro-D-alanine. *Proc. Natl. Acad. Sci. USA* **71**:417–421.

11. **Miyakawa, T., H. Matsuzawa, M. Matsuhashi, and Y. Sugino.** 1972. Cell wall peptidoglycan mutants of *Escherichia coli*: existence of two clusters of genes *mra* and *mrb* for cell wall peptidoglycan biosynthesis. *J. Bacteriol.* **112**:950–958.

11a. **Morrison, J. F., and C. T. Walsh.** 1987. The behavior and significance of slow-binding enzyme inhibitors. *Adv. Enzymol.* **57**:201–301.

12. **Neuhaus, F. C.** 1962. The enzymatic synthesis of D-alanyl-D-alanine synthetase. I. Purification and properties. *J. Biol. Chem.* **237**:778–786.

13. **Neuhaus, F. C.** 1962. The enzymatic synthesis of D-alanyl-D-alanine. II. Kinetic studies of D-alanyl-D-alanine synthetase. *J. Biol. Chem.* **237**:3128–3135.

14. **Neuhaus, F. C.** 1967. D-Cycloserine and O-carbamyl-D-serine, p. 40–83. *In* D. Gottlieb and P. D. Shaw (ed.), *Antibiotics: Mechanism of Action*, vol. 1. Springer Verlag, Heidelberg.

15. **Neuhaus, F. C., and W. P. Hammes.** 1981. Inhibition of cell wall biosynthesis by analogues of alanine. *Pharmacol. Ther.* **14**:265–319.

16. **Neuhaus, F. C., and J. Lynch.** 1964. The enzymatic synthesis of D-alanyl-D-alanine. III. On the inhibition of D-alanyl-D-alanine synthetase by the antibiotic D-cycloserine. *Biochemistry* **3**:471–480.

17. **Richaud, C., W. Higgins, D. Mengin-Lecreulx, and P. Stragier.** 1987. Molecular cloning, characterization and chromosomal location of *dapF*, the *Escherichia coli* gene for diaminopimelate epimerase. *J. Bacteriol.* **169**:1454–1459.

18. **Robinson, A., D. Kenan, J. Sweeney, and W. Donachie.** 1986. Further evidence for overlapping transcriptional units in an *Escherichia coli* cell envelope-cell division cluster: DNA sequence and transcriptional organization of the *ddl-ftsQ* region. *J. Bacteriol.* **167**:809–817.

19. **Soper, T. S., and J. M. Manning.** 1978. β-Elimination of β-halo substrates by D-amino acid transaminase associated with inactivation of the enzyme. *Biochemistry* **17**:3377–3384.

20. **Soper, T. S., J. M. Manning, P. A. Marcotte, and C. Walsh.** 1977. Inactivation of bacterial D-amino acid transaminases by the olefenic amino acid D-vinylglycine. *J. Biol. Chem.* **252**:1571–1575.

21. **Wang, E., and C. Walsh.** 1978. Suicide substrates for alanine racemase of *E. coli* B. *Biochemistry* **17**:1313–1321.
22. **Wang, E., and C. Walsh.** 1981. Characterizations of β,β-difluoroalanine and β,β,β-trifluoroalanine as suicide substrates for *E. coli* B alanine racemase. *Biochemistry* **20**:7539–7546.
23. **Wasserman, S., E. Daub, D. Grisafi, D. Botstein, and C. Walsh.** 1984. Catabolic alanine racemase from *Salmonella typhimurium*: DNA sequence, enzyme purification and characterization. *Biochemistry* **23**:5182–5187.
24. **Wasserman, S., C. Walsh, and D. Botstein.** 1983. Two alanine racemase genes in *Salmonella typhimurium* that differ in structure and function. *J. Bacteriol.* **153**:1439–1450.

Chapter 55

Phosphinic Acid Inhibitors of D-Alanyl-D-Alanine Ligase

Arthur A. Patchett

D-Alanyl-D-alanine ligase is one of the essential cytoplasmic bacterial enzymes required for the incorporation of D-alanine into the bacterial cell wall. It accomplishes the synthesis of a peptide bond while hydrolyzing one molecule of ATP to yield ADP and a molecule of inorganic phosphate. This enzyme and the stoichiometry of the reaction which it catalyzes were characterized in the early 1960s by Neuhaus (6–8) and by Ito and Strominger (4, 5). After years of relative inactivity, interest in D-alanyl-D-alanine ligase has intensified in the past several years. The enzyme from *Escherichia coli* has recently been cloned and sequenced by Donachie's group (11), and Walsh and colleagues (this volume) have described their progress in the cloning and sequencing of the *Salmonella* ligase. Thus, prospects for the rational design of potent inhibitors of this important enzyme have now been greatly enhanced.

The only reasonably effective inhibitor of D-alanyl-D-alanine ligase is cycloserine. However, its K_i using the *Streptococcus faecalis* ligase is only 10^{-5} M (10). The antibiotic potency of cycloserine is low by current standards, its antibacterial spectrum is broad but with serious deficiencies, and it can produce side effects in the central nervous system. Thus, it is little used today. Several years ago interest developed at Merck Sharp & Dohme Research Laboratories in a combination of fluoro-D-alanine with a prodrug of cycloserine (MK-641/MK-642) which displays potent, broad-spectrum antibacterial activity and is orally active (12; H. Kropp, F. M. Kahan, and H. B. Woodruff, *Program Abstr. Intersci. Conf. Antimicrob. Agents Chemother.*, abstr. no. 101, 1975). In this combination ligase inhibitory activity of cycloserine blocks the incorporation of fluoro-D-alanine into peptidoglycan and, thereby, prevents self-reversal of fluoro-D-alanine's antibacterial activity. Unfortunately, safety assess-

---

Arthur A. Patchett • Merck Sharp & Dohme Research Laboratories, Rahway, New Jersey 07065.

ment issues unrelated to MK-641/642's mechanisms of action hindered its clinical development. We were, however, encouraged to attempt the synthesis of more potent, structurally distinct inhibitors of alanine racemase and ligase.

For initial screening, a discontinuous assay was utilized by Mellin and R. D. Busch of the Merck laboratories which employed dialyzed, ammonium sulfate-precipitated enzyme from *S. faecalis* (ATCC 8043). Assay mixtures buffered at pH 7.9 consisting of 0.01 M KCl, 0.008 M MgCl$_2$ and 5 μl of enzyme (0.05 U) were preincubated with inhibitor for 40 min at room temperature. The ligase reaction was initiated by adding 0.005 M ATP and 0.005 M D-[1^{14}C]alanine. After incubation at 37°C for 30 min, 20-μl samples were chromatographed on thin-layer chromatography plates, and the radioactive zone of labeled D-alanyl-D-alanine was quantitated with a Berthold linear analyzer to determine 50% inhibitory concentrations (IC$_{50}$s).

The design of new inhibitors was based on several mechanistic assumptions. One of them was that the stoichiometry of the synthesis implied the intermediacy of an acyl phosphate. Another was that active-site inhibitors of an established acyl phosphate-dependent enzyme such as glutamine synthetase might be adapted to D-alanyl-D-alanine ligase by making appropriate structural changes to reflect the specificity of D-alanyl-D-alanine ligase. If an acyl phosphate mechanism were involved, inhibition might be observed in nondisplaceable or poorly displaceable variants of this functionality. This was indeed the case when $-O-PO(OH)_2$ was replaced by $-CH_2-PO(OH)_2$ and $NH-PO(OH)_2$ groups (P. K. Chakravarty, W. J. Greenlee, W. H. Parsons, P. Combs, A. Roth, A. A. Patchett, R. D. Busch, and T. Mellin, *Program Abstr. Am. Chem. Soc. 193rd Natl. Meet.*, MEDI 15) (Table 1). The observed modest inhibitions lent support to a hypothesized transition state for amide bond formation (Fig. 1). For the design of new inhibitors phosphinic acid functionality was chosen to mimic approximately the charge distribution and geometry of this transition state. In addition, the target phosphinic acids are analogous to the $-PO(OH)CH_3$

Table 1. Inhibition of D-alanyl-D-alanine ligase by D-alanyl phosphate analogs

X–Y[a]	IC$_{50}$ (M)[b]	Antibacterial activity[c]
$-CH_2-P(O)(OH)_2$	5.5×10^{-4}	Nil
$-NH-P(O)(OH)_2$	1.2×10^{-4}	Nil
Cycloserine	2.5×10^{-4}	Good

[a]Variables in the compound:

$$NH_2-CH-\overset{\overset{\displaystyle CH_3}{|}}{}\,\overset{\overset{\displaystyle O}{\|}}{C}-X-Y$$

[b]Determined by the discontinuous assay described in the text.
[c]As described in footnote *b* of Table 5.

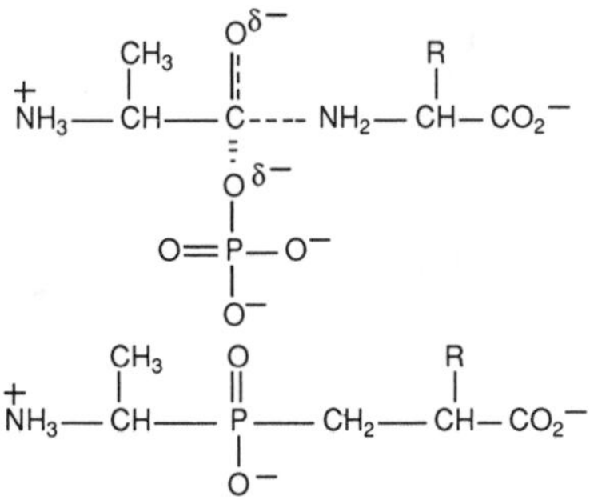

```
     CH₃    O δ⁻                    R
 +    |      ‖                      |
NH₃—— CH—— C ---- NH₂— CH— CO₂⁻
            ⁝
            O δ⁻
            |
       O= P— O⁻
            |
            O⁻
```

```
     CH₃    O                      R
 +    |     ‖                      |
NH₃—— CH—— P —— CH₂—— CH— CO₂⁻
            |
            O⁻
```

Figure 1. Proposed transition state for the reaction of D-alanyl phosphate with a second molecule of D-alanine. A generic formula of the phosphinic acids reported in this study is also shown.

functionality of the glutamine synthetase inhibitor phosphinothricin whose CH_3 presumably is an isosteric replacement for NH_3, whereas $-CH_2-CHRCO_2H$ is considered to be an alanine replacement. Investigators at the Merck laboratories (W. H. Parsons, A. A. Patchett, J. Davidson, W. Schoen, P. Combs, D. Taub, J. P. Springer, T. Mellin, R. Busch, N. Thornberry, H. Gadebusch, B. Weissberger, and M. Valiant, *Program Abstr. Am. Chem. Soc. 193rd Natl. Meet.*, MEDI 63) thus set out to synthesize phosphinic acids of the general formula

Table 2. Ligase inhibitory potencies of phosphinic acids related to D-alanine and D-alanyl-D-alanine

Compound		IC_{50} (μM)[a]
CH₃ O | ‖ NH₂—CH — P —H | DL– OH		1,000
CH₃ O CH₃ | ‖ | NH₂—CH — P —CH₂—CH — CO₂H | OH		
L–	DL–	>1,000
D–	DL–	115
D–	D–	35
CH₃ O CH₃ | ‖ | NH₂—CH — C —NH—CH — CO₂H D– D–		1,200
Cycloserine		250

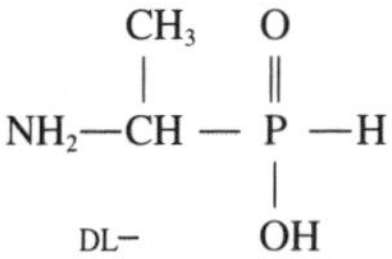

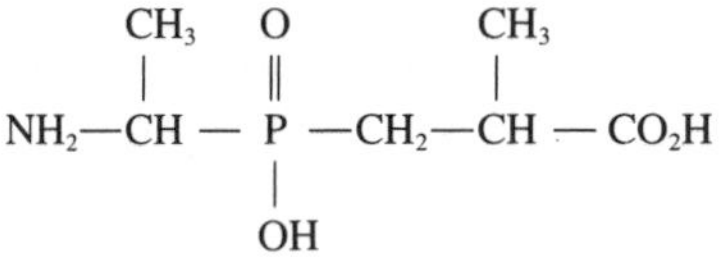

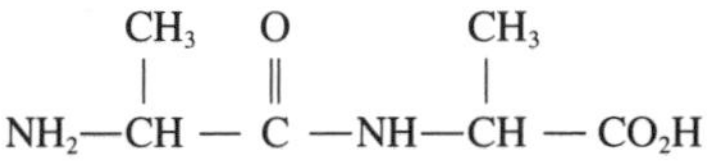

[a]Determined by the discontinuous assay described in the text.

Table 3. Ligase-inhibitory potencies of phosphinic acids bearing selected R_2 substituents

$R_2{}^a$	IC$_{50}$ (μM)^b
H–	535
CH$_3$–	155
CH$_3$S–	4
C$_2$H$_5$– (DL–; DL–)	45
C$_7$H$_{15}$–	4
1-Napthylethyl–	5
D-Cycloserine	250

^{a}Variable in the compound:

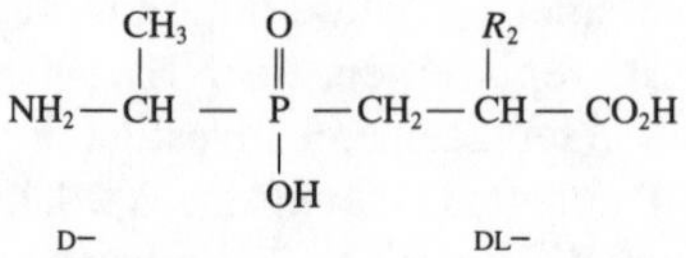

$$NH_2-\underset{\underset{D-}{|}}{CH} - \underset{\underset{OH}{|}}{\overset{\overset{CH_3}{|}}{\underset{\overset{\|}{O}}{P}}} -CH_2-\underset{\underset{DL-}{|}}{\overset{\overset{R_2}{|}}{CH}} - CO_2H$$

^{b}Determined by the discontinuous assay described in the text.

shown in Fig. 1. The most active phosphinic acids directly related to D-alanyl-D-alanine are also of DD absolute stereochemistry (Table 2). They are considerably more potent than D-alanyl-D-alanine, which is a product inhibitor (9), and both of the D-alanine binding sites have to be occupied for good inhibition. It was known from the work of Neuhaus (7) that D-alanyl-D-alanine ligase shows little specifity in its S_1' (acceptor) site, and indeed (Table 3), quite large R_2 groups confer excellent inhibitory activities. In the discontinuous assay the best of these inhibitors appear to be about two orders of magnitude more potent than cycloserine when correction is made for DL stereochemistry at R_2. This assay was used to provide a ranking of inhibitor potencies but was incapable of providing mechanistic insights. At his request we sent C. Walsh a sample

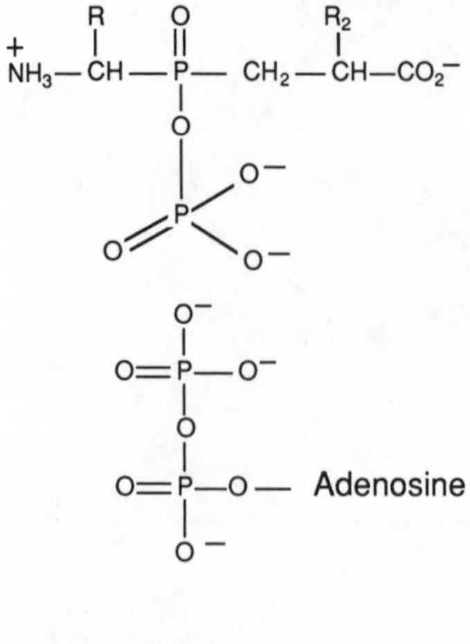

Figure 2. Proposed structure of tightly bound phosphorylated inhibitor in association with ADP in the active site of D-alanyl-D-alanine ligase.

Table 4. Activities of phosphinic acids bearing selected R_1 and R_2 substituents against a spectrum of gram-positive and gram-negative organisms

Organism[a]	MIC[b] (μg/ml) with:					
	R_1 DL-CH$_3$ R_2 DL-CH$_3$	R_1 D-CH$_3$ R_2 D-CH$_3$	R_1 D-CH$_3$ R_2 DL-C$_2$H$_5$	R_1 D-CH$_3$ R_2 SCH$_3$	R_1 D-CH$_3$ R_2 DL-C$_7$H$_{15}$	Cycloserine
Staphylococcus aureus 2865	>256	>256	>256	>256	128	8
Streptococcus faecalis 2864	128	8	16	4	64	16
Escherichia coli Tem 2$^+$ 4351	128	128	64	16	128	2
E. coli DC2 4353	128	256	32	64	32	8
E. coli 2891	>256	>256	128	128	128	4
Salmonella typhimurium 3860	>256	>256	256	128	128	8
Enterobacter cloacae 2647	>256	>256	128	128	256	4
Klebsiella pneumoniae 4005	>256	128	256	128	64	16
Proteus vulgaris 2829	>256	64	128	4	128	32
Pseudomonas aeruginosa 4279	>256	128	32	64	>256	16
Serratia marcescens 2840	>256	>256	>256	>256	>256	>256

[a] Numbers indicate Merck culture collection number.
[b] Determined in agar dilution assays using antagonist-free medium. R_1, R_2, variables in the compound:

$$NH_2-\underset{\underset{R_1}{|}}{CH}-\underset{\underset{OH}{|}}{\overset{\overset{O}{\|}}{P}}-CH_2-\underset{\underset{R_2}{|}}{CH}-CO_2H$$

Table 5. Combination studies of a phosphinic acid inhibitor with fluoro-D-alanine

Test culture[a]	MIC[b] (μg/ml) of:			Concn (μg/ml) of the following that no longer inhibits	
	Fluoro-D-alanine	R_1 D-CH$_3$ R_2 D-CH$_3$	Combination[c]	Fluoro-D-alanine	Combination[c]
Staphylococcus aureus 2865	2.0	>256.0	2.0 + 2.0	8	32
Streptococcus faecalis 2864	8.0	8.0	8.0 + 8.0	>256	>256
Escherichia coli Tem 2⁺ 4351	4.0	128.0	4.0 + 4.0	128	>256
E. coli DC2 4353	8.0	256.0	4.0 + 4.0	32	>256
E. coli 2891	4.0	>256.0	8.0 + 8.0	64	>256
Samonella typhimurium 3860	8.0	>256.0	8.0 + 8.0	32	>256
Enterobacter cloacae 2647	2.0	>256.0	4.0 + 4.0	16	128
Klebsiella pneumoniae 4005	1.0	128.0	1.0 + 1.0	32	256
Proteus vulgaris 2829	16.0	64.0	4.0 + 4.0	128	>256
Pseudomonas aeruginosa 4279	>256.0	128.0	4.0 + 4.0		>256
Serratia marcesens 2840	8.0	>256.0	8.0 + 8.0	32	>256

[a]Numbers indicate Merck culture number.
[b]Determined by agar dilution assay using a multipoint inoculator with an inoculum of 10⁴ CFU per spot on antagonist-free medium (Roche) and incubation at 35°C for 20 h.
[c]A 1:1 combination of fluoro-D-alanine and the phosphinic acid inhibitor.

of the phosphinic acid in which R_1 = D-CH$_3$ and R_2 = DL-heptyl for kinetic studies with pure *Salmonella* D-alanyl-D-alanine ligase. Much to our delight he has told us that this compound is a time-dependent, irreversible inhibitor. This result apparently reinforces an analogy with phosphinothricin, whose mechanism of inhibition of glutamine synthetase has recently been similarly described, including the presentation of evidence supportive of inhibitor phosphorylation in the formation of the tight binding complex (3). H. Bull of the Merck laboratories showed that when crude *S. faecalis* ligase was incubated with the heptyl inhibitor at 10^{-4} M followed by dialysis overnight, greater than 95% of the enzyme activity was recoverable, whereas if the initial incubation mixture contained also ATP only about 50% of the enzyme activity was assayable after an overnight dialysis. These results suggest that phosphorylation is involved in making a tight binding inhibitor species whose partial dissociation from the *S. faecalis* enzyme may be observable during an overnight period. From these preliminary findings and from the results of Walsh's group with the *Salmonella* ligase, we are led to hypothesize that the tight binding inhibitor complex formed in the reaction is phosphorylated as depicted in Fig. 2. We believe that its dissociation from the enzyme will be accompanied by rapid hydrolysis of the mixed anhydride without phosphorylation of the enzyme protein.

The antibacterial results shown in Table 4 include activities against both gram-positive and gram-negative organisms, but potencies are disappointing despite what we presume will be shown to be tight binding properties with all of these inhibitors. The antibacterial data were obtained by H. Gadebusch, M. Valiant, and B. Weissberger of the Merck laboratories using a synthetic medium suitable for D-alanine antagonists (1, 2). They also determined MICs for 1:1 combinations with fluoro-D-alanine (Table 5). Activity against a broad spectrum of bacteria including *Pseudomonas aeruginosa* is obtained in combinations of each component at 8 μg or less per ml. In most cases, however, the phosphinic acid adds little to the antibacterial activity of fluoro-D-alanine. The latter's self-reversal at higher concentration, however, is inhibited by the phosphinic acid, as would be expected of a ligase inhibitor. We thus believe that the modest antibacterial activities of these phosphinic acids are due to their potent ligase inhibition. Presumably they would be quite good antibacterial agents if their uptake by bacteria could be improved.

LITERATURE CITED

1. **Allen, J. G., F. R. Atherton, M. J. Hall, C. H. Hassall, S. W. Holmes, R. W. Lambert, L. J. Nisbet, and P. S. Ringrose.** 1979. Phosphonopeptides as antibacterial agents: alafosfalin and related phosphonopeptides. *Antimicrob. Agents Chemother.* **15:**684–695.
2. **Atherton, F. R., M. J. Hall, C. H. Hassall, R. W. Lambert, and P. S. Ringrose.** 1979. Phosphonopeptides as antibacterial agents: rationale, chemistry, and structure-activity relationships. *Antimicrob. Agents Chemother.* **15:**677–683.

3. **Colanduoni, J. A., and J. J. Villafranca.** 1986. Inhibition of *Escherichia coli* glutamine synthetase by phosphinothricin. *Biorg. Chem.* **14**:163–169.
4. **Ito, E., and J. L. Strominger.** 1960. Enzymatic synthesis of the peptide in a uridine nucleotide from *Staphylococcus aureus*. *J. Biol. Chem.* **235**:PC5-7.
5. **Ito, E., and J. L. Strominger.** 1962. Enzymatic synthesis of the peptide in bacterial uridine nucleotides. *J. Biol. Chem.* **237**:2696–2703.
6. **Neuhaus, F. C.** 1960. Enzymic synthesis of D-alanyl-D-alanine. *Biochem. Biophys. Res. Commun.* **3**:401–405.
7. **Neuhaus, F. C.** 1962. The enzymatic synthesis of D-alanyl-D-alanine. I. Purification and properties of D-analyl-D-alanine synthetase. *J. Biol. Chem.* **237**:778-786.
8. **Neuhaus, F. C.** 1962. The enzymatic synthesis of D-alanyl-D-alanine. II. Kinetic studies on D-alanyl-D-alanine synthetase. *J. Biol. Chem.* **237**:3128–3135.
9. **Neuhaus, F. C., C. V. Carpenter, J. L. Miller, N. M. Lee, M. Gragg, and R. A. Stickgold.** 1969. Enzymatic synthesis of D-alanyl-D-alanine. Control of D-alanine:D-alanine ligase (ADP). *Biochemistry* **8**:5119–5124.
10. **Neuhaus, F. C., and J. L. Lynch.** 1964. The enzymatic synthesis of D-alanyl-D-alanine. III. On the inhibition of D-alanyl-D-alanine synthetase by the antibiotic D-cycloserine. *Biochemistry* **3**:471–480.
11. **Robinson, A. C., D. J. Kenan, J. Sweeney, and W. D. Donachie.** 1986. Further evidence for overlapping transcriptional units in an *Escherichia coli* envelope-cell division gene cluster: DNA sequence and transcriptional organization of the *ddl ftsQ* region. *J. Bacteriol.* **167**:809–817.
12. **Wise, R., and J. M. Andrews.** 1984. In vitro activity of fludalanine combined with pentizidone compared with those of other agents. *Antimicrob. Agents Chemother.* **25**:612–617.

Chapter 56

Moenomycin: Inhibitor of Peptidoglycan Polymerization in *Escherichia coli*

Jean van Heijenoort
Yveline van Heijenoort
Peter Welzel

Antibiotics of the moenomycin group are phosphoglycolipid compounds produced by various species of *Streptomyces* (2). They are mostly obtained as complexes of very similar components. Moenomycin A is the main constituent of the product isolated from *S. bambergiensis* (2). On the basis of degradation and spectroscopic studies, its structure (Fig. 1) was determined (18, 19). A very similar structure was reported for the related antibiotic pholipomycin (12). Moenomycin is used only as a growth promoter in animal nutrition (flavomycin). Moreover, not being absorbed, it is limited to the intestinal tract, and, not being related to antibiotics used in human medicine, there is no cross resistance with such antibiotics (2). In this chapter, we describe work carried out on the cellular effects of moenomycin on *Escherichia coli,* on its mechanism of action, and on a structure-activity relationship (15, 17).

General Effects of Moenomycin on E. coli

Moenomycin can very efficiently promote autolysis of *E. coli* K-12 (5, 15). This bacteriolytic effect is greatly dependent on drug concentration and on growth rate. With fast-growing cells a maximal rate of autolysis was ob-

Jean van Heijenoort and Yveline van Heijenoort • Biochimie Moléculaire et Cellulaire, Université Paris-Sud, Orsay, France. **Peter Welzel** • Fakultät für Chemie der Ruhr-Universität, Bochum, Federal Republic of Germany.

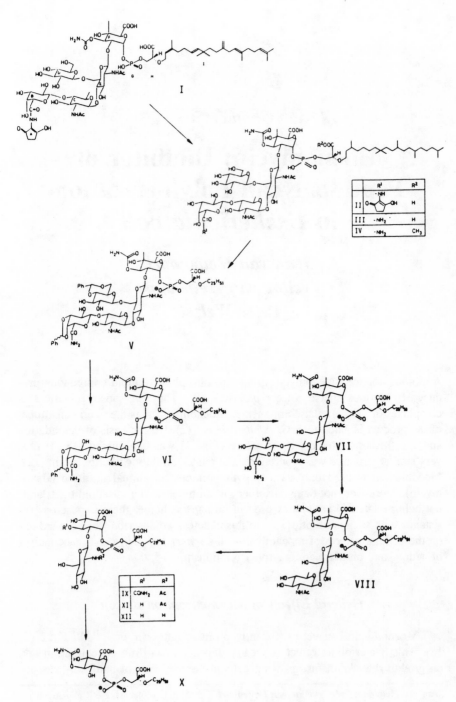

Figure 1. Degradation of moenomycin A.

served at 10 µg ml^{-1}. Mg^{2+} ions have been shown to inhibit autolysis in *E. coli* (6). Their addition at various times after treatment with moenomycin had an inhibitory effect on the rate of autolysis but none on the loss of viability. With cells growing in glucose minimal medium, the highest rate of autolysis was observed at 50 µg ml^{-1}. The effect of moenomycin was dependent on the time of its addition, early-exponential-phase cells being the most sensitive. Although a slow rate was still detectable with early-stationary-phase cells, no lysis could be promoted in overnight cultures. This dependence on growth phase and rate was consistent with the results described for other types of autolysis of *E. coli* (6). With fast-growing cells both rods and round cells were observed 15 min after addition of 10 µg of antibiotic ml^{-1}. After 30 min the remaining cells were mostly round, and after 45 min only ghosts and cell debris were observed. With cells growing in glucose minimal medium small normal rods were observed 60 min after addition of 25 µg of moenomycin ml^{-1}. After 90 min most rods were converted to swollen, irregularly shaped cells. When early-exponential-phase cells suspended in fresh medium containing sucrose were treated with moenomycin, they were readily converted to osmotically sensitive spheroplasts.

The MICs of moenomycin against several *E. coli* strains were determined in different media (2, 15). Values ranged from 4 to 512 µg ml^{-1}. Surprisingly, the highest values were observed in Penassay medium. This could perhaps be accounted for by association of the drug with components of the medium. Spontaneous moenomycin-resistant mutants of *E. coli* C600 were isolated by treating fast-growing, early-exponential-phase cells twice with moenomycin. Between the two treatments a period of regrowth without the antibiotic was allowed. After plating on agar containing moenomycin, 1,200 colonies were obtained. Owing to the duration of the regrowth period between moenomycin treatments, most of these colonies undoubtedly originated from a very limited number of individual variants. Assuming a fixed generation time for all cells, a frequency of mutation of 10^{-9} was estimated. Differences in the susceptibility to moenomycin in liquid medium were observed. Moenomycin led to very slow rates of autolysis with one-third of the colonies, to rates equivalent of that of the parental strain with another third, and to rates of intermediate value with the remaining third.

Interference with Peptidoglycan Synthesis

The interference of moenomycin with peptidoglycan synthesis has been studied by three approaches. With sublethal concentrations the accumulation of UDP-*N*-acetylmuramyl-pentapeptide was observed in *Staphylococcus aureus* (2) and in *E. coli* (Mengin-Lecreulx and van Heijenoort, in press). With cell-free systems which catalyze the in vitro synthesis of peptidoglycan from UDP-

N-acetylglucosamine and UDP-*N*-acetylmuramyl-pentapeptide, moenomycin was inhibitory and led to the accumulation of the lipid intermediate (7, 8). With a particulate fraction from *E. coli* and the lipid intermediate, an in vitro assay specific for the polymerization by transglycosylation was developed (14, 16). More specifically, penicillin-binding protein 1b (PBP 1b) is the polymerase responsible for this reaction (11). As a purified protein, with the lipid intermediate it can catalyze in vitro a transglycosylation reaction inhibited by moenomycin (10, 11). With both the cell-free systems and purified PBP 1b, moenomycin was inhibitory at concentrations between 10^{-8} and 10^{-7} M (10, 11, 16). The in vitro polymerization catalyzed by two other peptidoglycan polymerases of *E. coli,* PBP 1a and PBP 3, is also sensitive to moenomycin, or at least to a related antibiotic (3, 9).

Binding Assays

Decahydromoenomycin A (Table 1) was found to inhibit the in vitro polymerization by transglycosylation as efficiently as moenomycin. Therefore, [^{3}H]decahydromoenomycin A (6 Ci/mmol) was prepared and used in different binding assays.

A binding assay was developed with the particulate fraction from *E. coli* K-12 HfrH. When the concentration of free antibiotic in the assay was varied from 10^{-7} to 10^{-5} M, the amount of bound radioactive material increased steadily, and no well-defined plateau was observed. By adding unlabeled antibiotic (final concentration, 5×10^{-4} M), the bound radioactivity was completely chased in all cases. This suggested that the radioactive material may bind reversibly to the peptidoglycan polymerases present in the particulate fraction. Since the amount of antibiotic effectively bound exceeded the amounts of polymerases PBP 1a, PBP 1b, and PBP 3 present, presumably part of the bound [^{3}H]decahydromoenomycin was nonspecifically incorporated into membrane material in the same way as an amphipathic phospholipid would be, thus masking the specific binding to peptidoglycan polymerases.

The binding of [^{3}H]penicillin and [^{3}H]decahydromoenomycin to the membrane proteins of *E. coli* JA200(pLC19-19) was compared. With the [^{3}H]penicillin-treated samples, the well-known PBP pattern was observed (1), except that owing to the presence of plasmid pLC19-19 the overproduction of PBP 1b was quite conspicuous (13). With the [^{3}H]decahydromoenomycin-treated samples, examination of the autoradiographs of the electrophoretic gels revealed no radioactive protein band. However, in these samples a wide low-molecular-mass band corresponding to the radioactive antibiotic itself was clearly detectable. The same result was obtained when the experiment was carried out with purified PBP 1b isolated from membranes of *E. coli* JA200(pLC19-19).

Table 1. Effect of flavomycin, moenomycin A, and degradation products on the in vitro formation of peptidoglycan by transglycosylation[a]

Additive	Final concn (μg/ml)	Inhibition
Drug		
Flavomycin	10	100
	1	100
	0.1	23
Moenomycin A	10	100
	1	100
	0.1	54
Product		
II	10	100
	1	100
	0.1	0
III	10	100
	1	100
	0.1	8
IV	100	100
	10	64
	1	0
VII	10	100
	1	100
	0.1	75
VIII	10	100
	1	100
	0.1	30
IX	10	100
	1	100
	0.1	45
X	100	100
	10	100
	1	53
XI	100	100
	10	56
	1	0
XII	100	16
	10	0
	1	0

[a]Assays were carried out as described previously (14).

Minimal Structural Requirements for Inhibition

With the aim of defining the structural basis for the inhibitory effect of moenomycin A, its systematic chemical degradation was undertaken (Fig. 1). To avoid complications by the acid-sensitive allyl ether bond between units G and F, moenomycin A (I) was hydrogenated to give decahydromoenomycin (II). Next, unit A was removed by oxidation with $K_3Fe(CN)_6$ in K_2CO_3 solution to give compound III, which was further converted to compound V by the Barry degradation. The proposed structure for V rests on ^{13}C nuclear magnetic resonance and fast atom bombardment mass spectrometry results. Reaction of V with $NaIO_4$ in 50% acetic acid followed by purification and reaction with *N,N*-dimethylhydrazine led to VI. Liberation of another diol grouping was achieved by removing the benzylidene group from VI by hydrogenolysis to give VII. Degradation under the conditions described above converted VII to VIII, which was submitted to another degradation cycle to give IX. Under the same conditions IX was degraded to X. The urethane group was removed from IX by treatment with butylamine in methanol to give XI. Reaction with hydrazine hydrate converted IX into XII. The structural assignments of the moenomycin A degradation products rested mainly on ^{13}C nuclear magnetic resonance spectra and fast atom bombardment mass spectra. The various degradation products were assayed for their effect on the in vitro transglycosylation reaction (Table 1). The results enabled us to define the structural features in moenomycin A that are required for full in vitro biological activity: (i) units E, F, G, H, and I are essential; (ii) the moenocinol part I can be saturated; (iii) the carboxyl group in unit H must be free; and (iv) the carbamoyl group in the moenuronic acid moiety F must be present.

Conclusion

The effect of moenomycin on *E. coli* was followed by measuring the optical density of liquid culture, by determining viable counts, by microscopy, and by determining MIC values. Although these various approaches were not easily comparable, a certain consistency in the results was observed. In fast-growing cells the sharp decrease in optical density correlated well with the rapid loss of viability, the disappearance of whole cells, and the rapid formation of spheroplasts in sucrose-containing medium. The alterations in cell morphology, the formation of spheroplasts, and the proneness to cell lysis that were observed with moenomycin were in many respects similar to those described for other types of antibiotics that interfere with the biosynthesis of peptidoglycan in *E. coli* (1, 6). In particular, cephaloridine, which preferentially binds to PBP 1a and PBP 1b at low concentrations. also promotes fast cell lysis (1), in a manner similar to moenomycin (5). Both PBP 1a and PBP 1b appear to be bifunctional enzymes since they also catalyze an in vitro transpeptidation reaction sensitive to β-lactams (9, 10). In PBP 1b the transglycosylation and transpeptidation activities are on two separate domains (4, 10). Presumably, the target site for moenomycin is on the domain carrying the transglycosylation activity. The correlation between peptidoglycan metabolism and autolysis in *E. coli* has been discussed previously (6). Under conditions of normal growth, the polymerization reactions catalyzed by PBP 1a and PBP 1b are undoubtedly closely coupled in a well-balanced manner with the activity of specific endogenous peptidoglycan hydrolases. When the polymerization reactions are abruptly arrested by inhibitors such as moenomycin or cephaloridine, certain of these hydrolases are apparently not subject to any rapid control, and they continue unabated to catalyze the cleavage of peptidoglycan linkages, which will lead to autolysis. From the study of the structure-activity relationship it can be concluded that the inhibition of the transglycosylation reaction by moenomycin A and the active degradation products is based on the structural analogy between these and the membrane substrate. The structural analogy is especially apparent if compound IX and the membrane substrate are compared in the conformations

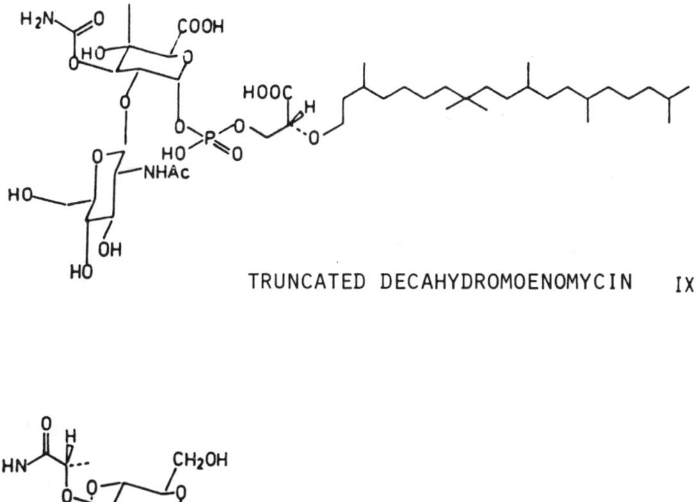

TRUNCATED DECAHYDROMOENOMYCIN   IX

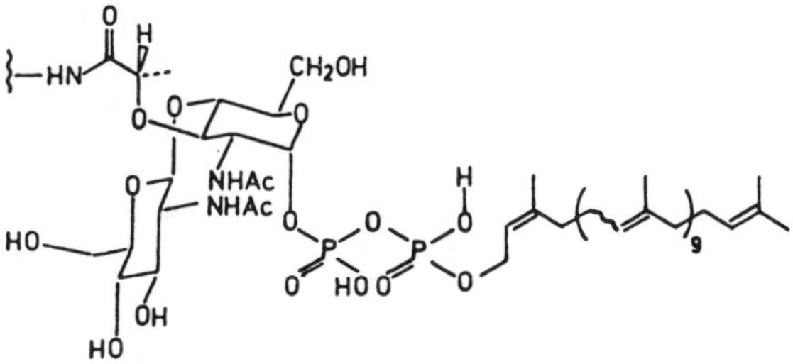

PEPTIDOGLYCAN PRECURSOR

Figure 2. Comparison of compound IX and the membrane substrate of peptidoglycan polymerization.

depicted in Fig. 2. This result, together with the reversible binding, strongly suggests that moenomycin is a competitive inhibitor of polymerase PBP 1b. This is in agreement with a previous suggestion (7).

LITERATURE CITED

1. **Gale, E. F., E. Cundliffe, P. E. Reynolds, M. H. Richmond, and M. J. Waring.** 1980. *The Molecular Basis of Antibiotic Action.* John Wiley & Sons, Inc., New York.
2. **Huber, G.** 1979. Moenomycin and related phosphorus containing antibiotics, p. 135–153. *In* F. E. Hahn (ed.), *Antibiotic V/I. Mechanism of Action of Antibacterial Agents.* Springer-Verlag, Berlin.
3. **Ishino, F., and M. Matsuhashi.** 1981. Peptidoglycan synthetic enzyme activities of highly purified penicillin-binding protein 3 in *Escherichia coli*: a septum-forming reaction sequence. *Biochem. Biophys. Res. Commun.* **101:**905–911.
4. **Kato, J., H. Suzuki, and Y. Hirota.** 1984. Overlapping of the coding regions for α and γ components of penicillin-binding protein lb in *Escherichia coli. Mol. Gen. Genet.* **196:**449–457.
5. **Leduc, M., C. Fréhel, and J. van Heijenoort.** 1985. Correlation between degradation and ultrastructure of peptidoglycan during autolysis of *Escherichia coli. J. Bacteriol.* **161:**627–635.
6. **Leduc, M., R. Kasra, and J. van Heijenoort.** 1982. Induction and control of the autolytic system of *Escherichia coli. J. Bacteriol.* **152:**26–34.
7. **Linnett, P. E., and J. L. Strominger.** 1973. Additional antibiotic inhibitors of peptidoglycan synthesis. *Antimicrob. Agents Chemother.* **4:**231–236.
8. **Lugtenberg, E. J. J., A. van Schijndel-van Dam, and T. H. M. van Bellegem.** 1971. In vivo and in vitro action of new antibiotics interfering with the utilization of *N*-acetyl-glucosamine-*N*-acetyl-muramyl-pentapeptide. *J. Bacteriol.* **108:**20–29.
9. **Matsuhashi, M., F. Ishino, J. Nakagawa, K. Mitsui, S. Nakajima-Iijima, and S. Tamaki.** 1981. Enzymatic activities of penicillin-binding proteins of *Escherichia coli* and their sensitivities to β-lactam antibiotics, p. 169–184. *In* M. Salton and G. D. Shockman (ed.), *β-Lactam Antibiotics.* Academic Press, Inc., New York.
10. **Nakagawa, J., S. Tamaki, S. Tomioka, and M. Matsuhashi.** 1984. Functional biosynthesis of cell wall peptidoglycan by polymorphic bifunctional polypeptides. Penicillin-binding protein 1Bs of *Escherichia coli* with activities of transglycosylase and transpeptidase. *J. Biol. Chem.* **259:**13937–13946.
11. **Suzuki, H., Y. van Heijenoort, T. Tamura, J. Mizoguchi, Y. Hirota, and J. van Heijenoort.** 1980. *In vitro* peptidoglycan polymerization catalysed by penicillin-binding protein lb of *Escherichia coli* K 12. *FEBS Lett.* **110:**245–249.
12. **Takahashi, S., K. Serita, and M. Aral.** 1983. Structure of pholipomycin. *Tetrahedron Lett.* **24:**499–502.
13. **Tamura, T., H. Suzuki, Y. Nishimura, J. Mizoguchi, and Y. Hirota.** 1980. On the process of cell division in *Escherichia coli*: isolation and characterization of penicillin-binding protein 1a, 1b and 3. *Proc. Natl. Acad. Sci. USA* **77:**4499–4503.
14. **van Heijenoort, Y., M. Derrien, and J. van Heijenoort.** 1979. Polymerization by transglycosylation in the biosynthesis of peptidoglycan of *Escherichia coli* K12 and its inhibition by antibiotics. *FEBS Lett.* **89:**141–144.
15. **van Heijenoort, Y., M. Leduc, H. Singer, and J. van Heijenoort.** 1987. Effects of moenomycin on *Escherichia coli. J. Gen. Microbiol.* **133:**667–674.
16. **van Heijenoort, Y., and J. van Heijenoort.** 1980. Biosynthesis of the peptidoglycan of

Escherichia coli K12. Properties of the *in vitro* polymerization by transglycosylation. *FEBS Lett.* **110**:241–244.

17. **Welzel, P., F. Kunisch, F. Kruggel, H. Stein, J. Scherkenbeck, A. Hiltmann, H. Duddeck, D. Müller, J. E. Maggio, H. W. Fehlhaber, G. Seibert, Y. van Heijenoort, and J. van Heijenoort.** 1987. Moenomycin A: minimum structural requirements for biological activity. *Tetrahedron* **43**:585–598.

18. **Welzel, P., B. Wietfeld, F. Kunisch, T. Schubert, K. Hobert, H. Duddeck, D. Müller, G. Huber, J. E. Maggio, and D. H. Williams.** 1983. Moenomycin A: further structural studies and preparation of simple derivatives. *Tetrahedron* **39**:1583–1591.

19. **Welzel, P., F. J. Witteler, D. Müller, and W. Riemer.** 1981. Structure of the antibiotic moenomycin A. *Angew. Chem. Int. Ed. Engl.* **20**:121–123.

Chapter 57

Effects of LY146032 on Cell Wall Precursors in Gram-Positive Bacteria

N. E. Allen

J. N. Hobbs, Jr.

W. E. Alborn, Jr.

LY146032 is a new calcium-requiring cyclic lipopeptide antibiotic which is active against gram-positive bacteria. The action of this compound has been characterized as an inhibition of cell wall biosynthesis with simultaneous effects on membrane function (N. Allen, W. Alborn, Jr., J. Hobbs, Jr., and H. Percifield, *Program Abstr. 24th Intersci. Conf. Antimicrob. Agents Chemother.*, abstr. no. 1081, 1984). The studies reported here help to elucidate how LY146032 inhibits cell wall formation and how the membrane effects may relate to this action.

Several peptide-containing antibiotics (e.g., amphomycin, vancomycin, and bacitracin) interfere with the membrane-associated steps in cell wall assembly, i.e., the formation and polymerization of disaccharide-peptide subunits. The chemical structure of LY146032 (Fig. 1) suggested that it may behave similarly. However, inhibition of [^{14}C]alanine incorporation into peptidoglycan of *Staphylococcus aureus* as well as inhibition of incorporation of [^{14}C]diaminopimelic acid or *N*-acetyl-[^{14}C]glucosamine into *Bacillus megaterium* peptidoglycan by LY146032 did not cause detectable accumulation of nucleotide-linked sugar-peptide precursors (Fig. 2). This suggested that LY146032 may inhibit an "early" stage in cell wall formation, since antibiotics inhibiting the membrane-mediated reactions often cause precursor accumulation (4, 6).

N. E. Allen, J. N. Hobbs, Jr., and W. E. Alborn, Jr. • Lilly Research Laboratories, Indianapolis, Indiana 46285.

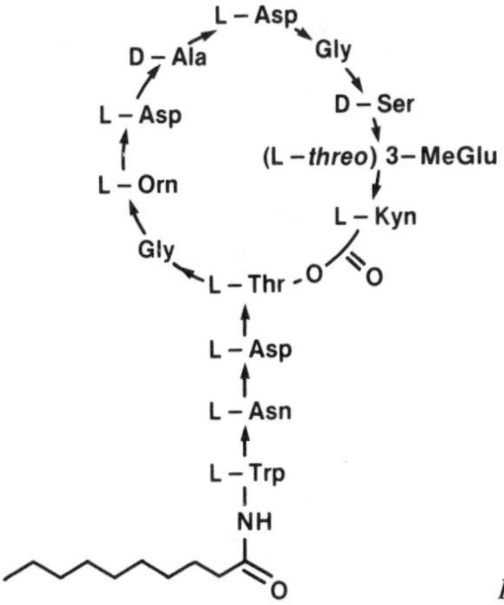

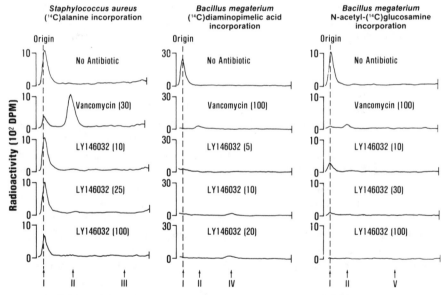

Figure 1. Structure of LY146032.

Figure 2. Effect of LY146032 on the accumulation of cell wall precursors: I, peptidoglycan; II, UDP-*N*-acetylmuramyl-pentapeptide; III, [^{14}C]alanine; IV, [^{14}C]diaminopimelic acid; V, *N*-acetyl-[^{14}C]glucosamine. Incorporation of [^{14}C]alanine, [^{14}C]diaminopimelic acid, and *N*-acetyl-[^{14}C]glucosamine was as described by Lugtenberg and DeHaan (5). The medium was supplemented with 1.25 mM CaCl$_2$. Reactions were run for 15 min (*S. aureus*) or 30 min (*B. megaterium*). Cultures were then centrifuged, and pellets were suspended in isobutyric acid–1 M ammonium hydroxide (5:3), spotted on Whatman 3MM paper, and chromatographed in the same solvent. Radioactivity was measured with a radiochromatogram scanner.

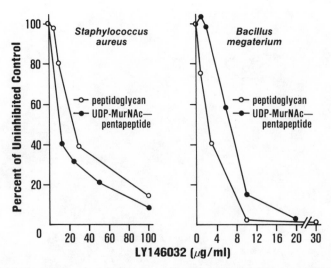

Figure 3. Inhibition of formation of UDP-*N*-acetylmuramyl-pentapeptide and peptidoglycan bio-synthesis by LY146032. The pentapeptide precursor was radioactively labeled with [¹⁴C]alanine (*S. aureus*) or [¹⁴C]diaminopimelic acid (*B. megaterium*). Accumulation of the precursor was facilitated by the addition of 100 μg of vancomycin per ml. Quantitation of precursor was as described in the legend to Fig. 2. Peptidoglycan biosynthesis was measured in separate reactions in the absence of added vancomycin. Peptidoglycan was detected as trichloroacetic acid-insoluble radioactivity (5).

Further investigation (Fig. 3) revealed that LY146032 had a potent inhib-itory effect on the biosynthesis of UDP-*N*-acetylmuramyl-pentapeptide in both *S. aureus* and *B. megaterium*. In this experiment, vancomycin (100 μg/ml) was used to induce accumulation of the radioactive pentapeptide precursor. LY146032 inhibited the formation of this precursor at concentrations essen-tially the same as those needed to inhibit the incorporation of radioactivity into peptidoglycan. In a separate experiment (data not shown), LY146032 was shown to inhibit the formation of the D-cycloserine-induced accumulation of UDP-*N*-acetylmuramyl-tripeptide. These findings strongly suggest that LY146032 in-hibits the formation of nucleotide-linked sugar-peptide precursors involved in peptidoglycan biosynthesis. Fosfomycin has been shown to inhibit this stage of cell wall assembly (3), but LY146032 had very little effect on the fosfo-mycin-sensitive phosphoenolpyruvate:UDP-*N*-acetylglucosamine enolpyruvyl-transferase from *B. megaterium* (data not shown).

The biosynthesis of peptidoglycan precursors occurs in the cytoplasm via soluble enzymes (2, 7). The effect of LY146032 on precursor formation implies that this antibiotic may require transit across the cytoplasmic membrane to reach its target site. Results from the experiment shown in Fig. 4 indicate that LY146032 causes leakage of intracellular potassium in *S. aureus* and *B. megaterium*. Moreover, this effect on membrane function occurred at the same concentra-

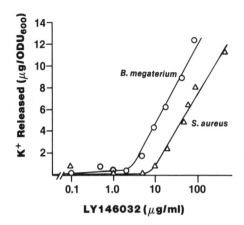

Figure 4. Effect of LY146032 on intracellular potassium in *S. aureus* and *B. megaterium*. Cells were suspended in a defined medium (1; prepared with sodium phosphate and supplemented with 1.25 mM $CaCl_2$) and exposed to various concentrations of LY146032 for 20 min. Potassium released into the medium was measured with an Orion ion-specific electrode.

tions required to inhibit peptidoglycan biosynthesis and precursor formation (compare Fig. 3 and 4). Although the membrane perturbation does not prove transit of LY146032, the effects are consistent with this notion. In spite of the potassium leakage, LY146032 had only a minor and transitory effect on the phosphotransferase carbohydrate transport system in *S. aureus* (data not shown). Compounds that specifically target the cytoplasmic membrane (e.g., gramicidin S and cetylpyridinium chloride) completely destroyed the functioning of this system. The data support the contention that LY146032 has a disruptive yet nonlethal effect on the cytoplasmic membrane.

Inhibition of peptidoglycan biosynthesis by LY146032 has been demonstrated in *S. aureus* and *B. megaterium* by [^{14}C]alanine, [^{14}C]diaminopimelic acid, and *N*-acetyl-[^{14}C]glucosamine incorporation experiments. The evidence favors a lethal target involved in the formation of peptidoglycan precursors. The observed membrane perturbation by LY146032 may result from transit of the antibiotic to its target site. It is tempting to speculate that the calcium requirement of this novel antibiotic may be related to its need to cross the formidable cytoplasmic membrane barrier.

LITERATURE CITED

1. **Allen, N. E.** 1977. Macrolide resistance in *Staphylococcus aureus*: induction of macrolide-resistant protein synthesis. *Antimicrob. Agents Chemother.* **11:**661–668.
2. **Gale, E. F., E. Cundliffe, P. E. Reynolds, M. H. Richmond, and M. J. Waring.** 1981. *The Molecular Basis of Antibiotic Action*, 2nd ed., p. 49–174. John Wiley & Sons, Inc., New York.
3. **Kahan, F. M., J. S. Kahan, P. J. Cassidy, and H. Kropp.** 1974. The mechanism of action of fosfomycin (phosphonomycin). *Ann. N.Y. Acad. Sci.* **235:**364–386.
4. **Linnett, P. E., and J. L. Strominger.** 1973. Additional antibiotic inhibitors of peptidoglycan synthesis. *Antimicrob. Agents Chemother.* **4:**231–236.

5. **Lugtenberg, E. J. J., and P. G. DeHaan.** 1971. A simple method for following the fate of alanine-containing components in murein synthesis in *Escherichia coli. Antonie van Leeuwenhoek J. Microbiol. Serol.* **37:**537–552.
6. **Lugtenberg, E. J. J., A. van Schijndel-Van Dam, and T. H. M. van Bellegen.** 1971. In vivo and in vitro action of new antibiotics interfering with the utilization of *N*-acetyl-glucosamine-*N*-acetyl-muramyl-pentapeptide. *J. Bacteriol.* **108:**20–29.
7. **Rogers, H. J., H. R. Perkins, and J. B. Ward.** 1980. *Microbial Cell Walls and Membranes*, p. 239–297. Chapman & Hall, Ltd., London.

Chapter 58

Vancomycin Prevents β-Lactam Antibiotics from Causing Soluble Peptidoglycan Secretion by *Staphylococcus aureus* Cells

Allen R. Zeiger

Study of the mode of action of antibiotics is essential for initiating new strategies to combat bacterial resistance. Among the most effective antibiotics, β-lactams and vancomycin act at the level of peptidoglycan biosynthesis. In vitro studies, however, indicate that they act differently. The β-lactam antibiotics, for example, inactivate enzymes important for cell wall formation (2, 11), whereas vancomycin binds to peptidoglycan precursors (1, 6). At least some precursors of peptidoglycan that are capable of reacting with vancomycin occur earlier in the biosynthetic pathway than the enzymes affected by β-lactam antibiotics. Nevertheless, the precursors to which vancomycin binds in vivo are not known (1, 3). Moreover, vancomycin has been shown to bind to cell walls (3), where the last steps in peptidoglycan synthesis occur.

The inhibition of a metabolic pathway often results in the buildup of product at the steps immediately preceding the site of blockage. One sequela of β-lactam treatment of some gram-positive bacteria, such as *Staphylococcus aureus*, is the secretion of a soluble, high-molecular-weight form of peptidoglycan (SPG) (4, 5, 7–10, 13). SPG is a transglycosylated but not transpeptidated product of nucleotide-containing peptidoglycan precursors. Vancomycin causes an accumulation of these precursors, but no SPG secretion occurs (5, 13). If vancomycin does in fact act in vivo at earlier biosynthetic steps than do the β-lactams, it should be able to prevent the secretion of SPG following β-lactam antibiotic treatment.

Allen R. Zeiger • Department of Biochemistry and Molecular Biology, Jefferson Medical College, Thomas Jefferson University, Philadelphia, Pennsylvania 19107.

SPG was prepared in a minimal cell wall growth medium containing 4 μCi of L-[^{14}C]alanine (0.36 μCi/μM) with *S. aureus* cells as described (13). The chromatography of the concentrated spent cell wall medium with Sephadex G-100 (13) and the enzyme-linked immunosorbent assay (ELISA) for SPG (12) have been reported. Briefly, polyvinyl wells were coated with 100 μg of vancomycin per ml in a carbonate buffer overnight at 4°C. The wells were washed with phosphate-buffered saline–0.05% Tween 20, and the spent minimal medium was added. After incubation for 1 h at 37°C, affinity-purified rabbit anti-Lys-D-Ala-D-Ala was added (5 μg/ml), and the mixture was incubated for another 1 h at 37°C. The wells were then washed, and goat anti-rabbit immunoglobulin that was covalently bound to alkaline phosphatase (Sigma Chemical Co., St. Louis, Mo.) was diluted according to the manufacturer's directions and added. After 2 h at 37°C, the substrate *p*-nitrophenylphosphate (Sigma) was added at 1 mg/ml in diethanolamine buffer, and the absorbance was read periodically at 405 nm.

SPG has usually been obtained from gram-positive bacteria incubated in a minimal cell wall growth medium containing a labeled SPG precursor, such as [^{14}C]alanine, and penicillin G (5, 10, 13). The amount of SPG produced is then determined by gel sieving chromatography of the spent medium. With Sephadex G-100, the SPG from a spent medium of *S. aureus* cells elutes at the void volume (peak 1). The amount of SPG has been previously shown to be relatively constant, from 10 to 250 μg of penicillin G per ml (13). Below 10 μg of penicillin G per ml, the amount of SPG decreases, until about 1 ng of penicillin G per ml, where there is no significant difference in peak 1 from that of untreated *S. aureus* cells (Table 1).

For experiments on the inhibition of SPG secretion by vancomycin, we chose a concentration of penicillin G that results in nonmaximal, but signifi-

Table 1. Secretion of SPG by *S. aureus* Rb as a function of penicillin G concentration[a]

Penicillin G (μg/ml)	A_{405} of 1:10 dilution	Peak 1 (μg/liter)[b]
10	0.796	1,615
1	0.708	1,210
0.1	0.527	730
0.01	0.358	450
0.001	0.064	58
0	0.075	28

[a]*S. aureus* Rb (2×10^8 CFU) was incubated in 20 ml of minimal cell wall growth medium containing [^{14}C]-alanine (0.357 mCi/mmol) and penicillin G for 60 min at 37°C. The spent medium was used in the ELISA at a 10-fold dilution (200 μl). Nonspecific hydrolysis of the substrate (0.083) was subtracted from the observed values.

[b]The contents of the high-molecular-weight peak (peak 1) from Sephadex G-100 chromatography of the spent minimal medium based on specific activity of 0.357 mCi/mmol and a molecular weight of SPG monomer per alanine of 411.

cant, SPG secretion, i.e., 100 ng/ml. Figure 1 shows that there was a peak at the void volume, although not as prominent as at higher penicillin G concentrations. The presence of 100 μg of vancomycin per ml resulted in an elution profile with virtually no peak 1 material that is indistinguishable from the elution profile of spent medium in the absence of either antibiotic. Thus, a high vancomycin concentration is able to prevent the penicillin-induced secretion of SPG and probably acts at an earlier stage in the biosynthetic pathway of peptidoglycan than does penicillin G.

Lower concentrations of vancomycin with 100 ng of penicillin G per ml give a more complex elution profile. At 1 μg of vancomycin per ml, some material elutes at the void volume. However, the peak maximum is shifted, indicating the presence of some less polymerized oligosaccharide-peptides. There also appears to be more material between peak 1 and the low-molecular-weight peak than is present with *S. aureus* treated only with penicillin G (Fig. 2). At 10 μg of vancomycin per ml, there is no more peak 1 material than with the cells treated with 100 μg/ml. Yet, as with the cells treated with 1 μg of vancomycin per ml, there does appear to be more material between peak 1 and the low-molecular-weight peak.

Interpretation of the results in Fig. 1 is complicated by the fact that vancomycin can combine with each peptide in the oligosaccharide-peptides and thereby appear to increase the size of the latter. Vancomycin alone migrates with the low-molecular-weight peak (Fig. 2) and would probably have to combine multivalently with the oligosaccharide-peptides to form a complex eluting at or close to peak 1. This multivalent combination certainly does not occur in the media from the cells treated with 100 or 10 μg of vancomycin per ml, since there is so little high-molecular-weight [^{14}C]alanine-containing material present in the medium. One does, however, have to consider this possibility with medium from cells treated with 1 μg of vancomycin per ml. On the basis of the reported equilibrium constant of 6×10^{-8} to 8×10^{-8} (6), the amount of vancomycin bound to peptide will not be close to maximal, and thus this possibility is unlikely.

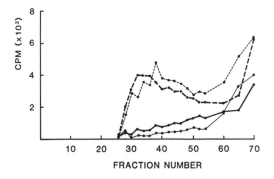

Figure 1. Fractionation through a Sephadex G-100 column of 5 ml of concentrated L-[^{14}C]alanine-labeled spent minimal cell wall medium (4 μCi/20 ml) from *S. aureus*, incubated with 100 ng of penicillin G per ml and various amounts of vancomycin. Elution was with distilled water. Elution profiles with vancomycin present at 100 (•——•), 10 (•———•), and 1 (•·····•) μg/ml and with no vancomycin (•----•).

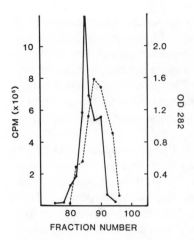

Figure 2. Low-molecular-weight peak from the elution of the Fig. 1 sample incubated with 1 μg of vancomycin per ml (•——•) and optical density (280 nm) profile of vancomycin (13.1 ng) applied to the same column (•-----•).

Nevertheless, to make certain that the material in the high-molecular-weight fractions of the medium treated with 1 μg of vancomycin per ml is multivalent, an ELISA requiring at least divalency of peptide was used. The steps in the ELISA utilize vancomycin as the coating antigen, the spent minimal medium, an affinity-purified rabbit antiserum that is specific for the Lys-D-Ala-D-Ala sequence present in SPG and many of its precursors, an antirabbit immunoglobulin linked to alkaline phosphatase, and *p*-nitrophenylphosphate. Monovalent peptides and vancomycin will both act as inhibitors of the ELISA. De-

Table 2. Effects of vancomycin on the secretion of SPG by *S. aureus* Rb in the presence of penicillin G[a]

Expt no.	Penicillin G (μg/ml)	Vancomycin (μg/ml)	A_{405} of 1:10 dilution
1	10	0	0.613
	10	0.01	0.579
	10	0.1	0.549
	10	1	0.487
	10	10	0.196
	10	100	0.011
	0	0	0.052
2	0.1	0	0.513
	0.1	0.01	0.446
	0.1	0.1	0.439
	0.1	1	0.449
	0.1	10	0.079
	0.1	100	0.042
	0	0	0.035

[a]Conditions as described for Table 1, except that the vancomycin was added 1 min before the penicillin G. The nonspecific lysis was 0.069 and 0.104 for experiments 1 and 2, respectively.

spite this, the absorption value from the ELISA of the spent minimal medium with 1 μg of vancomycin and 100 ng of penicillin G per ml was 0.449, which was about the same as for lower vancomycin concentrations and slightly lower than 0.513, which was the value obtained with spent medium having only 100 ng of penicillin G per ml (Table 2). These values were higher than for spent medium with neither antibiotic (0.035). The values for spent media with high vancomycin concentrations were close to the latter, but this could be due to inhibition of the ELISA by vancomycin in the media.

It seems apparent that high vancomycin concentrations were able to shut down the secretion of high-molecular-weight SPG by *S. aureus* cells. At lower concentrations, however, vancomycin may not have been able to bind completely to the earliest peptidoglycan precursors, and in the presence of penicillin G the size of secreted fragments appeared to increase. It thus may be that the concentration of vancomycin dictates the steps in the biosynthetic pathway to peptidoglycan at which synthesis ceases.

LITERATURE CITED

1. **Barna, J. C. J., and D. H. Williams.** 1984. The structure and mode of action of glycopeptide antibiotics of the vancomycin group. *Annu. Rev. Microbiol.* **38:**339–357.
2. **Blumberg, P. M., and J. L. Strominger.** 1974. Interaction of penicillin with the bacterial cell: penicillin-binding proteins and penicillin-sensitive enzymes. *Bacteriol. Rev.* **38:**291–335.
3. **Gale, E. F., E. Cundliffe, P. E. Reynolds, M. H. Richmond, and M. J. Waring.** 1981. Inhibitors of bacterial and fungal cell wall synthesis. *The Molecular Basis of Antibiotic Action*, 2nd ed., p. 49–174. John Wiley & Sons, Inc., New York.
4. **Keglevic, D., B. Ladesic, O. Hadzija, J. Tomasic, A. Valinger, and M. Pokorny.** 1974. Isolation and study of the composition of a peptidoglycan complex excreted by the biotin-requiring mutant of *Brevibacterium divaricatum* NRRL-2311 in the presence of penicillin. *Eur. J. Biochem.* **42:**389–400.
5. **Mirelman, D., R. Bracha, and N. Sharon.** 1974. Penicillin-induced secretion of a soluble, uncross-linked peptidoglycan by *Micrococcus luteus* cells. *Biochemistry* **13:**5045–5053.
6. **Perkins, H. R.** 1969. Specificity of combination between mucopeptide precursors and vancomycin or ristocetin. *Biochem. J.* **111:**195–205.
7. **Rosenthal, R. S., D. Jungkind, L. Daneo-Moore, and G. D. Shockman.** 1975. Evidence for the synthesis of soluble peptidoglycan fragments by protoplasts of Streptococcus faecalis. *J. Bacteriol.* **125:**398–409.
8. **Seidl, P. H., and K. H. Schleifer.** 1985. Secretion of fragments from bacterial cell wall peptidoglycan, p. 443–450. *In* I. S. Kulaev, E. A. Dawes, and D. W. Tempest (ed.), *Environmental Regulation of Microbial Metabolism*. Academic Press, Inc., London.
9. **Tynecka, A., and J. B. Ward.** 1975. Peptidoglycan synthesis in *Bacillus licheniformis*: the inhibition of cross-linking by benzylpenicillin and cephaloridine *in vivo* accompanied by the formation of soluble peptidoglycan. *Biochem. J.* **146:**253–267.
10. **Waxman, D. J., W. Yu, and J. L. Strominger.** 1980. Linear, uncross-linked peptidoglycan secreted by penicillin-treated *Bacillus subtilis*. *J. Biol. Chem.* **255:**11577–11587.
11. **Yocum, R. R., D. J. Waxman, and J. L. Strominger.** 1980. Interaction of penicillin with its receptors in bacterial membranes. *Trends Biochem. Sci.* **4:**97–101.

12. **Zeiger, A. R.** 1986. Antibodies against a synthetic peptidoglycan-precursor pentapeptide containing diaminopimelic acid, p. 95–98. *In* P. H. Seidl and K. H. Schleifer (ed.), *Biological Properties of Peptidoglycan*. Walter deGruyter & Co., Berlin.
13. **Zeiger, A. R., W. Wong, A. N. Chatterjee, F. E. Young, and C. U. Tuazon.** 1982. Evidence for the secretion of soluble peptidoglycans by clinical isolates of *Staphylococcus aureus. Infect. Immun.* **37:**1112–1118.

Chapter 59

Bicyclic Pyrazolidinones: A New Class of γ-Lactam Compounds Inhibiting Cell Wall Biosynthesis

N. E. Allen

J. N. Hobbs, Jr.

E. Wu

The β-lactam antibiotics comprise a diverse array of chemical structure types (Fig. 1). The penicillins, penems, carbapenems, cephalosporins, and oxacephems all contain a classic β-lactam ring which is fused to a secondary ring structure. The clinically useful β-lactam antibiotics contain different substitutions on the β-lactam ring and demonstrate a wide diversity in the structure of the secondary ring. In spite of major variations in the secondary ring system, all of these compounds share the ability to interfere with cell wall formation by acylating certain proteins associated with the bacterial cytoplasmic membrane (8, 9, 11). The most recently discovered β-lactam antibiotics (the monobactams) have a monocyclic structure without any fused secondary ring, yet they too bind to the same membrane proteins and inhibit cell wall assembly (1). Apparently, considerable diversity in the non-β-lactam ring portion of these molecules can be tolerated without causing major changes in target specificity.

Although the acylating activity of β-lactam antibiotics is fundamental to the antibacterial properties of these compounds, the effects of β-lactams on bacteria are manifested in different ways. The antibiotics bind to penicillin-binding proteins (PBPs) and inhibit transpeptidase, D-alanine carboxypeptidase, and endopeptidase reactions in vitro (11). In intact cells, these effects can lead to reduced numbers of peptide cross-linkages in newly synthesized peptido-

N. E. Allen, J. N. Hobbs, Jr., and E. Wu • Lilly Research Laboratories, Indianapolis, Indiana 46285.

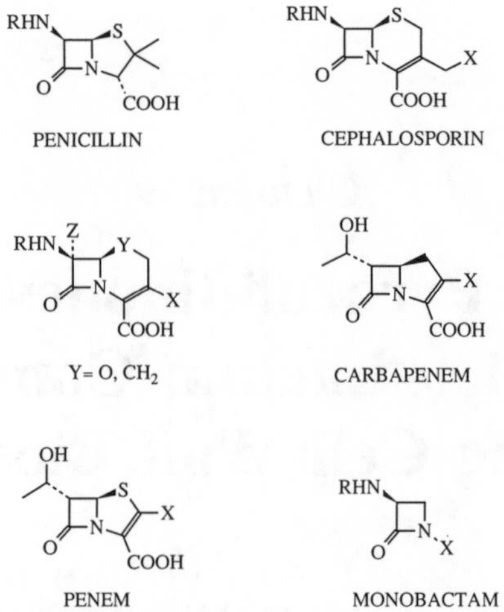

Figure 1. Structures of typical β-lactam antibiotics.

glycan (6) and ultimately to cell lysis as a result of the action of endogenous autolytic enzymes (10).

The magnitude of these effects is dependent on the specific antibiotic tested, but generally speaking, the effects are common to all of the β-lactams. How much chemical diversity can be introduced into these molecules without having major deleterious effects on biological specificity? In particular, how critical is the β-lactam ring to the activity of these compounds? This latter question has been asked by many investigators but until recently has remained largely unanswered.

Recent reports from the Lilly Research Laboratories (2, 3; L. N. Jungheim and S. K. Sigmund, submitted for publication) have described the synthesis of bicyclic pyrazolidinone compounds containing a novel γ-lactam ring in place of a β-lactam ring and having broad-spectrum antibacterial activity. LY186826 (Fig. 2) is a bicyclic pyrazolidinone containing a five-membered aza-γ-lactam ring substituted with a classical acylamino side chain at the C-7 position. This molecule has antibacterial activity against a variety of gram-positive and gram-negative bacteria (Table 1). We have investigated the mechanism of action of this compound against *Providencia rettgeri* C24.

LY186826 inhibited the incorporation of [^{14}C]diaminopimelic acid into acid-

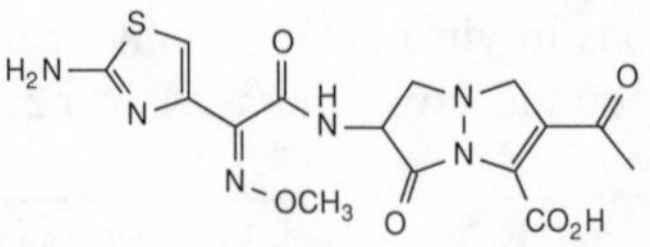

Figure 2. Structure of LY186826.

Table 1. Agar dilution MICs[a]

Strain	MIC (μg/ml)
Staphylococcus aureus X1.1	32
S. aureus V41	64
S. aureus X400	>128
S. epidermidis 222	32
Streptococcus pyogenes C203	0.5
S. pneumoniae Park	1
S. faecalis X66	>128
Haemophilus influenzae C.L.	8
H. influenzae 76	8
Escherichia coli EC14	2
E. coli TEM	4
Klebsiella pneumoniae KAE	16
K. pneumoniae X68	2
Enterobacter aerogenes C32	4
E. cloacae EB5	16
Salmonella typhi X514	2
Serratia marcescens SE3	4
Morganella morganii PR15	4
Providencia alcalifacius PR33	0.5
P. rettgeri C24	0.25

[a]Data of J. Ott, Lilly Research Laboratories.

insoluble peptidoglycan but had little to no effect on the incorporation of the appropriate radiolabeled precursors into protein, RNA, or DNA (Table 2). Activity against *Staphylococcus aureus* X1 but not against an isogenic L form of this strain supported the idea that LY186826 could be a selective inhibitor of cell wall biosynthesis.

The effects of LY186826 were compared with effects of cefamandole on cell wall biosynthesis in *P. rettgeri*. Both compounds inhibited incorporation of [^{14}C]diaminopimelic acid into sodium dodecyl sulfate (SDS)-insoluble as well as trichloroacetic acid (TCA)-insoluble cell wall (Table 3). Moreover, both compounds lowered the ratio of detergent-insoluble to acid-insoluble incorporation compared with uninhibited controls. This ratio estimates the amount of cross-linked to uncross-linked peptidoglycan, and a reduction in the ratio indicates inhibition of the final stage (formation of cross-linkages) in cell wall assembly (5).

Incorporation of UDP-*N*-acetyl-[^{14}C]glucosamine (UDP-[^{14}C]GlcNAc) into peptidoglycan by permeabilized *P. rettgeri* likewise was inhibited by both LY186826 and cefamandole (Table 4). Fosfomycin had no effect in this system, as was expected since this antibiotic inhibits one of the early stages in peptidoglycan biosynthesis (data not shown). The results with LY186826 and cefamandole are consistent with a mode of action affecting the final steps in

Table 2. Effect of LY186826 on macromolecular biosynthesis
in *P. rettgeri*[a]

Antibiotic	Concn (μg/ml)	% Inhibition of precursor incorporation into:			
		Cell wall	Protein	RNA	DNA
LY186826	500	80	<0	<0	<0
	100	54	<0	33	<0
	10	38	5	13	<0
	1	36	<0	<0	<0
Chloramphenicol	500		98		
	100		88		
	10		48		
	1		6		
Rifampin	500			79	
	100			68	
	10			54	
	1			43	
Nalidixic acid	500				100
	100				94
	10				75
	1				57

[a]Cell wall, protein, RNA, and DNA biosyntheses were measured by following incorporation of ^{14}C-labeled diaminopimelic acid, leucine, uracil, and thymidine, respectively, into TCA-insoluble material. Cell wall was measured in cells suspended in CWSM (cell wall synthesis medium; 4). Other precursors were measured in cells suspended in phosphate-buffered saline supplemented with 0.1 mM glucose.

wall assembly. Both compounds appeared to affect cross-linkage formation.

β-Lactam antibiotics block reactions in permeabilized bacterial cells that release alanine (6). C-terminal D-alanine is released by the action of transpeptidase and DD-alanine carboxypeptidase enzymes which are sensitive to penicillins, cephalosporins, and related antibiotics. In experiments in vitro with permeabilized *P. rettgeri*, LY186826 and cefamandole inhibited the release of [14C]alanine from UDP-*N*-acetylmuramyl-pentapeptide (labeled in the C-ter-

Table 3. Effects of LY186826 on cross-linkage formation in intact *P. rettgeri*[a]

Compound	Concn (μg/ml)	SDS-insoluble incorporation: dpm (%)	TCA-insoluble incorporation: dpm (%)	SDS/TCA
None		11,157	12,073	0.92
LY186826	250	2,004 (82)	3,668 (69)	0.54
	50	5,527 (50)	7,994 (36)	0.69
Cefamandole	25	5,302 (52)	8,393 (30)	0.63
	5	6,474 (42)	9,892 (18)	0.65

[a]Effects on cross-linkage formation were estimated by measuring SDS-insoluble and TCA-insoluble incorporation of [14C]diaminopimelic acid (Table 2) and comparing SDS/TCA incorporation ratios (see text).

Table 4. Effects of LY186826 on cross-linkage formation in permeabilized *P. rettgeri*[a]

Compound	Concn (μg/ml)	SDS-insoluble incorporation: dpm (%)	TCA-insoluble incorporation: dpm (%)	SDS/TCA
None		723	5,837	0.124
LY186826	500	208 (71)	2,441 (58)	0.085
	100	633 (12)	4,566 (22)	0.137
Cefamandole	500	105 (85)	1,975 (66)	0.053
	100	511 (29)	4,026 (31)	0.126

[a]Cells were permeabilized by ether treatment, and UDP-[^{14}C]GlcNAc incorporation was performed as described (6). Effects on cross-linkage formation were estimated as described in Table 3 footnote *a*.

minal alanine). Omission of UDP-GlcNAc in this system gives a measure of alanine released as a result of DD-alanine carboxypeptidase activity because release of alanine from the pentapeptide in the presence of UDP-GlcNAc is caused by carboxypeptidase plus transpeptidase activities. The difference indicates transpeptidase activity. LY186826 appeared to inhibit alanine release by both enzymes (Table 5).

Taken together, these findings suggest that LY186826 inhibits cell wall formation in a manner typical of β-lactam antibiotics. Further investigation has allowed us to look more closely at the specific type of inhibition by this compound. Growth of *P. rettgeri* in the presence of 10 and 100 μg of LY186826 per ml had little effect on growth rate for the first 4 h (Fig. 3). Increases in optical density stopped after 4 h, but there was little indication of significant cell lysis. However, microscopic observation of these cells showed formation of filaments at 10 μg/ml, suggesting inhibition of cross wall formation. Filaments with bulges appeared at the higher concentration. Similar observations have been made with *Escherichia coli* (D. Preston and E. Wu, unpublished

Table 5. Effect of LY186826 on [^{14}C]alanine release by permeabilized *P. rettgeri*[a]

Compound	Concn (μg/ml)	Estimate of enzyme activity [pmol (%)]		
		CPase (without UDP-GlcNAc)	CPase + TPase (with UDP-GlcNAc)	TPase (difference)
None		2,231	2,940	709
LY186826	500	1,148 (49)	1,445 (51)	297 (58)
	100	1,222 (45)	1,588 (46)	366 (48)
	10	1,744 (22)	2,230 (24)	484 (32)
Cefamandole	10	1,860 (17)	2,096 (29)	236 (67)

[a]Enzymatic release of [^{14}C]alanine from UDP-*N*-acetylmuramyl-[^{14}C]alanine-pentapeptide (C-terminal labeled) by permeabilized bacteria was measured by the procedure of Mirelman et al. (6). Reactions were incubated with and without unlabeled UDP-GlcNAc. Free [^{14}C]alanine generated in the reaction mixture was determined by Dowex cation-exchange chromatography of the TCA-soluble material. CPase, DD-Carboxypeptidase; TPase, transpeptidase.

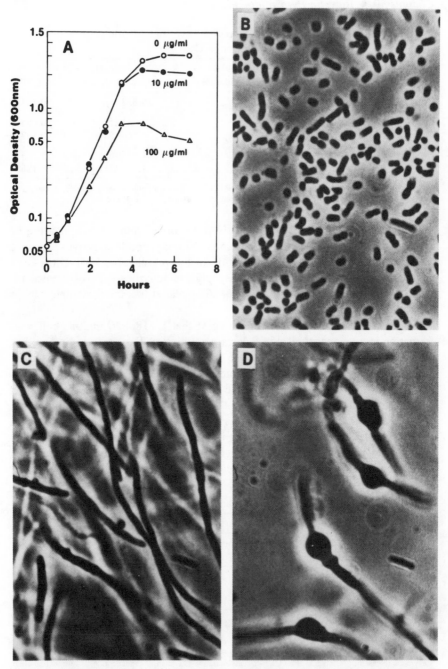

Figure 3. (A) Optical density of *P. rettgeri* grown in Trypticase soy broth in the presence and absence of LY186826. Photomicrographs of cells exposed for 4 h to 0 (B), 10 (C), and 100 (D) μg of LY186826 per ml. Magnification, ×1,600.

574

data). β-Lactam antibiotics that induce filament formation (e.g., cephalexin) often bind preferentially to PBP 3.

LY186826 was tested for its ability to compete with [^{3}H]penicillin G for binding to PBPs in membranes isolated from *P. rettgeri* and *E. coli* K-12. Gel profiles are shown in Fig. 4 and 5. LY186826 bound predominantly to PBP 3 of both *P. rettgeri* and *E. coli*. In *E. coli*, binding is clearly specific for PBP 3. Although LY186826 binds mainly to PBP 3 in *P. rettgeri*, we cannot completely rule out binding to other PBPs (e.g., PBP 2) in this organism.

Conclusions

The studies reported here demonstrate that the antibacterial action of LY186826 is due to an effect on cell wall formation. This action appears to be specific for cell wall biosynthesis. LY186826 inhibits the in vitro polymerization of peptidoglycan from nucleotide precursors in permeabilized cells. Inhibition in this in vitro system as well as in intact cells leads to a reduction

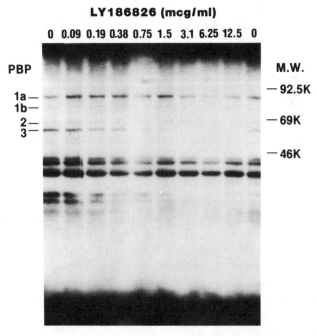

Figure 4. Binding of LY186826 to PBPs of *P. rettgeri* C24. Cells were grown in brain heart infusion and broken in an X-press. Membranes were purified by discontinuous sucrose gradient centrifugation. Binding specificity of LY186826 at the indicated concentrations was determined by competition with [^{3}H]penicillin G as described by Spratt (9). PBPs were separated by SDS-polyacrylamide gel electrophoresis and detected by fluorography.

LY186826 (mcg/ml)

0   25 12.5 6.25 3.1   0   1.5 0.78 0.39 0.2 0.1   0

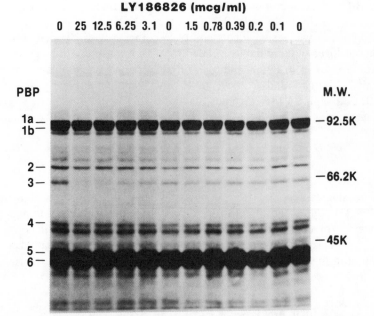

PBP   M.W.

1a —   — 92.5K
1b —

2 —   — 66.2K
3 —

4 —   — 45K

5 —
6 —

Figure 5. Binding of LY186826 to PBPs of *E. coli* K-12. Membranes were prepared from cells grown in antibiotic medium no. 3, and PBP binding was determined as described in the legend of Fig. 4.

in cross-linkage formation and an inhibition of the carboxypeptidase- and transpeptidase-catalyzed release of alanine. LY186826 binds to PBPs and shows a strong preference for PBP 3, which is consistent with the observation of filament formation.

Despite the nonclassical chemical structure of this compound, LY186826 behaves biologically in an identical manner to β-lactam antibiotics without having a β-lactam ring. As mentioned previously, the β-lactams are a structurally diverse group of compounds having the fundamental property of acylating bacterial PBPs. Assumption that the β-lactam ring is crucial for this activity has been doctrinaire. The findings reported here on LY186826, an aza-γ-lactam-containing compound, and the recent report (7) describing the β-lactam-like activity of lactivicin (a natural product lacking a classic β-lactam ring) suggest that the lactam bond rather than the β-lactam ring may be the common structural denominator that determines the specialized mechanism of action shown by all of these compounds.

ACKNOWLEDGMENTS. We thank Mimi Andonov-Roland, Lowell Hatfield, Louis Jungheim, Fred Mertz, John Ott, and David Preston for assistance and encouragement with this work.

LITERATURE CITED

1. **Georgopapadakou, N. H., S. A. Smith, and R. B. Sykes.** 1982. Mode of action of azthreonam. *Antimicrob. Agents Chemother.* **21**:950–956.
2. **Jungheim, L. N., S. K. Sigmund, and J. W. Fisher.** 1987. Bicyclic pyrazolidinones, a new class of antibacterial agent based on the β-lactam model. *Tetrahedron Lett.* **28**:285–288.
3. **Jungheim, L. N., S. K. Sigmund, N. D. Jones, and J. K. Swartzendruber.** 1987. Bicyclic pyrazolidinones, steric and electronic effects on antibacterial activity. *Tetrahedron Lett.* **28**:289–292.
4. **Lugtenberg, E. J. J., and P. B. DeHaan.** 1971. A simple method for following the fate of alanine-containing components in murein synthesis in *Escherichia coli. Antonie van Leeuwenhoek J. Microbiol. Serol.* **37**:537–552.
5. **Mirelman, D., and Y. Nuchamowitz.** 1979. Biosynthesis of peptidoglycan in *Pseudomonas aeruginosa*. 1. The incorporation of peptidoglycan into the cell wall. *Eur. J. Biochem.* **94**:541–548.
6. **Mirelman, D., Y. Yashouv-Gan, and U. Schwarz.** 1976. Peptidoglycan biosynthesis in a thermosensitive division mutant of *Escherichia coli. Biochemistry* **15**:1781–1790.
7. **Nozaki, Y., N. Katayama, H. Ono, S. Tsubotani, S. Harada, H. Okazaki, and Y. Nakao.** 1987. Binding of a non-β-lactam antibiotic to penicillin-binding proteins. *Nature* (London) **325**:179–180.
8. **Spratt, B. G.** 1975. Distinct penicillin binding proteins involved in the division, elongation, and shape of *Escherichia coli* K12. *Proc. Natl. Acad. Sci. USA* **72**:2999–3003.
9. **Spratt, B. G.** 1977. Properties of the penicillin-binding proteins of *Escherichia coli* K12. *Eur. J. Biochem.* **72**:341–352.
10. **Tomasz, A., and S. Waks.** 1975. Mechanism of action of penicillin: triggering of the pneumococcal autolytic enzyme by inhibitors of cell wall synthesis. *Proc. Natl. Acad. Sci. USA* **72**:4162–4166.
11. **Waxman, D. J., and J. L. Strominger.** 1982. β-Lactam antibiotics: biochemical modes of action, p. 209–285. *In* R. B. Morin and M. Gorman (ed.), *Chemistry and Biology of β-Lactam Antibiotics.* Academic Press, Inc., New York.

Chapter 60

Antibacterial Activities of the Lipid Dodecylglycerol

Ronald A. Pieringer
H. S. Ved
Janice L. Brissette
Evan Gustow
Erlinda A. Cabacungan

Although the antibacterial nature of certain lipids has been known for over half a century (1, 3, 7), relatively little is known about how a lipid, especially at low concentrations (a few micrograms per milliliter), exerts its antibacterial action. Our interest in attempting to explain some aspects of this unknown led us to the discovery that the alkylglycerol ether lipid dodecylglycerol (DDG) is an excellent antibacterial agent (8) and is at least twofold more potent than the corresponding acylglycerol ester lipid dodecanoylglycerol (monolaurin). The greater potency of dodecylglycerol can probably be attributed to the greater metabolic and chemical stability of its ether bond, which was in part verified by the fact that it is retained for longer periods in bacteria than is the ester-containing monolaurin (8).

The fact that DDG can exert its antibacterial activity at relatively low concentrations (the MIC can be as low as 4 μg/ml of medium) suggested that its mode of antibacterial action must include an antimetabolic as well as a physical phenomenon (8). This view was substantiated by showing that heat-treated (pasteurized) cells (*Streptococcus faecium* ATCC 9790) did not lyse upon exposure to 50 μg of DDG per ml, whereas untreated cells readily lysed at 6 μg of DDG per ml (8). The lysis of *S. faecium* by DDG is brought about by its

Ronald A. Pieringer, H. S. Ved, Janice L. Brissette, Evan Gustow, and Erlinda A. Cabacungan • Department of Biochemistry, Temple University School of Medicine, Philadelphia, Pennsylvania 19140.

stimulation of the degradative, autolytic peptidoglycan hydrolase (*S. faecium* muramidase) (8), a β-1,4-*N*-acetylmuramyl hydrolase (5). The stimulation of this autolytic enzyme by DDG was indirect and occurred only when whole cells were exposed to the lipid (8). It had no direct stimulative effect on preparations of the enzyme isolated from broken cells (8). The indirect stimulation of autolytic activity by DDG in whole cells was shown to be caused by its release or activation of an endogenous proteinase (9), which in turn converted the autolysin from a latent, inactive form to an active state. The activation of the autolysin by an exogenous proteolytic enzyme, such as trypsin, had been demonstrated previously (5).

The proteinase, which was localized primarily in a 25,000 × *g* supernatant fraction of broken cells of *S. faecium* ATCC 9790, had a specific activity threefold greater in fractions isolated from DDG-treated cells than in those from untreated cells (9). The DDG stimulation of autolysin activity was additive with trypsin treatment and was reversed by proteinase inhibitors. Specific proteinase inhibitors also reversed DDG-stimulated proteinase activity of *S. faecium* ATCC 9790 (9). These data demonstrate at least one mode of antibacterial activity of DDG. Our hypothesis is that the hydrophobic environment of the membrane is sufficiently perturbed by DDG to release or activate a proteinase which in turn partially degrades the autolysin to an active enzyme.

Since DDG stimulates the autolysin enzyme, which hydrolyzes cell wall peptidoglycan, the question of whether it might interfere with peptidoglycan synthesis was also considered (10). This appeared to be a particularly pertinent question because the bacterium *S. mutans* BHT has little autolytic activity (6) but is still very susceptible to DDG even though it does not lyse. In this organism DDG drastically curtailed the incorporation of [^{14}C]lysine into peptidoglycan, suggesting that peptidoglycan synthesis is also inhibited (10). In *S. faecium* ATCC 9790, the synthesis of peptidoglycan was differentiated from the degradation of peptidoglycan by inhibiting autolytic degradation with chloramphenicol. In the presence of chloramphenicol the stimulation of autolysin by DDG in *S. faecium* ATCC 9790 is prevented. Under these conditions DDG inhibits the incorporation of [^{14}C]lysine into peptidoglycan, which again demonstrates that peptidoglycan synthesis is inhibited by DDG (10). This, the second known mode of action of DDG, prevents the synthesis of a key component of the bacterial cell wall and thus stops growth.

Since DDG is taken up rather readily into the membrane of certain bacteria, its potential for altering the hydrophobic environment of the membrane and of the enzyme activities associated with the cell membrane is rather high. Certainly one of the obvious enzymatic systems likely to be disturbed is the enzymes involved in the synthesis of glycerolipids. This proved to be the case. Exposure of *S. mutans* BHT to growth-inhibitory concentrations of racemic *sn*-l(3)-DDG inhibited the incorporation of [^{14}C]glycerol into lipids and lipoteichoic acids and altered the percent composition of the glycerolipids (2).

Increases in phosphatidic acid and diphosphatidylglycerol, and decreases in phosphatidylglycerol, contributed the most to the change in lipid composition. No cellular lysis occurred under these conditions, mainly because *S. mutans* BHT lacks significant autolysin activity (6). Radioactive racemic *sn*-l(3)-DDG was readily taken up by the cell and was metabolized primarily to lysophosphatidic acid and phosphatidic acid, with smaller amounts converted to phosphatidylglycerol and diacylglycerol. The accumulation of phosphatidic acid and the loss of viability responded in parallel to different concentrations of DDG. An increase in the concentration of CTP was also observed in response to DDG treatment. The buildup of this substrate together with the increase in phosphatidic acid suggested a possible impairment in the synthesis of CDP-diacylglycerol (2). The CDP-diacylglycerol synthase appeared not to readily accept a DDG-derived phosphatidic acid as a substrate.

The use of pure *sn*-1-, *sn*-2-, and *sn*-3-DDG stereoisomers has produced some unexpected results. Because most glycerol-containing phospholipids found in nature have the phosphate in the *sn*-3 position of the glycerol moiety of these lipids (the major exception is the archaebacteria [4]), it was thought that *sn*-3-DDG would not be metabolized. However, *S. mutans* BHT cells readily take it up and convert it to lysophosphatidic acid, phosphatidic acid, and diacylglycerol, but no other lipid. The surprising aspect of these results is that the monoacylglycerol kinase (DDG → lysophosphatidic acid) and the acyltransferase (lysophosphatidic acid → phosphatidic acid) are obviously not stereospecific. However, the CDP-diacylglycerol synthase is highly stereospecific and will not accept an unnatural phosphatidic acid (with its phosphate moiety in the *sn*-1 position) as a substrate. Phosphatidic acids derived from *sn*-1-DDG and *sn*-2-DDG are partially acceptable as substrates to this enzyme (R. A. Pieringer, E. Gustow, and E. Cabacungan, *Fed. Proc.* **46:**637, 1987). The diglyceride derived from the *sn*-3-DDG was all of the *sn*-1,3 configuration, whereas the diglyceride derived from *sn*-1-DDG was one-third *sn*-1,3- and two-thirds *sn*-1-dodecyl-2-acylglycerol. These results suggest that the phosphatidic acid phosphatase is stereospecific in that no *sn*-2-acyl-3-alkylglycerol is formed from unnatural phosphatidic acid. The diglyceride of the *sn*-1,3-configuration is probably formed from the direct acylation of DDG.

Although only three modes have been uncovered as yet, it is obvious that DDG can have a profound detrimental effect on the metabolism of the bacterial cell. Thus, it is also clear that DDG is not acting as an antibacterial agent solely through a physical phenomenon.

ACKNOWLEDGMENTS. This work was supported in part by Public Health Service research grant AI-05730 from the National Institute of Allergy and Infectious Diseases and by Biomedical Research Support Grant S07RR05417 from the Division of Research Resources, National Institutes of Health.

LITERATURE CITED

1. **Avery, O. T., and G. E. Cullen.** 1923. Studies on the enzymes of pneumococcus IV. Bacteriolytic enzyme. *J. Exp. Med.* **38**:199–206.
2. **Brissette, J. L., E. A. Cabacungan, and R. A. Pieringer.** 1986. Studies on the antibacterial activity of dodecylglycerol: its limited metabolism and inhibition of glycerolipid and lipoteichoic acid biosynthesis in *Streptococcus mutans* BHT. *J. Biol. Chem.* **261**:6338–6345.
3. **Cornett, J. B., and G. D. Shockman.** 1978. Cellular lysis of *Streptococcus faecalis* induced with Triton X-100. *J. Bacteriol.* **135**:153–160.
4. **DeRosa, M., A. Gambacorta, and A. Gliozzi.** 1986. Structure, biosynthesis, and physicochemical properties of archaebacterial lipids. *Microbiol. Rev.* **50**:70–80.
5. **Kawamura, T., and G. D. Shockman.** 1983. Purification and some properties of the endogenous, autolytic N-acetylmuramoylhydrolase of *Streptococcus faecium*, a bacterial glycoenzyme. *J. Biol. Chem.* **258**:9514–9521.
6. **Shockman, G. D., R. Kessler, J. B. Cornett, and M. Mychajlonka.** 1978. Turnover and excretion of streptococcal surface components, p. 803–814. *In* J. R. McGhee, J. Mestecky, and J. L. Babb (ed.), *Secretory Immunity and Infection.* Plenum Publishing Corp., New York.
7. **Valko, E. I.** 1946. Surface active agents in biology and medicine. *Ann. N.Y. Acad. Sci.* **46**:451–478.
8. **Ved, H. S., E. Gustow, V. Mahadevan, and R. A. Pieringer.** 1984. Dodecylglycerol: a new type of antibacterial agent which stimulates autolysin activity in *Streptococcus faecium* ATCC 9790. *J. Biol. Chem.* **259**:8115–8121.
9. **Ved, H. S., E. Gustow, and R. A. Pieringer.** 1984. The involvement of the proteinase of *Streptococcus faecium* ATCC 9790 in the stimulation of its autolysin activity by dodecylglycerol. *J. Biol. Chem.* **259**:8122–8124.
10. **Ved, H. S., E. Gustow, and R. A. Pieringer.** 1984. Inhibition of peptidoglycan synthesis of *Streptococcus faecium* ATCC 9790 and *Streptococcus mutans* BHT by the antibacterial agent dodecyl glycerol. *Biosci. Rep.* **4**:659–664.

Chapter 61

Binding of Polycationic Antibiotics to Lipopolysaccharides of *Pseudomonas aeruginosa*

Mildred Rivera
Estelle J. McGroarty

Lipopolysaccharide (LPS) is a major component of the outer membrane of gram-negative bacteria. The structure of LPS is critical in forming a penetration barrier to amphipathic and hydrophobic compounds (10). LPS contains a variety of ionic groups, with acidic phosphates and carboxyl moieties concentrated within the core and lipid A regions (16). Therefore, LPS is associated with a variety of cations, including the metal ions Mg^{2+} and Ca^{2+} (3, 10). Polycationic antibiotics such as polymyxin B and gentamicin C interact with the outer membrane of gram-negative bacteria and destabilize the membrane structure (5, 10). This mechanism involves the displacement of divalent cations from LPS by polycationic compounds, thus destroying the LPS cross-bridging and destabilizing the outer membrane (5).

We measured the relative affinities of various polycationic antibiotics and polyamines for purified preparations of *Pseudomonas aeruginosa* LPS by using the cationic spin probe CAT_{12}. We found that polycationic antibiotics bind and rigidify MgLPS and CaLPS complexes from *P. aeruginosa* H215 (antibiotic-sensitive mutant), H216 (revertant), and H219 (parent) to different extents. The chemical basis for differences seen in LPS-cation interactions among the three isolates studied here may result from substoichiometric substitution of the core-lipid A. No other chemical difference was noted in the LPS isolates.

Mildred Rivera and Estelle McGroarty • Department of Biochemistry, Michigan State University, East Lansing, Michigan 48824.

Sample Preparations

LPS Samples

LPS from PAO1 strains of *P. aeruginosa* H215 (aminoglycoside-super-sensitive mutant), H216 (revertant), the H219 (parent) were a gift from R. E. W. Hancock (Vancouver, Canada). The magnesium (MgLPS) and calcium (CaLPS) salts of the LPS isolates were prepared as described elsewhere (3). Samples were lyophilized and weighed to quantitate LPS concentrations. Elemental analysis of LPS was done as described previously (3).

SDS-PAGE

Sodium dodecyl sulfate-polyacrylamide gel electrophoresis (SDS-PAGE) gels were prepared as described by Peterson and McGroarty (13). Separating gels were formed with either 11 or 15% acrylamide. The gels were either silver stained (4) or electrotransferred to nitrocellulose (150 mA, 18 h) and were visualized by using anti-503 (O-antigen specific) monoclonal antibody (11).

Column Chromatography

LPS samples were fractionated with a Sephadex G-200 column (64 cm by 25 mm) by using the column system of Peterson and McGroarty (13). Fractions were analyzed for amino sugars, phosphate, and 2-keto-3-deoxyoctulosonic acid as described previously (13).

Partitioning of Spin Probe

Titrations of the calcium and magnesium salts of LPS (10 mg/ml in KOH-HEPES [*N*-2-hydroxyethylpiperazine-*N'*-2-ethanesulfonic acid], 50 mM, pH 7.2) with polycationic antibiotics (polymyxin B and gentamicin C), polyamines (spermine), $CaCl_2$, and $MgCl_2$ were analyzed by competitive displacement of the spin probe 4-dodecyl dimethyl ammonium-1-oxyl-2,2,6,6-tetramethyl piperidine bromide (CAT_{12}) (1).

Chemical Characterization

Approximately 11 phosphates were bound per LPS molecule (Table 1). All of the samples studied were in the magnesium or calcium forms.

The silver stain procedure was used to visualize different bands on SDS-PAGE. Three regions of bands were observed for the parent (H219), revertant

Table 1. Elemental composition of purified *P. aeruginosa* LPS isolates[a]

Strain	Ca/P	Mg/P	+/P	Ca/LPS[c]	Mg/LPS[c]	P/LPS
			Ratio[b] (mol/mol)			
MgLPS						
H219	<0.01	0.76 ± 0.01	1.55 ± 0.04	—[d]	10.20 ± 0.13	13.45 ± 0.07
H216	<0.01	0.75 ± 0.01	1.52 ± 0.02	—[d]	9.30 ± 0.56	12.40 ± 0.57
H215	<0.01	0.72 ± 0.01	1.48 ± 0.01	—[d]	8.39 ± 2.14	11.66 ± 2.74
CaLPS						
H219	0.78 ± 0.04	<0.01	1.60 ± 0.09	7.64 ± 1.51	—[d]	9.80 ± 1.48
H216	0.81 ± 0.04	<0.01	1.69 ± 0.04	7.84 ± 0.71	—[d]	9.78 ± 1.25
H215	0.77 ± 0.04	<0.01	1.57 ± 0.09	7.66 ± 0.01	—[d]	10.00 ± 0.44

[a]The metal and phosphate contents of the magnesium and calcium salts from the three LPS isolates were determined by inductively coupled plasma emission spectroscopy. There was less than 0.01 metal per phosphate of Na, K, Fe, Al, and Zn ions associated with the samples.
[b]Expressed as the mean ± SD. Comparisons of the values were made by using the randomized block ANOVA statistical analysis at $P < 0.05$.
[c]Assumes an average molecular mass of 9,000.
[d]There was less than 0.01 metal per phosphate of Ca or Mg ions associated with the sample. The thiobarbituric acid assay showed two reactive 2-keto-3-deoxyoctulosonic acid residues per molecule of LPS.

(H216), and mutant (H215) (Fig. 1A, lanes 1–3). At low concentrations, there was no difference in the migration pattern of the short-chain LPS of the three strains (Fig. 1A, lanes 4–6). Analysis of Western blots of LPS isolates with anti-O-antigen-specific monoclonal antibodies revealed the ladder pattern of O-antigen-containing molecules with O antigens of increasing length (Fig. 1B). This blot indicates that the ladder consists of doublet bands, and the level of one of the bands in the doublet is in much lower amounts in the isolate from the mutant, H215, than from the parent, H219. Presumably, this reflects a lack of substoichiometric substitution within the core lipid A region of the molecules (10, 15).

The size heterogeneity of the molecules was characterized by separating the samples on a Sephadex G-200 column (13). The three major amino sugar-containing peaks, very long O-antigen chain (peak 1), long chain (peak 2), and short chain (peak 3), corresponded to the three major sets of bands seen on SDS-PAGE (Fig. 1A). Analyses of the levels of these peaks revealed no differences in size heterogeneity between the sample from H215 (mutant) and H216 (revertant) (Table 2). SDS-PAGE of the column fractions of samples from strain H215 and H216 revealed two distinct ladder patterns of apparently different sizes, the A bands (slow-moving ladder bands) and the B bands (faster-moving ladder bands) (results not shown). Amino sugar analysis of the column fractions containing mainly A bands lacked reactive amino sugars, and reaction with monoclonal anti-503 LPS antibody indicated that only the B bands and not the A bands were antigenically active (data not shown). These data support the hypothesis that *Pseudomonas* spp. have the ability to synthesize more than one type of LPS with different O-antigenic side chains (16).

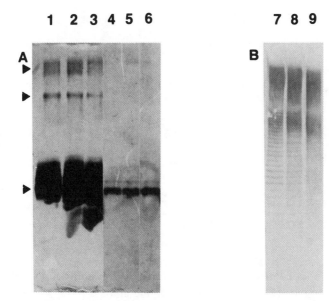

Figure 1. (A) Silver-stained SDS-PAGE of LPS from *P. aeruginosa* H219 (lanes 1 and 4), H216 (lanes 2 and 5), and H215 (lanes 3 and 6). Samples of either 5 μg (lanes 1–3) or 0.10 μg (lanes 4–6) were applied to a 15% polyacrylamide gel which had polymerized overnight with a butanol overlay. Arrowheads indicate the three intensively stained regions of the *P. aeruginosa* samples. (B) Western blots of LPS from *P. aeruginosa* H215 (lane 7), H216 (lane 8), and H219 (lane 9), allowed to react with monoclonal anti-503 antibody.

Spin Probe Analysis

Scatchard analysis of the binding of spin probe to LPS suggests that the three LPS isolates possess specific and nonspecific binding sites for CAT_{12} with different affinities (Table 3). A decrease in probe binding was observed for the CaLPS complexes of strains H219 and H215. This difference between the partitioning of the spin probe in the MgLPS and in CaLPS complexes may result from the physicochemical properties of Mg^{2+} and Ca^{2+} ions (6, 10). Ca^{2+} ions

Table 2. Chemical analysis of size fractions of LPS from strains H215 and H216

Strain	Peak no.	% Total phosphate	Amino sugar/phosphate
H216	1	3.5	21.4
	2	2.0	16.4
	3	93.0	1.0
H215	1	4.0	23.3
	2	2.0	17.5
	3	94.0	1.0

Table 3. Scatchard analysis of CAT_{12} binding to LPS
isolates of *P. aeruginosa*[a]

Strain	High-affinity site		Low-affinity site	
	K_D (μM)	N[b]	K_D (μM)	N[b]
MgLPS				
H219	0.16	0.9	2.07	2.6
H216	0.40	0.9	2.19	2.6
H215	0.73	1.0	2.51	2.4
CaLPS				
H219	0.59	0.8	2.49	2.0
H215	0.31	0.8	2.53	2.7

[a]Concentrations of 5 μM of the calcium and magnesium salts of the LPS isolates in 50 mM KOH-HEPES (pH 7.2), 25°C, were mixed with different concentrations of CAT_{12}. The levels of bound and free probe were determined from the resultant electron spin resonance spectra. Data represent averages of three samples.
[b]N, Number of probe-binding sites per LPS molecule.

form higher coordination complexes with irregular geometry compared with Mg^{2+} ions (6). Therefore, Ca^{2+} can induce the anionic groups in the core and lipid A region to form a more "closed" structure, whereas Mg^{2+} forms a more "open" type of structure.

Titration of LPS samples containing CAT_{12} with various cations displaced different amounts of the probe, reflected in the magnitude of the partitioning parameter Ψ_i. Displacement of the spin probe from the MgLPS complex of the mutant strain (H215) occurred at much lower levels of added cation per LPS than for the parent (H219) or revertant (H216) strains (Fig. 2, left panels). LPS from strains H216 and H219 exhibit similar antibiotic binding abilities. The binding preference of the compounds for LPS of strains H215, H216, and H219 was as follows: polymyxin B $>>$ gentamicin C $\simeq$ spermine. The affinity of cations for LPS results, in part, from ionic attractions. The strength of binding will depend on the net positive charge of the cation and on the arrangement of these charges within the cationic molecule (12). Cations with higher net charge are better competitors because of the large number of negative charges they can sequester. In addition, as any given cation binds to LPS, the LPS aggregate may alter its conformation to maximize ionic interactions, which, in turn, may change the binding sites for other cations (12). The data also suggest that cations displaced CAT_{12} more readily from strain H215 samples than from strains H219 and H216. This difference in binding to the MgLPS complexes from the various strains of *P. aeruginosa* is probably related to differences in the phosphate substituents in the 2-keto-3-deoxyoctulosonic acid-lipid A region (10, 15).

Samples of the calcium salts of the H215 and H219 isolates were also titrated with the various polycations (Fig. 2, right panel). The competitive displacement curves showed a pattern of binding preference similar to that of the

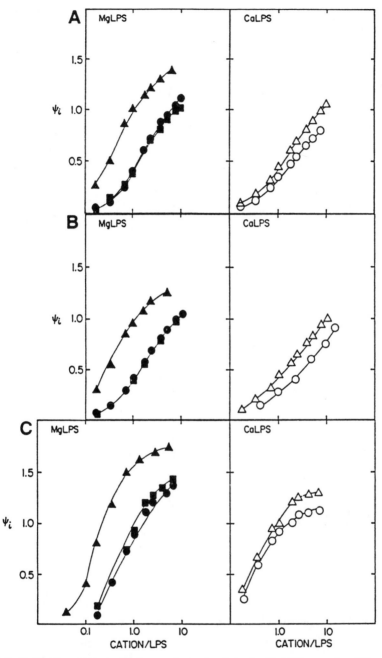

Figure 2. Partitioning parameter Ψ_i measured as a function of added (A) spermine, (B) genta-micin C, and (C) polymyxin B. Partitioning of the spin probe CAT_{12} was determined upon ad-dition of cation to MgLPS (closed symbols) and CaLPS (open symbols) from *P. aeruginosa* H219 (parent, $\bigcirc$), H215 (antibiotic-supersensitive mutant, $\triangle$), and H216 (revertant, $\square$). The amount of added cation is plotted as the number of cations per LPS (moles per mole).

587

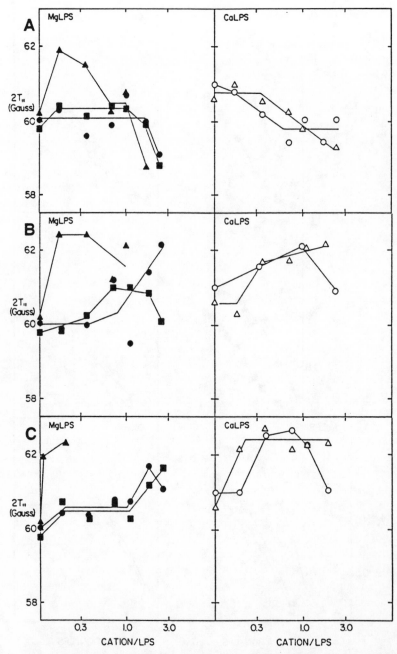

Figure 3. Head group mobility of LPS from *P. aeruginosa* H219 (O), H216 (□), and H215 (△), measured by the hyperfine splitting parameter ($2T_{11}$), in gauss, of bound CAT_{12}. This parameter was measured as a function of increasing concentrations of (A) spermine, (B) gentamicin C, and (C) polymyxin B to the magnesium (closed symbols) and calcium (open symbols) complexes of LPS.

588

MgLPS complexes. However, the Ψ_i values calculated for both strains were dramatically decreased when compared with the MgLPS complexes, with the mutant strain being the most affected. A decrease in Ψ_i might suggest that a rearrangement in the LPS aggregate structure had occurred when changing the counterion from Mg^{2+} to Ca^{2+}. Perhaps what we observed when using the Ca^{2+} salts was an alteration in the LPS aggregate and, thus, differential alteration of the barrier function of the intact outer membrane to polycationic compounds (2, 9).

The relative fluidity of the bound CAT_{12} probe was measured by calculating $2T_{11}$ as a function of cation concentration (Fig. 3). It can be seen from our data that (i) after cationic titration with polycations, the onset of MgLPS structural change for the H215 sample occurs at lower cation concentrations than for strain H219 (Fig. 3, left panel), and (ii) in the absence of added cations, the CaLPS aggregate has a more rigid structure than the MgLPS aggregate (Fig. 3, right panel). The antibiotics polymyxin B and gentamicin C rigidified the MgLPS of strain H215 to a greater extent than that of the parent strain (Fig. 3A and 3B, left panel). On the other hand, spermine fluidized the MgLPS complexes of the mutant as well as the parent strain. It has been proposed that the fluidity depends on the class of cation added, rather than on the relative affinity (12).

A different structural alteration was observed in the CaLPS complexes from the mutant and parent strain when polycations were added (Fig. 3, right panel). The perturbation induced at low concentrations of cations was different between the MgLPS and CaLPS complexes of strains H215 and H219. However, similar values of $2T_{11}$ were calculated for the Ca^{2+} salts of both strains. The antibiotic-induced alteration in the structure of LPS aggregates is more pronounced in the MgLPS preparations than in the CaLPS, which, again, supports the idea that Ca^{2+} protects against the binding or structural alterations, or both, induced by antibiotics.

The increase in rigidity of LPS aggregates upon addition of cationic antibiotics indicates an antibiotic-induced alteration of the LPS packing. The antibiotics rearranged the LPS packing which might have resulted in cracks or nonlamellar structure of LPS within the aggregate (7, 8). In contrast, polyamines, which are naturally present in the outer membrane, bind tightly to LPS, yet do not immobilize the LPS head groups or increase the permeability of the outer membrane (12, 14). The small size of polyamines may allow binding to LPS aggregates, neutralizing some of the charges without disrupting the LPS packing.

LITERATURE CITED

1. **Coughlin, R. T., C. R. Caldwell, A. Haug, and E. J. McGroarty.** 1981. A cationic electron spin resonance probe used to analyze cation interactions with lipopolysaccharide. *Biochem. Biophys. Res. Commun.* **100**:1137–1142.

2. **Coughlin, R. T., A. Haug, and E. J. McGroarty.** 1983. Physical properties of defined lipopolysaccharide salts. *Biochemistry* **22**:2007–2013.
3. **Coughlin, R. T., S. Tonsager, and E. J. McGroarty.** 1983. Quantitation of metal cations bound to membranes and extracted lipopolysaccharide of *Escherichia coli. Biochemistry* **22**:2002–2007.
4. **Dubray, G., and G. Bezard.** 1982. A highly sensitive periodic acid-silver stain for 1,2-diol groups of glycoproteins and polysaccharides in polyacrylamide gels. *Anal. Biochem.* **119**:325–329.
5. **Hancock, R. E. W.** 1984. Alterations in outer membrane permeability. *Annu. Rev. Microbiol.* **38**:237–264.
6. **Hughes, M. N. (ed.).** 1981. *The Inorganic Chemistry of Biological Processes*, p. 256–295. John Wiley & Sons, Inc., New York.
7. **Lounatmaa, K., P. H. Mäkelä, and M. Sarvas.** 1976. Effect of polymyxin on the ultrastructure of the outer membrane of wild-type and polymyxin-resistant strains of *Salmonella. J. Bacteriol.* **127**:1400–1407.
8. **Martin, N. L., and T. J. Beveridge.** 1986. Gentamicin interaction with *Pseudomonas aeruginosa* cell envelope. *Antimicrob. Agents Chemother.* **29**:1079–1087.
9. **Newton, B. A.** 1954. Site of action of polymyxin on *Pseudomonas aeruginosa*: antagonism by cations. *J. Gen. Microbiol.* **10**:491–499.
10. **Nikaido, H., and M. Vaara.** 1985. Molecular basis of bacterial outer membrane permeability. *Microbiol. Rev.* **49**:1–32.
11. **Otten, S., S. Iyer, W. Johnson, and R. Montgomery.** 1986. Serospecific antigens of *Legionella pneumophila. J. Bacteriol.* **167**:893–904.
12. **Peterson, A. A., R. E. W. Hancock, and E. J. McGroarty.** 1985. Binding of polycationic antibiotics and polyamines to lipopolysaccharides of *Pseudomonas aeruginosa. J. Bacteriol.* **164**:1256–1261.
13. **Peterson, A. A., and E. J. McGroarty.** 1985. High-molecular-weight components in lipopolysaccharides of *Salmonella typhimurium, Salmonella minnesota,* and *Escherichia coli. J. Bacteriol.* **162**:738–745.
14. **Stevens, L.** 1967. Studies on the interaction of homologues of spermine with deoxyribonucleic acid and with bacterial protoplasts. *Biochem. J.* **103**:811–815.
15. **Vaara, M., T. Vaara, M. Jensen, I. Helander, M. Nurminen, E. T. Rietschel, and P. H. Mäkelä.** 1981. Characterization of the lipopolysaccharide from the polymyxin-resistant pmrA mutants of *Salmonella typhimurium. FEBS Lett.* **129**:145–149.
16. **Wilkinson, S. G.** 1983. Composition and structure of lipopolysaccharides from *Pseudomonas aeruginosa. Rev. Infect. Dis.* **5**:S941–S949.

Chapter 62

Can Synthesis of Cell Wall Anionic Polymers in *Bacillus subtilis* Be a Target for Antibiotics?

H. M. Pooley
D. Karamata

The rapid increase over recent years in the proportion of hospital infections caused by β-lactam-resistant gram-positive cocci (6) underlines the interest in exploring new potential antibiotic targets. There are grounds, we believe, for considering the synthesis of cell wall anionic polymers found in many gram-positive organisms as one such target.

The chemistry and biosynthesis of these polymers, and the coordination of their synthesis with that of peptidoglycan (PG), have been extensively investigated in vitro (1, 3, 4, 12). However, until recently, little was known about their genetics; in *Bacillus subtilis,* the only genes mapped were *gtaA gtaB* (16) involved in glucosylation of the major wall teichoic acid (TA) and *tag-1* which, at the nonpermissive temperature, leads to a reduced concentration of this polymer (2).

The role(s) and importance of this class of polymers remain essentially the subject of speculation (9). Could the relatively restricted number of publications concerned with these polymers in recent years be a reflection of a generally held view that they are of secondary importance and may not be essential for growth? Nevertheless, in *B. subtilis* there is evidence—genetic as well as that from antibiotic and biosynthetic studies—which points to the opposite conclusion.

Data presented recently (5, 7) enable us to assign most, if not all, TA genes to a small region, around 310 degrees on the *B. subtilis* map. Genetic

H. M. Pooley and D. Karamata • Institut de Génétique et Biologie Microbiennes, 1005, Lausanne, Switzerland.

determinants for the poly(ribitol phosphate) TA [poly(rit*P*)] of strain W23 were transferred to strain 168, replacing those coding for the poly(glycerol phosphate) polymer [poly(gro*P*)] of the recipient. In these heterologous (168 × W23) crosses the frequency of genetic exchange in the region of TA markers was two orders of magnitude lower than that of reference markers. In all of about 60 rare hybrid recombinants at the *tag-1* locus, the presence of ribitol-containing TAs of the W23 type was shown by phage resistance patterns, and in a limited number of strains this result was confirmed by cell wall analysis. The noteworthy absence, among these recombinants, of strains devoid of both major anionic polymers [poly(gro*P*) and poly(rit*P*)] was taken as indicating their nonviability. Indeed, the latter class of recombinants, involving distinct sets of genes that occupy overlapping positions on their respective chromosomes, is, statistically, the one most frequently expected. This provides, we believe, part of the explanation for the low frequency of hybrid recombinants.

This observation puts in a new light the extremely limited nature of defects in TA synthesis among a large number (>200) of *B. subtilis* mutants resistant to phage ϕ25 or ϕ29 (15; H. M. Pooley, D. Paschoud, and D. Karamata, *J. Gen. Microbiol.*, in press). Particularly significant is the absence of mutants with cell walls devoid of phosphate. Since poly(gro*P*) forms part of the ϕ29 receptor, mutants lacking this polymer would be expected to be resistant; their absence is consistent with nonviability.

The essentiality of a cell component can also be deduced from the existence of conditional lethal mutants. It has been advanced that one mutation associated with such a phenotype, *tag-1*, is affected in TA synthesis (2). Further studies on *tag-1* and other mutations with a conditional lethal phenotype, including *rodC1*, revealed, under restrictive conditions, a fivefold increase in the rate of PG synthesis relative to that of protein. This led to the conclusion that these mutations affected primarily PG synthesis (10, 11). However, recent results from heterologous transformations (5) give strong support for believing that *tag-1* affects one of the steps in the synthesis of poly(gro*P*). Further support comes from genetic and cell wall analyses of a new collection of temperature-sensitive mutants obtained when selecting for resistance to ϕ29 at 37°C (M. M. Briehl, H. M. Pooley, N. H. Mendelson, and D. Karamata, *Abstr. Annu. Meet. Am. Soc. Microbiol. 1986*, J17, p. 191). The majority of the temperature-sensitive mutations are linked to *tag-1* by transformation and form a linkage group large enough to include several genes, consistent with the number expected to be involved in the synthesis of poly(gro*P*) (5).

Most of the temperature-sensitive TA mutants, including those carrying *tag-1*, have a somewhat leaky phenotype. Transfer to the restrictive temperature is accompanied by an important residual mass increase, continuing for five generation times or longer. Nevertheless, in many cases mass increase eventually stops. Residual growth at the restrictive temperature is associated with dramatic cellular deformations, some of which have been seen previously (2).

Further work is required to establish whether the massive selective increase in PG synthesis by *tag-1* mutants at the restrictive temperature is characteristic of all new temperature-sensitive mutations belonging to this linkage group. Nevertheless, a subgroup of temperature-sensitive markers genetically distant from *tag-1*, apparently affected in TA synthesis, do not produce thickened walls at the restrictive temperature and are associated with distinctive deformations (not presented).

Now that *rodC1* has been shown to map close to *tag-1* (Pooley et al., in press), it is clear that many mutations in this organism associated with morphological deformations map within, or close to, the TA gene region. It appears increasingly likely that blocking synthesis of anionic wall polymers interferes with orderly cell surface expansion, leading to deformations and, ultimately, to arrest of cell growth. This last implies that TA synthesis should be a potential target for antibiotics.

Indeed, in vitro studies have revealed that the synthesis of the linkage unit joining the main TA polymer to muramic acid and more specifically that of the first enzyme involved, the UDP-*N*-acetylglucosamine translocase, are exquisitely sensitive to the antibiotic tunicamycin (4, 13, 14). The concentration required to block 50% of the translocase activity, 0.2 µg/ml (10-fold lower than that required to inhibit PG synthesis), was, remarkably, virtually identical to the MIC for growth of *B. subtilis*. We believe that at 0.2 µg/ml TA synthesis could be the target responsible for stopping growth, although previously (12, 13) this was considered unlikely. Furthermore, cells treated with tunicamycin developed deformations closely similar to those characteristic of *tag-1* at high temperature (14).

Identification of the target of an antibiotic can be made through genetic and biochemical characterization of resistant mutants. If resistance to the MIC of tunicamycin could be achieved through structural modification of the UDP-*N*-acetylglucosamine translocase, the marker responsible should be in the structural gene for this enzyme. Although the location of this gene is unknown, most, if not all, TA genes map in a limited region near the *hisA* marker. To test this hypothesis, tunicamycin-resistant mutants forming colonies on L agar containing 0.4 µg of tunicamycin per ml were isolated.

Screening of mutants thus obtained by PBS1 transduction revealed resistance mutations (*tmr*) linked to the *hisA* marker; three point transduction crosses with *hisA* and *hag-4* markers gave results consistent with *tmr* markers being localized within the same segment of the genome as the TA genes. Further characterization is needed to establish whether or not modification in the translocase sensitivity to tunicamycin is associated with these new *tmr* markers.

It is noteworthy that two previously described *tmr* markers associated with resistance to about 50 and 250 times the MIC of tunicamycin map a long way from the TA region (8); furthermore, the translocase isolated from strains containing these markers was reported to have unchanged sensitivity to this antibiotic (14). For one of these markers, resistance was correlated with a modification in a protein increasing excretion of α-amylase (14).

Results obtained so far encourage a belief that a reevaluation of the importance of wall anionic polymers for the growth of gram-positive organisms is necessary and timely. While this communication is voluntarily restricted to one genetically well-characterized organism, it is evident that any findings and conclusions cannot be extended to other species, e.g., the clinically important cocci (6), without further study.

LITERATURE CITED

1. **Archibald, A. R.** 1974. The structure, biosynthesis and function of teichoic acid. *Adv. Microb. Physiol.* **11**:53–95.
2. **Boylan, R. J., N. H. Mendelson, D. Brooks, and F. E. Young.** 1972. Regulation of the bacterial cell wall: analysis of a mutant of *Bacillus subtilis* defective in biosynthesis of teichoic acid. *J. Bacteriol* **110**:281–290.
3. **Bracha, R., and L. Glaser.** 1976. In vitro system for the synthesis of teichoic acid linked to peptidoglycan. *J. Bacteriol.* **125**:872–879.
4. **Hancock, I. C., G. Wiseman, and J. Baddiley.** 1976. Biosynthesis of a unit that links teichoic acid to the bacterial wall: inhibition by tunicamycin. *FEBS Lett.* **69**:75–80.
5. **Karamata, D., H. M. Pooley, and M. Monod.** 1987. Expression of heterologous genes for wall teichoic acid in *Bacillus subtilis* 168. *Mol. Gen. Genet.* **297**:73–81.
6. **Lyon, B. R., and R. Skurray.** 1987. Antimicrobial resistance of *Staphylococcus aureus*: genetic basis. *Microbiol. Rev.* **51**:88–134.
7. **Monod, M., H. M. Pooley, and D. Karamata.** 1983. *Bacillus subtilis* W23/168 recombinants with hybrid cell walls and phage resistance patterns, p. 285–290. *In* R. Hakenbeck, J.-V. Höltje, and H. Labischinski (ed.), *The Target of Penicillin.* Walter de Gruyter, Berlin.
8. **Nomura, S., K. Yamane, T. Sasaki, M. Yamasaki, G. Tamura, and B. Maruo.** 1978. Tunicamycin-resistant mutants and chromosomal locations of mutational sites in *Bacillus subtilis. J. Bacteriol.* **136**:818–821.
9. **Rogers, H. J., H. R. Perkins, and J. B. Ward.** 1980. *Microbial Cell Walls and Membranes,* p. 219–221. Chapman & Hall, Ltd., London.
10. **Rogers, H. J., and C. Taylor.** 1979. Autolysins and shape change in *rodA* mutants of *Bacillus subtilis. J. Bacteriol.* **135**:1032–1042.
11. **Rogers, H. J., P. F. Thurman, C. Taylor, and J. R. Reeve.** 1974. Mucopeptide synthesis by *rod* mutants of *Bacillus subtilis. J. Gen. Microbiol.* **85**:335–350.
12. **Ward, J. B.** 1981. Teichoic and teichuronic acids: biosynthesis, assembly, and location. *Microbiol. Rev.* **45**:211–243.
13. **Ward, J. B., A. W. Wyke, and C. A. M. Curtis.** 1980. The effect of tunicamycin on wall synthesis in *Bacilli. Biochem. Soc. Trans.* **8**:164–166.
14. **Yamasaki, M., A. Takatsuki, and G. Tamura.** 1981. Tunicamycin in microbiology, p. 73–96. *In* G. Tamura (ed.), *Tunicamycin.* Japan Scientific Societies Press, Tokyo.
15. **Young, F. E.** 1967. Requirement of glucosylated teichoic acid for adsorption of phage in *Bacillus subtilis* 168. *Proc. Natl. Acad. Sci. USA* **58**:2377–2384.
16. **Young, F. E., C. Smith, and B. E. Reilly.** 1969. Chromosomal location of genes regulating resistance to bacteriophage in *Bacillus subtilis. J. Bacteriol.* **98**:1087–1097.

IX. RESISTANCE

Chapter 63

Overproduction of Penicillin-Binding Protein 4 in *Staphylococcus aureus* Is Associated with Methicillin Resistance

Nafsika H. Georgopapadakou
Lisa M. Cummings
Eugene R. LaSala
Joel Unowsky
David L. Pruess

Stapyhylococcus aureus, a major human pathogen, is normally susceptible to β-lactam antibiotics, which act by inhibiting enzymes essential for cell wall biosynthesis (4, 9). These enzymes are most likely peptidoglycan transpeptidases. Both they and the related DD-carboxypeptidases covalently bind penicillin and are detected as penicillin-binding proteins (PBPs) (9). Of the four or five PBPs normally found in *S. aureus* (PBP 1, 87 kilodaltons [kDa]; PBP 2, 80 kDa; PBP 3, 75 kDa; PBP 3′, 70 kDa; and PBP 4, 41 kDa), PBPs 2 and 3 are essential and may be involved, respectively, in cell wall growth and septation (2).

There are three mechanisms of resistance to β-lactam antibiotics in *S. aureus* (6): (i) β-lactamase-mediated inactivation through hydrolysis of the β-lactam nucleus, (ii) PBP-associated intrinsic resistance due to the lowering of the affinity of PBPs or the acquisition of new PBPs, and (iii) tolerance to the bactericidal effect of β-lactam antibiotics as a result of inhibition of autolysins.

Nafsika H. Georgopapadakou, Lisa M. Cummings, Eugene R. LaSala, Joel Unowsky, and David L. Pruess • Roche Research Center, Nutley, New Jersey 07110.

597

Our studies of the mechanism of methicillin resistance in a laboratory strain of *S. aureus* show overproduction of PBP 4 and suggest that methicillin resistance can occur by changes other than the well-documented appearance of a new PBP, 2a or 2' (7, 10).

Organisms and Reference Antibiotics

S. aureus 67 and Evans were generous gifts from, respectively, A. Tomasz, Rockefeller University, and G. Best, Medical College of Georgia. *S. aureus* ATCC 25923 and ATCC 6538P were obtained from the American Type Culture Collection (Rockville, Md.).

Benzylpenicillin and vancomycin were obtained from Eli Lilly & Co. (Indianapolis, Ind.); methicillin, from Bristol Laboratories (Syracuse, N.Y.); cefoxitin and cycloserine, from Merck Sharp & Dohme (Rahway, N.J.); tetracycline, from Pfizer Inc. (Groton, Conn.); gentamicin, from Schering-Plough Corp. (Bloomfield, N.J.); and fleroxacin, from Roche Laboratories.

Assays

MICs were determined by the agar diffusion or agar dilution method in Mueller-Hinton agar or antibiotic seed agar.

The PBP assay was carried out with solubilized membranes from sonicated bacteria (3, 8). PBP binding was measured as inhibition of binding of [^{14}C]benzylpenicillin.

The DD-carboxypeptidase-transpeptidase assay was performed with solubilized *S. aureus* membranes and [^{14}C]diac-L-Lys-D-Ala-D-Ala as substrate (carboxypeptidase activity) or diac-L-Lys-D-Ala-D-Ala and [^{14}C]glycine as substrates (transpeptidase activity) (5).

The autolysin assay was performed with freeze-thaw extracts of *S. aureus* and [^{14}C]glycine-labeled *S. aureus* ATCC 25923 treated with trichloroacetic acid and tryspin (1). The liberation of the ^{14}C label from the insoluble substrate was monitored.

Antibiotic Profile of Methicillin-Resistant Strain 3037B

The methicillin-resistant strain 3037B was obtained from *S. aureus* ATCC 6538P by serial subculturing on increasing concentrations of benzylpenicillin (Fig. 1). The antibiotic profiles of these two strains are shown in Table 1. Also shown are the antibiotic profiles of *S. aureus* ATCC 25923, a standard susceptible strain; *S. aureus* Evans, a standard tolerant strain (1); and *S. aureus*

<u>S. aureus</u> 6538P

(penicillin sensitive)

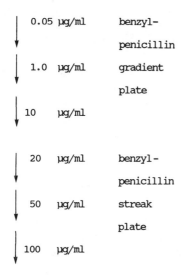

| 0.05 μg/ml | benzyl-
| ↓ | penicillin
| 1.0 μg/ml | gradient
| ↓ | plate
| 10   μg/ml |

| 20   μg/ml | benzyl-
| ↓ | penicillin
| 50   μg/ml | streak
| ↓ | plate
| 100   μg/ml |

<u>S. aureus</u> 3037B

(penicillin resistant)

Figure 1. Isolation of *S. aureus* 3037B. For each passage, the highest concentration of benzylpenicillin allowing growth is indicated.

Table 1. Antibiotic susceptibility profile of methicillin-resistant *S. aureus* 3037B and some reference strains

	MIC (μg/ml)[a] for strain:				
Antibiotic	3037B (mutant)	6538P (parent)	25923	67	Evans
Benzylpenicillin	>128	<0.02	0.03	16	32
Methicillin	32	1	1	>128	4
Cefoxitin	1	2	2	>128	4
Vancomycin	0.2	2	1	2	1
Cycloserine	32	32	16	16	
Tetracycline	0.06	0.12	0.12	0.12	0.12
Gentamicin	0.5	0.5	0.5	0.5	0.25
Fleroxacin	0.5	0.5	0.5	0.5	0.5

[a]Agar dilution in Mueller-Hinton agar.

67, a typical methicillin-resistant strain. Strains 3037B and 67 were both resistant to methicillin and benzylpenicillin; however, 3037B was susceptible to cefoxitin while 67 was resistant. The Evans strain had normal sensitivity to all antibiotics tested, with the exception of benzylpenicillin. Thus, under assay conditions, tolerance did not generally manifest itself as resistance.

As expected, the tolerant Evans strain did not show an increase in autolytic activity after exposure to 10 μg of methicillin per ml. Neither did 3037B or ATCC 6538P (data not shown), although when grown in the absence of methicillin, their autolysin levels were comparable to literature values for other strains (1). Surprisingly, no autolytic activity could be detected in *S. aureus* ATCC 25923 under assay conditions, casting some doubt on the usefulness of this assay.

Resistance of *S. aureus* 3037B to methicillin was atypical; that is, it was independent of growth conditions such as pH, temperature, and salt (Table 2).

PBP Profile of Methicillin-Resistant Strain 3037B

The PBP profile of *S. aureus* 3037B showed a 25-fold increase in PBP 4 and the appearance of a ~100-kDa PBP (Fig. 2A). Both PBP 4 and the 100-kDa PBP were relatively resistant to β-lactam antibiotics except for cefoxitin (Fig. 2B). Both PBPs possessed β-lactamase activity.

PBP 4 is a DD-carboxypeptidase with secondary peptidoglycan cross-linking activity (11). Consistent with PBP 4 overproduction, there was a marked increase in DD-carboxypeptidase-transpeptidase activity in *S. aureus* 3037B (Table 3). Activity was strongly inhibited by submicromolar concentrations of cefoxitin. In preliminary experiments, PBP 4 was partially purified from solubilized membranes by affinity chromatography on 6-aminopenicillanic acid-Sepharose. Preincubation with 2 μg of cephalothin per ml blocked the binding of all other PBPs with the exception of the 100-kDa PBP.

Table 2. MICs of methicillin for methicillin-resistant *S. aureus* 3037B and the methicillin-susceptible parent strain ATCC 6538P under various conditions of growth

| | MIC (μg/ml) under the growth conditions indicated: | | | | | | | |
| | pH[a] | | | Temp[a] | | | NaCl[b] | |
Strain	7.1	5.9	5.1	36°C	32°C	26°C	0%	4%
ATCC 6538P	4	2	2	2	2	2	0.5	0.2
3037B	125	125	NG[c]	125	125	62	16	8

[a]Agar diffusion in antibiotic seed agar.
[b]Agar dilution in antibiotic seed agar.
[c]NG, No growth.

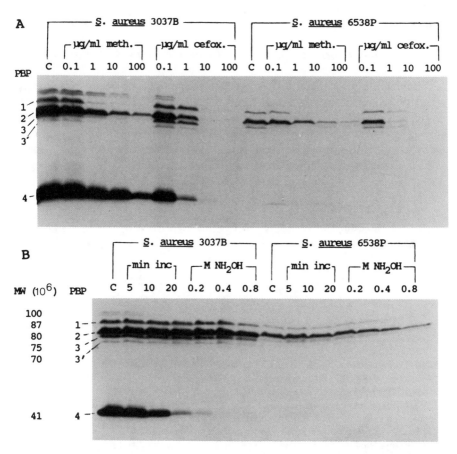

Figure 2. (A) Binding of methicillin (meth.) and cefoxitin (cefox.) to *S. aureus* 3037B and ATCC 6538P PBPs. (B) Release of bound penicillin by PBPs in the absence (min inc) and presence of hydroxylamine (M NH₂OH).

Table 3. DD-Carboxypeptidase and transpeptidase activity in *S. aureus* 3037B and ATCC 6538P

| Incubation time (min) | Sp act (cpm/25 µg of protein) in strain: | | | |
| | 3037B | | ATCC 6538P | |
	Hydrolysis	Transpeptidase[a]	Hydrolysis	Transpeptidase
60	9,974	1,742	308	445
90	9,492	2,767	<100	747
120	10,046	3,312	<100	443
150	ND[b]	4,174	ND	698
180	ND	4,673	ND	553

[a]Cefoxitin inhibited activity by 75% at 0.1 µM.
[b]ND, Not determined.

601

Comparison with Other Methicillin-Resistant Strains

Methiçillin-resistant *S. aureus* and *S. epidermidis* have been isolated from a variety of clinical settings. In all cases, resistance has been attributed to overproduction of PBP 2'. Since our strain was isolated in the laboratory and is a slow grower with a prolonged lag phase, it is unlikely that this form of resistance will be clinically significant. However, the relatively accessible PBP 4 is a most plausible model for the penicillin-sensitive peptidoglycan transpeptidases, and its 25-fold increase in strain 3037B should facilitate its isolation and characterization.

In conclusion, the resistance of *S. aureus* 3037B to most β-lactam antibiotics is probably due to the overproduction of PBP 4 or the appearance of the 100-kDa PBP, which may be related (a precursor?) to PBP 4.

LITERATURE CITED

1. **Best, G. K., N. H. Best, and A. V. Koval.** 1974. Evidence for participation of autolysins in bactericidal action of oxacillin on *Staphylococcus aureus*. *Antimicrob. Agents Chemother.* **6:**825–830.
2. **Georgopapadakou, N. H., B. A. Dix, and Y. R. Mauriz.** 1986. Possible physiological functions of penicillin-binding proteins in *Staphylococcus aureus*. *Antimicrob. Agents Chemother.* **29:**333–336.
3. **Georgopapadakou, N. H., and F. Y. Liu.** 1980. Binding of β-lactam antibiotics to penicillin-binding proteins of *Staphylococcus aureus* and *Streptococcus faecalis*: relation to antibacterial activity. *Antimicrob. Agents Chemother.* **18:**148–157.
4. **Georgopapadakou, N. H., and F. Y. Liu.** 1980. Penicillin-binding proteins in bacteria. *Antimicrob. Agents Chemother.* **18:**834–836.
5. **Kozarich, J. W., and J. L. Strominger.** 1978. A membrane enzyme from *Staphylococcus aureus* which catalyzes transpeptidase, carboxypeptidase, and penicillinase activities. *J. Biol. Chem.* **253:**1272–1278.
6. **Lyon, B. R., and R. Skurray.** 1987. Antimicrobial resistance of *Staphylococcus aureus*: genetic basis. *Microbiol. Rev.* **51:**88–134.
7. **Rossi, L., E. Tonin, Y. R. Cheng, and R. Fontana.** 1985. Regulation of penicillin-binding protein activity: description of a methicillin-inducible penicillin-binding protein in *Staphylococcus aureus*. *Antimicrob. Agents Chemother.* **27:**828–831.
8. **Spratt, B. G.** 1977. Properties of the penicillin-binding proteins of *Escherichia coli*. *Eur. J. Biochem.* **72:** 341–352.
9. **Spratt, B. G.** 1983. Penicillin-binding proteins and the future of β-lactam antibiotics. *J. Gen. Microbiol.* **129:**1247–1260.
10. **Ubukata, K., N. Yamashita, and M. Konno.** 1985. Occurrence of a β-lactam-inducible penicillin-binding protein in methicillin-resistant staphylococci. *Antimicrob. Agents Chemother.* **27:**851–857.
11. **Wyke, A. W., J. B. Ward, M. V. Hayes, and N. A. C. Curtis.** 1981. A role in vivo for methicillin-binding protein 4 of *Staphylococcus aureus*. *Eur. J. Biochem.* **119:**389–393.

Chapter 64

Mechanisms and Significance of Antimicrobial Resistance in Enterococci

S. Zighelboim-Daum
R. C. Moellering, Jr.

Enterococci are unique among streptococci in their resistance to a variety of antimicrobial agents. In this chapter we discuss a number of features of this resistance, emphasizing in particular recent findings on the mechanism of intrinsic resistance to β-lactams.

Enterococci belong to the Lancefield group D streptococci, all of which contain a glycerol lipoteichoic acid in their membrane (group D-specific antigen). The group D streptococci have traditionally been divided into two physiologically distinct groups as follows: (i) group D enterococci, including *Streptococcus faecalis, S. faecium*, and *S. durans*, all of which have an unusual capacity to survive in the presence of β-lactam antibiotics, and (ii) group D nonenterococci, encompassing organisms like *S. bovis* and *S. equinus*, which are usually very susceptible to penicillin.

Recent biochemical, immunological, and DNA hybridization studies have demonstrated that *S. faecalis* and *S. faecium* are only distantly related to the nonenterococcal streptococci of group D, and a new genus, *Enterococcus*, has been proposed to include these organisms (30). In addition to *E. faecalis* and *E. faecium* the following streptococci would be included in this group: *E. avium, E. casseliflavus, E. durans, E. gallinarum, E. hirae*, and *E. mundtii*. However, only *E. faecalis, E. faecium, E. durans*, and probably *E. avium* are clinically important in humans.

Strain ATCC 9790 has been extensively used in the study of enterococcal

S. Zighelboim-Daum and R. C. Moellering, Jr. • New England Deaconess Hospital, Harvard Medical School, Boston, Massachusetts 02215.

resistance to β-lactam antibiotics, and its taxonomic status should be reviewed briefly. This strain was originally thought to be *S. faecalis*, but was reclassified by Smith in 1970 (31) as *S. faecium* var. *durans*. In 1983, Williamson et al. (37) confirmed that the penicillin-binding protein (PBP) pattern of this strain resembled the pattern seen in other clinical isolates of *S. faecium*, even though the MIC of penicillin for this strain was relatively low when compared with those for other *S. faecium* strains. In 1985, Farrow and Collins (11), carrying out DNA-DNA hybridization, fatty acid analysis, and detailed biochemical tests, found that *S. faecium* ATCC 9790 and other atypical *S. faecium* strains from chicken and hog intestines form a phenotypically distinct taxon which was named *Enterococcus hirae*, with strain ATCC 9790 as the type strain. Furthermore, *E. hirae* ATCC 9790 was demonstrated to be different from *S. durans* homology group I (now named *E. durans*), but identical to *S. durans* homology group II (now included with *E. hirae*) and to have less than 30% homology with other strains of enterococci (20).

Enterococci are normal inhabitants of the gastrointestinal tract, from which, under the appropriate circumstances, they can invade the host to cause a variety of diseases: endocarditis, urinary tract infections, intra-abdominal abscesses, soft tissue infections, and bacteremia.

The treatment of enterococcal endocarditis, bacteremia, and other infection for which bactericidal therapy is needed poses a real therapeutic challenge, since enterococci have a unique ability to survive in the presence of antibiotics.

Antibiotic Resistance in Enterococci

The remarkable capacity of enterococci to evade antimicrobial agents is due not only to their innate resistance to antibiotics (24), but also to the fact that they are able to acquire new resistance traits with great facility by transfer of plasmids and transposons, mainly by conjugation (19). Not only are they able to conjugate with one another, but conjugal transfer of genetic information has been demonstrated between enterococci and other streptococci, *Staphylococcus aureus*, and *Staphylococcus epidermidis* (10), thus providing an almost limitless potential for the acquisition and dissemination of new genetic determinants among these organisms.

The antibiotic resistance of enterococci appears to be of two types, intrinsic and acquired. Intrinsic resistance is found in all strains regardless of previous exposure to antimicrobial agents and includes (i) low-level resistance to lincomycin, clindamycin (MIC, ≤40 μg/ml), and aminoglycosides (MIC, ≤1,000 μg/ml) and (ii) the ability to survive in the presence of β-lactams. The low-level resistance to aminoglycosides appears to be due to the inability of these agents to penetrate effectively into the cell (22, 25). For the aminoglycosides, it has been proposed that the cell envelope is an important determinant of low-level resistance, since uptake of these agents is enhanced specifically by antimicrobial agents that alter the cell envelope (27). In addition, low-level resistance to kanamycin, tobramycin, and netilmicin in *S. faecium* has been shown to be related to the production of an aminoglycoside-modifying enzyme (AAC 6′), which may be chromosomally mediated (35). The ability to survive in the presence of β-lactams is due to at least two different mechanisms which may or may not be superimposed. The first is intrinsic resistance, which results in the need for high antibiotic concentrations to prevent growth (i.e., the MIC is increased); the mechanism of this intrinsic resistance to β-lactams is reviewed in detail in later sections of this chapter. The second mechanism is tolerance, which results in a slow rate of killing even when the bacteria are exposed to penicillin concentrations equal to or greater than the MIC; the problem of tolerance is reviewed in another chapter.

Acquired resistance is present in some strains of enterococci and includes (i) resistance to tetracycline, erythromycin, and chloramphenicol; (ii) high-level resistance to lincomycin, clindamycin, and aminoglycosides (25); and (iii) production of β-lactamases, a property recently acquired by some strains of *S. faecalis* (28).

Enterococcal resistance to tetracycline, erythromycin, and high levels of lincomycin and clindamycin is mediated by plasmids or transposons, but the mechanisms by which plasmid-containing strains are made resistant to the corresponding drugs has not been determined (22, 24, 25). Resistance to chloramphenicol is due to plasmids that code for an enzyme (chloramphenicol acetyltransferase) capable of inactivating the drug (3).

In general, it appears that high-level resistance to aminoglycosides other than streptomycin is plasmid mediated and is due to the production of enzymes capable of acetylating, phosphorylating, or adenylylating these agents (3, 21, 35). High levels of resistance to streptomycin can be obtained via two different mechanisms. (i) Some strains produce a plasmid-mediated adenylyltransferase, an enzyme seen in both *S. faecium* and *S. faecalis* (21). This plasmid usually codes for a phosphotransferase (APH 3′) that mediates resistance to kanamycin and amikacin, and most, but not all, clinical isolates with high-level resistance to streptomycin produce this enzyme. (ii) Other strains show a ribosomal alteration that prevents streptomycin from inhibiting protein synthesis. This mechanism was originally described in laboratory mutants (41), but is now known to be present in up to 5% of clinical isolates with high resistance to streptomycin (8).

High-level resistance to kanamycin, neomycin, and amikacin in *S. faecalis* and *S. faecium* is mediated by a phosphotransferase (APH 3′), which may be coded for by the same plasmid that carries the gene for the production of streptomycin-inactivating enzymes (21). High-level resistance to kanamycin, netilmicin, sisomicin, and tobramycin appears to be mediated by the production of an acetyltransferase (AAC 6′), which is plasmid mediated in *S. faecalis*. This enzyme appears to be chromosomally determined in *S. faecium*, which is resistant to synergistic killing by combinations of penicillin with kanamycin, netilmicin, sisomicin, or tobramycin, but is not highly resistant to the aminoglycosides (35).

Acquired resistance to gentamicin can be achieved in two ways: (i) by production of a plasmid-mediated phosphotransferase (APH 2″) in *S. faecalis*, which results in high-level resistance to gentamicin (38), or (ii) by a specific transport defect for gentamicin that leads to decreased uptake of the drug (26). Only one clinical isolate of *S. faecalis* has so far been described with this latter type of resistance, and the MIC of gentamicin was not very high (MIC/MBC, 8/500 µg/ml).

It is clear that enterococci cannot be killed by either β-lactams or aminoglycosides when used alone at concentrations achievable in serum. Therefore, synergistic combinations of β-lactams plus an aminoglycoside are required for the treatment of infections against which bactericidal therapy is needed. In general, concentrations of aminoglycosides well below their MICs produce synergistic killing when combined with agents such as penicillin or ampicillin (23). The molecular basis for penicillin-aminoglycoside synergism against enterococci has been extensively studied. It is believed that low-level resistance to the aminoglycosides is due to the inability of the drugs to penetrate the bacterial cell to bind their ribosomal targets. Thus, it has been proposed that the β-lactam antibiotic inhibits peptidoglycan synthesis, causing an alteration in the cell envelope that leads to enhanced uptake of the aminoglycoside (27).

Given the above observations, it is reasonable to conclude that β-lactam-aminoglycoside synergism against enterococci will not occur in the following circumstances: (i) organisms that acquire high-level resistance to the aminoglycoside will also become resistant to penicillin-aminoglycoside synergism (21, 22, 25, 26), and (ii) any increase in the MIC of the β-lactam beyond levels achievable in vivo would also render the combination ineffective. The resistance of enterococci to antistaphylococcal penicillins and cephalosporins appears to be an example of the latter phenomenon.

Major research efforts have been and are under way to try to understand the biochemical basis of the intrinsic resistance of enterococci to β-lactams and the factors that may contribute to make them even more resistant to penicillin.

Mechanisms of β-Lactam Resistance in Enterococci

It has been proposed that penicillin kills bacteria in a biphasic process. In the primary phase, the attachment of penicillin to PBPs in the plasma membrane causes inhibition of penicillin-sensitive enzymes, and this inhibition results in an interference with cell wall synthesis, which is not lethal in itself but only causes a reversible inhibition of growth. In the secondary phase, the signal(s) generated by the inhibition of peptidoglycan synthesis causes the dissociation of an autolysin-inhibitor complex, with secretion of the inhibitor into the media and the activation of the autolysin, which is then free to cause various degrees of structural damage to the cell wall that result in death or lysis, or both (irreversible effects) (32, 33). In general, it can be said that any factors that interfere with the primary phase of penicillin action will lead to development of resistance (higher MIC), while interference with the secondary phase will result in tolerance (decreased rate of killing).

Resistance to β-lactam antibiotics may occur by any of the following mechanisms. (i) β-Lactamases may be produced that inactivate the drugs before they reach their targets in the plasma membrane; recently, a few strains of *S. faecalis* have been found to produce a plasmid-mediated β-lactamase (28). (ii) Permeability to the antibiotics may decrease; the outer membrane of gram-negative bacteria plays a significant role in increased resistance in these organisms (29), but a substantial barrier to the penetration of β-lactams in gram-positive bacteria, including enterococci, has yet to be demonstrated. (iii) Alterations may be made in the target proteins (penicillin-sensitive enzymes, PBPs). Theoretically, these may involve a decrease in the affinity of the PBPs for the drug, in which case the antibiotic would still be effective but higher concentrations would be required; the affinity could be so low that the PBP(s) would no longer bind the drugs but would still carry out its biological functions, thus ensuring the survival of the bacteria. Changes in the quantity of PBPs present in the membrane may also be involved: increased resistance could conceivably be paralleled by an increase in the amount of enzyme available for the interaction with the antibiotic. Other possible alterations are an increase in the turnover rate of the enzymes, so that the inactivated PBPs are rapidly replaced by new, active proteins, and the appearance of new or modified proteins with altered affinity for the β-lactam, whose presence bypasses the effect of the drug.

PBP alterations associated with intrinsic resistance have been demonstrated in a number of bacterial species. In the penicillin-resistant gonococci, the PBP changes seem to be restricted to lower antibiotic affinities (5–7). Overproduction of one PBP has been described for *Bacillus megaterium* (16) and *Pseudomonas aeruginosa* (17). In *Staphylococcus aureus* resistance seems to be associated with the acquisition of a new PBP with extremely low affinity for the β-lactams (2, 18). *Streptococcus pneumoniae* exhibits one of the most complex mechanisms described so far: low and intermediate levels of resistance are associated with a gradual decrease in the penicillin affinity of several of the pneumococcal PBPs (similar to resistant gonococci), while high levels of penicillin resistance are accompanied by acquisition of a new PBP (similar to staphylococci) (34, 39, 40).

The biochemical basis for the intrinsic penicillin resistance of enterococci has been the object of intensive study in recent years. With the exception of a few strains of *S. faecalis* (28), enterococci do not produce β-lactamases, and their cell wall poses no barrier to the penetration of β-lactams. Therefore, most of the research has centered on the enterococcal PBPs, and a review of our current knowledge in this field follows.

Mechanisms of Intrinsic β-Lactam Resistance in S. faecium

Given that *S. faecium* is the most resistant of the enterococci (MICs of penicillin ranging between 16 and 64 μg/ml), it is not surprising that most of the research on intrinsic penicillin resistance of enterococci has focused on this organism and its PBPs. The basic tool for the study of PBPs is the labeling of these penicillin-sensitive enzymes by incubating either membranes (in vitro) or whole cells (in vivo) with radiolabeled (^{14}C or ^{3}H) penicillin for a defined period of time. This time has traditionally been 10 to 15 min, and the first studies on intrinsic resistance of *S. faecium* were carried out with this incubation period.

One of these early studies was the original characterization of the PBPs of strain ATCC 9790 by Coyette et al. (4). This strain was found to contain seven PBPs of molecular weights ranging from 43,000 to 140,000, with PBP 3 having the highest affinity for penicillin and PBP 5 having the lowest affinity. None of the PBPs could be identified as the lethal target since (i) inactivation of PBP 3 (the most penicillin-susceptible PBP) was not sufficient to inhibit growth, and (ii) inhibition of growth could be obtained under circumstances in which any of the other PBPs (PBP 1, 2, 4, 5, or 6) escapes inactivation. Therefore, these authors suggested that several PBPs may fulfill similar physiological functions. If so, inactivation of any one protein could be compensated for by another PBP.

A subsequent attempt to identify the killing target came from Fontana et al. (12, 13). Again using strain ATCC 9790, these authors demonstrated that changes in the growth conditions (media, temperature of growth) were associated with alterations in penicillin susceptibility and PBP patterns. For cells growing at a maximal rate (45°C, in synthetic medium), the MIC of penicillin was lower (0.04 μg/ml), and growth could be inhibited through binding to PBP 3 alone. In contrast, the MIC of penicillin for cells growing at a slower rate (32°C, in synthetic medium) was higher (4.0 μg/ml), and at that concentration all PBPs bound radioactivity and PBP 3 was not the lethal target. They proposed that the physiological importance of PBPs was heavily influenced by growth conditions and that PBP 3 could have an essential role at 45°C but not at 32°C, as a result of changes in the physiological status of the PBPs.

Williamson et al. (37) tried to explain the difference in penicillin susceptibility of clinical isolates of *S. faecium* (MIC, 37.2 μg/ml), *S. faecalis* (MIC, 4.8 μg/ml), and *S. bovis* (MIC, 0.036 μg/ml) by comparing the penicillin affinities of their PBPs. The PBPs of *S. bovis* appeared to have higher affinities than those of *S. faecalis*, which in turn were higher than those of *S. faecium*. However, no quantitative correlation could be established between the binding of penicillin to any one PBP or groups of PBPs and the MIC for the corresponding organisms. In addition, their studies appear to suggest that the intrinsic resistance of enterococci to cephalosporins is related to altered affinity of the lower-molecular-weight PBPs for these drugs.

A further attempt to explain the intrinsic resistance of *S. faecium* to penicillin was carried out by Eliopoulos et al. (9), who compared the PBP patterns of clinical isolates of *S. faecium* with those of hypersusceptible mutants derived from them by novobiocin treatment. No difference in the PBPs of the resistant and hypersusceptible mutants could be detected. It was concluded that the marked differences seen in the MICs of penicillin for the mutants could not be attributed to differences in their PBPs.

A major breakthrough in the understanding of penicillin resistance in enterococci occurred when Fontana et al. (14) modified the PBP assay by extending the time the cells are exposed to [^{3}H]penicillin from 10 min to 60 min. When this protocol was applied to strain ATCC 9790, it was found that this organism contains a PBP (PBP 5) that (i) reacts very slowly with penicillin, taking it 20 min to bind enough radioactivity to be detected and 60 min to become saturated, and (ii) has a very low affinity for penicillin, being eightfold lower than the second most resistant PBPs (PBPs 4 and 6) and 100-fold lower than that of the most susceptible PBP (PBP 3).

It was proposed that PBP 5 is responsible for both the natural low susceptibility of *S. faecium* to penicillin and the increase in penicillin resistance seen in mutants derived from strain ATCC 9790 by serial passage on increasing penicillin concentrations. Thus, PBP 5 is overproduced in mutants with increased penicillin resistance (14). When PBP 5 is overproduced, it becomes the only essential PBP; i.e., it takes over the function of all the other PBPs and fully compensates for their activity when they are inhibited. Therefore, these cells can grow normally (whether at slow or fast rate) in the presence of penicillin concentrations that saturate all PBPs except PBP 5, and since the affinity of PBP 5 for penicillin is so low, the cells are very resistant to penicillin. Growth inhibition will occur only at concentrations saturating the overproduced PBP 5.

In addition to its presence in strain ATCC 9790, PBP 5 is found in all streptococcal strains that show intrinsic resistance to penicillin and is absent from all susceptible species studied (14). In strains such as ATCC 9790 in which PBP 5 is present in normal amounts (not overproduced), this PBP is the target for growth inhibition when the cells are growing at a slow rate, and growth is inhibited in the presence of penicillin concentrations fully saturating three or four different PBPs and partially saturating PBP 5. It is assumed that the capability of such a PBP to take over the functions of other PBPs is proportional to its amount and could be lost by its partial inactivation by penicillin. It has been proposed that in cells growing at a fast rate (for which the MIC of penicillin is lower) the target of inhibition shifts to PBP 3, which then becomes the crucial target. Since PBP 3 has a high affinity for penicillin, it can be saturated with lower concentrations of the drug, and the cells are more sensitive to penicillin (1).

When PBP 5 is lost or substantially reduced by mutation, the cells become hypersusceptible to penicillin. In the mutants derived by Fontana et al. (15) from strain ATCC 9790, PBP 5 is absent, and the hypersusceptible mutants obtained by Eliopoulos et al. (9) from clinical isolates show a two-thirds reduction in the amount of PBP 5 (36; G. M. Eliopoulos, personal communication). These latter mutants were originally reported to have the same PBP pattern as the penicillin-resistant parental strain, but when retested by exposing the cells to [³H]penicillin for 60 min (instead of the original 10 min), the substantial reduction in the amount of PBP 5 was found. In the absence of PBP 5 there is no PBP able to compensate for the inactivation of other PBPs by penicillin, and the target for growth inhibition appears to be PBP 3. Even a partial saturation of this high-affinity PBP will cause growth inhibition. It therefore appears that PBP 5 is dispensable in undisturbed cells but plays an important role in protecting bacteria against the action of penicillin (1, 14).

More recently, Williamson et al. (36) reported some hypersusceptible mutants derived from clinical isolates of S. faecium in which PBP 5 is present in significantly reduced amounts, with the concomitant appearance of a new higher-affinity PBP (PBP 5*). A survey of PBPs of various clinical isolates revealed that the strains could be divided into six groups on the basis of the amounts of PBP 5, the presence or absence of PBP 5*, and the affinity of PBP 5* for benzylpenicillin. It was proposed that in S. faecium the level of resistance to benzylpenicillin is related to the relative amounts and the relative affinities of these two essential PBPs (PBP 5 and PBP 5*), which are by themselves capable of sustaining growth in the presence of penicillin concentrations below the MIC.

Mechanism of Intrinsic β-Lactam Resistance in S. faecalis

With very few exceptions (14), most of our knowledge on intrinsic penicillin resistance in enterococci has been derived from the study of strain ATCC 9790 (1, 4, 12–15) and some clinical isolates of S. faecium (9, 36, 37), and very little is known about S. faecalis, the most common enterococcus. Even if E. hirae ATCC 9790 had really been an S. faecium strain, there is no reason to believe that results obtained with S. faecium can be extrapolated to S. faecalis. These two species are both enterococci, but they are completely different in their biochemical properties, PBP patterns, and peptidoglycan structure (the end result of PBP activity).

Studying resistant mutants derived from *S. faecalis*, Fontana et al. (14) suggested that increased resistance to penicillin in these organisms was also due to an overproduction of PBP 5.

We are currently studying the mechanisms of intrinsic penicillin resistance in *S. faecalis*. For this purpose, we have selected a group of human isolates collected 20 years ago in the highlands of Malaita, one of the Solomon Islands (R. C. Moellering, Jr., personal communication). These strains are unique in that they have never been exposed to antibiotics and thus they represent a penicillin-"virgin" population, which offers an opportunity to study the evolution of penicillin resistance in enterococci. Tests of the penicillin susceptibility of seven of the Solomon Islands strains indicated that the MIC of benzylpenicillin was 3.1 μg/ml for strains SI-E1, SI-E39, SI-E45, SI-E47, and SI-E51 and 1.6 μg/ml for strains SI-E38 and SI-E44. These MICs are in a range very similar to that seen in U.S. isolates.

We studied the PBPs of *S. faecalis* SI-E39 by labeling them in vivo with [³H]penicillin. In these experiments, 1-ml amounts of log-phase cultures of SI-E39 (approximately 2×10^8 bacteria per ml) were rapidly chilled in ice and centrifuged at 5,000 rpm for 10 min at 2°C. The pellet was suspended in 100 μl of growth medium (dextrose-phosphate broth) containing different concentrations of [³H]penicillin, and samples were incubated at 37°C for 60 min. The samples were then placed on ice, and excess unlabeled penicillin was added to stop the reaction. The cells were recovered by centrifugation (5,000 rpm, 10 min at 2°C), suspended in 50 μl of potassium phosphate buffer (10 mM, pH 6.8) containing 10 μg of lysozyme, 10 μl of M1-muramidase, and 0.1% (wt/vol) Triton X-100, and incubated for 20 min at 37°C. Subsequently, 2.5 μl of 20% Sarkosyl was added, and the incubation was continued for 5 min at 37°C, when the lysis was completed. The PBPs in the lysates were analyzed by sodium dodecyl sulfate-polyacrylamide gel electrophoresis as previously described (39), except that the final concentrations of acrylamide and *N,N*-bis-acrylamide in the separating gels were reduced to 6% (wt/vol) and 0.08% (wt/vol), respectively, to enable better resolution of the individual PBPs. The results of these experiments show that *S. faecalis* SI-E39 has six PBPs, of molecular weights ranging from 43,000 to 115,000 (Fig. 1), and that the PBP

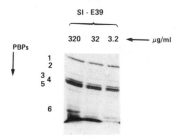

Figure 1. PBPs of *S. faecalis* SI-E39.

pattern is remarkably similar to that of American isolates for which the MIC of penicillin is the same (data not shown).

The fact that the penicillin MICs and the PBP patterns from the pre-antibiotic-era Solomon Islands strains are similar to those of current isolates of *S. faecalis* suggests that the particular complement of PBPs present in all *S. faecalis* strains confers the intermediate level of resistance which is commonly found in *S. faecalis* and that this level has not increased despite extensive exposure to antibiotics in the clinical setting.

In an attempt to better understand the mechanism of the intrinsic penicillin resistance of *S. faecalis*, the level of penicillin resistance of strain SI-E39 was increased in a stepwise process by serial passage of this strain on plates containing increasing concentrations of penicillin. A series of isogenic strains that differ only in their level of penicillin resistance was obtained (pen 3.1, pen 6.2, pen 12.5, and pen 31.0). The PBPs of these isogenic strains were labeled as described above and compared with those of the parental strain SI-E39 (Fig. 2). Two major changes parallel the increase in β-lactam resistance. (i) As the level of resistance increases, there is a parallel increase in the quantity

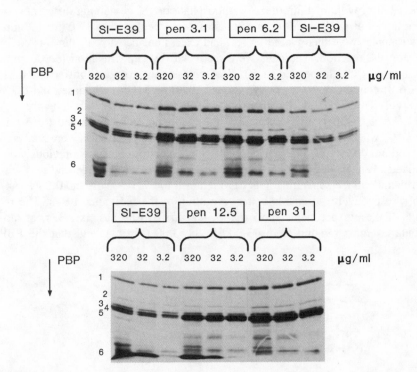

Figure 2. Comparison of PBPs of *S. faecalis* SI-E39 and a series of isogenic penicillin-resistant strains.

of PBPs 4 and 5. This change seems to be analogous to the findings of Fontana et al. (14) for strain ATCC 9790 and other streptococci, that mutants with increased levels of resistance overproduce the low-affinity PBP 5. It remains to be seen whether PBP 4 and PBP 5 of the isogenic penicillin-resistant Solomon Islands strains are indeed low-affinity PBPs. (ii) A novel change, not previously described for any enterococcal species, occurs in PBP 1, whose quantity appears to be increased in all the penicillin-resistant mutants. The significance of this finding is currently under investigation.

Whatever turns out to be the relevant explanation for the increased resistance in the isogenic strains, the important question to be asked is why the base-line level of resistance of *S. faecalis* has remained the same from the preantibiotic era (as exemplified by the Solomon Islands strains) to the present time, despite extensive use of antibiotics in clinical practice, and especially since it is relatively easy to increase the MIC for the antibiotic-"virgin" SI-E39 by simply exposing it to antibiotic pressure.

LITERATURE CITED

1. **Canepari, P., M. del Mar Lleo, G. Cornaglia, R. Fontana, and G. Satta.** 1986. In *Streptococcus faecium* penicillin binding protein 5 alone is sufficient for growth at sub-maximal but not at maximal rate. *J. Gen Microbiol.* **132:**625–631.
2. **Chambers, H. F., B. J. Hartman, and A. Tomasz.** 1985. Increased amounts of a novel penicillin binding protein in a strain of methicillin-resistant *Staphylococcus aureus. J. Clin. Invest.* **76:**325–331.
3. **Courvalin, P. M., M. V. Shaw, and A. E. Jacob.** 1978. Plasmid-mediated mechanism of resistance to aminoglycoside aminocyclitol antibiotics and to chloramphenicol in a group D streptococcus. *Antimicrob. Agents Chemother.* **13:**716–720.
4. **Coyette, J., J. M. Ghuysen, and R. Fontana.** 1980. The penicillin binding proteins in *Streptococcus faecalis* ATCC 9790. *Eur. J. Biochem.* **110:**445–456.
5. **Dougherty, T. J.** 1985. Involvement of a change in penicillin target and peptidoglycan structure in low-level resistance to beta-lactam antibiotics in *Neisseria gonorrhoeae. Antimicrob. Agents Chemother.* **28:**90–95.
6. **Dougherty, T. J., A. E. Koller, and A. Tomasz.** 1980. Penicillin binding proteins of penicillin susceptible and intrinsically resistant *Neisseria gonorrhoeae. Antimicrob. Agents Chemother.* **18:**730–737.
7. **Dougherty, T. J., A. E. Koller, and A. Tomasz.** 1981. Competition of beta-lactam antibiotics for the penicillin-binding proteins of *Neisseria gonorrhoeae. Antimicrob. Agents Chemother.* **20:**109–114.
8. **Eliopoulos, G. M., B. F., Farber, B. E. Murray, C. Wennersten, and R. C. Moellering, Jr.** 1984. Ribosomal resistance of clinical enterococcal isolates to streptomycin. *Antimicrob. Agents Chemother.* **25:**398–399.
9. **Eliopoulos, G. M., C. Wennersten, and R. C. Moellering, Jr.** 1982. Resistance to beta-lactam antibiotics in *Streptococcus faecium. Antimicrob. Agents Chemother.* **22:**295–301.
10. **Engel, H. W. B., N. Soedirman, J. A. Rost, W. Van Leuwen, and J. D. A van Embden.** 1980. Transferability of macrolide, lincomycin, and streptogramin resistance between groups A, B, and D streptococci, *Streptococcus pneumoniae*, and *Staphylococcus aureus. J. Bacteriol.* **142:**407–413.

11. **Farrow, J. A., and M. D. Collins.** 1985. *Enterococcus hirae,* a new species that includes amino acid assay strain NCDO 1258 and strains causing growth depression in young chickens. *Int. J. Syst. Bacteriol.* **35**:73–75.

12. **Fontana, R., P. Canepari, G. Satta, and J. Coyette.** 1980. Identification of the lethal target of benzylpenicillin in *Streptococcus faecalis* by *in vivo* penicillin binding studies. *Nature* (London) **287**:70–72.

13. **Fontana, R., P. Canepari, G. Satta, and J. Coyette.** 1983. *Streptococcus faecium* ATCC 9790 penicillin binding proteins and penicillin sensitivity are heavily influenced by growth conditions: proposal for an indirect mechanism of growth inhibition by beta-lactams. *J. Bacteriol.* **154**:916–923.

14. **Fontana, R., R. Cerini, P. Longoni, A. Grossato, and P. Canepari.** 1983. Identification of a streptococcal binding protein that reacts very slowly with penicillin. *J. Bacteriol.* **155**:1343–1350.

15. **Fontana, R., A. Grossato, L. Rossi, Y. R. Cheng, and G. Satta.** 1985. Transition from resistance to hypersusceptibility to beta-lactam antibiotics associated with loss of a low-affinity penicillin-binding protein in a *Streptococcus faecium* mutant highly resistant to penicillin. *Antimicrob. Agents Chemother.* **28**:678–683.

16. **Giles, A. F., and P. E. Reynolds.** 1979. *Bacillus megaterium* resistance to oxacillin accompanied by a compensatory change in penicillin binding proteins. *Nature* (London) **280**:167–169.

17. **Godfrey, A. J., E. Bryan, and H. R. Rabin.** 1981. Beta lactam resistant *Pseudomonas aeruginosa* with modified penicillin-binding proteins emerging during cystic fibrosis treatment. *Antimicrob. Agents Chemother.* **19**:705–711.

18. **Hartman, B. J., and A. Tomasz,** 1984. Low affinity penicillin-binding protein associated with beta lactam resistance in *Staphylococcus aureus. J. Bacteriol.* **158**:513–516.

19. **Jacob, A. E., and S. J. Hobbs.** 1974. Conjugal transfer of plasmid-borne multiple antibiotic resistance in *Streptococcus faecalis* var. *zymogenes. J. Bacteriol.* **117**:360–365.

20. **Knight, R. G., and D. M. Shlaes.** 1986. Deoxyribonucleic acid relatedness of *Enterococcus hirae* and "*Streptococcus durans*" homology group II. *Int. J. Syst. Bacteriol.* **36**:111–113.

21. **Krogstad, D. J., T. R. Korfhagen, R. C. Moellering, Jr., C. Wennersten, M. N. Swartz, S. Persynski, and J. Davies.** 1978. Aminoglycoside-inactivating enzymes in clinical isolates of *Streptococcus faecalis*: an explanation for resistance to antibiotic synergism. *J. Clin. Invest.* **62**:480–486.

22. **Krogstad, D. J., and R. C. Moellering, Jr.** 1982. Genetic and physiological determinants of the enterococcal response to antimicrobial synergism, p. 195–198. *In* D. Schlessinger (ed.), *Microbiology—1982.* American Society for Microbiology, Washington, D.C.

23. **Moellering, R. C., Jr.** 1981. Antimicrobial susceptibility of enterococci; *in vitro* studies of the action of antibiotics alone and in combination, p. 81–96. *In* A. L. Bisno (ed.), *Treatment of Infective Endocarditis.* Grune and Stratton, New York.

24. **Moellering, R. C., Jr.** 1984. Treatment of enterococcal endocarditis, p. 113–133. *In* M. E. Sande, D. Kaye, and R. K. Root (ed.), *Endocarditis.* Churchill Livingston, Ltd., Edinburgh.

25. **Moellering, R. C., Jr., and D. J. Krogstad.** 1979. Antibiotic resistance in enterococci, p. 293–298. *In* D. Schlessinger (ed.), *Microbiology—1979.* American Society for Microbiology, Washington, D.C.

26. **Moellering, R. C., Jr., B. E. Murray, S. C. Shoenbaum, J. Adler, and C. Wennersten.** 1980. A novel mechanism of resistance to penicillin-gentamicin synergism in *Streptococcus faecalis. J. Infect. Dis.* **141**:81–86.

27. **Moellering, R. C., Jr., and A. N. Weinberg.** 1971. Studies on antibiotic synergism against enterococci. II. Effects of various antibiotics on the uptake of ^{14}C-labeled streptomycin by enterococci. *J. Clin. Invest.* **50**:2580–2584.

28. **Murray, B. E., D. A. Church, A. Wanger, K. Zscheck, M. E. Levison, M. J. Ingerman, E. Abrutyn, and B. Mederski-Samoraj.** 1986. Comparison of two beta-lactamase-producing strains of *Streptococcus faecalis. Antimicrob. Agents Chemother.* **30:**861–864.

29. **Nikaido, H.** 1981. Outer membrane permeability of bacteria: resistance and accessibility of targets, p. 249–260. *In* M. Salton and G. D. Shockman (ed.), *Beta-Lactam Antibiotics.* Academic Press, Inc., New York.

30. **Schleifer, K. H., and R. Kilpper-Balz.** 1984. Transfer of *Streptococcus faecalis* and *Streptococus faecium* to the genus *Enterococcus* nom. rev. as *Enterococcus faecalis* comb. nov. and *Enterococcus faecium* comb. nov. *Int. J. Syst. Bacteriol.* **34:**31–34.

31. **Smith, D. G.** 1970. The identity of *Streptococcus faecalis* ATCC 9790. *J. Appl. Bacteriol.* **33:**474–477.

32. **Tomasz, A.** 1979. The mechanism of irreversible antimicrobial effects of penicillin: how the beta-lactam antibiotics kill and lyse bacteria. *Annu. Rev. Microbiol.* **33:**113–137.

33. **Tomasz, A., and S. Waks.** 1975. Mechanism of action of penicillin: triggering of the pneumococcal autolytic enzyme by inhibitors of cell wall synthesis. *Proc. Natl. Acad. Sci. USA* **72:**4162–4166.

34. **Tomasz, A., S. Zighelboim-Daum, S. Handwerger, H. Liu, and H. Qian.** 1984. Physiology and genetics of intrinsic beta-lactam resistance in pneumococci, p. 393–397. *In* L. Leive and D. Schlessinger (ed.), *Microbiology—1984.* American Society for Microbiology, Washington, D.C.

35. **Wennersten, C. B., and R. C. Moellering, Jr.** 1980. Mechanism of resistance to penicillin-aminoglycoside synergism in *Streptococcus faecium,* p. 710–712. *In* J. Nelson and C. Grassi (ed.), *Current Chemotherapy and Infectious Disease,* vol. 1. American Society for Microbiology, Washington, D.C.

36. **Williamson, R., C. L. Bouguenec, L. Gutmann, and T. Horaud.** 1985. One or two low affinity penicillin binding proteins may be responsible for the range of susceptibility of *Enterococcus faecium* to benzylpenicillin. *J. Gen. Microbiol.* **131:**1933–1940.

37. **Williamson, R., S. B. Calderwood, R. C. Moellering, Jr., and A. Tomasz.** 1983. Studies on the mechanism of intrinsic resistance to beta-lactam antibiotics in Group D streptococci. *J. Gen. Microbiol.* **129:**813–822.

38. **Zervos, M. J., T. S. Mikesell, and D. R. Shaberg.** 1980. Heterogenecity of plasmid determining high-level resistance to gentamicin in clinical isolates of *Streptococcus faecalis. Antimicrob. Agents Chemother.* **30:**78–81.

39. **Zighelboim, S., and A. Tomasz.** 1980. Penicillin-binding proteins of multiply antibiotic resistant South African strains of *Streptococcus pneumoniae. Antimicrob. Agents Chemother.* **17:**434–442.

40. **Zighelboim, S., and A. Tomasz.** 1981. Multiple antibiotic resistance in South African strains of *Streptococcus pneumoniae*: mechanism of resistance to beta-lactam antibiotics. *Rev. Infect. Dis.* **3:**267–276.

41. **Zimmerman, R. A., R. C. Moellering, Jr., and A. N. Weinberg.** 1971. Mechanism of resistance to antibiotic synergism in enterococci. *J. Bacteriol.* **105:**873–879.

Chapter 65

Resistance and Tolerance to β-Lactam Antibiotics in Pneumococci and in *Staphylococcus aureus*

Alexander Tomasz

Previous chapters have presented discussions of antibacterial agents targeted on various enzymatic steps in the biosynthesis of bacterial cell walls and their potentials as chemotherapeutic agents. The prompt emergence of bacterial strains resistant to these drugs shows that the ingenuity of clinical chemists is more than matched by the resourcefulness of bacteria in finding countermeasures that allow them to evade the inhibitory effects of these antibiotics. My purpose in this chapter is to provide a brief update on two such relatively novel bacterial mechanisms, intrinsic penicillin resistance and penicillin tolerance, in pneumococci and in staphylococci.

Penicillin resistance, by definition, results in an increase in the MIC of the antibiotic, and one of the most recently observed mechanisms for this involves several different kinds of alterations in the enzymatic targets of β-lactam antibiotics, the penicillin-binding proteins (PBPs). In contrast, antibiotic tolerance does not alter susceptibility to the drug (i.e., the MIC remains unchanged), but it improves bacterial survival during antibiotic treatment. Tolerance appears to have a variety of mechanisms, and it has been proposed that all of these collectively be referred to as survival mutations (18).

The fact that bacterial susceptibility and survival can change independently of each other is unique to the class of antibiotics that inhibit cell wall synthesis, and the basis of this duality of mechanisms is the indirect nature of the antibacterial effects of these antibiotics (13). In the laboratory setting, one can

---

Alexander Tomasz • The Rockefeller University, New York, New York 10021.

select either for penicillin resistance (6, 17) or for penicillin tolerance (8, 20), depending on the specific kind of antibiotic selection applied. Resistant mutants isolated in this manner have elevated MICs and altered (lower antibiotic reactivity) PBPs. On the other hand, tolerant mutants are either defective in autolysin activity or have a variety of defects, some of which appear to be in the regulation of autolytic activity (5). In the clinical setting, β-lactam resistance linked to altered PBPs has by now been detected in many natural isolates, including most of the major human invasive pathogens (15), and it seems likely— although it has not yet been rigorously proved—that strains with improved survival (tolerance) are also being selected by antibiotic pressure in the in vivo environments of bacteria. More detailed recent critical reviews are available for the interested reader on both mechanistic as well as clinical-epidemiological aspects of penicillin resistance and tolerance (4, 15, 18).

In the limited space available here I will concentrate on describing three intriguing observations recently made in our laboratory and take the liberty of presenting some speculations concerning their possible mechanisms. The first of these observations is the apparent "orderliness" in the sequential PBP alterations that accompany gradually increasing levels of penicillin resistance in pneumococci. The second observation concerns the unusually frequent association of penicillin resistance and defective autolysis in clinical isolates of pneumococci. Finally, possible mechanisms for the heterogeneous or homogeneous expression of methicillin resistance in *Staphylococcus aureus* are considered.

Penicillin Resistance in Pneumococci

Figure 1 shows a sketch of the PBP profiles of the penicillin-susceptible pneumococcal laboratory strain R6 (MIC, 0.006 μg/ml) and the highly penicillin-resistant South African strain 8249 (MIC, 6.0 μg/ml), as well as some intermediate-level transformants that have been studied in some detail in my laboratory. Recent studies with antisera directed against PBPs 1a and 2b of the susceptible pneumococci provided some clarification for two of the most peculiar features of the PBP profiles of the resistant strain: the apparent absence of two PBPs (corresponding to PBPs 1a and 2b of the susceptible bacteria) and the presence of a "new" 92-kilodalton PBP 1c in the fluorograms of the resistant cells (21). Development of the membrane gels by immunoblotting showed that PBP 1c gave strong cross-reaction with antibody prepared against PBP 1a, while no reactive material could be detected at the position normally occupied by PBP 1a. Immunoblotting with antisera against PBP 2b of the sensitive cells produced a single band corresponding to the position of PBP 2b in spite of the fact that the penicillin-binding technique could not detect this PBP. The simplest interpretation of these findings is that (i) PBP 1c of the resistant cells is

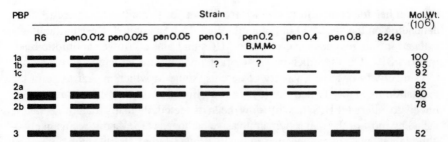

Figure 1. PBP patterns of penicillin-susceptible and -resistant pneumococci. The thickness of bars indicates intensity of the bands in the fluorograms. In addition to a series of isogenic transformants, the penicillin-sensitive recipient strain (R6) and the DNA donor penicillin-resistant strain (8249) are included as controls. Strains M, Mo, and B are independent clinical isolates from the Oklahoma Children's Hospital. (Reproduced with permission of Springer-Verlag, from A. L. Demain and N. A. Solomon, ed., *Handbook of Experimental Pharmacology*, vol. 67/I, 1983.)

in fact a remodeled (or, possibly, truncated) form of PBP 1a and (ii) that the PBP 2b molecules are present in the resistant cells (and are, possibly, performing their physiological function in cell wall synthesis), but they are undetectable by the penicillin-binding technique as a result of a mutationally decreased antibiotic binding capacity (3). The "disappearance" of PBP 1b from the fluorograms of the PBPs from resistant cells may have a similar mechanism.

Additional insights into the relationship between altered PBPs and penicillin resistance were provided by genetic analysis. Introduction of the high-level penicillin resistance of strain 8249 into susceptible cells required several (at least five) rounds of genetic transformation, and in this manner a series of isogenic transformants were constructed that differed from one another in their MICs. Examination of the PBP profiles and penicillin-binding capacities of these low- and intermediate-level transformants revealed a gradual shift in PBP pattern along with the gradually increasing MICs from the five PBPs of the susceptible recipient to the trio of PBPs characteristic of the DNA donor (Fig. 1). During this shift, four of the five PBPs of the recipient cells (i.e., every PBP except PBP 3) were shown to undergo stepwise changes in their reactivity to penicillin (decreased rates of acylation or affinity, or both, and in some cases, a change in the total amounts of PBPs) (5). These findings suggest that in pneumococci four of the five PBPs are physiologically important targets of penicillin, and the multiple rounds of transformation are needed for the stepwise introduction of the genetic determinants of these four mutationally altered proteins. Presumably, these determinants contain multiple point mutations leading to amino acid substitutions at some critical sites in the PBPs that affect the kinetic properties of the proteins. Indeed, peptide mapping indicated that at least some of these amino acid replacements were in the vicinity of the β-lactam-binding site of PBP 1a (3).

The order of penicillin reactivity of the PBPs in the susceptible pneumococci is 1a > 2a > 1b > 2b. Thus, one would expect that at the first, lowest level of resistance the MIC increase is due to the introduction of a lower-affinity form of PBP 1a. This in fact appears to be the case (6). However, the next higher levels of resistance appear to involve simultaneous changes (lower reactivity) in more than one PBP, and this fact may, in part, explain why transformation frequencies were found to decline rapidly with increasing levels of resistance, in spite of the fact that the recipients used in each round of transformation were always transformants with the nearest lower level of penicillin resistance isolated in the previous cross (21). It is most likely that the series of stepwise transformants with their gradually increasing MICs and gradually shifting PBP patterns reconstruct the mutational steps that originally led to the emergence of the highly penicillin-resistant strain 8249 in South Africa. If this interpretation is correct, then the number of mutations that have accumulated in this strain is remarkable indeed, considering the low incidence (10^{-6} or less) of spontaneous mutations even to low levels of penicillin resistance (6).

A survey of the epidemiological literature indicates that natural populations of pneumococci have undergone a substantial upward shift in penicillin resistance levels worldwide since the 1950s, and this resistance involves the same type of PBP-linked mutations described in the highly resistant South African strains (6). One should also remember that some of the South African strains, including strain 8249, are also resistant to a large number of mechanistically diverse antibacterial agents (22), indicating the capacity of the originally "docile" pneumococcus to reemerge as a potentially much more difficult-to-control pathogen.

None of the speculations outlined above can fully explain the apparent sequential order of PBP alterations observed in the series of isogenic transformants that span the range of MICs from that of the recipient cell to the nearly 1,000 times higher MIC of the DNA donor cells (Fig. 1). To put it differently, one of the basic unanswered questions is why the most extensively changed forms of a given PBP show up only above certain MIC levels. For instance, if PBP 1a undergoes multiple stepwise changes in penicillin affinity all the way to its most extensively changed form (which is PBP 1c), why is it that PBP 1c is not detectable in low-level transformants also? Given the reasonable size of the DNA determinant of such a protein (92 kilodaltons), it is extremely unlikely that the DNA determinants of PBP 1c are not taken up and are not fully integrated in at least some of the competent recipients used in the first round of transformation of penicillin resistance. Yet, analysis of a reasonable number of low- and intermediate-level transformants showed no detectable PBP 1c. PBP 1c has become detectable only at higher levels of resistance, and we could never detect PBPs 1a and 1c in the same cells.

A possible explanation of these findings is that recombinants that have integrated the most extensively changed form of the PBP 1a gene (which carries multiple point mutations and codes for PBP 1c) do actually form but they are not viable. One reason for this may be that the extensively low-affinity form of PBP 1a cannot perform its physiological function in cell wall synthesis in a cell in which the other PBPs are present in unaltered, highly reactive forms. This proposal assumes that a PBP with lower reactivity to the penicillin molecule (which is a structural analog of the cell wall building blocks) will also show abnormalities in its physiological function in catalyzing the incorporation of wall building blocks during cell wall synthesis. Experimental evidence indicating this has recently been obtained in our laboratory. This proposition leads to the interesting prediction that PBPs in pneumococci, and perhaps in other bacteria as well, function in a cooperative manner, as a kind of "assembly line" in the synthesis of cell wall. In this model the relative catalytic activities of individual PBPs must be tuned to one another, and extensive alterations in only a single PBP (without an appropriate parallel tune-down in the reactivities of other PBPs) may make cooperative functioning impossible.

Figure 2 shows a model in which the donor DNA prepared from the highly resistant strain 8249 contains DNA molecules carrying the genetic determinant of PBP 1a in a form that is assumed to have several point mutations (see heavy horizontal bar with three solid squares representing the mutated sites). When fully expressed, this DNA codes for the lowest affinity form of PBP 1a ($\equiv$ PBP 1c). The model assumes that both kinds of recombinants (1 and 2) actually form between this donor DNA and its allelic determinant present in the

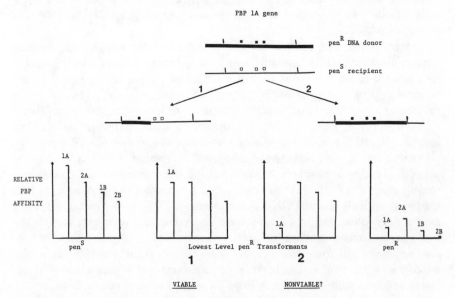

Figure 2. Model for the apparent orderliness of PBP alterations in genetic transformants to penicillin resistance.

penicillin-susceptible recipient cell (thin horizontal line; sites corresponding to the mutated sites in the donor DNA are represented by empty squares above the horizontal lines). (The different connotations for donor and recipient DNA [heavy versus thin lines] do not mean to imply lack of homology but serve strictly as illustrative devices.) Both recombinants 1 and 2 are formed already in the first round of transformation which yields only low-level penicillin-resistant transformants. The fact that such transformants do not contain the highly mutated form of PBP 1a is explained by assuming that such a low-affinity protein could not perform its normal physiological function in the company of unaltered PBPs 2a, 1b, and 2b, implying a cooperative functioning of PBPs.

The lower portion of Fig. 2 depicts the relative drug (and, presumed, substrate) affinities of four pneumococcal PBPs in the penicillin-susceptible recipient (left-side barogram, pen^S) and in the two low-level transformants (1 and 2), of which 2 is assumed to be nonviable, and the relative PBP affinities in the DNA donor strain (right-side barogram, pen^R).

Experiments testing the validity of the model proposed in Fig. 2 are in progress. Given the topological complexity of the process of wall assembly, the existence of an organized unit of cooperatively functioning catalysts (PBPs) is an attractive concept.

Autolytic Defect and Penicillin Resistance in Clinical Isolates of Pneumococci

Detailed examination of the antibiotic response of some penicillin-resistant South African isolates has revealed that not only are the MICs of these strains elevated but an autolytic defect is present as well: when treated with penicillin concentrations above the MIC for these strains, cultures of five of six strains failed to undergo autolysis, and their rate of loss of viability was also slow compared with that of typical penicillin-susceptible strains (10). Our initial suspicion that the autolytic defect was an aspect of the resistance mechanism was not borne out by experimental tests. First of all, the autolytic defect was not related to the level of resistance, but was observed both in highly resistant isolates (MIC, ≥ 6 μm/ml) and in low-resistance strains (MIC, 0.05). Furthermore, the resistant and autolysis-defect traits could be easily separated by genetic transformation. More recent observations with clinical isolates from the United States, New Guinea, and Switzerland have confirmed the frequent association between autolytic defect and penicillin resistance.

For a possible explanation of such a frequent association between these two mechanistically unrelated traits, one may consider the nature of antibiotic pressure applied in the clinical environment (Fig. 3). During many therapeutic courses of antibiotic use, bacteria at an infection site are exposed to cycles of high antibiotic concentration (far above the MIC) followed by drug clearance and a period of sub-MIC drug levels, during which bacteria that were not killed can resume growth. This type of antibiotic exposure selects for bacterial survival rather than resistance, and survivor (tolerant) mutants of staphylococci (16), *E. coli* (8), and pneumococci (20) have been isolated by such drug exposures in the laboratory.

If such a process leads to the enrichment of a pathogen population in autolysis-defective mutants, then there should be a higher probability for resistant mutants to emerge from such autolysis-defective cells. This (secondary) selection for resistant strains involves a completely different type of drug treatment: prolonged exposure to a low, constant concentration of penicillin slightly above the MIC. This method was used in the laboratory for the isolation of β-lactam-resistant mutants with altered PBPs, both in pneumococci (6) and in staphylococci (17). If similar antibiotic pressure were responsible for the selection of resistant strains in vivo, then this process is likely to be restricted to the tail end ("trough") of the pharmacokinetic antibiotic dose (Fig. 3).

Homogeneous and Heterogeneous Methicillin Resistance in Staphylococci

The PBP-linked β-lactam resistance mutation found in clinical isolates of *S. aureus* has a completely different mechanism from that of the penicillin-resistant pneumococci. Methicillin-resistant *S. aureus* strains contain the same

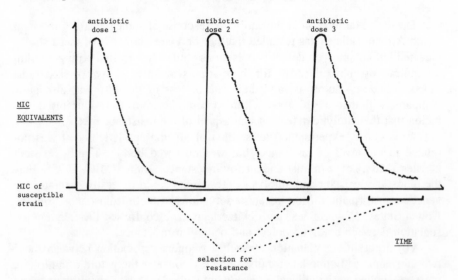

Figure 3. Schematic illustration of the type of antibiotic pressure operating in the clinical environment. The curves represent the changing drug concentrations at an idealized infection site during three doses of antibiotic administration.

set of PBPs with normal antibiotic affinities as the methicillin-sensitive strains, but the resistant strains also have acquired an extra PBP of approximately 78 kilodaltons, referred to in the literature either as PBP 2′ or PBP 2a (for review, see reference 15). This PBP has extremely low antibiotic reactivity, and all experimental evidence to date suggests that it is the biochemical correlate of methicillin resistance in all the methicillin-resistant *S. aureus* isolates examined so far, whether the strains come from the United States, Japan, Australia, or Europe (12). It is not clear whether PBP 2a takes part in the catalysis of cell wall synthesis in methicillin-resistant *S. aureus* growing in the absence of antibiotics. In some of the resistant homogeneous strains studied in our laboratory, growth rates are normal. In addition, in some of these strains the production of PBP 2a is inducible by β-lactam antibiotics (2, 11, 19).

These observations suggest that methicillin-resistant *S. aureus* may primarily use the normal, high-affinity PBPs for wall synthesis in drug-free media, and the shift to the use of the low-affinity PBP 2a occurs when methicillin is added to the medium. Such a strategy appears to be a "smart" one since it would circumvent the problems penicillin-resistant pneumococci may encounter by decreasing the penicillin affinity of their normal PBPs and thus, possibly, introducing problems into the normal functioning of these enzymes as well. On the other hand, the presence of the normal set of PBPs in methicillin-resistant *S. aureus* raises a dilemma. In the susceptible staphylococci acylation of the normal PBPs by the β-lactam antibiotics quickly leads to the triggering of irreversible effects: cell death and autolysis. If the methicillin-resistant *S. aureus* isolates contain the same set of normal PBPs, one wonders what prevents in these cells the initiation of the same irreversible effects upon the acylation of the normal PBPs. One has to think about some mechanism that prevents these irreversible effects from occurring before the switchover to the functioning of PBP 2a can take effect.

Examination of the physiology of methicillin-resistant *S. aureus* isolates indicates that the great majority of the clinical strains are in fact not protected from the irreversible effects of the antibiotics (in spite of the fact that they contain PBP 2a); upon addition of methicillin to such so-called heterogeneous methicillin-resistant *S. aureus* isolates, rapid cell death ensues which proceeds to kill all but a small minority (10^{-4} or less) of the population. In some of these heterogeneous isolates, cell killing and lysis can be prevented if the bacteria are shifted to a lower temperature (30°C instead of 37°C), if NaCl is added to the medium at the time of drug addition, or if both changes are made. The important (and initially very surprising) observation relevant for the interpretation of these facts is that all cells in heterogeneous cultures appear to produce normal quantities of PBP 2a, whether grown at the lower temperature (30°C, at which all cells can grow in the presence of high concentrations of methicillin) or the higher temperature (at which only a tiny fraction of the population survives) (7).

To explain these paradoxical observations, it was proposed that in addition to the production of PBP 2a, methicillin-resistant *S. aureus* also has to produce another factor X to enable all cells to utilize PBP 2a for continued growth in the presence of methicillin. Factor X was, presumably, made (or was present) constitutively in homogeneous strains, while its production was defective (or thermolabile) in the heterogeneous strains (7). The need for more than one gene product for the full expression of resistance in methicillin-resistant *S. aureus* is consistent with genetic evidence (1, 9). However, nothing is known about the nature of factor X. It is conceivable that it represents a process or a product that is required for the normal functioning or membrane incorporation of PBP 2a. It is also plausible to think of factor X as controlling the rate of transcription or translation of PBP 2a (2). However, a completely different and more intriguing hypothesis has also been proposed to explain the nature of factor X and the mechanism of expression of resistance in heterogeneous and homogeneous strains. This proposal originates in the question posed at the beginning of this discussion: if the normal set of PBPs are present in methicillin-resistant *S. aureus,* what prevents the triggering of cell death and lysis upon the acylation of these proteins which is known to occur with great rapidity upon the addition of methicillin to the medium?

We proposed that the "success" of PBP 2a in the homogeneously resistant cells may be due to a tuning down of the lethal pathway and autolysis-triggering events in these bacteria (7, 14). In this model factor X may be involved with the regulation of autolytic activity or some other aspect of methicillin lethality rather than with a modification of PBP 2a (14). In heterogeneous methicillin-resistant *S. aureus,* factor X may be defective (or thermolabile), and thus acylation of the normal PBPs rapidly activates the same lethal pathway that kills susceptible bacteria before an effective switchover to the functioning of PBP 2a could be accomplished. In other words, effective, homogeneous expression of resistance may require that the PBP 2a gene be expressed in a tolerant cell.

The model presented in Fig. 4 represents the response of homogeneous and heterogeneous methicillin-resistant *S. aureus* cultures to the addition of methicillin to the media. In both homogeneous and heterogeneous strains addition of methicillin results in the rapid acylation (inactivation) of the complement of "normal" PBPs 1, 2, and 3; only the low-affinity PBP 2a remains unacylated and functional (see upper part of Fig. 4). Up to this point the two types of cultures are assumed to behave in an identical manner. The rapid killing of most of the heterogeneous cells and continued growth of the homogeneous bacteria are explained by the assumption that homogeneous cells may also carry an additional trait, a tuned-down autolytic system, and thus acylation of the normal PBPs initiates killing and lysis only after a longer lag time than in the heterogeneous cells. The slower triggering of the autolytic system is assumed to allow the cells to switch over to a cell wall synthesis catalyzed by PBP 2a.

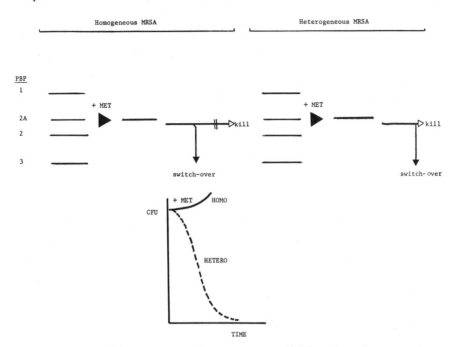

Figure 4. Response of homogeneous and heterogeneous methicillin-resistant *S. aureus* cultures to methicillin. The addition of methicillin to the medium is indicated by + MET above the arrowheads. The longer and shorter horizontal arrows pointing to "kill" represent the longer and shorter lag times before killing in the homogeneous and heterogeneous bacteria, respectively.

The intriguing feature of this model is that it brings together β-lactam resistance and tolerance. The frequent association of these two traits has already been observed in clinical isolates of pneumococci. Cyclic exposure of non-mutagenized penicillin-susceptible staphylococci to β-lactam antibiotics in the laboratory leads to the emergence of autolysis-defective mutants which are also substantially less sensitive to the bactericidal effect of penicillin (16). Our model predicts that the combination of some type of autolytic defect with the expression of the PBP 2a gene may be an essential feature of methicillin-resistant *S. aureus* strains.

ACKNOWLEDGMENTS. Work described in this paper was supported by Public Health Service grant RO1 AI 16794 from the National Institute of Allergy and Infectious Diseases.

I thank Elaine Tuomanen for numerous stimulating discussions of the models described in this chapter.

LITERATURE CITED

1. **Berger-Bachi, B.** 1983. Insertional inactivation of staphylococcal methicillin resistance by Tn*551*. *J. Bacteriol.* **154:**479–487.

2. **Chambers, H. F., B. J. Hartman, and A. Tomasz.** 1985. Increased amounts of a novel penicillin binding protein in a strain of methicillin-resistant *Staphylococcus aureus* exposed to nafcillin. *J. Clin. Invest.* **76:**325–331.

3. **Hakenbeck, R., H. Ellerbrok, T. Briese, S. Handwerger, and A. Tomasz.** 1986. Penicillin-binding proteins of penicillin-susceptible and -resistant pneumococci: immunological relatedness of altered proteins and changes in peptides carrying the beta-lactam binding site. *Antimicrob. Agents Chemother.* **30:**553–558.

4. **Handwerger, S., and A. Tomasz.** 1985. Antibiotic tolerance among clinical isolates of bacteria. *Rev. Infect. Dis.* **7:**368–386.

5. **Handwerger, S., and A. Tomasz.** 1986. Alterations in kinetic properties of penicillin-binding proteins of penicillin-resistant *Streptococcus pneumoniae*. *Antimicrob. Agents Chemother.* **30:**57–63.

6. **Handwerger, S., and A. Tomasz.** 1986. Alterations in penicillin-binding proteins of clinical and laboratory isolates of pathogenic *Streptococcus pneumoniae* with low levels of penicillin resistance. *J. Infect. Dis.* **153:**83–89.

7. **Hartman, B. J., and A. Tomasz.** 1986. Expression of methicillin resistance in heterogeneous strains of *Staphylococcus aureus*. *Antimicrob. Agents Chemother.* **29:**85–92.

8. **Kitano, K., and A. Tomasz.** 1979. *Escherichia coli* mutants tolerant to beta-lactam antibiotics. *J. Bacteriol.* **140:**955–962.

9. **Kornblum, J., B. J. Hartman, R. P. Novick, and A. Tomasz.** 1986. Conversion of a homogeneously methicillin-resistant strain of *Staphylococcus aureus* to heterogeneous resistance of Tn551-mediated insertional inactivation. *Eur. J. Clin. Microbiol.* **5:**714–718.

10. **Liu, H., and A. Tomasz.** 1985. Penicillin tolerance in multiply drug-resistant natural isolates of *Streptococcus pneumoniae*. *J. Infect. Dis.* **152:**365–372.

11. **Rossi, L., E. Tonin, Y. R. Cheng, and R. Fontana.** 1985. Regulation of penicillin-binding protein activity: description of a methicillin-inducible penicillin-binding protein in *Staphylococcus aureus*. *Antimicrob. Agents Chemother.* **27:**828–831.

12. **Shimada, K., and A. Tomasz (ed.).** 1986. *Methicillin-Resistant Staphylococcus aureus*. University of Tokyo Press, Japan.

13. **Tomasz, A.** 1979. The mechanism of the irreversible antimicrobial effects of penicillins: how the beta-lactam antibiotics kill and lyse bacteria. *Annu. Rev. Microbiol.* **33:**113–137.

14. **Tomasz, A.** 1986. Beta-lactam antibiotic resistant bacteria with altered penicillin binding proteins, p. 55–60. *In* K. Shimada, and A. Tomasz (ed.), *Methicillin-Resistant Staphylococcus aureus*. University of Tokyo Press, Japan.

15. **Tomasz, A.** 1986. Penicillin-binding proteins and the antibacterial effectiveness of beta-lactam antibiotics. *Rev. Infect. Dis.* **8**(Suppl. 3):260–278.

16. **Tomasz, A., and M.-L. de Vegvar.** 1986. Construction of a penicillin-tolerant laboratory mutant of *Staphylococcus aureus*. *Eur. J. Clin. Microbiol.* **5:**710–713.

17. **Tonin, E., and A. Tomasz.** 1986. Beta-lactam-specific resistant mutants of *Staphylococcus aureus*. *Antimicrob. Agents Chemother.* **30:**577–583.

18. **Tuomanen, E., D. T. Durack, and A. Tomasz.** 1986. Antibiotic tolerance among clinical isolates of bacteria. *Antimicrob. Agents Chemother.* **30:**521–527.

19. **Ubukata, K., N. Yamashita, and M. Konno.** 1985. Occurrence of a beta-lactam-inducible penicillin-binding protein in methicillin-resistant staphylococci. *Antimicrob. Agents Chemother.* **27:**851–857.

20. **Williamson, R., and A. Tomasz.** 1980. Antibiotic-tolerant mutants of *Streptococcus pneumoniae* that are not deficient in autolytic activity. *J. Bacteriol.* **144:**105–113.

21. **Zighelboim, S., and A. Tomasz.** 1980. Penicillin-binding proteins of multiply antibiotic-resistant South African strains of *Streptococcus pneumoniae*. *Antimicrob. Agents Chemother.* **17:**434–442.
22. **Zighelboim, S., and A. Tomasz.** 1981. Multiple antibiotic resistance in South African strains of *Streptococcus pneumoniae*: mechanism of resistance to beta-lactam antibiotics. *Rev. Infect. Dis.* **3:**267–276.

Chapter 66

Penicillin Tolerance in *Enterococcus hirae* ATCC 9790

L. Daneo-Moore
I. Said
H. Fletcher
O. Massidda
F. Pittaluga

Distinguishing Autolytic Deficiency and Tolerance

Exposure of susceptible enterococci to penicillin G can cause one of three responses. First, addition of penicillin G at the MIC can cause lysis and killing. Autolysis-defective organisms show a second type of response consisting of inhibition of growth, a slow rate of lysis, and a slow rate of killing. A third type response is obtained in many fresh clinical isolates of enterococci, as well as some laboratory strains. Here addition of penicillin G results in a slow rate of lysis and reasonably rapid killing, but a fraction of the population survives. The three responses shown in Fig. 1 are from derivatives of the same ATCC 9790 strain of *Enterococcus hirae* (formerly *Streptococcus faecium*). Figure 1A shows the response of a highly autolytic strain which we call ATCC 9790-S. It has been maintained in the laboratory of G. D. Shockman for some 20 years and is stably susceptible to penicillin G. Figure 1B shows the response of Aut-1, a lysis-defective derivative of ATCC 9790-S obtained by nitrosoguanidine mutagenesis (10). Figure 1C shows the response of a strain of *E. hirae* ATCC 9790 obtained in 1980 from the American Type Culture Collection. This organism (designated ATCC 9790-80T) lyses slowly with penicillin G and has some survivors at 24 h. Therefore, poorly lytic organisms shown in Fig. 1C differ from autolysis-deficient organisms of the type shown in Fig. 1B in that,

---

L. Daneo-Moore, I. Said, H. Fletcher, O. Massidda, and F. Pittaluga • Temple University School of Medicine, Philadelphia, Pennsylvania 19140.

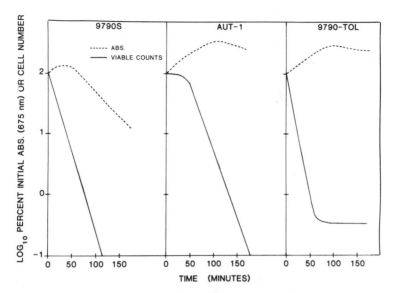

Figure 1. Schematic pattern of effects of penicillin G (5 μg/ml) addition on viable cell counts and absorbance at 675 nm. (A) *E. hirae* ATCC 9790S. (B) *E. hirae* Aut-1. (C) *E. hirae* ATCC 9790-80T. Solid lines, Viable cell counts; dashed lines, absorbance at 675 nm (1).

at concentrations around the MIC, a fraction of the cells (on the order of 0.1 to 1%) survive the penicillin G treatment (1).

When the MIC is compared with the MBC, ATCC 9790-S and the Aut-1 derivative are inhibited and killed by about the same penicillin G concentrations, but ATCC 9790-80T requires 16 times more penicillin G to be killed than to be inhibited (Table 1). A much higher MBC than MIC is one of the criteria used to define tolerance (5, 8). The important point made by these data is that a slow rate of lysis is not necessarily an index of survival to penicillin G exposure. Some other factor (e.g., tolerance) permits survival of ATCC 9790-80T but not of other autolysis-defective organisms. A distinction between autolytic deficiency and tolerance in pneumococci was made by Williamson and Tomasz on entirely different grounds (11).

Clinical Relevance of Tolerance

Table 2 shows the frequency of enterococci for which the MBCs are at least 32 times higher than the MICs in fresh clinical isolates. In all the tolerant clinical isolates the MBC/MIC ratio was greater than 32. A consequence of a definition of tolerance based on an MBC/MIC ratio is that tolerance occurs over a limited antibiotic concentration range. At high enough antibiotic concentrations there are no survivors from tolerant cultures.

Table 1. MICs and MBCs of penicillin G for derivatives
of *E. hirae* ATCC 9790

Strain	MIC (μg/ml)	MBC (μg/ml)
ATCC 9790-S	6.25	12.5
ATCC 9790-Lyt14	6.25	12.5
ATCC 9790-80T	12.5	200.0

Tolerance Is Switched On and Off

Since only a fraction of the ATCC 9790-80T cultures survive exposure to penicillin G at or above the MIC, we tried to isolate these organisms from single colonies. We used a modification of a method originally described by Kim and Anthony (4), which is shown diagrammatically in Fig. 2. The idea was to isolate organisms from single colonies, patch them on a plate that does not contain penicillin G, and test these patches for ability to survive exposure to penicillin G on a second plate. The third plate was used to distinguish survivors (T colonies) from nonsurvivors (S colonies).

To our surprise, about 95% of the colony patches isolated from ATCC 9790-80T were survivors, and about 5% were susceptible in this test. Even more surprising was the observation that on further isolation (including repeated streaking for single-colony isolation), T colonies always produced about 5% S colonies. The S colonies derived from T isolates always produced about 5% T colonies. The detailed lineage analysis of this phenomenon has been published elsewhere (2). The clear indication of that study was that T colonies and S colonies can switch spontaneously and reversibly from one state to the other (Fig. 3). As mentioned above, S strains can become stably susceptible on prolonged laboratory subculture (7). The ATCC 9790-S strain maintained in G. D. Shockman's laboratory appears to be an example of this.

The T colonies isolated from this analysis had an MBC/MIC ratio of 32 for penicillin G and had high MBC/MIC ratios also for bacitracin, vanco-

Table 2. Frequency of tolerance in 45 unselected, freshly
isolated streptococci from the Clinical Microbiology
Laboratory, Temple University School of Medicine

Serotype	No. isolated	No. tolerant[a]
Group A	5	1
Group B	12	6
Group D		
Enterococcus	23	20
Nonenterococcus	2	1
Unclassified	3	2
Total	45	30

[a]Tolerance determined by an MBC/MIC ratio of 32 or greater.

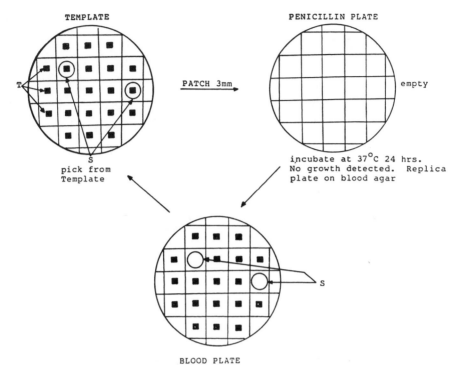

Figure 2. Scheme for the identification of S and T derivatives. Note that both S and T derivatives identified on the blood plate can be recovered from the template.

mycin, and D-cycloserine (2) but not for fosfomycin (unpublished data). Rates of lysis of T and S derivatives were examined at various concentrations of penicillin, bacitracin, vancomycin, and D-cycloserine. For each of these antibiotics, at some, but not at all, concentrations, T derivatives lysed more slowly than S derivatives (9). In the case of fosfomycin, lysis rates were the same for S and T derivatives over a range of concentrations. The cellular autolytic activity of the stably susceptible strain (ATCC-S) and that of three S derivatives of ATCC 9790-80T were examined and compared with the cellular autolytic activity of ATCC 9790-80T and three of its T derivatives (Fig. 4). All the tolerant bacteria had severely reduced cellular autolytic activity.

Quantitative Differences in Autolytic Enzyme Levels between Tolerant and Susceptible Bacteria

Since *E. hirae* has only two detectable autolytic enzymes (E-1 and E-2; see Shockman, Dolinger, and Daneo-Moore, this volume), we examined T and S derivatives for absence of one or the other enzyme. Plates containing strep-

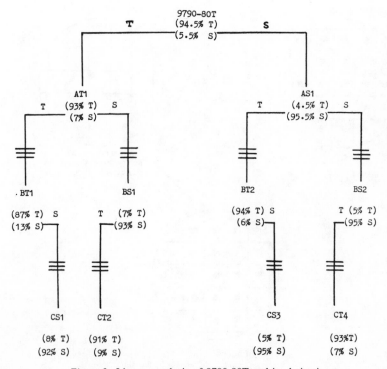

Figure 3. Lineage analysis of 9790-80T and its derivatives.

tococcal cells (specific substrates for E-1) or micrococcal cells (specific substrates for E-2) were inoculated with S and T derivatives and were examined for the possible absence of zones of clearing around the colonies. Of 380 T colonies and 383 S colonies examined, all had autolytic halos on both types of plates. However, the S colonies had visibly smaller halos that T colonies on micrococcal plates, suggesting a higher level of E-2 in T than in S strains. Also, the T colonies had smaller halos than S colonies on streptococcal plates, suggesting a higher level of E-1 in S than in T strains. These differences were statistically significant (Table 3).

To confirm that T colonies have higher levels of E-2 and S colonies have higher levels of E-1, we extracted the enzymes by using the method of Kawamura and Shockman (3). Since the results (Table 4) indicate quantitative rather than qualitative differences in muramidase levels between S and T derivatives, the tentative conclusion is that some factor other than expression of a muramidase may be responsible for the autolytic deficiency of T strains. This conclusion is reinforced by the relatively narrow range of antibiotic concentrations at which T derivatives are lysed more slowly and killed less than S bacteria.

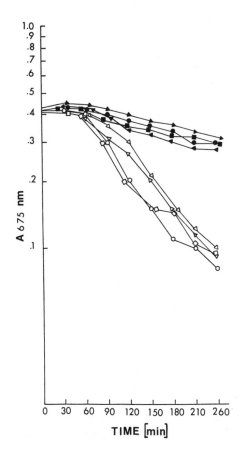

Figure 4. Cellular autolytic activity. Cells in the exponential phase of growth were harvested by filtration through a 0.65-μm-pore-size membrane filter (Millipore Corp.) and washed three times with ice-cold distilled water. The filter was suspended in 6 ml of 0.3 M potassium phosphate buffer (pH 6.8), and the cell suspension was examined for subsequent lysis at 37°C. Symbols: ○, 9790S; ●, 9790-80T; □, AS1; ■, AT1; △, BS1; ▲, BT1; ▽, BS2; ▼, BT2.

Cloning and Expression of Gene Fragment Conferring Tolerance in *S. sanguis*

To identify the hypothetical gene responsible for tolerance, we partially digested the DNA of a T derivative for 30 min with *Eco*RI and ligated this DNA into pVA736, a streptococcal vector with an erythromycin resistance determinant (6). The ligated plasmid was used to transform penicillin-susceptible *S. sanguis* Challis 685 (6). After 2 h the transformation mixture received penicillin G at 4 μg/ml (about 300 times the MIC for this organism) for 2 more h. The survivors were plated on erythromycin, and the colonies were then replica plated by the method shown in Fig. 2, but with penicillin at 4 μg/ml. In two separate experiments tolerant *S. sanguis* was found with frequencies of 3.5×10^{-3} and 6.5×10^{-3} per erythromycin-resistant transformant. When *S. sanguis*

Table 3. Zones of lysis on *Micrococcus luteus* plates (specific for E-2) and *E. hirae* plates (specific for E-1)[a]

Plates	Derivative	Mean zone (mm) Susceptible	Tolerant	*t* value
M. luteus	A1	9.09 ± 0.19	9.95 ± 0.21	8.95
	B1	9.10 ± 0.37	10.00 ± 0.29	5.73
	B2	8.73 ± 0.33	9.63 ± 0.41	5.11
E. hirae	A1	7.79 ± 0.14	7.01 ± 0.09	26.05
	B1	7.97 ± 0.09	6.91 ± 0.08	26.17
	B2	8.05 ± 0.08	6.44 ± 0.01	33.47

[a]On each plate, 10 colonies selected at random were measured at a 10× magnification. The *t* tests indicated a significant difference at the 99.5% level.

Table 4. Muramidase activities in S and T derivatives

Strain	E-1 (U)[a] Total[b] ($n = 10$)	Active ($n = 7$)	Activatable[c]	E-2 (U)[a] active ($n = 7$)
Susceptible	18.5 + 5.2	5.8 + 4.5	11.8 ± 6.7	6.8 ± 4.9
Tolerant	11.3 + 4.6	5.7 + 3.8	5.2 ± 4.1	13.3 ± 7.8

[a]A unit is defined as the amount which reduces the turbidity of the walls by 0.001 optical density unit per min at 37°C.
[b]Determined after addition of trypsin, 0.3 μg/ml.
[c]Determined as the difference between total and active for each determination.

transformed with pVA736 was analyzed by the same method, there were no penicillin-tolerant transformants. Transformation of *S. sanguis* with plasmid T32 containing one of these cloned genes gave tolerant *S. sanguis* (Table 5). Tolerant transformants were found at a frequency of 3×10^{-2} to 6×10^{-2} per erythromycin-resistant transformant. In this experiment there was no penicillin enrichment of the transformation mixture. Table 5 shows that the MBC/MIC ratio is elevated in the original transformant and in an *S. sanguis* strain transformed with purified T32. Restriction analysis of the *E. hirae* insert indicates that it is 1.6 kilobases in size and contains one *Ava*I and one *Cla* site. Current

Table 5. MICs and MBCs of penicillin for *S. sanguis* transformants

Strain	MBC (μg/ml)	MIC (μg/ml)	MBC/MIC
S. sanguis 685	0.015	0.015	1.00
S. sanguis T[a]	0.313	0.020	15.65
S. sanguis T32[b]	0.313	0.020	15.65

[a]Tolerant transformant of *S. sanguis* 685.
[b]Transformed with one of the plasmids isolated from *S. sanguis* T.

studies are directed at elucidation of the nature of the product of the DNA insert and of the molecular basis of the mechanism responsible for the switch between the susceptible and tolerant states.

ACKNOWLEDGMENTS. This research was supported by Biomedical Research Support Grant S07 RR05417 and Public Health Service grant AI 23394-01 from the National Institutes of Health.

LITERATURE CITED

1. **Daneo-Moore, L., and M. Pucci.** 1982. Penicillin tolerance and resistance in *Streptococcus faecium*, p. 199–203. *In* D. Schlessinger (ed.), *Microbiology—1982*. American Society for Microbiology, Washington, D.C.
2. **Daneo-Moore, L., A. Volpe, and I. Said.** 1985. Variation between penicillin-tolerant and penicillin-sensitive states in *S. faecium* ATCC 9790. *FEMS Microbiol. Lett.* **30:**319–323.
3. **Kawamura, T., and G. D. Shockman.** 1983. Purification and some properties of the endogenous, autolytic N-acetylmuramoylhydrolase of *Streptococcus faecium*, a bacterial glycoenzyme. *J. Biol. Chem.* **258:**9514–9521.
4. **Kim, K. S., and B. F. Anthony.** 1983. Use of penicillin-gradient and replicate plates for demonstrating penicillin tolerance in streptococci. *J. Infect. Dis.* **148:**488–491.
5. **Krogstadt, D. J., and A. R. Parquette.** 1980. Defective killing of enterococci: a common property of antimicrobial agents acting on a cell wall. *Antimicrob. Agents Chemother.* **17:**965–968.
6. **Macrina, F. L., K. R. Jones, and P. H. Wood.** 1980. Chimeric streptococcal plasmids and their use as molecular cloning vehicles in *Streptococcus sanguis* (Challis). *J. Bacteriol.* **143:**1425–1435.
7. **Mayhall, C. G., and E. Apollo.** 1980. Effect of storage and changes in bacterial growth phase and antibiotic concentrations on antimicrobial tolerance in *Staphylococcus aureus*. *Antimicrob. Agents Chemother.* **18:**784–788.
8. **Sabath, L. D., N. Wheeler, M. Laverdiere, D. Blazevic, and B. J. Wilkinson.** 1977. A new type of penicillin resistance of *Staphylococcus aureus*. *Lancet* **i:**443–447.
9. **Said, I., H. Fletcher, A. Volpe, and L. Daneo-Moore.** 1987. Penicillin tolerance in *Streptococcus faecium* ATCC 9790. *Antimicrob. Agents Chemother.* **31:**1150–1152.
10. **Shungu, D. L., J. B. Cornett, and G. D. Shockman.** 1979. Morphological and physiological study of autolytic-defective *Streptococcus faecium* strains. *J. Bacteriol.* **138:**598–608.
11. **Williamson, R., and A. Tomasz.** 1980. Antibiotic-tolerant mutants of *Streptococcus pneumoniae* that are not deficient in autolytic activity. *J. Bacteriol.* **144:**105–113.

X.   CONCLUDING REMARKS

Chapter 67

The Bacterial Surface—Where Does It Begin and End?

H. J. Rogers

I thought it appropriate in the last chapter of a book devoted to the most up-to-date investigations of peptidoglycan and the inhibitors of its synthesis to consider some aspects of bacterial surfaces that have received less attention than those in the mainstream of advance. Progress to date has necessarily involved disassembly of the envelope components, and we now talk, often with unconsidered confidence, of walls, cytoplasmic membranes, outer membranes, sacculi, and so on. The concept of the envelope thus developed has depended on the design of methods for the separation of the components and growing suppositions as to what these ought to be or to do. In the living cell the structure and function of the components of the envelope must be so integrated that the flow of macromolecules remains responsive to the environment. The biosynthesis of these starts in the cytoplasm but for some must finish far outside the outer limits of the cell as seen by ordinary electron microscopic techniques. What follows is an attempt to look for hints as to the nature of this necessary integration.

When Dawson (7) and Salton and Horne (37, 38) turned their electron microscopes on bacteria that they had ruptured in various ways, the long era of guesswork about the nature of the skins of the little beasties, begun by Leeuwenhoek, had ended. Workers were no longer entitled to discard the insoluble residue from ruptured bacterial suspensions as debris. So well did Salton and Horne (35, 39) design the methods for rupturing bacterial cells and preparing walls, apparently free from cytoplasmic material, that for gram-positive bacteria they have remained in use to the present day some 35 years later, a rare achievement. When hydrolyzed with acid, the extraordinary nature of this ma-

H. J. Rogers • Biological Laboratory, University of Kent, Canterbury, Kent CT2 7NJ, United Kingdom.

terial became apparent (36), and one organism in particular yielded only three amino acids, amino sugars, and glucose. This was *Micrococcus lysodeikticus* or, if we must, *M. luteus*. The other five gram-positive organisms Salton examined yielded predominantly 3 of the 4 amino acids, but also sizable amounts of 9 or 10 others. The early recognition of a probable basal mucocomplex alias mucopeptide murein, glycopeptide, and peptidoglycan started the search for "undamaged clean" walls and founded the subject which ultimately brought us together here. Varieties of treatments, ranging from incubation with proteolytic enzymes (6) to treatment with strong solutions of anionic detergents (4, 42), were designed to rid the walls of their unwanted contaminating amino acids. The walls of gram-positive bacteria have hence come to be regarded by many as the insoluble residue made of covalently linked molecules resisting solution by these treatments. From the start (36) it was clear that the walls of gram-negative bacteria were more complex than those of gram-positive bacteria, the former containing more equal proportions of a wider range of amino acids as well as considerable amounts of phospholipid. To avoid undue elaboration of the historical aspect of this topic, let me say that each layer of these walls has had its own specific methods for isolation along with its own empirical standards of "purity," and each has its own separate well-developed portrait. My long familiarity dictates that most of this chapter should deal with the walls of gram-positive organisms, but at this point it would seem appropriate to point out that evidence might already question entire reliance on the study of separated "purified" entities from the outer layers of gram-negative envelopes. Isolation of peptidoglycan sacculi as thin, flexible tubes from *Escherichia coli* (48, 49) fitted well with critical ultrastructural work of the time (8, 23). More recent work, however (15, 18), using much gentler methods of preparation has at a stroke made dubious the reality of a periplasmic space and a thin monolayer of peptidoglycan. The erstwhile periplasmic space may be filled by an expanded three-dimensional network of peptidoglycan, and the holes in the net may be filled with the solutions of oligosaccharides and proteins, thus making much more plausible close interactions of function between inner and outer membranes through a periplasmic "gel."

Proteins and Walls

To return to the walls of gram-positive bacteria, we may ask whether the cytoplasmic membrane is related to the "clean" walls of our common parlance only by the necessity of its carrying out essential biosynthetic functions. The first unambiguous answer to this was provided by the observation that in some lactobacilli lipoteichoic acid rooted in the membrane could be seen on the cell surface by antibodies to it (45). A much more general answer was provided by Owen and Salton (25), who in their elegant work showed the penetration of the wall of *M. lysodeikticus* by a number of membrane antigens. Of the 27 antigens recognized in membrane preparations by crossed immunoelectrophoresis, 12 resided on the outer face of the membrane, since antibodies to them were absorbed by intact protoplasts. More surprisingly, these same 12 antibodies were also absorbed by whole intact bacteria. This strongly suggests that a considerable number of antigens can penetrate from the outer region of the cytoplasmic membrane through a considerable thickness of the wall. It is very improbable that antibody molecules can penetrate any great distance from the outside. One of the strongest reacting antigens was the succinylated mannan (24, 26, 31), to be compared with the penetration of walls of lactobacilli by lipoteichoic acid. Of the 11 other surface antigens, 5 reacted to lectin specific for sugar groups; the nature of none of them apart from the mannan is known.

Wall preparations from organisms other than *M. lysodeikticus* when made by the original method were infamous for retaining what were regarded as contaminating amino acids. Indeed, those of *Bacillus licheniformis* and *B. subtilis* failed to part with them even when repeatedly incubated with proteolytic enzymes; only warm sodium dodecyl sulfate (SDS) solutions persuaded them into solution, leaving insoluble peptidoglycan-teichoic acid structures. Reexamination of walls of *B. subtilis* 168 prepared by rupturing exponentially growing cells, removing unbroken cells by differential centrifugation, and washing them much as Salton and Horne did showed that about 2% of the preparation was protein extractable by 2 to 5 M LiCl. This process was originally introduced to extract autolysins from walls (30). By SDS-polyacrylamide gel electrophoresis of the extracts, 20 to 25 polypeptides could be distinguished. Two of these are autolysins that have been purified from such extracts (14, 33). The nature of the others is unknown. Some may simply be absorbed from the cytoplasm after cell rupture, but others may be wall-membrane interactive proteins. This situation could be readily elucidated by the immunological techniques employed so profitably by Owen and Salton (25).

Membrane-Wall Cooperation in Biosynthesis

Glycoside polymerization to give so called uncross-linked peptidoglycan is undertaken by membrane preparations alone from gram-positive bacteria, when supplied by the appropriate nucleotide precursors (44). To obtain cross-linked material from in vitro preparations from some species, the transpeptidases need a more complex system. Either wall fragments with membrane attached (21, 22) or permeabilized whole bacteria (40) must be used. The need for continued close association between wall and membrane does not appear to be limited to the action of the transpeptidases. For example, good synthesis of polyribitol phosphate was achieved only by membrane-wall preparations from *Lactobacillus plantarum,* whereas in preparations from *Staphylococcus aureus* membrane alone was fully competent (46). Preliminary examination of the association of membrane proteins with walls in preparations from a mutant of *B. subtilis* made by grinding the bacteria in the cold with alumina, the method used originally to produce biosynthetically active preparations from *S. aureus* (22), gave interesting results. The mutant used was the conditional morphological *rodA* mutant, grown at 25°C as rods or at 45°C as cocci. The membrane-wall particles were extracted successively with 1 mM EDTA at 0°C for 5 min, 0.1% Triton X-100 at 0°C for 20 min or at 37°C for 20 min, and finally with 0.1% SDS at 37°C for 90 min. At the end of each extraction a sample of the insoluble material was removed and extracted with 2% SDS at 100°C. The proteins in these final extracts were examined by SDS-polyacrylamide gel electrophoresis, alongside an isolated membrane preparation. Individual membrane proteins were removed at different rates by the various extraction procedures, but the most interesting result was for the residue from extraction by warm 0.1% SDS. A few prominent bands were seen, and the patterns were quite different for the bacteria grown at 25°C as rods compared with those grown at 45°C as cocci. The very strong band from bacteria grown at 45°C but not at 25°C, having a molecular weight of 38,000, one might suppose to be the autolysin *N*-acetylmuramyl-L-alanine amidase, except that less enzyme activity can be extracted from *rodA* mutants grown at 45°C than from cells grown at lower temperatures (32). The strongly differential association of the different membrane proteins with the wall would seem to be worth further investigation by the sort of methods used by Owen and Salton (25).

The necessity for close association between wall and membrane for the in vitro synthesis of cross-linked peptidoglycan must be set against the well-known ability of protoplasts of many species of gram-positive bacteria to regenerate walls to form normal vegetative cells when spread on solid media (10, 13, 16, 17, 39). Careful examination has shown that no wall polymers can be detected chemically or ultrastructurally on the surface of such protoplasts, yet not only is the peptidoglycan of the fully reverted cells normal, but even the very small amounts of material formed a few hours after plating the protoplasts already have a significant degree of cross-linking (9). It must therefore be that cross-linking can occur in the absence of wall. At this early stage of reversion, material recognized as peptidoglycan by ferritin-labeled antibodies prepared against it appeared as a loose fibrillar, microcapsular material extending some distance from the cytoplasmic membrane of the protoplast. Within a few more hours during the regeneration process, the degree of cross-linking of the peptides was indistinguishable from that in either the original or the fully reverted bacilli. During this latter period the wall started to assume its normal ultrastructural appearance. So far as is known, reversion of protoplasts will occur only on solid media. The obvious explanation that comes to mind is that on agar and gelatin media the rate of diffusion of larger molecules away from the protoplast is limited. This may involve, in theory, diffusion of either the transpeptidases or the substrate. Local concentration of substrate at the protoplast boundary might favor cross-linking by increased contiguity of donor and acceptor molecules. At first sight the transpeptidases, inasmuch as only penicillin-binding proteins (PBPs) are active, do not seem to be likely candidates for diffusion because they are thought to be integral membrane proteins exposed on the outer face of the membrane but fixed to it by short hydrophobic tails (47). However, it is of great interest that Linder and Salton (19) reported the presence of a soluble, β-lactam-sensitive D-alanine carboxypeptidase that was liberated during protoplast formation from *M. lysodeikticus*. This raises the question of whether transpeptidases can also under some circumstances either become solubilized from the outer surface of membranes or otherwise penetrate the wall. Whether or not this occurs, it would seem that the wall in the membrane-wall preparations found necessary for in vitro biosynthesis and the solid supporting media necessary for protoplast reversion may have the common function of keeping the biosynthetic system concentrated around the cytoplasmic membrane; therefore, the porosity of the wall may regulate continued cross-linking.

Effects of Inhibition of Wall Synthesis upon Membrane Function

When wall synthesis is inhibited, irrespective of the means used the autolytic enzymes formed by the bacteria lead to bacterial dissolution or lysis. This, however, does not happen with some species such as, for example, *Streptococcus mutans* and *S. sanguis,* or with organisms in which autolysins have been greatly reduced in activity either phenotypically or genotypically. Growth stops and the biomass remains constant, but in rather short times macromolecular biosynthesis is inhibited and at least in some examples the organisms slowly die (34, 41). Two types of effect upon the membrane have been shown. There is a leakage from the bacteria of phospholipids (2, 12) and a temporary increase in their rate of turnover (34). In some species membrane proteins, including PBPs (11), leak out of the cell, and in another instance (34) there is damage to membrane transport. These effects begin rapidly after the addition of cell wall-inhibiting antibiotics to exponentially growing cultures, within less than the equivalent of one generation time for the uninhibited culture. It is, of course, not possible to exclude dogmatically sublytic action of residual autolysin in these effects, but so far the only functional change in the walls that has been measured is increased permeability to the fluorescent probe 8-anilino-1-naphthalene sulfonate. However, the increase in fluorescence of the probe during inhibition of wall synsthesis might also be due to penetration through the wall of hydrophobic membrane components to meet the probe (34a). The presence of 0.5 M sucrose in the medium does not stop the effects on membrane of inhibiting wall synthesis in *lyt* mutants of *B. subtilis* (unpublished data).

Evidence of Interplay between Wall Synthesis and Cytoplasmic Events

Present orthodoxy would say that those membrane proteins which bind β-lactams to a major extent are wholly concerned with peptidoglycan synthesis, even if the exact way in which some such as PBPs of *E. coli* or the DD-alanylcarboxypeptidases of this same organism are involved is not entirely clear. Yet, very old and to my knowledge neglected evidence (27–29)—neglected, that is, until the present book, in which one paper may render this section of the present contribution obsolete before it appears—suggests that acylation of PBPs with β-lactams may affect not only wall synthesis but in a specific manner protein synthesis too. When vegetative cells of *B. cereus* were treated with very low concentrations of benzylpenicillin at 0°C, thoroughly washed, and resuspended in medium not containing the antibiotic, extracellular β-lactamase formation was induced at a linear rate after a short lag. The rate of enzyme formation remained unaltered for a long time. Penicilloic acid was quite inactive when used in place of benzylpenicillin for pretreatment. The optimum concentration of β-lactam required for induction was that necessary to saturate what was then called the "penicillin binding component of the cell." We now know that it is the PBPs. Much obviously needs to be done to apply the advances of 37 years to this observation. The weak β-lactamase activity of some DD-alanylcarboxypeptidases and the amino acid sequence analogy between the —NH$_2$-terminal ends of these enzymes and some β-lactamases (47) are tantalizing hints. The induction of penicillinase, however, appeared to be a true derepression phenomenon, and it is difficult to see how an integral membrane protein could act either directly or indirectly as a repressor molecule unless it first occurs as a cytoplasmic protein. If it were with such a soluble precursor of the membrane protein that the β-lactam reacted, it is difficult to see how β-lactamase induction conforms to the kinetics for saturation of the PBPs in the membrane.

Another equally tantalizing piece of evidence of interrelations between cytoplasmic events and the apparatus of wall synthesis is to be found in the latest (1) of the large number of studies of attachment of the chromosome to the cytoplasmic membrane. For example, a fragment of cytoplasmic membrane was isolated from *B. subtilis* which amounted to only 3.9% of the total membrane protein but was associated with 13.2% of the total PBPs and 25% of the chromosome-origin marker which represented only 2.5% of the total bacterial DNA. This concentration of PBPs at the chromosome origin may be principally related to cell division and chromosome segregation, although it should be noted that the concentration of one DD-alanylcarboxypeptidase was some three times higher than that of the two higher-molecular-weight PBPs which are generally thought to represent the biosynthetic transpeptidases. The presence of such a high concentration of elements of the peptidoglycan biosynthetic apparatus in close proximity to the chromosome origin inevitably raises speculation as to the role of the complex in cell division. The relation between this membrane fragment and the ribosome-deficient membrane fraction isolated from *B. subtilis,* which also has a high proportion of PBPs (20), is also of great interest.

Beyond the Wall

To succeed in their natural environment, microorganisms need many properties not demanded by laboratory growth situations. They may need to adhere to surfaces in a specific or nonspecific way, move, or repel phagocytic cells or marauding bacteriophages. To do this, bacteria possess various appendages such as flagella and fimbriae, coats of lectins, proteins, and polysaccharide capsules. To reach their destination, these molecules or structures must pass through the membrane and wall but be retained at the outermost layers. Some proteins, such as the M substance of *Streptococcus pyogenes* or the A protein of *Staphylococcus aureus*, appear to solve their problems by becoming covalently attached to the peptidoglycan. Others form patterned coats attached by noncovalent bonds (43). Polysaccharides forming well-defined capsules must presumably have some mechanism for keeping them in place, since in other instances the same polysaccharides are formed in an uncontrolled manner, appearing as slimes leaving no visible capsule at all on the organism. Do these many substances pass through the peptidoglycan network leaving no trace behind, or is the wall a highly tensile three-dimensional network, its spaces filled with proteins and polysaccharides, some as membrane extensions, some en route to the exterior surface, and some covalently joined. In other words, is the wall of a gram-positive organism really so different from that of a gram-negative organism with its periplasmic gel (15; J. Stephenson, C. R. Harwood, and A. R. Archibald, *Abstr. 107th Meet. Soc. Gen. Microbiol.*, p. 18, 1986)? Perhaps we should return to reexamine the origin of those "nonpeptidoglycan" amino acids observed by Salton but cheerfully disregarded by the rest of us.

There may be no clear beginning and certainly there is no clear end to the cell surface. Rather, the envelope is an organ which for analytical convenience we have separated into membranes, walls, and glycocalyx (5), but in the living cell one shades into another and they are all interdependent in function and formation.

LITERATURE CITED

1. **Bone, E. J., J. A. Todd, D. J. Ellar, M. G. Sargent, and A. W. Wyke.** 1985. Membrane particles from *Escherichia coli* and *Bacillus subtilis* containing penicillin-binding proteins enriched for chromosomal-origin DNA. *J. Bacteriol.* **164:**192–200.
2. **Brissete, J. L., G. D. Shockman, and R. A. Pieringer.** 1982. Effects of penicillin on

synthesis and excretion of lipid and lipoteichoic acid from *Streptococcus mutans* BH7. *J. Bacteriol.* **151:**838–824.

3. **Caulfield, M. P. P., P. C. Tai, and B. D. Davis.** 1983. Association of penicillin-binding proteins and other proteins with the ribosome-free membrane fraction of *Bacillus subtilis. J. Bacteriol.* **156:**1–5.

4. **Conover, M. J., J. S. Thompson, and G. D. Shockman.** 1966. Autolytic enzyme of *Streptococcus faecalis*: release of soluble enzyme from cell walls. *Biochem. Biophys. Res. Commun.* **23:**713–719.

5. **Costerton, J. W., and R. T. Irwin.** 1981. The bacterial glycocalyse in nature and disease. *Annu. Rev. Microbiol.* **35:**299–324.

6. **Cummins, C. S., and H. Harris.** 1956. The chemical compositions of the cell wall in some gram-positive bacteria and its possible value as a taxonomic character. *J. Gen. Microbiol.* **14:**583–600.

7. **Dawson, I. M.** 1949. *Nature of the Bacterial Surface*, p. 842. Oxford University Press, Oxford.

8. **DePetris, S.** 1967. Ultrastructure of the cell wall of *Escherichia coli* and the chemical nature of its constituent layers. *J. Ultrastruct. Res.* **19:**45–83.

9. **Elliott, T. S. J., J. B. Ward, and H. J. Rogers.** 1975. Formation of cell wall polymers by reverting protoplasts of *Bacillus licheniformis. J. Bacteriol.* **124:**623–632.

10. **Elliot, T. S. J., J. B. Ward, P. B. Wyrick, and H. J. Rogers.** 1975. Ultrastructural study of the reversion of protoplasts of *Bacillus licheniformis* to bacilli. *J. Bacteriol.* **124:**905–917.

11. **Hackenbeck, R., C. Martin, and G. Morelli.** 1983. *Streptococcus pneumoniae* proteins released in the medium upon inhibition of cell wall biosynthesis. *J. Bacteriol.* **155:**1372–1381.

12. **Hackenbeck, R., S. Waks, and A. Tomasz.** 1978. Characterization of wall polymers secreted into the growth medium of lysis-defective pneumococci during treatment with penicillin and other inhibitors of cell wall synthesis. *Antimicrob. Agents Chemother.* **13:**302–311.

13. **Hadlaczky, G., K. Fodor, and L. Alfoldi.** 1976. Morphological study of the reversion to the bacilliary form of *Bacillus megaterium* protoplasts. *J. Bacteriol.* **125:**1172–1179.

14. **Herbold, D. R., and L. Glaser.** 1975. The purification of *Bacillus subtilis* autolytic amidase. *J. Biol. Chem.* **250:**1676–1682.

15. **Hobot, J. A., E. Carlemalm, W. Villiger, and E. Kellenberger.** 1984. Periplasmic gel: new concept resulting from the reinvestigation of bacterial cell ultrastructure by new methods. *J. Bacteriol.* **160:**143–152.

16. **King, J. R., and H. Gooder.** 1970. Reversion to the streptococcal state of enterococcal protoplasts, spheroplasts and L-forms. *J. Bacteriol.* **103:**692–696.

17. **Landman, O. E., and A. Forman.** 1969. Gelatin-induced reversion of protoplasts of *Bacillus subtilis* to the bacillary form-biosynthesis of macromolecules and wall during successive steps. *J. Bacteriol.* **99:**576–589.

18. **Leduc, M., C. Frehel, and J. van Heijenoort.** 1985. Correlation between degradation and ultrastructure of peptidoglycan during autolysis of *Escherichia coli. J. Bacteriol.* **161:**627–635.

19. **Linder, R., and M. R. J. Salton.** 1975. D-alanine carboxypeptidase activity of *Micrococcus lysodeikticus* released into the protoplasting medium. *Eur. J. Biochem.* **95:**291–297.

20. **Marty-Mazars, D., S. Horluchi, P. C. Tai, and B. D. Davis.** 1983. Proteins of ribosome-bearing and free-membrane domains in *Bacillus subtilis. J. Bacteriol.* **154:**1381–1388.

21. **Mirelman, D., R. Bracha, and N. Sharon.** 1972. Role of the penicillin-sensitive transpeptidation reaction in attachment of newly synthesized peptidoglycan to cell walls of *Micrococcus lutens. Proc. Natl. Acad. Sci. USA* **69:**3355–3359.

22. **Mirelman, D., R. Bracha, and N. Sharon.** 1979. Biosynthesis of peptidoglycan by a cell wall preparation of *Staphylococcus aureus* and its inhibition by penicillin. *Biochemistry* **13:**5045–5053.

23. **Murray, R. G. E., P. Steed, and H. E. Elson.** 1965. The location of the mucopeptide in sections of the cell wall of *Escherichia coli* and other Gram negative bacteria. *Can. J. Microbiol.* **11:**547–560.

24. **Owen, P., and M. R. J. Salton.** 1975. A succinylated mannan in the membrane system of *Micrococcus lysodeikticus*. *Biochem. Biophys. Res. Commun.* **63:**875–880.

25. **Owen, P., and M. R. J. Salton.** 1977. Membrane asymmetry and expression of cell surface antigens of *Micrococcus lysodeikticus* established by crossed immunoelectrophoresis. *J. Bacteriol.* **132:**974–985.

26. **Pess, D. D., A. S. Schmit, and W. J. Lennarz.** 1975. The characterization of mannan of *Micrococcus lysodeikticus* as an acidic lipopolysaccharide. *J. Biol. Chem.* **250:**1319–1327.

27. **Pollock, M. R.** 1950. Penicillinase adaptation in *Bacillus cereus*. Adaptive enzyme formation in the absence of free substrate. *Br. J. Exp. Pathol.* **31:**739–753.

28. **Pollock, M. R.** 1953. Penicillinase adaptation and fixation of penicillin sulphur by *Bacillus cereus* spores. *J. Gen. Microbiol.* **8:**186–197.

29. **Pollock, M. P., and C. J. Perret.** 1951. The reaction between fixation of penicillin sulphur and penicillinase adaptation in *Bacillus cereus*. *Br. J. Exp. Pathol.* **32:**387–395.

30. **Pooley, H. M., J. M. Porres-Juan, and G. D. Shockman.** 1970. Dissociation of an autolytic enzyme-cell wall complex by treatment with unusually high concentrations of salt. *Biochem. Biophys. Res. Commun.* **38:**1134–1140.

31. **Powell, D. A., M. Duckworth, and J. Baddiley.** 1985. A membrane-associated lipomannan and micrococci. *Biochem. J.* **151:**387–397.

32. **Rogers, H. J., and C. Taylor.** 1978. Autolysine and shape change in *rodA* mutants of *Bacillus subtilis*. *J. Bacteriol.* **135:**1032–1042.

33. **Rogers, H. J., C. Taylor, S. Rayter, and J. B. Ward.** 1984. Purification and properties of autolytic endo-β-N-acetylglucosaminidase and the N-acetylmuranyl-L-alanine amidase from *Bacillus subtilis* strain 168. *J. Gen. Microbiol.* **130:**2395–2402.

34. **Rogers, H. J., P. F. Thurman, and J. D. J. Burdett.** 1983. The bactericidal action of β-lactam antibiotics on an autolysin deficient strain of *Bacillus subtilis*. *J. Gen. Microbiol.* **129:**465–478.

34a. **Rogers, H. J., and G. Wright.** 1987. Early changes in bacterial envelopes as shown by the use of a fluorescent probe. *J. Gen. Microbiol.* **133:**2567–2572.

35. **Salton, M. R. J.** 1952. Studies of the bacterial cell wall. III. Preliminary investigation of the chemical constitution of the cell wall of *Streptococcus faecalis*. *Biochim. Biophys. Acta* **8:**510–519.

36. **Salton, M. R. J.** 1953. Studies of the bacterial cell wall. IV. The composition of the cell walls of some Gram positive and Gram negative bacteria. *Biochim. Biophys. Acta* **10:**512–523.

37. **Salton, M. R. J., and R. W. Horne.** 1951. Studies of the bacterial cell wall. I. Electron microscopical observations on heated bacteria. *Biochim. Biophys. Acta* **7:**177–191.

38. **Salton, M. R. J., and R. W. Horne.** 1951. Studies of the bacterial cell wall. II. Methods of preparation and some properties of cell walls. *Biochim. Biophys. Acta* **7:**177–191.

39. **Schonfeld, J. K., and W. C. De Bruijen.** 1973. Ultrastructure of the intermediate stages in the reverting L-phase organisms of *Staphylococcus aureus* and *Streptococcus faecalis*. *J. Gen. Microbiol.* **77:**261–271.

40. **Schrader, W. P., and D. P. Fan.** 1974. Synthesis of cross-linked peptidoglycan attached to previously formed cell wall by Toluene-treated cells of *Bacillus inegaterium*. *J. Biol. Chem.* **249:**4815–4818.

41. **Shockman, G. D., L. Daneo-Moore, T. D. McDowell, and W. Wong.** 1981. Function and structure of the cell wall. Its importance in the life and death of bacteria, p. 31–65. *In* M. Salton and G. D. Shockman (ed.), β-*Lactam Antibiotics Mode of Action.* Academic Press, Inc., New York.

42. **Shockman, G. D., J. S. Thompson, and M. J. Conover.** 1967. Autolytic enzyme system of Streptococcus faecalis. II. Partial characterization of the autolysin and its substrate. *Biochemistry* **6:**1054–1064.

43. **Sleytr, U. B.** 1978. Regular arrays of macromolecules on bacterial cell walls, structure, chemistry, assembly and function. *Int. Dev. Cytol.* **53:**1–64.

44. **Strominger, J. L.** 1970. Penicillin-sensitive enzymatic reactions in bacterial cell wall synthesis. *Harvey Lect.* (69) *Ser.* **64:**179–213.

45. **Van Driel, D., A. J. Wicken, M. R. Dickson, and K. W. Knox.** 1973. Cellular location of the lipoteichoic acids of *Lactobacillus fermenti* NCTC 6991 and *Lactobacillus casei* NCTC 6375. *J. Ultrastruct. Res.* **43:**483–497.

46. **Ward, J. B.** 1981. Teichoic and teichuronic acids: biosynthesis, assembly and location. *Microbiol. Rev.* **45:**211–243.

47. **Waxman, D. J., and J. L. Strominger.** 1982. Penicillin binding proteins and the mechanism of action of β-lactam antibiotics. *Annu. Rev. Biochem.* **52:**825–869.

48. **Weidel, W., H. Frank, and W. Leutgeb.** 1963. Autolytic enzymes as a source of error in the preparation of gram-negative cell walls. *J. Gen. Microbiol.* **30:**127–130.

49. **Weidel, W., H. Frank, and H. H. Martin.** 1960. The rigid layer of the cell wall of *Escherichia coli* strain B. *J. Gen. Microbiol.* **22:**158–166.

Index

Access index, *see* Target access index
N-Acetylmuramyl-L-alanine amidase, nascent peptidoglycan processing in *G. homari,* 168–176
Acyl enzymes, formation with β-lactams, 270–271, 276–277
Aglycones (glycopeptide antibiotics), structure, 517–521, 525–527
D-Alanine, macrofiber twist states, 110–120
Alanine racemases
    genes, *S. typhimurium,* 531–538
    inhibitors, 542–547
    suicide substrates (β-halo-D-alanines), 536–537
    time-dependent inactivation, 533
D-Alanyl-D-alanine, pool size and twist state, 120
D-Alanyl-D-alanine ligase
    design of inhibitors, 542–547
    E. coli genes (*ddl*), 531–532, 535
    inhibition, 506, 536–538, 541–547
    phosphinic acid inhibition, 541–547
    S. faecalis, 534, 542
    S. typhimurium
        gene cloning, 534–535
        inhibitor studies, 536–538
        substrate specificity, 535–536
        time-dependent inhibition, 537–538
Aminoethyl phosphinodecanoate, ligase inhibitor, 537–538
AmpC, β-lactamase regulation, 489–490, 492, 500
AmpD, β-lactamase regulation, 490–492, 500
Ampicillin
    A. aquaticus, turgor pressure effects, 157–163
    S. faecium, surface growth effects, 94–95
AmpR, β-lactamase regulation, 489–492, 500
Ancylobacter aquaticus, turgor pressure
    ampicillin effect, 157–163
    cell components I and II, 159, 162–163
    measurement in gas vesicles, 157–163

Anionic polymers, cell wall, as target for antibiotic action, 591–594
Archaebacteria, Gram stain response, 7–9
Aridicin
    peptidoglycan synthesis inhibitor, mechanism of action, 510–511
    structure, 519–521
Assembly, cell wall
    B. subtilis, 98–106
    inhibition by LY186826 (γ-lactam), 570–576
    β-lactam effect, three-dimensional studies, 261–267
    S. faecium, 92
p-Azidophenacylbromide, use in photo-cross-linking, 321–322, 325

Bacillus licheniformis
    β-lactamase genes, 482–486
    β-lactamase induction, 481–486
    β-lactamase repressor (BlaI)
        sequence, 484
        structure, 484–485
Bacillus subtilis
    anionic polymers of cell wall as target for antibiotic action, 591–594
    bacterial threads, 110, 120–124
    cell wall
        anionic polymers, 591–594
        proteins, functions, 146–149
        turnover, 99–100, 104
    elongation, 98–106
    macrofiber twist, 109–120
    mutants
        autolysin deficient (*flaD* [*lyt*]), 146–149
        filamentous, 48–49
        multiple PBP, 337–340
        PBP 4 deficient, 335–337
        phosphoglucomutase (*gtaC*), 100–102
        teichoic acid synthesis, 592–593
        tunicamycin resistant (*tmr*), 593–594
    PBPs
        interstrain variability, 334–335

649